INTERACTION DESIGN

beyond human-computer interaction

John Wiley & Sons, Inc.

ACQUISITIONS EDITOR	Gaynor Redvers-Mutton/Paul Crockett
MARKETING MANAGER	Katherine Hepburn
SENIOR PRODUCTION EDITOR	Ken Santor
COVER DESIGNER	Madelyn Lesure
ILLUSTRATION EDITOR	Anna Melhorn
ILLUSTRATIONS	Tech-Graphics, Inc.
COVER IMAGE	"Thoughts in Passage II" by Michael Jon March. Courtesy of Grand Image Publishing

This book was set in 10/12 Times Ten by UG / GGS Information Services, Inc., and printed and bound by R. R. Donnelley/Crawfordsville. The cover and the color insert were printed by Phoenix Color Corporation.

This book is printed on acid free paper. ∞

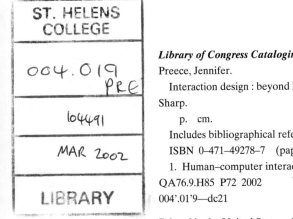
Library of Congress Cataloging in Publication Data:

Preece, Jennifer.
   Interaction design : beyond human– computer interaction/ Jennifer Preece, Yvonne Rogers, Helen Sharp.
      p.   cm.
   Includes bibliographical references and index.
   ISBN 0–471–49278–7   (paper : alk. paper)
   1. Human–computer interaction.   I. Rogers, Yvonne.   II. Sharp, Helen.   III. Title.
QA76.9.H85 P72 2002
004'.01'9—dc21

Printed in the United States of America   2001006730

10 9 8 7 6 5 4 3 2 1

Preface

Welcome to *Interaction Design: Beyond Human-Computer Interaction*, and our interactive website at *ID-Book.com*

This textbook is for undergraduate and masters students from a range of backgrounds studying classes in human-computer interaction, interaction design, web design, etc. A broad range of professionals and technology users will also find this book useful, and so will graduate students who are moving into this area from related disciplines.

Our book is called *Interaction Design: Beyond Human-Computer Interaction* because it is concerned with a broader scope of issues, topics, and paradigms than has traditionally been the scope of human-computer interaction (HCI). This reflects the exciting times we are living in, when there has never been a greater need for interaction designers and usability engineers to develop current and next-generation interactive technologies. To be successful they will need a mixed set of skills from psychology, human-computer interaction, web design, computer science, information systems, marketing, entertainment, and business.

What exactly do we mean by interaction design? In essence, we define interaction design as:

"designing interactive products to support people in their everyday and working lives".

This entails creating user experiences that enhance and extend the way people work, communicate, and interact. Now that it is widely accepted that HCI has moved beyond designing computer systems for one user sitting in front of one machine to embrace new paradigms, we, likewise, have covered a wider range of issues. These include ubiquitous computing and pervasive computing that make use of wireless and collaborative technologies. We also have tried to make the book up-to-date with many examples from contemporary research.

The book has 15 chapters and includes discussion of how cognitive, social, and affective issues apply to interaction design. A central theme is that design and evaluation are interleaving, highly iterative processes, with some roots in theory but which rely strongly on good practice to create usable products. The book has a 'hands-on' orientation and explains how to carry out a variety of techniques. It also has a strong pedagogical design and includes many activities (with detailed comments), assignments, and the special pedagogic features discussed below.

The style of writing is intended to be accessible to students, as well as professionals and general readers, so it is conversational and includes anecdotes, cartoons, and case studies. Many of the examples are intended to relate to readers' own experiences. The book and the associated website encourage readers to be active when reading and to think about seminal issues. For example, one feature we have included in the book is the "dilemma," where a controversial topic is aired. The aim is for readers to understand that much of interaction design needs consideration of the issues, and that they need to learn to weigh-up the pros and cons and be prepared to make trade-offs. We particularly want readers to realize that there is rarely a right or wrong answer although there are good designs and poor designs.

This book is accompanied by a website, which provides a variety of resources and interactivities. The website offers a place where readers can learn how to design websites and other kinds of multimedia interfaces. Rather than just provide a list of guidelines and design principles, we have developed various interactivities, including online tutorials and step-by-step exercises, intended to support learning by doing.

Special features

We use both the textbook and the web to teach about interaction design. To promote good pedagogical practice we include the following features:

Chapter design

Each chapter is designed to motivate and support learning:

- *Aims* are provided so that readers develop an accurate model of what to expect in the chapter.
- *Key points* at the end of the chapter summarize what is important.
- *Activities* are included throughout the book and are considered an essential ingredient for learning. They encourage readers to extend and apply their knowledge. Comments are offered directly after the activities, because pedagogic research suggests that turning to the back of the text annoys readers and discourages learning.
- An *assignment* is provided at the end of each chapter. This can be set as a group or individual project. The aim is for students to put into practice and consolidate knowledge and skills either from the chapter that they have just studied or from several chapters. Some of the assignments build on each other and involve developing and evaluating designs or actual products. Hints and guidance are provided on the website.
- *Boxes* provide additional and highlighted information for readers to reflect upon in more depth.
- *Dilemmas* offer honest and thought-provoking coverage of controversial or problematic issues.
- *Further reading* suggestions are provided at the end of each chapter. These refer to seminal work in the field, interesting additional material, or work that has been heavily drawn upon in the text.
- *Interviews* with nine practitioners and visionaries in the field enable readers to gain a personal perspective of the interviewees' work, their philosophies, their ideas about what is important, and their contributions to the field.
- *Cartoons* are included to make the book enjoyable.

ID-Book.com website

The aim of the website is to provide you with an opportunity to learn about inter-action design in ways that go "beyond the book." Additional in-depth material, hands-on interactivities, a student's corner and informal tutorials will be provided. Specific features planned include:

- Hands-on interactivities, including designing a questionnaire, customizing a set of heuristics, doing a usability analysis on 'real' data, and interactive tools to support physical design.
- Recent case studies.
- Student's corner where you will be able to send in your designs, thoughts, written articles which, if suitable, will be posted on the site at specified times during the year.
- Hints and guidance on the assignments outlined in the book.
- Suggestions for additional material to be used in seminars, lab classes, and lectures.
- Key terms and concepts (with links to where to find out more about them).

Readership

This book will be useful to a wide range of readers with different needs and aspirations.

Students from Computer Science, Software Engineering, Information Systems, Psychology, Sociology, and related disciplines studying courses in Interaction Design and Human-Computer Interaction will learn the knowledge, skills, and tech-niques for designing and evaluating state-of-the-art products, and websites, as well as traditional computer systems.

Web and *Interaction Designers*, and *Usability Professionals* will find plenty to satisfy their need for immediate answers to problems as well as for building skills to satisfy the demands of today's fast moving technical market.

Users, who want to understand why certain products can be used with ease while others are unpredictable and frustrating, will take pleasure in discovering that there is a discipline with practices that produce usable systems.

Researchers and *developers* who are interested in exploiting the potential of the web, wireless, and collaborative technologies will find that, as well as offering guid-ance, techniques, and much food for thought, a special effort has been made to in-clude examples of state-of-the-art systems.

In the next section we recommend various routes through the text for different kinds of readers.

How to use this book

Interaction Design is not a linear design process but is essentially iterative and some readers and experienced instructors will want to find their own way through the chapters. Others, and particularly those with less experience, may prefer to work through chapter by chapter. Readers will also have different needs. For example, students in Psychology will come with different background knowledge and needs from those in Computer Science. Similarly, professionals wanting to learn the fundamentals in a one-week course have different needs. This book and the website are designed for using in various ways. The following suggestions are provided to help you decide which way is best for you.

From beginning to end

There are fifteen chapters so students can study one chapter per week during a fifteen-week semester course. Chapter 15 contains design and evaluation case studies. Our intention is that these case studies help to draw together the contents of the rest of the book by showing how design and evaluation are done in the real world. However, some readers may prefer to dip into them along the way.

Getting a quick overview

For those who want to get a quick overview or just the essence of the book, we suggest you read Chapters 1, 6, and 10. *These chapters are recommended for everyone.*

Suggestions for computer science students

In addition to reading Chapters 1, 6, and 10, Chapters 7 and 8 contain the material that will feel most familiar to any students who have been introduced to software development. These chapters cover the process of interaction design and the activities it involves, including establishing requirements, conceptual design, and physical design. The book itself does not include any coding exercises, but the website will provide tools and widgets with which to interact.

For those following the ACM-IEEE Curriculum (2001)*, you will find that this text and website cover most of this curriculum. The topics listed under each of the following headings are discussed in the chapters shown:

- HC1 Foundations of Human-Computer Interaction (Chapters 1–5, 14, website).

- HC2 Building a simple graphical user interface (Chapters 1, 6, 8, 10 and the website).

- HC3 Human-Centered Software Evaluation (Chapters 1, 10–15, website).

- HC4 Human-Centered Software Design (Chapters 1, 6–9, 15).

- HC5 Graphical User-Interface Design (Chapters 2 and 8 and the website. Many relevant examples are discussed in Chapters 1–5 integrated with discussion of cognitive and social issues).

---

*ACM–IEEE Curriculum (2001) [computer.org/education/cc2001/] is under development at the time of writing this book.

- HC6 Graphical User-Interface Programming (touched upon only in Chapters 7–9 and on the website).
- HC7 HCI Aspects of Multimedia Information Systems and the web (integrated into the discussion of Chapters 1–5, and in examples throughout the text, and on the website).
- HC8 HCI Aspects of Group Collaboration and Communication Technology (discussed in 1–5, particularly in Chapter 4. Chapters 6–15 discuss design and evaluation and some examples cover these systems, as does the website.)

Suggestions for information systems students

Information systems students will benefit from reading the whole text, but instructors may want to find additional examples of their own to illustrate how issues apply to business applications. Some students may be tempted to skip Chapters 3–5 but we recommend that they should read these chapters since they provide important foundational material. This book does not cover how to develop business cases or marketing.

Suggestions for psychology and cognitive science students

Chapters 3–5 cover how theory and research findings have been applied to interaction design. They discuss the relevant issues and provide a wide range of studies and systems that have been informed by cognitive, social, and affective issues. Chapters 1 and 2 also cover important conceptual knowledge, necessary for having a good grounding in interaction design.

Practitioner and short course route

Many people want the equivalent of a short intensive 2–5 day course. The best route for them is to read Chapters 1, 6, 10 and 11 and dip into the rest of the book for reference. For those who want practical skills, we recommend Chapter 8.

Plan your own path

For people who do not want to follow the "beginning-to-end" approach or the suggestions above, there are many ways to use the text. Chapters 1, 6, 10 and 11 provide a good overview of the topic. Chapter 1 is an introduction to key issues in the discipline and Chapters 6 and 10 offer introductions to design and evaluation. Then go to Chapters 2–5 for user issues, then on to the other design chapters, 2–9, dipping into the evaluation chapters 10–14 and the case studies in 15. Another approach is to start with one or two of the evaluation chapters after first reading Chapters 1, 6, 10 and 11, then move into the design section, drawing on Chapters 2–5 as necessary.

Web designer route

Web designers who have a background in technology and want to learn how to design usable and effective websites are advised to read Chapters 1, 7, 8, 13 and 14.

These chapters cover key issues that are important when designing and evaluating the usability of websites. A worked assignment runs through these chapters.

Usability professionals' route

Usability professionals who want to extend their knowledge of evaluation techniques and read about the social and psychological issues that underpin design of the web, wireless, and collaborative systems are advised to read Chapter 1 for an overview, then select from Chapters 10–14 on usability testing. Chapters 3, 4, and 5 provide discussion of seminal user issues (cognitive, social, and affective aspects). There is new material throughout the rest of the book, which will also be of interest for dipping into as needed. This group may also be particularly interested in Chapter 8 which, together with material on the book website, provides practical design examples.

Acknowledgements

Many people have helped to make this book a reality. We have benefited from the advice and support of our many professional colleagues across the world, our students, friends, and families and we thank you all. We also warmly thank the following people for reviewing the manuscript and making many helpful suggestions for improvements: Liam Bannon, Sara Bly, Penny Collings, Paul Dourish, Jean Gasen, Peter Gregor, Stella Mills, Rory O'Connor, Scott Toolson, Terry Winograd, Richard Furuta, Robert J.K. Jacob, Blair Nonnecke, William Buxton, Carol Traynor, Blaise Liffich, Jan Scott, Sten Hendrickson, Ping Zhang, Lyndsay Marshall, Gary Perlman, Andrew Dillon, Michael Harrison, Mark Crenshaw, Laurie Dingers, David Carr, Steve Howard, David Squires, George Weir, Marilyn Tremaine, Bob Fields, Frances Slack, Ian Graham, Alan O'Callaghan, Sylvia Wilbur, and several anonymous reviewers. We also thank Geraldine Fitzpatrick, Tim and Dirk from DSTC (Australia) for their feedback on Chapters 1 and 4, Mike Scaife, Harry Brignull, Matt Davies, the HCCS Masters students at Sussex University (2000–2001), Stephanie Wilson and the students from the School of Informatics at City University and Information Systems Department at UMBC for their comments.

We are particularly grateful to Sara Bly, Karen Holtzblatt, Jakob Nielsen, Abigail Sellen, Suzanne Robertson, Gitta Salomon, Ben Shneiderman, Gillian Crampton Smith, and Terry Winograd for generously contributing in-depth interviews.

Lili Cheng and her colleagues allowed us to use the *HutchWorld* case study. Bill Killam provided the *TRIS* case study. Keith Cogdill supplied the *MEDLINEplus* case study. We thank Lili, Bill, and Keith for supplying the basic reports and commenting on various drafts. Jon Lazar and Dorine Andrews contributed material for the section on questionnaires, which we thank them for.

We are grateful to our Editors Paul Crockett and Gaynor Redvers-Mutton and the production team at Wiley: Maddy Lesure, Susannah Barr, Anna Melhorn, Gemma Quilter, and Ken Santor. Without their help and skill this book would not have been produced. Bill Zobrist played a significant role in persuading us to work with Wiley and we thank him too.

About the authors

The authors are all senior academics with a background in teaching, researching, and consulting in the UK, USA, Canada, Australia, and Europe. Having worked together on two other successful text books, they bring considerable experience in curriculum development, using a variety of media for distance learning as well as face-to-face teaching. They have considerable knowledge of creating learning texts and websites that motivate and support learning for a range of students.

All three authors are specialists in interaction design and human-computer interaction (HCI). In addition they bring skills from other disciplines. Yvonne Rogers is a cognitive scientist, Helen Sharp is a software engineer, and Jenny Preece works in information systems. Their complementary knowledge and skills enable them to cover the breadth of concepts in interaction design and HCI to produce an interdisciplinary text and website. They have collaborated closely, supporting and commenting upon each other's work to produce a high degree of integration of ideas with one voice. They have shared everything from initial concepts, through writing, design and production.

Contents

Chapter 1   **What is interaction design?**   1

1.1   Introduction   1

1.2   Good and poor design   2

   1.2.1   What to design   4

1.3   What is interaction design?   6

   1.3.1   The makeup of interaction design   6

   1.3.2   Working together as a multidisciplinary team   9

   1.3.3   Interaction design in business   10

1.4   What is involved in the process of interaction design?   12

1.5   The goals of interaction design   13

   1.5.1   Usability goals   14

   1.5.2   User experience goals   18

1.6   More on usability: design and usability principles   20

   1.6.1   Heuristics and usability principles   26

   Interview with Gitta Salomon   31

Chapter 2   **Understanding and conceptualizing interaction**   35

2.1   Introduction   35

2.2   Understanding the problem space   36

2.3   Conceptual models   39

   2.3.1   Conceptual models based on activities   41

   2.3.2   Conceptual models based on objects   51

   2.3.3   A case of mix and match?   54

2.4   Interface metaphors   55

2.5   Interaction paradigms   60

2.6   From conceptual models to physical design   64

   Interview with Terry Winograd   70

Chapter 3   **Understanding users**   73

3.1   Introduction   73

3.2   What is cognition?   74

3.3   Applying knowledge from the physical world to the digital world   90

3.4   Conceptual frameworks for cognition   92

   3.4.1   Mental models   92

3.4.2   Information processing   96

3.4.3   External cognition   98

3.5   Informing design: from theory to practice   101

Chapter 4   **Designing for collaboration and communication**   105

4.1   Introduction   105

4.2   Social mechanisms used in communication and collaboration   106

4.2.1   Conversational mechanisms   107

4.2.2   Designing collaborative technologies to support conversation   110

4.2.3   Coordination mechanisms   118

4.2.4   Designing collaborative technologies to support coordination   122

4.2.5   Awareness mechanisms   124

4.2.6   Designing collaborative technologies to support awareness   126

4.3   Ethnographic studies of collaboration and communication   129

4.4   Conceptual frameworks   130

4.4.1   The language/action framework   130

4.4.2   Distributed cognition   133

Interview with Abigail Sellen   138

Chapter 5   **Understanding how interfaces affect users**   141

5.1   Introduction   141

5.2   What are affective aspects?   142

5.3   Expressive interfaces   143

5.4   User frustration   147

5.4.1   Dealing with user frustration   152

5.5   A debate: the application of anthropomorphism to interaction design   153

5.6   Virtual characters: agents   157

5.6.1   Kinds of agents   157

5.6.2   General design concerns   160

Chapter 6   **The process of interaction design**   165

6.1   Introduction   165

6.2   What is interaction design about?   166

6.2.1   Four basic activities of interaction design   168

6.2.2   Three key characteristics of the interaction design process   170

6.3   Some practical issues   170

6.3.1   Who are the users?   171

6.3.2   What do we mean by "needs"?   172
6.3.3   How do you generate alternative designs?   174
6.3.4   How do you choose among alternative designs?   179
6.4   Lifecycle models: showing how the activities are related   182
6.4.1   A simple lifecycle model for interaction design   186
6.4.2   Lifecycle models in software engineering   187
6.4.3   Lifecycle models in HCI   192
Interview with Gillian Crampton Smith   198

Chapter 7   **Identifying needs and establishing requirements**   201
7.1   Introduction   201
7.2   What, how, and why?   202
7.2.1   What are we trying to achieve in this design activity?   202
7.2.2   How can we achieve this?   202
7.2.3   Why bother? The importance of getting it right   203
7.2.4   Why *establish* requirements?   204
7.3   What are requirements?   204
7.3.1   Different kinds of requirements   205
7.4   Data gathering   210
7.4.1   Data-gathering techniques   211
7.4.2   Choosing between techniques   215
7.4.3   Some basic data-gathering guidelines   216
7.5   Data interpretation and analysis   219
7.6   Task description   222
7.6.1   Scenarios   223
7.6.2   Use cases   226
7.6.3   Essential use cases   229
7.7   Task analysis   231
7.7.1   Hierarchical Task Analysis (HTA)   231
Interview with Suzanne Robertson   236

Chapter 8   **Design, prototyping and construction**   239
8.1   Introduction   239
8.2   Prototyping and construction   240
8.2.1   What is a prototype?   240
8.2.2   Why prototype?   241
8.2.3   Low-fidelity prototyping   243
8.2.4   High-fidelity prototyping   245
8.2.5   Compromises in prototyping   246

      8.2.6   Construction: from design to implementation   248

8.3   Conceptual design: moving from requirements to first design   249

      8.3.1   Three perspectives for developing a conceptual model   250

      8.3.2   Expanding the conceptual model   257

      8.3.3   Using scenarios in conceptual design   259

      8.3.4   Using prototypes in conceptual design   262

8.4   Physical design: getting concrete   264

      8.4.1   Guidelines for physical design   266

      8.4.2   Different kinds of widget   268

8.5   Tool support   275

Chapter 9   **User-centered approaches to interaction design**   279

9.1   Introduction   279

9.2   Why is it important to involve users at all?   280

      9.2.1   Degrees of involvement   281

9.3   What is a user-centered approach?   285

9.4   Understanding users' work: applying ethnography in design   288

      9.4.1   Coherence   293

      9.4.2   Contextual Design   295

9.5   Involving users in design: Participatory Design   306

      9.5.1   PICTIVE   307

      9.5.2   CARD   309

      Interview with Karen Holtzblatt   313

Chapter 10   **Introducing evaluation**   317

10.1   Introduction   317

10.2   What, why, and when to evaluate   318

      10.2.1   What to evaluate   318

      10.2.2   Why you need to evaluate   319

      10.2.3   When to evaluate   323

10.3   HutchWorld case study   324

      10.3.1   How the team got started: early design ideas   324

      10.3.2   How was the testing done?   327

      10.3.3   Was it tested again?   333

      10.3.4   Looking to the future   334

10.4   Discussion   336

Chapter 11   **An evaluation framework**   339

11.1   Introduction   339

11.2   Evaluation paradigms and techniques   340
    11.2.1   Evaluation paradigms   341
    11.2.2   Techniques   345
11.3   D E C I D E: A framework to guide evaluation   348
    11.3.1   Determine the goals   348
    11.3.2   Explore the questions   349
    11.3.3   Choose the evaluation paradigm and techniques   349
    11.3.4   Identify the practical issues   350
    11.3.5   Decide how to deal with the ethical issues   351
    11.3.6   Evaluate, interpret, and present the data   355
11.4   Pilot studies   356

Chapter 12   **Observing users**   359
12.1   Introduction   359
12.2   Goals, questions and paradigms   360
    12.2.1   What and when to observe   361
    12.2.2   Approaches to observation   363
12.3   How to observe   364
    12.3.1   In controlled environments   365
    12.3.2   In the field   368
    12.3.3   Participant observation and ethnography   370
12.4   Data collection   373
    12.4.1   Notes plus still camera   374
    12.4.2   Audio recording plus still camera   374
    12.4.3   Video   374
12.5   Indirect observation: tracking users' activities   377
    12.5.1   Diaries   377
    12.5.2   Interaction logging   377
12.6   Analyzing, interpreting and presenting data   379
    12.6.1   Qualitative analysis to tell a story   380
    12.6.2   Qualitative analysis for categorization   381
    12.6.3   Quantitative data analysis   384
    12.6.4   Feeding the findings back into design   384
    Interview with Sara Bly   387

Chapter 13   **Asking users and experts**   389
13.1   Introduction   389
13.2   Aking users: interviews   390
    13.2.1   Developing questions and planning an interview   390

13.2.2   Unstructured interviews   392

13.2.3   Structured interviews   394

13.2.4   Semi-structured interviews   394

13.2.5   Group interviews   396

13.2.6   Other sources of interview-like feedback   397

13.2.7   Data analysis and interpretation   398

13.3   Asking users: Questionnaires   398

13.3.1   Designing questionnaires   398

13.3.2   Question and response format   400

13.3.3   Administering questionnaires   404

13.3.4   Online questionnaires   405

13.3.5   Analyzing questionnaire data   407

13.4   Asking experts: Inspections   407

13.4.1   Heuristic evaluation   408

13.4.2   Doing heuristic evaluation   410

13.4.3   Heuristic evaluation of websites   412

13.4.4   Heuristics for other devices   419

13.5   Asking experts: walkthroughs   420

13.5.1   Cognitive walkthroughs   420

13.5.2   Pluralistic walkthroughs   423

Interview with Jakob Nielsen   426

Chapter 14   **Testing and modeling users**   429

14.1   Introduction   429

14.2   User testing   430

14.2.1   Testing MEDLINEplus   432

14.3   Doing user testing   438

14.3.1   Determine the goals and explore the questions   439

14.3.2   Choose the paradigm and techniques   439

14.3.3   Identify the practical issues: Design typical tasks   439

14.3.4   Identify the practical issues: Select typical users   440

14.3.5   Identify the practical issues: Prepare the testing conditions   441

14.3.6   Identify the practical issues: Plan how to run the tests   442

14.3.7   Deal with ethical issues   443

14.3.8   Evaluate, analyze, and present the data   443

14.4   Experiments   443

14.4.1   Variables and conditions   444

14.4.2   Allocation of participants to conditions   445

14.4.3   Other practical issues   446
14.4.4   Data collection and analysis   446
14.5   Predictive models   448
14.5.1   The GOMS model   449
14.5.2   The Keystroke level model   450
14.5.3   Benefits and limitations of GOMS   453
14.5.4   Fitts' Law   454
Interview with Ben Shneiderman   457

Chapter 15   **Design and evaluation in the real world: communicators and advisory systems**   461
15.1   Introduction   461
15.2   Key Issues   462
15.3   Designing mobile communicators   463
15.3.1   Background   463
15.3.2   Nokia's approach to developing a communicator   464
15.3.3   Philip's approach to designing a communicator for children   474
15.4   Redesigning part of a large interactive phone-based response system   482
15.4.1   Background   483
15.4.2   The redesign   483

Reflections from the Authors   491

References   493

Credits   503

Index   509

Foreword

by Gary Perlman

As predicted by many visionaries, devices everywhere are getting *"smarter."* My camera has a multi-modal hierarchical menu and form interface. Even my toaster has a microprocessor. Computing is not just for computers anymore. So when the authors wrote the subtitle *"beyond human-computer interaction,"* they wanted to convey that the book generalizes the human side to people, both individuals and groups, and the computer side to desktop computers, handheld computers, phones, cameras . . . maybe even toasters.

My own interest in this book is motivated by having been a software developer for 20 years, during which time I was a professor and consultant for 12. Would the book serve as a textbook for students? Would it help bring software development practice into a new age of human-centered interaction design?

A textbook for students . . .

More than anything, I think students need to be motivated, inspired, challenged, and I think this book, particularly Chapters 1–5, will do that. Many students will not have the motivating experience of seeing projects and products fail because of a lack of attention, understanding, and *zeal* for the user, but as I read the opening chapters, I imagined students thinking, *"This is what I've been looking for!"* The interviews will provide students with the wisdom of well-chosen experts: what's important, what worked (or didn't), and why. I see students making career choices based on this motivating material.

The rest of the book covers the art and some of the science of interaction design, the basic knowledge needed by practitioners and future innovators. Chapters 6–9 give a current view of analysis, design, and prototyping, and the book's website should add motivating examples. Chapters 10–14 cover evaluation in enough depth to facilitate understanding, not just rote application. Chapter 15 brings it all together, adding more depth. For each topic, there are ample pointers to further reading, which is important because interaction design is not a one-book discipline.

Finally, the book itself is pedagogically well designed. Each chapter describes its aims, contains examples and subtopics, and ends with key points, assignments, and an annotated bibliography for more detail.

A guide for development teams . . .

When I lead or consult on software projects, I face the same problem over and over: many people in marketing and software development–these are the people who have the most input into design, but it applies to any members of multidisciplinary teams–have little knowledge or experience building systems with a user-centered focus. A user-centered focus requires close work with users (not just customer–buyers), from analysis through design, evaluation, and maintenance. A lack of user-centered focus results in products and services that often do not meet the needs of their intended users. Don Norman's design books have convinced many that these problems are not unique to software, so this book's focus on interaction design feels right.

To help software teams adopt a user-centered focus, I've searched for books with end-to-end coverage from analysis, to design, to implementation (possibly of prototypes), to evaluation (with iteration). Some books have tried to please all audiences and have become encyclopedias of user interface development, covering topics worth knowing, but not in enough detail for readers to understand them. Some books have tried to cover theory in depth and tried to appeal to developers who have little interest in theory. Whatever the reasons for these choices, the results have been lacking. This book has chosen fewer topics and covered them in more depth; enough depth, I think, to put the ideas into practice. I think the material is presented in a way that is understandable by a wide audience, which is important in order for the book to be useful to whole multidisciplinary teams.

A recommended book . . .

I've been waiting for this book for many years. I think it's been worth the wait.

As the director of the HCI Bibliography project (www.hcibib.org), a free-access HCI portal receiving a half-million hits per year, I receive many requests for suggestions for books, particularly from students and software development managers. To answer that question, I maintain a list of recommended readings in ten categories (with 20,000 hits per year). Until now, it's been hard to recommend just one book from that list. I point people to some books for motivation, other books for process, and books for specific topics (e.g., task analysis, ergonomics, usability testing). This book fits well into half the categories in my list and makes it easier to recommend one book to get started and to have on hand for development.

I welcome the commitment of the authors to building a website for the book. It's a practice that has been adopted by other books in the field to offer additional information and keep the book current. The site also presents interactive content to aid in tasks like conducting surveys and heuristic evaluations. I look forward to seeing the book's site present new materials, but as director of www.hcibib.org, I hope they use links to instead of re-inventing existing resources.

Gary Perlman
Columbus
October 2001

About Gary Perlman

Gary Perlman is a consulting research scientist at the OCLC–Online Computer Library Center (www.oclc.org) where he works on user interfaces for bibliographic and full-text retrieval. His research interests are in making information technology more useful and usable for people.

He has also held research and academic positions at Bell Labs in Murray Hill, New Jersey; Wang Institute of Graduate Studies; Massachusetts Institute of Technology; Carnegie-Mellon University; and The Ohio State University. Dr. Perlman's Ph.D. is in experimental psychology from the University of California, San Diego. He is the author of over 75 publications in the areas of mathematics education, statistical computing, hypertext, and user interface development. He has lectured and consulted internationally since 1980.

He is best known in the HCI community as the director of the HCI Bibliography (www.hcibib.org), a free-access online resource of over 20,000 records searched hundreds of thousands of times each year.

A native of Montreal, Canada, Gary now lives in Columbus, Ohio with his wife and two sons.

Chapter 1

What is interaction design?

1.1   Introduction
1.2   Good and poor design
    1.2.1   What to design
1.3   What is interaction design?
    1.3.1   The makeup of interaction design
    1.3.2   Working together as a multidisciplinary team
    1.3.3   Interaction design in business
1.4   What is involved in the process of interaction design?
1.5   The goals of interaction design
    1.5.1 Usability goals
    1.5.2 User experience goals
1.6. More on usability: design and usability principles

1.1   Introduction

How many interactive products are there in everyday use? Think for a minute about what you use in a typical day: cell phone, computer, personal organizer, remote control, soft drink machine, coffee machine, ATM, ticket machine, library information system, the web, photocopier, watch, printer, stereo, calculator, video game . . . the list is endless. Now think for a minute about how usable they are. How many are actually easy, effortless, and enjoyable to use? All of them, several, or just one or two? This list is probably considerably shorter. Why is this so?

Think about when some device caused you considerable grief—how much time did you waste trying to get it to work? Two well-known interactive devices that cause numerous people immense grief are the photocopier that doesn't copy the way they want and the VCR that records a different program from the one they thought they had set or none at all. Why do you think these things happen time and time again? Moreover, can anything be done about it?

Many products that require users to interact with them to carry out their tasks (e.g., buying a ticket online from the web, photocopying an article, pre-recording a TV program) have not necessarily been designed with the users in mind. Typically, they have been engineered as systems to perform set functions. While they may work effectively from an engineering perspective, it is often at the expense of how the system will be used by real people. The aim of interaction design is to redress this concern by

1

bringing usability into the design process. In essence, it is about developing interactive products[1] that are easy, effective, and enjoyable to use—from the users' perspective.

In this chapter we begin by examining what interaction design is. We look at the difference between good and poor design, highlighting how products can differ radically in their usability. We then describe what and who is involved in interaction design. In the last part of the chapter we outline core aspects of usability and how these are used to assess interactive products. An assignment is presented at the end of the chapter in which you have the opportunity to put into practice what you have read, by evaluating an interactive product using various usability criteria.

The main aims of the chapter are to:

- Explain the difference between good and poor interaction design.
- Describe what interaction design is and how it relates to human-computer interaction and other fields.
- Explain what usability is.
- Describe what is involved in the process of interaction design.
- Outline the different forms of guidance used in interaction design.
- Enable you to evaluate an interactive product and explain what is good and bad about it in terms of the goals and principles of interaction design.

1.2 Good and poor design

A central concern of interaction design is to develop interactive products that are usable. By this is generally meant easy to learn, effective to use, and provide an enjoyable user experience. A good place to start thinking about how to design usable interactive products is to compare examples of well and poorly designed ones. Through identifying the specific weaknesses and strengths of different interactive systems, we can begin to understand what it means for something to be usable or not. Here, we begin with an example of a poorly designed system—voice mail—that is used in many organizations (businesses, hotels, and universities). We then compare this with an answering machine that exemplifies good design.

Imagine the following scenario. You're staying at a hotel for a week while on a business trip. You discover you have left your cell (mobile) phone at home so you have to rely on the hotel's facilities. The hotel has a voice-mail system for each room. To find out if you have a message, you pick up the handset and listen to the tone. If it goes "beep beep beep" there is a message. To find out how to access the message you have to read a set of instructions next to the phone.

You read and follow the first step:

"1. Touch 491".
The system responds, "You have reached the Sunny Hotel voice message center. Please enter the room number for which you would like to leave a message."

---

[1]We use the term *interactive products* generically to refer to all classes of interactive systems, technologies, environments, tools, applications, and devices.

You wait to hear how to listen to a recorded message. But there are no further instructions from the phone. You look down at the instruction sheet again and read:

"2. Touch*, your room number, and #". You do so and the system replies,

"You have reached the mailbox for room 106. To leave a message type in your password."

You type in the room number again and the system replies, "Please enter room number again and then your password."

You don't know what your password is. You thought it was the same as your room number. But clearly not. At this point you give up and call reception for help. The person at the desk explains the correct procedure for recording and listening to messages. This involves typing in, at the appropriate times, the room number and the extension number of the phone (the latter is your password, which is different from the room number). Moreover, it takes six steps to access a message and five steps to leave a message. You go out and buy a new cell phone.

What is problematic with this voice-mail system?

- It is infuriating.
- It is confusing.
- It is inefficient, requiring you to carry out a number of steps for basic tasks.
- It is difficult to use.
- It has no means of letting you know at a glance whether any messages have been left or how many there are. You have to pick up the handset to find out and then go through a series of steps to listen to them.
- It is not obvious what to do: the instructions are provided partially by the system and partially by a card beside the phone.

Now consider the following phone answering machine. Figure 1.1 shows two small sketches of an answering machine phone. Incoming messages are represented using physical marbles. The number of marbles that have moved into the pinball-like chute indicates the number of messages. Dropping one of these marbles into a slot in the machine causes the recorded message to play. Dropping the same marble into another slot on the phone dials the caller who left the message.

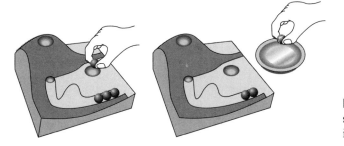

Figure 1.1 Two small sketches showing answering phone.

How does the "marble" answering machine differ from the voice-mail system?

- It uses familiar physical objects that indicate visually at a glance how many messages have been left.
- It is aesthetically pleasing and enjoyable to use.
- It only requires one-step actions to perform core tasks.
- It is a simple but elegant design.
- It offers less functionality and allows anyone to listen to any of the messages.

The marble answering machine was designed by Durrell Bishop while a student at the Royal College of Art in London (described by Crampton-Smith, 1995). One of his goals was to design a messaging system that represented its basic functionality in terms of the behavior of everyday objects. To do this, he capitalized on people's everyday knowledge of how the physical world works. In particular, he made use of the ubiquitous everyday action of picking up a physical object and putting it down in another place. This is an example of an interactive product designed with the users in mind. The focus is on providing them with an enjoyable experience but one that also makes efficient the activity of receiving messages. However, it is important to note that although the marble answering machine is a very elegant and usable design, it would not be practical in a hotel setting. One of the main reasons is that it is not robust enough to be used in public places, for example, the marbles could easily get lost or taken as souvenirs. Also, the need to identify the user before allowing the messages to be played is essential in a hotel setting. When considering the usability of a design, therefore, it is important to take into account *where* it is going to be used and *who* is going to use it. The marble answering machine would be more suited in a home setting—provided there were no children who might be tempted to play with the marbles!

1.2.1   What to design

Designing usable interactive products thus requires considering who is going to be using them and where they are going to be used. Another key concern is understanding the kind of *activities* people are doing when *interacting* with the products. The appropriateness of different kinds of interfaces and arrangements of input and output devices depends on what kinds of activities need to be supported. For example, if the activity to be supported is to let people communicate with each other at a distance, then a system that allows easy input of messages (spoken or written) that can be readily accessed by the intended recipient is most appropriate. In addition, an interface that allows the users to interact with the messages (e.g., edit, annotate, store) would be very useful.

The range of activities that can be supported is diverse. Just think for a minute what you can currently do using computer-based systems: send messages, gather information, write essays, control power plants, program, draw, plan, calculate, play games—to name but a few. Now think about the number of interfaces and interactive devices that are available. They, too, are equally diverse:

multimedia applications virtual-reality environments, speech-based systems, personal digital assistants and large displays—to name but a few. There are also many ways of designing the way users can interact with a system (e.g., via the use of menus, commands, forms, icons, etc.). Furthermore, more and more novel forms of interaction are appearing that comprise physical devices with embedded computational power, such as electronic ink, interactive toys, smart fridges, and networked clothing (See Figure 1.2 on Color Plate 1). What this all amounts to is a multitude of choices and decisions that confront designers when developing interactive products.

A key question for interaction design is: how do you optimize the users' interactions with a system, environment or product, so that they match the users' activities that are being supported and extended? One could use intuition and hope for the best. Alternatively, one can be more principled in deciding which choices to make by basing them on an understanding of the users. This involves:

- taking into account what people are good and bad at
- considering what might help people with the way they currently do things
- thinking through what might provide quality user experiences
- listening to what people want and getting them involved in the design
- using "tried and tested" user-based techniques during the design process

The aim of this book is to cover these aspects with the goal of teaching you how to carry out interaction design. In particular, it focuses on how to identify users' needs, and from this understanding, move to designing usable, useful, and enjoyable systems.

ACTIVITY 1.1 How does making a phone call differ when using:

- a public phone box
- a cell phone?

How have these devices been designed to take into account (a) the kind of users, (b) type of activity being supported, and (c) context of use?

Comment (a) Public phones are designed to be used by the general public. Many have Braille embossed on the keys and speaker volume control to enable people who are blind and hard of hearing to use them.

Cell phones are intended for all user groups, although they can be difficult to use for people who are blind or have limited manual dexterity.

(b) Most phone boxes are designed with a simple mode of interaction: insert card or money and key in the phone number. If engaged or unable to connect the money or card is returned when the receiver is replaced. There is also the option of allowing the caller to make a follow-on call by pressing a button rather than collecting the money and reinserting it again. This function enables the making of multiple calls to be more efficient.

Cell phones have a more complex mode of interaction. More functionality is provided, requiring the user to spend time learning how to use them. For example, users can save phone numbers in an address book and then assign these to "hotkeys," allowing them to be called simply through pressing one or two keys.

(c) Phone boxes are intended to be used in public places, say on the street or in a bus station, and so have been designed to give the user a degree of privacy and noise protection through the use of hoods and booths.

Cell phones have have been designed to be used any place and any time. However, little consideration has been given to how such flexibility affects others who may be in the same public place (e.g., sitting on trains and buses).

1.3   What is interaction design?

By interaction design, we mean

designing interactive products to support people in their everyday and working lives.

In particular, it is about creating user experiences that enhance and extend the way people work, communicate and interact. Winograd (1997) describes it as "the design of spaces for human communication and interaction." In this sense, it is about finding ways of supporting people. This contrasts with software engineering, which focuses primarily on the production of software solutions for given applications. A simple analogy to another profession, concerned with creating buildings, may clarify this distinction. In his account of interaction design, Terry Winograd asks how architects and civil engineers differ when faced with the problem of building a house. Architects are concerned with the people and their interactions with each other and within the house being built. For example, is there the right mix of family and private spaces? Are the spaces for cooking and eating in close proximity? Will people live in the space being designed in the way it was intended to be used? In contrast, engineers are interested in issues to do with realizing the project. These include practical concerns like cost, durability, structural aspects, environmental aspects, fire regulations, and construction methods. Just as there is a difference between designing and building a house, so too, is there a distinction between interaction design and software engineering. In a nutshell, interaction design is related to software engineering in the same way as architecture is related to civil engineering.

1.3.1   The makeup of interaction design

It has always been acknowledged that for interaction design to succeed many disciplines need to be involved. The importance of understanding how users act and react to events and how they communicate and interact together has led people from a variety of disciplines, such as psychologists and sociologists, to become involved. Likewise, the growing importance of understanding how to design different kinds of interactive media in effective and aesthetically pleasing ways has led to a diversity of other practitioners becoming involved, including graphic designers, artists, animators, photographers, film experts, and product designers. Below we outline a brief history of interaction design.

In the early days, engineers designed hardware systems for engineers to use. The computer interface was relatively straightforward, comprising various switch panels and dials that controlled a set of internal registers. With the advent of monitors (then referred to as visual display units or VDUs) and personal workstations in the late '70s and early '80s, interface design came into being (Grudin, 1990). The new concept of the user interface presented many challenges:

> *Terror. You have to confront the documentation. You have to learn a new language. Did you ever use the word 'interface' before you started using the computer?*

—Advertising executive Arthur Einstein (1990)

One of the biggest challenges at that time was to develop computers that could be accessible and usable by other people, besides engineers, to support tasks involving human cognition (e.g., doing sums, writing documents, managing accounts, drawing plans). To make this possible, computer scientists and psychologists became involved in designing user interfaces. Computer scientists and software engineers developed high-level programming languages (e.g., BASIC, Prolog), system architectures, software design methods, and command-based languages to help in such tasks, while psychologists provided information about human capabilities (e.g., memory, decision making).

The scope afforded by the interactive computing technology of that time (i.e., the combined use of visual displays and interactive keyboards) brought about many new challenges. Research into and development of graphical user interfaces (GUI for short, pronounced "goo-ee") for office-based systems took off in a big way. There was much research into the design of widgets (e.g., menus, windows, palettes, icons) in terms of how best to structure and present them in a GUI.

In the mid '80s, the next wave of computing technologies—including speech recognition, multimedia, information visualization, and virtual reality—presented even more opportunities for designing applications to support even more people. Education and training were two areas that received much attention. Interactive learning environments, educational software, and training simulators were some of the main outcomes. To build these new kinds of interactive systems, however, required a different kind of expertise from that of psychologists and computer programmers. Educational technologists, developmental psychologists, and training experts joined in the enterprise.

As further waves of technological development surfaced in the '90s—networking, mobile computing, and infrared sensing—the creation of a diversity of applications for *all* people became a real possibility. All aspects of a person's life—at home, on the move, at school, at leisure as well as at work, alone, with family or friends—began to be seen as areas that could be enhanced and extended by designing and integrating various arrangements of computer technologies. New ways of learning, communicating, working, discovering, and living were envisioned.

BOX 1.1   The Relationship between Interaction Design, Human-Computer Interaction, and Other Approaches

We view interaction design as fundamental to all disciplines, fields, and approaches that are concerned with researching and designing computer-based systems for people (see Figure 1.3). The best-known interdisciplinary field is human-computer interaction (HCI), which is "concerned with the design, evaluation, and implementation of interactive computing systems for human use and with the study of major phenomena surrounding them" (ACM SIGCHI, 1992, p. 6). Until the early '90's, the focus of HCI was primarily designing interfaces for single users. In response to a growing concern for the need also to support multiple individuals working together using computer systems, the interdisciplinary field of computer-supported cooperative work (CSCW) emerged (Greif, 1988). Information systems is another area concerned with the application of computing technology in domains like business, health, and education. Other fields related to interaction design include human factors, cognitive ergonomics, and cognitive engineering. All are concerned with designing systems to match users' goals, however, each has a different focus and methodology.

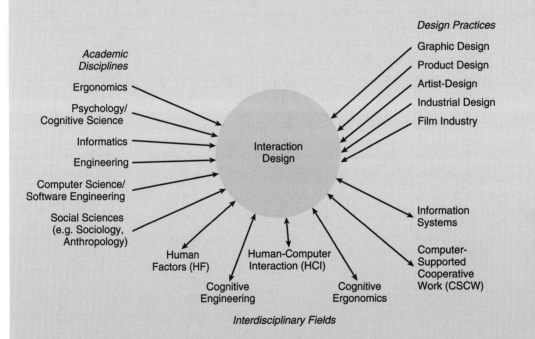

Figure 1.3 Relationship among contributing academic disciplines, design practices, and interdisciplinary fields concerned with interaction design.

In the mid '90s, many companies realized it was necessary again to extend their existing multidisciplinary design teams to include professionals trained in media and design, including graphical design, industrial design, film, and narrative. Sociologists, anthropologists, and dramaturgists were also brought on board, all having quite a different take on human interaction from psychologists. This wider set of people were thought to have the right mix of skills and understanding of the different application areas necessary to design the new generation of interactive systems. For example, designing a reminder application for the family requires understanding how families interact; creating an interactive story kit for children requires understanding how children write and understand narrative, and developing an interactive guide for art-gallery visitors requires appreciating what people do and how they move through public spaces.

Now in the '00s, the possibilities afforded by emerging hardware capabilities—e.g., radio-frequency tags, large interactive screens, and information appliances—has led to a further realization that engineers, who know about hardware, software, and electronics are needed to configure, assemble, and program the consumer electronics and other devices to be able to communicate with each other (often referred to as middleware).

1.3.2 Working together as a multidisciplinary team

Bringing together so many people with different backgrounds and training has meant many more ideas being generated, new methods being developed, and more creative and original designs being produced. However, the down side is the costs involved. The more people there are with different backgrounds in a design team, the more difficult it can be to communicate and progress forward the designs being generated. Why? People with different backgrounds have different perspectives and ways of seeing and talking about the world (see Figure 1.4). What one person values as important others may not even see (Kim, 1990). Similarly, a computer scientist's understanding of the term representation is often very different from a graphic designer's or a psychologist's.

Figure 1.4 Four different team members looking at the same square, but each seeing it quite differently.

What this means in practice is that confusion, misunderstanding, and communication breakdowns can often surface in a team. The various team members may have different ways of talking about design and may use the same terms to mean quite different things. Other problems can arise when a group of people is "thrown" together who have not worked as a team. For example, the Philips Vision of the Future Project found that its multidisciplinary teams—who were responsible for developing ideas and products for the future—experienced a number of difficulties, namely, that project team members did not always have a clear idea of who needed what information, when, and in what form (Lambourne et al., 1997).

ACTIVITY 1.2   In practice, the makeup of a given design team depends on the kind of interactive product being built. Who do you think would need to be involved in developing:

(a) a public kiosk providing information about the exhibits available in a science museum?

(b) an interactive educational website to accompany a TV series?

Comment   Each team will need a number of different people with different skill sets. For example, the first interactive product would need:

(a) graphic and interaction designers, museum curators, educational advisors, software engineers, software designers, usability engineers, ergonomists

The second project would need:

(b) TV producers, graphic and interaction designers, teachers, video experts, software engineers, software designers, usability engineers

In addition, as both systems are being developed for use by the general public, representative users, such as school children and parents, should be involved.

In practice, design teams often end up being quite large, especially if they are working on a big project to meet a fixed deadline. For example, it is common to find teams of fifteen people or more working on a website project for an extensive period of time, like six months. This means that a number of people from each area of expertise are likely to be working as part of the project team.

1.3.3   Interaction design in business

Interaction design is now big business. In particular, website consultants, start-up companies, and mobile computing industries have all realized its pivotal role in successful interactive products. To get noticed in the highly competitive field of web products requires standing out. Being able to say that your product is easy and effective to use is seen as central to this. Marketing departments are realizing how branding, the number of hits, customer return rate, and customer satisfaction are greatly affected by the usability of a website. Furthermore, the presence or absence of good interaction design can make or break a company.

One infamous dot.com fashion clothes company that failed to appreciate the importance of good interaction design paid heavily for its oversight, becoming bankrupt within a few months of going public.[2] Their approach had been to go for an "all singing and all dancing," glossy 3D graphical interface. One of the problems with this was that it required several minutes to download. Furthermore, it often took more than 20 minutes to place an order by going through a painfully long and slow process of filling out an online form—only to discover that the order was not successful. Customers simply got frustrated with the site and never returned.

In response to the growing demand for interaction design, an increasing number of consultancies are establishing themselves as interaction design experts. One such company is Swim, set up by Gitta Salomon to assist clients with the design of interactive products (see the interview with her at the end of this chapter). She points out how often companies realize the importance of interaction design but don't know how to do it themselves. So they get in touch with companies, like Swim, with their partially developed products and ask them for help. This can come in the form of an expert "crit" in which a detailed review of the usability and design of the product is given (for more on expert evaluation, see Chapter 13). More extensively, it can involve helping clients create their products.

Another established design company that practices interaction design is IDEO, which now has many branches worldwide. Drawing on over 20 years of experience in the area, they design products, services, and environments for other companies, pioneering new user experiences (Spreenberg et al., 1995). They have developed

---

BOX 1.2 What's In a Name? From Interface Designers to Information Architects

Ten years ago, when a company wanted to develop an interface for an interactive product it advertised for interface designers. Such professionals were primarily involved in the design and evaluation of widgets for desktop applications. Now that the potential range of interactive products has greatly diversified, coupled with the growing realization of the importance of getting the interface right, a number of other job descriptions have begun to emerge. These include:

- interactive/interaction designers (people involved in the design of all the interactive aspects of a product, not just the graphic design of an interface)

- usability engineers (people who focus on evaluating products, using usability methods and principles)

- web designers (people who develop and create the visual design of websites, such as layouts)

- information architects (people who come up with ideas of how to plan and structure interactive products, especially websites)

- user-experience designers (people who do all the above but who may also carry out field studies to inform the design of products)

---

[2]This happened before the dot.com crash in 2001.

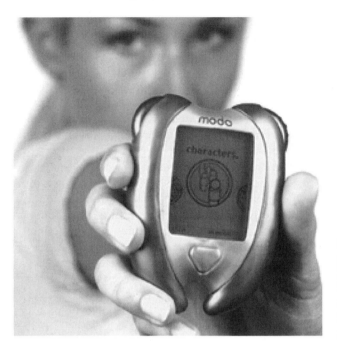

Figure 1.5 An innovative product developed by IDEO: Scout Modo, a wireless handheld device delivering up-to-date information about what's going on in a city.

thousands of products for numerous clients, each time following their particular brand of user-centered design (see Figure 1.5).

1.4 What is involved in the process of interaction design?

Essentially, the process of interaction design involves four basic activities:

1. identifying needs and establishing requirements
2. developing alternative designs that meet those requirements
3. building interactive versions of the designs so that they can be communicated and assessed
4. evaluating what is being built throughout the process

These activities are intended to inform one another and to be repeated. For example, measuring the usability of what has been built in terms of whether it is easy to use provides feedback that certain changes must be made or that certain requirements have not yet been met.

Evaluating what has been built is very much at the heart of interaction design. Its focus is on ensuring that the product is usable. It is usually addressed through a user-centered approach to design, which, as the name suggests, seeks to involve users throughout the design process. There are many different ways of achieving this: for example, through observing users, talking to them, interviewing them, testing them using performance tasks, modeling their performance, asking them to fill in questionnaires, and even asking them to become co-designers. The findings from the different ways of engaging and eliciting knowledge from users are then interpreted with respect to ongoing design activities (we give more detail about all these aspects of evaluation in Chapters 10–14).

Equally important as involving users in evaluating an interactive product is understanding what people currently do. This form of research should take place before building any interactive product. Chapters 3, 4, and 5 cover a lot of this ground by explaining in detail how people act and interact with one another, with information, and with various technologies, together with describing their strengths and weaknesses. Such knowledge can greatly help designers determine which solutions to choose from the many design alternatives available and how to develop and test these further. Chapter 7 describes how an understanding of users' needs can be translated to requirements, while Chapter 9 explains how to involve users effectively in the design process.

A main reason for having a better understanding of users is that different users have different needs and interactive products need to be designed accordingly. For example, children have different expectations about how they want to learn or play from adults. They may find having interactive quizzes and cartoon characters helping them along to be highly motivating, whereas most adults find them annoying. Conversely, adults often like talking-heads discussions about topics, but children find them boring. Just as everyday objects like clothes, food, and games are designed differently for children, teenagers, and adults, so, too, must interactive products be designed to match the needs of different kinds of users.

In addition to the four basic activities of design, there are three key characteristics of the interaction design process:

1. Users should be involved through the development of the project.
2. Specific usability and user experience goals should be identified, clearly documented, and agreed upon at the beginning of the project.
3. Iteration through the four activities is inevitable.

We have already mentioned the importance of involving users and will return to this topic throughout the book. Iterative design will also be addressed later when we talk about the various design and evaluation methods by which this can be achieved. In the next section we describe usability and user experience goals.

1.5 The goals of interaction design

Part of the process of understanding users' needs, with respect to designing an interactive system to support them, is to be clear about your primary objective. Is it to design a very efficient system that will allow users to be highly productive in their work, or is it to design a system that will be challenging and motivating so that it supports effective learning, or is it something else? We call these top-level concerns usability goals and user experience goals. The two differ in terms of how they are operationalized, i.e., how they can be met and through what means. Usability goals are concerned with meeting specific usability criteria (e.g., efficiency) and user experience goals are concerned with explicating the quality of the user experience (e.g., to be aesthetically pleasing).

1.5.1 Usability goals

To recap, usability is generally regarded as ensuring that interactive products are easy to learn, effective to use, and enjoyable from the user's perspective. It involves optimizing the interactions people have with interactive products to enable them to carry out their activities at work, school, and in their everyday life. More specifically, usability is broken down into the following goals:

- effective to use (effectiveness)
- efficient to use (efficiency)
- safe to use (safety)
- have good utility (utility)
- easy to learn (learnability)
- easy to remember how to use (memorability)

For each goal, we describe it in more detail and provide a key question.

Effectiveness is a very general goal and refers to how good a system is at doing what it is supposed to do.

Question: Is the system capable of allowing people to learn well, carry out their work efficiently, access the information they need, buy the goods they want, and so on?

Efficiency refers to the way a system supports users in carrying out their tasks. The answering machine described at the beginning of the chapter was considered efficient in that it let the user carry out common tasks (e.g., listening to messages) through a minimal number of steps. In contrast, the voice-mail system was considered inefficient because it required the user to carry out many steps and learn an arbitrary set of sequences for the same common task. This implies that an efficient way of supporting common tasks is to let the user use single button or key presses. An example of where this kind of efficiency mechanism has been effectively employed is in e-tailing. Once users have entered all the necessary personal details on an e-commerce site to make a purchase, they can let the site save all their personal details. Then, if they want to make another purchase at that site, they don't have to re-enter all their personal details again. A clever mechanism patented by Amazon.com is the one-click option, which requires users only to click a single button when they want to make another purchase.

Question: Once users have learned how to use a system to carry out their tasks, can they sustain a high level of productivity?

Safety involves protecting the user from dangerous conditions and undesirable situations. In relation to the first ergonomic aspect, it refers to the external conditions where people work. For example, where there are hazardous conditions—like X-ray machines or chemical plants—operators should be able to interact with and control computer-based systems remotely. The second aspect refers to helping any kind of user in any kind of situation avoid the dangers of carrying out unwanted actions accidentally. It also refers to the perceived fears users might have of the consequences of making errors and how this affects their behavior. To make computer-based systems safer in this sense involves (i) preventing the user from making serious errors by reducing the risk of wrong keys/buttons being mistakenly activated (an example is *not* placing the quit or delete-file command right next to the save command on a menu) and (ii) providing users with various means of recovery should they make errors. Safe interactive systems should engender confidence and allow the user the opportunity to explore the interface to try out new operations (see Figure 1.6a). Other safety mechanisms include undo facilities and

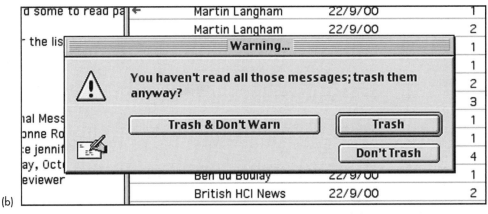

Figure 1.6 (a) A safe and an unsafe menu. Which is which and why? (b) Warning dialog message from Eudora.

confirmatory dialog boxes that give users another chance to consider their intentions (a well-known example used in e-mail applications is the appearance of a dialog box, after the user has highlighted messages to be deleted, saying: "Are you sure you want to delete all these messages?" See Figure 1.6(b)).

Question: Does the system prevent users from making serious errors and, if they do make an error, does it permit them to recover easily?

Utility refers to the extent to which the system provides the right kind of functionality so that users can do what they need or want to do. An example of a system with high utility is an accounting software package providing a powerful computational tool that accountants can use to work out tax returns. A example of a system with low utility is a software drawing tool that does not allow users to draw freehand but forces them to use a mouse to create their drawings, using only polygon shapes.

Question: Does the system provide an appropriate set of functions that enable users to carry out all their tasks in the way they want to do them?

Learnability refers to how easy a system is to learn to use. It is well known that people don't like spending a long time learning how to use a system. They want to get started straight away and become competent at carrying out tasks without too much effort. This is especially so for interactive products intended for everyday use (e.g., interactive TV, email) and those used only infrequently (e.g., videoconferencing). To a certain extent, people are prepared to spend longer learning more complex systems that provide a wider range of functionality (e.g., web authoring tools, word processors). In these situations, CD-ROM and online tutorials can help by providing interactive step-by-step material with hands-on exercises. However, many people find these tedious and often difficult to relate to the tasks they want to

---

BOX 1.3   The Ten-Minute Rule

A criterion for assessing whether a system is easy to learn is to apply the "ten-minute rule" (Nelson, 1980). It proposes that novice users should be able to learn how to use a system in under 10 minutes. If not the system fails. As pointed out by Rubinstein and Hersh (1984), many computer systems do not meet this criterion. To make systems easy to learn, they suggest that designers capitalize on people's existing knowledge: "A computer system for architects is not expected to teach architecture. Quite the reverse: the ten-minute rule requires that what an architect already knows be helpful in learning to use the architecture system," (Rubinstein and Hersh, 1984 p. 9).

When is the ten-minute rule not appropriate?

The ten-minute rule is a useful rule of thumb for evaluating many kinds of systems. However, it is inappropriate for using with complex systems, where it would be difficult and reckless to think that a user could learn how to use them in under ten minutes. For example, would you feel safe knowing that the pilots flying your plane had only spent ten minutes learning how to use the various devices in the cockpit? You would expect them to have spent considerable time (in addition to the years of training to become a pilot) thoroughly learning how to use the array of controls and displays for that particular kind of plane and what to do if any of them malfunction. Likewise, it is unrealistic to assume that ten minutes is enough to learn a system that provides diverse functionality (e.g., a word processor) or that needs high levels of skill to use (e.g., a video game).

accomplish. A key concern is determining how much time users are prepared to spend learning a system. There seems little point in developing a range of functionality if the majority of users are unable or not prepared to spend time learning how to use it.

Question: How easy is it and how long does it take (i) to get started using a system to perform core tasks and (ii) to learn the range of operations to perform a wider set of tasks?

Memorability refers to how easy a system is to remember how to use, once learned. This is especially important for interactive systems that are used infrequently. If users haven't used a system or an operation for a few months or longer, they should be able to remember or at least rapidly be reminded how to use it. Users shouldn't have to keep relearning how to carry out tasks. Unfortunately, this tends to happen when the operations required to be learned are obscure, illogical, or poorly sequenced. Users need to be helped to remember how to do tasks. There are many ways of designing the interaction to support this. For example, users can be helped to remember the sequence of operations at different stages of a task through meaningful icons, command names, and menu options. Also, structuring options and icons so they are placed in relevant categories of options (e.g., placing all the drawing tools in the same place on the screen) can help the user remember where to look to find a particular tool at a given stage of a task.

Question: What kinds of interface support have been provided to help users remember how to carry out tasks, especially for systems and operations that are used infrequently?

ACTIVITY 1.3   How long do you think it *should* take to learn how to use the following interactive products and how long does it *actually* take most people to learn them? How memorable are they?

    (a) using a VCR to play a video

    (b) using a VCR to pre-record two programs

    (c) using an authoring tool to create a website

Comment   (a) To play a video should be as simple as turning the radio on, should take less than 30 seconds to work out, and then should be straightforward to do subsequently. Most people are able to fathom how to play a video. However, some systems require the user to switch to the "video" channel using one or two remote control devices, selecting from a choice of 50 or more channels. Other settings may also need to be configured before the video will play. Most people are able to remember how to play a video once they have used a particular VCR.

    (b) This is a more complex operation and should take a couple of minutes to learn how to do and to check that the programming is correct. In reality, many VCRs are so poorly designed that 80% of the population is unable to accomplish this task, despite several attempts. Very few people remember how to pre-record a program, largely because the interaction required to do this is poorly designed, with poor or no feedback, and is often illogical from the user's perspective. Of those, only a few will bother to go through the manual again.

(c) A well-designed authoring tool should let the user create a basic page in about 20 minutes. Learning the full range of operations and possibilities is likely to take much longer, possibly a few days. In reality, there are some good authoring tools that allow the user to get started straight away, providing templates that they can adapt. Most users will extend their repertoire, taking another hour or so to learn more functions. However, very few people actually learn to use the full range of functions provided by the authoring tool. Users will tend to remember frequently used operations (e.g., cut and paste, inserting images), especially if they are consistent with the way they are carried out in other software applications. However, less frequently used operations may need to be relearned (e.g., formatting tables).

The usability goals discussed so far are well suited to the design of business systems intended to support working practices. In particular, they are highly relevant for companies and organizations who are introducing or updating applications running on desktop and networked systems—that are intended to increase productivity by improving and enhancing how work gets done. As well as couching them in terms of specific questions, usability goals are turned into *usability criteria*. These are specific objectives that enable the usability of a product to be assessed in terms of how it can improve (or not) a user's performance. Examples of commonly used usability criteria are time to complete a task (efficiency), time to learn a task (learnability), and the number of errors made when carrying out a given task over time (memorability).

1.5.2 User experience goals

The realization that new technologies are offering increasing opportunities for supporting people in their everyday lives has led researchers and practitioners to consider further goals. The emergence of technologies (e.g., virtual reality, the web, mobile computing) in a diversity of application areas (e.g., entertainment, education, home, public areas) has brought about a much wider set of concerns. As well as focusing primarily on improving efficiency and productivity at work, interaction design is increasingly concerning itself with creating systems that are:

- satisfying
- enjoyable
- fun
- entertaining
- helpful
- motivating
- aesthetically pleasing
- supportive of creativity
- rewarding
- emotionally fulfilling

The goals of designing interactive products to be fun, enjoyable, pleasurable, aesthetically pleasing and so on are concerned primarily with the user experience. By this we mean what the interaction with the system *feels* like to the users. This involves explicating the nature of the user experience in subjective terms. For example, a new software package for children to create their own music may be designed with the primary objectives of being fun and entertaining. Hence, user experience goals differ from the more objective usability goals in that they are concerned with how users experience an interactive product from their perspective, rather than assessing how useful or productive a system is from its own perspective. The relationship between the two is shown in Figure 1.7.

Much of the work on enjoyment, fun, etc., has been carried out in the entertainment and computer games industry, which has a vested interest in understanding the role of pleasure in considerable detail. Aspects that have been described as contributing to pleasure include: attention, pace, play, interactivity, conscious and unconscious control, engagement, and style of narrative. It has even been suggested that in these pleasure contexts, it might be interesting to build systems that are *non-easy* to use, providing opportunities for quite different user experiences from those designed based on usability goals (Frohlich and Murphy, 1999). Interacting with a virtual representation using a physical device (e.g., banging a plastic

Figure 1.7 Usability and user experience goals. Usability goals are central to interaction design and are operationalized through specific criteria. User experience goals are shown in the outer circle and are less clearly defined.

hammer to hit a virtual nail represented on the computer screen) compared with using a more efficient way to do the same thing (e.g., selecting an option using command keys) may require *more effort* but could, conversely, result in a *more enjoyable* and *fun* experience.

Recognizing and understanding the trade-offs between usability and user experience goals is important. In particular, this enables designers to become aware of the consequences of pursuing different combinations of them in relation to fulfilling different users' needs. Obviously, not all of the usability goals and user experience goals apply to every interactive product being developed. Some combinations will also be incompatible. For example, it may not be possible or desirable to design a process control system that is both safe and fun. As stressed throughout this chapter, what is important depends on the use context, the task at hand, and who the intended users are.

ACTIVITY1.4   Below are a number of proposed interactive products. What do you think are the key usability goals and user experience goals for each of them?

(a) a mobile device that allows young children to communicate with each other and play collaborative games

(b) a video and computer conferencing system that allows students to learn at home

(c) an Internet application that allows the general public to access their medical records via interactive TV

(d) a CAD system for architects and engineers

(e) an online community that provides support for people who have recently been bereaved

Comment

(a) Such a collaborative device should be easy to use, effective, efficient, easy to learn and use, fun and entertaining.

(b) Such a learning device should be easy to learn, easy to use, effective, motivating and rewarding.

(c) Such a personal system needs to be safe, easy to use and remember how to use, efficient and effective.

(d) Such a tool needs to be easy to learn, easy to remember, have good utility, be safe, efficient, effective, support creativity and be aesthetically pleasing.

(e) Such a system needs to be easy to learn, easy to use, motivating, emotionally satisfying and rewarding.

1.6   More on usability: design and usability principles

Another way of conceptualizing usability is in terms of design principles. These are generalizable abstractions intended to orient designers towards thinking about different aspects of their designs. A well-known example is feedback: systems should be designed to provide adequate feedback to the users to ensure they know what to do next in their tasks. Design principles are derived from a mix of theory-based knowledge, experience, and common sense. They tend to be written in a prescriptive manner, suggesting to designers what to provide and what to avoid at the interface—if you like, the do's and don'ts of interaction design. More specifically, they are intended to help designers explain and improve the design (Thimbleby, 1990). However, they are not intended to specify how to design an actual interface (e.g., telling the designer how to design a particular icon or how to structure a web portal) but act more like a set of reminders to designers, ensuring that they have provided certain things at the interface.

A number of design principles have been promoted. The best known are concerned with how to determine what users should see and do when carrying out their tasks using an interactive product. Here we briefly describe the most common ones: visibility, feedback, constraints, mapping, consistency, and affordances. Each of these has been written about extensively by Don Norman (1988) in his bestseller *The Design of Everyday Things*.

Visibility   The importance of visibility is exemplified by our two contrasting examples at the beginning of the chapter. The voice-mail system made the presence and number of waiting messages invisible, while the answer machine made both aspects highly visible. The more visible functions are, the more likely users will be able to know what to do next. In contrast, when functions are "out of sight," it makes them more difficult to find and know how to use. Norman (1988) describes the controls of a car to emphasize this point. The controls for different operations are clearly visible (e.g., indicators, headlights, horn, hazard warning lights), indicating what can be done. The relationship between the way the controls have been positioned in the car and what they do makes it easy for the driver to find the appropriate control for the task at hand.

Feedback   Related to the concept of visibility is feedback. This is best illustrated by an analogy to what everyday life would be like without it. Imagine trying to play a guitar, slice bread using a knife, or write using a pen if none of the actions produced any effect for several seconds. There would be an unbearable delay before the music was produced, the bread was cut, or the words appeared on the paper, making it almost impossible for the person to continue with the next strum, saw, or stroke.

Feedback is about sending back information about what action has been done and what has been accomplished, allowing the person to continue with the activity. Various kinds of feedback are available for interaction design—audio, tactile, verbal, visual, and combinations of these. Deciding which combinations are appropriate for different kinds of activities and interactivities is central. Using feedback in the right way can also provide the necessary visibility for user interaction.

Constraints   The design concept of constraining refers to determining ways of restricting the kind of user interaction that can take place at a given moment. There are various ways this can be achieved. A common design practice in graphical user interfaces is to deactivate certain menu options by shading them, thereby restrict-

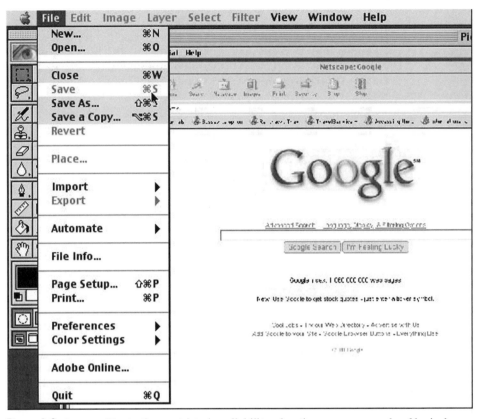

Figure 1.8 A menu illustrating restricted availability of options as an example of logical constraining. Shaded areas indicate deactivated options.

ing the user to only actions permissible at that stage of the activity (see Figure 1.8). One of the advantages of this form of constraining is it prevents the user from selecting incorrect options and thereby reduces the chance of making a mistake. The use of different kinds of graphical representations can also constrain a person's interpretation of a problem or information space. For example, flow chart diagrams show which objects are related to which, thereby constraining the way the information can be perceived.

Norman (1999) classifies constraints into three categories: physical, logical, and cultural. Physical constraints refer to the way physical objects restrict the movement of things. For example, the way an external disk can be placed into a disk drive is physically constrained by its shape and size, so that it can be inserted in only one way. Likewise, keys on a pad can usually be pressed in only one way.

Logical constraints rely on people's understanding of the way the world works (cf. the marbles answering machine design). They rely on people's common-sense reasoning about actions and their consequences. Picking up a physical marble and placing it in another location on the phone would be expected by most people to

◀▷▷▷   ▷▷▷◀

(a)   (b)

Figure 1.9 (a) Natural mapping between rewind, play, and fast forward on a tape recorder device. (b) An alternative arbitrary mapping.

trigger something else to happen. Making actions and their effects obvious enables people to logically deduce what further actions are required. Disabling menu options when not appropriate for the task in hand provides logical constraining. It allows users to reason why (or why not) they have been designed this way and what options are available.

Cultural constraints rely on learned conventions, like the use of red for warning, the use of certain kinds of audio signals for danger, and the use of the smiley face to represent happy emotions. Most cultural constraints are arbitrary in the sense that their relationship with what is being represented is abstract, and could have equally evolved to be represented in another form (e.g., the use of yellow instead of red for warning). Accordingly, they have to be learned. Once learned and accepted by a cultural group, they become universally accepted conventions. Two universally accepted interface conventions are the use of windowing for displaying information and the use of icons on the desktop to represent operations and documents.

Mapping This refers to the relationship between controls and their effects in the world. Nearly all artifacts need some kind of mapping between controls and effects, whether it is a flashlight, car, power plant, or cockpit. An example of a good mapping between control and effect is the up and down arrows used to represent the up and down movement of the cursor, respectively, on a computer keyboard. The mapping of the relative position of controls and their effects is also important. Consider the various musical playing devices (e.g., MP3, CD player, tape recorder). How are the controls of playing, rewinding, and fast forward mapped onto the desired effects? They usually follow a common convention of providing a sequence of buttons, with the play button in the middle, the rewind button on the left and the fast-forward on the right. This configuration maps directly onto the directionality of the actions (see Figure 1.9a). Imagine how difficult it would be if the mappings in Figure 1.9b were used. Look at Figure 1.10 and determine from the various mappings which is good and which would cause problems to the person using it.

Figure 1.10 Four possible combinations of arrow-key mappings. Which is the most natural mapping?

Consistency   This refers to designing interfaces to have similar operations and use similar elements for achieving similar tasks. In particular, a consistent interface is one that follows rules, such as using the same operation to select all objects. For example, a consistent operation is using the same input action to highlight any graphical object at the interface, such as always clicking the left mouse button. Inconsistent interfaces, on the other hand, allow exceptions to a rule. An example of this is where certain graphical objects (e.g., email messages presented in a table) can be highlighted only by using the right mouse button, while all other operations are highlighted using the left button. A problem with this kind of inconsistency is that it is quite arbitrary, making it difficult for users to remember and making the users more prone to mistakes.

One of the benefits of consistent interfaces, therefore, is that they are easier to learn and use. Users have to learn only a single mode of operation that is applicable to all objects. This principle works well for simple interfaces with limited operations, like a mini CD player with a small number of operations mapped onto separate buttons. Here, all the user has to do is learn what each button represents and select accordingly. However, it can be more problematic to apply the concept of consistency to more complex interfaces, especially when many different operations need to be designed for. For example, consider how to design an interface for an application that offers hundreds of operations (e.g. a word-processing application). There is simply not enough space for a thousand buttons, each of which maps onto an individual operation. Even if there were, it would be extremely difficult and time-consuming for the user to search through them all to find the desired operation.

A much more effective design solution is to create categories of commands that can be mapped into subsets of operations. For the word-processing application, the hundreds of operations available are categorized into subsets of different menus. All commands that are concerned with file operations (e.g., save, open, close) are placed together in the same file menu. Likewise, all commands concerned with formatting text are placed in a format menu. Selecting an operation then becomes a matter of homing in on the right category (menu) of options and scanning it for the desired one, rather than scrolling through one long list. However, the consistency rule of having a visible one-to-one mapping between command and operation is broken. Operations are not immediately visible at the interface, but are instead hidden under different categories of menus. Furthermore, some menu items are immediately visible, when a top-level menu is first pulled down, while others remain hidden until the visible items are scrolled over. Thus, users need to learn what items are visible in each menu category and which are hidden in submenus.

The way the items are divided between the categories of menu items can also appear inconsistent to users. Various operations appear in menus where they do not belong. For example, the sorting operation (very useful for listing references or names in alphabetical order) in Microsoft Word 2001 is in the Table menu (the Mac Version). In the previous Word 98 version, it was in both the Tools and Table menus. I always thought of it as a Tool operation (like Word Count), and became very frustrated to discover that as a default for Word 2001 it is only in the Table menu. This makes it inconsistent for me in two ways: (i) with the previous version and (ii) in the category it has been placed. Of course, I can customize the new version so that the menus are structured in the way I think they should be, but this all takes considerable time (especially when I use different machines at work, home, and when travelling).

Another problem with consistency is determining what aspect of an interface to make consistent with what else. There are often many choices, some of which can be inconsistent with other aspects of the interface or ways of carrying out actions. Consider the design problem of developing a mechanism to let users lock their files on a shared server. Should the designer try to design it to be consistent with the way people lock things in the outside world (called external consistency) or with the way they lock objects in the existing system (called internal consistency)? However, there are many different ways of locking objects in the physical world (e.g., placing in a safe, using a padlock, using a key, using a child safety lock), just as there are different ways of locking electronically (e.g., using PIN numbers, passwords, permissions, moving the physical switches on floppy disks). The problem facing designers is knowing which one to be consistent with.

Affordance is a term used to refer to an attribute of an object that allows people to know how to use it. For example, a mouse button invites pushing (in so doing activating clicking) by the way it is physically constrained in its plastic shell. At a very simple level, to afford means "to give a clue" (Norman, 1988). When the affordances of a physical object are perceptually obvious it is easy to know how to interact with it. For example, a door handle affords pulling, a cup handle affords grasping, and a mouse button affords pushing. Norman introduced this concept in the late '80s in his discussion of the design of everyday objects. Since then, it has been much popularized, being used to describe how interface objects should be designed so that they make obvious what can be done to them. For example, graphical elements like buttons, icons, links, and scroll bars are talked about with respect to how to make it appear obvious how they should be used: icons should be designed to afford clicking, scroll bars to afford moving up and down, buttons to afford pushing.

Unfortunately, the term affordance has become rather a catch-all phrase, losing much of its potency as a design principle. Norman (1999), who was largely responsible for originally promoting the concept in his book *The Design of Everyday Things* (1988), now despairs at the way it has come to be used in common parlance:

> *"I put an affordance there," a participant would say, "I wonder if the object affords clicking ... " affordances this, affordances that. And no data, just opinion. Yikes! What had I unleashed upon the world?* Norman's (1999) reaction to a recent CHI-Web discussion.

He has since tried to clarify his argument about the utility of the concept by saying there are two kinds of affordance: perceived and real. Physical objects are said to have real affordances, like grasping, that are perceptually obvious and do not have to be learned. In contrast, user interfaces that are screen-based are virtual and do not have these kinds of real affordances. Using this distinction, he argues that it does not make sense to try to design for real affordances at the interface—except when designing physical devices, like control consoles, where affordances like pulling and pressing are helpful in guiding the user to know what to do. Alternatively, screen-based

---

BOX 1.4   Can You Afford a Screen?

A problem in applying the concept of affordance to interface objects is that virtual objects have quite different properties from physical objects. A physical door handle affords pulling because its physical properties constrain what can be done with it in relation to the person and environment. It results in it opening (if it is closed) and it closing (if it is open). It is obvious to the person how to interact with it. However, a virtual object like an icon button invites clicking on only because a user has learned initially that the graphical element on the screen is a representation that, when clicked on, makes something else happen (such as moving to another page). It could equally trigger other system responses, like a window closing down. Hence, the mapping between a virtual representation and its behavior is arbitrary, relying on the user learning the accepted conventions.

A problem in using the concept of affordance in this context is that it can be misleading. Designers can be misled into thinking that virtual objects should be designed to look and behave like physical objects because people know intuitively how to interact with these. This may lead them into thinking that interfaces that exhibit this kind of realism will be easier to learn and use. These assumptions,

however, are incorrect for the reason stated above. To illustrate this point further, consider the design of screen buttons. A number of them have been designed to have a 3D look, giving the appearance of protruding. An assumption is that this kind of illusion will give the buttons the *affordance* of pushing, inviting the user to click on them, as they would do with actual physical buttons. While users may readily learn this association, it is equally the case that they will be able to learn how to interact with a simple, 2D representation of a button on the screen. The effort to learn the association is similar. However, the effort to design 3D buttons is likely to be greater than simple 2D buttons.

A danger with trying to design graphical interfaces to afford in the way physical objects do is that it can inadvertently lead to poor design. The use of shadowing and other perceptual illusions to give the effect of 3D can have the undesirable effect of cluttering up an interface, often making it more difficult to find particular objects. Simple, plain 2D abstract shapes (e.g., a square or a circle) used to represent objects like buttons, on the other hand, can be easier to perceive and recognize at the interface (See Figure 1.11 on Color Plate 1).

---

interfaces are better conceptualized as *perceived* affordances, which are essentially learned conventions. In conclusion, Norman argues that other design concepts—conventions, feedback and cultural and logical constraints—are far more useful for helping designers develop graphical user interfaces.

1.6.1   Heuristics and usability principles

When design principles are used in practice they are commonly referred to as heuristics. This term emphasizes that something has to be done with them when they are applied to a given problem. In particular, they need to be interpreted in the design context, drawing on past experience of, for example, how to design feedback and what it means for something to be consistent.

Another form of guidance is usability principles. An example is "speak the user's language." These are quite similar to design principles, except that they tend to be more prescriptive. In addition, whereas design principles tend to be used mainly for informing a design, usability principles are used mostly as the basis for evaluating prototypes and existing systems. In particular, they provide the framework for heuristic evaluation (see Chapter 13). They, too, are called heuristics when used as part of an evaluation. Below are the ten main usability principles, developed by Nielsen (2001) and his colleagues. Note how some of them overlap with the design principles.

1. *Visibility of system status*—always keep users informed about what is going on, through providing appropriate feedback within reasonable time

2. *Match between system and the real world*—speak the users' language, using words, phrases and concepts familiar to the user, rather than system-oriented terms

3. *User control and freedom*—provide ways of allowing users to easily escape from places they unexpectedly find themselves, by using clearly marked 'emergency exits'

4. *Consistency and standards*—avoid making users wonder whether different words, situations, or actions mean the same thing

5. *Help users recognize, diagnose, and recover from errors*—use plain language to describe the nature of the problem and suggest a way of solving it

6. *Error prevention*—where possible prevent errors occurring in the first place

7. *Recognition rather than recall*—make objects, actions, and options visible

8. *Flexibility and efficiency of use*—provide accelerators that are invisible to novice users, but allow more experienced users to carry out tasks more quickly

9. *Aesthetic and minimalist design*—avoid using information that is irrelevant or rarely needed

10. *Help and documentation*—provide information that can be easily searched and provides help in a set of concrete steps that can easily be followed

ACTIVITY 1.5   One of the main design principles which Nielsen has proselytized, especially for website design, is simplicity. He proposes that designers go through all of their design elements and remove them one by one. If a design works just as well without an element, then remove it. Do you think this is a good design principle? If you have your own website, try doing this and seeing what happens. At what point does the interaction break down?

Comment   Simplicity is certainly an important design principle. Many designers try to cram too much into a screenful of space, making it unwieldy for people to find what they are interested in. Removing design elements to see what can be discarded without affecting the overall function of the website can be a salutary lesson. Unnecessary icons, buttons, boxes, lines, graphics, shading, and text can be stripped, leaving a cleaner, crisper, and easier-to-navigate website. However, a certain amount of graphics, shading, coloring, and formatting can make a site aesthetically pleasing and enjoyable to use. Plain vanilla sites with just lists of text and a few hyperlinks may not be as appealing and may put certain visitors off returning. The key is getting the right balance between aesthetic appeal and the right amount and kind of information per page.

Design and usability principles have also been operationalized into even more specific prescriptions called rules. These are guidelines that should be followed. An example is "always place the quit or exit button at the bottom of the first menu list in an application."

BOX 1.5   Usable Usability: Which Terms Do I Use?

The various terms proposed for describing the different aspects of usability can be confusing. They are often used interchangeably and in different combinations. Some people talk about usability design principles, others usability heuristics, and others design concepts. The key is understanding how to use the different levels of guidance. Guidelines is the most general term used to refer to *all forms* of guidance. Goals refer to the high-level usability aims of the system (e.g., it should be efficient to use). Principles refer to general guidance intended to inform the design and evaluation of a system. Rules are low-level guidance that refer to a particular prescription that must be followed. Heuristics is a general term used to refer to design and usability principles when applied to a particular design problem.

Concept	Level of guidance	Also sometimes called	How To Use
Usability goals	General		Setting up usability criteria for assessing the acceptability of a system (e.g., "How long does it take to perform a task?").
User experience goals	General	Pleasure factors	Identifying the important aspects of the user experience (e.g., "How can you make the interactive product fun and enjoyable?").
Design principles	General	Heuristics when used in practice. Design concepts	As reminders of what to provide and what to avoid when designing an interface (e.g., "What kind of feedback are you going to provide at the interface?").
Usability principles	Specific	Heuristics when used in practice	Assessing the acceptability of interfaces, used during heuristic evaluation (e.g., "Does the system provide clearly marked exits?").
Rules	Specific		To determine if an interface adheres to a specific rule when being designed and evaluated (e.g., "Always provide a backwards and forwards navigation button on a web browser").

Assignment

This assignment is intended for you to put into practice what you have read about in this chapter. Specifically, the objective is to enable you to define usability and user experience goals and to use design and usability principles for evaluating the usability of an interactive product.

Find a handheld device (e.g. remote control, handheld computer, or cell phone) and examine how it has been designed, paying particular attention to how the user is meant to interact with it.

(a) From your first impressions, write down what first comes to mind as to what is good and bad about the way the device works. Then list (i) its functionality and (ii) the range of tasks a typical user would want to do using it. Is the functionality greater, equal, or less than what the user wants to do?

(b) Based on your reading of this chapter and any other material you have come across, compile your own set of usability and user experience goals that you think will be

DILEMMA   Usability Trade-Offs

One of the problems of applying more than one of the design principles in interaction design is that trade-offs can arise between them. For example, the more you try to constrain an interface, the less visible information becomes. The same can also happen when trying to apply a single design principle. For example, we saw how the more an interface is designed to "afford" through trying to resemble the way physical objects look, the more cluttered and difficult to use it can become. Consistency is another design principle that can be problematic to apply. As we saw earlier, trying to design an interface to be consistent with something can make it inconsistent with something else. Furthermore, sometimes inconsistent interfaces are actually easier to use than consistent interfaces. A trade-off, however, is that it can take longer to learn such an interface.

Grudin (1989) illustrates the consistency dilemma with an analogy to where knives are stored in a house. Knifes have a variety of forms—butter knives, steak knives, table knifes, fish knives. An easy place to put them all and subsequently locate them is in the top drawer by the sink. This makes it easy for everyone to find them and follows

a simple consistent rule. But what about the knives that don't fit or are too sharp to put in the drawer, like carving knives and bread knives? They are placed in a wooden block. And what about the best knives kept only for special occasions? Another exception, as they are placed in the cabinet in the other room for safekeeping. And what about other knives like putty knives and paint-scraping knives used in home projects (kept in the garage) and jack knives (kept in one's pockets or backpack)? Very quickly the consistency rule begins to break down.

Grudin notes how in extending the number of places where knives are kept inconsistency is introduced, which in turn increases the time needed *to learn* where they are all stored. However, the placement of the knives in different places often makes it *easier* to find them because they are at hand for the context in which they are used and also next to the other objects used for a specific task (e.g. all the home project tools are stored together in a box in the garage). The same is true when designing interfaces: introducing inconsistency can make it more difficult to learn an interface but in the long run can make it easier to use.

most useful in evaluating the device. Decide which are the most important ones and explain why.

(c) Translate the core usability and user experience goals you have selected into two or three questions. Then use them to assess how well your device fares (e.g., *Usability goals.* What specific mechanisms have been used to ensure safety? How easy is it to learn? *User experience goals:* Is it fun to use? Does the user get frustrated easily? If so, why?).

(d) Repeat (b) and (c) for design concepts and usability principles (again choose a relevant set).

(e) Finally, discuss possible improvements to the interface based on your usability evaluation.

Summary

In this chapter we have looked at what interaction design is and how it has evolved. We examined briefly its makeup and the various processes involved. We pointed out how the notion of usability is fundamental to interaction design. This was explained in some detail, describing what it is and how it is operationalized to assess the appropriateness, effectiveness, and quality of interactive products. A number of high-level design principles were also introduced that provide different forms of guidance for interaction design.

Key points

- Interaction design is concerned with designing interactive products to support people in their everyday and working lives.

- Interaction design is multidisciplinary, involving many inputs from wide-reaching disciplines and fields.

- Interaction design is now big business: many companies want it but don't know how to do it.

- Optimizing the interaction between users and interactive products requires taking into account a number of interdependent factors, including context of use, type of task, and kind of user.

- Interactive products need to be designed to match usability goals like ease of use and learning.

- User experience goals are concerned with creating systems that enhance the user experience in terms of making it enjoyable, fun, helpful, motivating, and pleasurable.

- Design and usability principles, like feedback and simplicity, are useful heuristics for analyzing and evaluating aspects of an interactive product.

Further reading

Here we recommend a few seminal readings. A more comprehensive list of useful books, articles, websites, videos, and other material can be found at our website.

WINOGRAD, T. (1997) From computing machinery to interaction design. In P. Denning and R. Metcalfe (eds.) *Beyond Calculation: the Next Fifty Years of Computing*. New York: Springer-Verlag, 149–162. Terry Winograd provides an overview of how interaction design has emerged as a new area, explaining how it does not fit into any existing design or computing fields. He describes the new demands and challenges facing the profession.

NORMAN, D. *The Design of Everyday Things*. New York: Doubleday, 1988 (especially Chapter 1). Norman's writing is highly accessible and enjoyable to read. He writes extensively about the design and usability of everyday objects like doors, faucets, and fridges. These examples provide much food for thought in relation to designing interfaces. The Voyager CD-ROM (sadly, now no longer published) of his collected works provides additional videos and animations that illustrate in an entertaining way many of the problems, design ideas and issues raised in the text.

NORMAN, D. (1999) *ACM Interactions Magazine*, May/June, 38–42. Affordances, conventions and design. This is a short and thought-provoking critique of design principles.

GRUDIN, J. (1990) The computer reaches out: the historical continuity of interface design. In *CHI'90 Proc.* 261–268. GRUDIN, J. (1989) The case against user interface consistency. *Communications of the ACM*, 32(10), 1164–1173 Jonathan Grudin is a prolific writer and many of his earlier works provide thought-provoking and well documented accounts of topical issues in HCI. The first paper talks about how interface design has expanded to cover many more aspects in its relatively short history. The second paper, considered a classic of its time, discusses why the concept of consistency—which had been universally accepted as good interface design up until then—was in fact highly problematic.

Interactions, January/February 2000, ACM. This special issue provides a collection of visions, critiques, and sound bites on the achievements and future of HCI from a number of researchers, designers, and practitioners.

IDEO provides a well illustrated online archive of a range of interactive products it has designed. (see *www.ideo.com*)

INTERVIEW   with Gitta Salomon

Gitta Salomon is a consultant interaction designer. She founded Swim Interaction Design Studio (swimstudio.com) in 1996 as a consultancy company to assist clients with the design of interactive products. Recently, many of her clients have included start-up companies, developing web-based and other products, who realize the importance of interaction design in ensuring their products are successful but don't know how to do this. Often they get in touch with Swim with partially developed products and ask for help with their interaction design. Swim has consulted for a range of clients, including Apple Computer, Nike, IBM, DoubleClick, Webex, and RioPort.

YR: What is your approach to interaction design?

GS: I've devised my own definition: interaction design is the design of products that reveal themselves over time. Users don't necessarily see all the functionality in interactive products when they first look at them. For example, the first screen you see on a cell phone doesn't show you everything you can do with it. As you use it, additional functionality is revealed to you. Same thing with a web-based application or a Window's application—as you use them you find yourself in different states and suddenly you can do different things. This idea of revealing over time is possible because there is a microprocessor behind the product and usually there is also a dynamic display. I believe this definition characterizes the kind of products we work on—which is a very wide range, not just web products.

YR: How would you say interaction design has changed in the years since you started Swim?

GS: I don't think what we do has changed fundamentally, but the time frame for product development is much shorter. And seemingly more people think they want interaction design assistance. That has definitely changed. There are more people who don't necessarily know what interaction design is, but they are calling us and saying "we need it." All of a sudden there is a great deal of focus and money on all of these products that are virtual and computationally based, which require a different type of design thinking.

YR: So what were the kinds of projects you were working on when you first started Swim?

GS: They were less web-centric. There was more software application design and a few hardware/software type things. For the last year and a half the focus shifted to almost exclusively web-based applications. However, these are quite similar to software applications—they just have different implementation constraints. Right at the moment, the hardware/software products are starting to pick up again—it does seem that information appliances are going to take off. The nature of the problems we solve hasn't changed much; it's the platform and associated constraints that change.

YR: What would you say are the biggest challenges facing yourself and other consultants doing interaction design these days?

GS: One of the biggest challenges is remembering that half of what we do is the design work and the other half is the communication of that design work. The clients almost never bridge the gap for us: we need to bridge it. We always have to figure out how to deliver the work so it is going to have impact. We are the ones who need to ensure that the client is going to understand it and know what to do with it. That part of the work is oftentimes the most difficult. It means we've got to figure out what is going on internally with the client and decide how what we deliver will be effective. In some cases you just start seeing there is no place to engage with the client. And I think that is a very difficult problem. Most people right now don't have a product development process. They are just going for it. And we have to figure out how to fit into what is best described as a moving train.

YR: And what do you use when you try to communicate with them? Is it a combination of talking, meetings, and reports?

GS: We do a number of different things. Usually we will give them a written document, like a report or a critique of their product. Sometimes we will give them interactive prototypes in Director or HTML, things that simulate what the product experience would feel like. In the written materials, I

Figure 1 Steelcase Worklife New York retail showroom. One of the projects Gitta Salomon was involved in was to develop an interactive sales showroom for the company called Steelcase, based in New York. The sales environment was developed to provide various sales tools, including an interactive device allowing salespeople to access case-study videos that can be projected onto the large screens in the background.

often name the things that we all need to be talking about. Then at least we all have a common terminology to discuss things. It is a measure of our success if they start using the words that we gave them, because we truly have influenced their thinking. A lot of times we'll give them a diagram of what their system is like, because nobody has ever visualized it. We serve as the visualizers, taking a random assortment of vaguely defined concepts and giving some shape to them. We'll make an artifact, which allows them to say "Yes, it is like that" or "No, it's not like that, it's like this. . . ." Without something to point to they couldn't even say to each other "No, that is not what I mean" because they didn't know if they were talking about the same thing. Many times we'll use schematic diagrams to represent system behavior. Once they have these diagrams then they can say "Oh no, we need all this other stuff in there, we forgot to tell you." It seems that nobody is writing complete lists of functionality, requirements specifications, or complete documentation anymore. This means the product ideas stay in somebody's head until we make them tangible through visualization.

YR: So this communication process is just as important as the ideas?

GS: I think it is, a lot of times.

YR: So, how do you start with a client?

GS: For clients who already have something built, I find that usually the best way for us to get started, is to begin with the client doing a comprehensive demo of their product for us. We will usually spend a whole day collecting information. Besides the demo, they tell us about their target market, competitors, and a whole range of things. It then takes a longer period of time for us to use the product and observe other people using it to get a much broader picture. Because the client's own vision of their product is so narrow, we really have to step back from what they initially tell us.

YR: So do you write notes, and then try and put it together afterwards, or—what?

GS: We use all kinds of things. We use notes and video, and we sit around with tracing paper and marker pens. When reviewing the materials, I often try and bring them together in some sort of thematic way. It's often mind-boggling to bring a software product that's been thrown together into any kind of coherent framework. It's easy to write a shopping list of observations, but we want to assemble a larger structure and framework and that takes several weeks to construct. We need time to reflect and stew on what was done and what maybe should have been done. We need to highlight the issues and put them into some kind of larger order. If you always operate at a low level of detail, like worrying and critiquing the size of a button, you end up solving only local issues. You never really get to the big interaction design problems of the product, the ones that should be solved first.

YR: If you're given a prototype or product to evaluate and you discover that it is really bad, what do you do?

GS: Well, I never have the guts to go in and say something is fundamentally flawed. And that's maybe not the best strategy anyway, because it's your word against theirs. Instead, I think it is always about making the case for why something is wrong or flawed. Sometimes I think we are like lawyers. We have to assemble the case for what's wrong with the product. We have to make a convincing argument. A lot of times I think the kind of argumentation we do is very much like what lawyers do.

YR: Finally, how do you see interaction design moving in the next five years? More of the same kind of problems with new emerging technologies? Or do you think there are going to be more challenges, especially with the hardware/software integration?

GS: I think there will be different constraints as new technologies arise. No matter what we are designing, we have to understand the constraints of the implementation. And yes, different things will happen when we get more into designing hardware/software products. There are different kinds of cost constraints and different kinds of interactions you can do when there is special purpose hardware involved. Whereas designing the interaction for applications requires visual design expertise, designing information appliances or other hardware products requires experience with product design. Definitely, there will be some new challenges.

Hopefully, in the next few years, people will stop looking for interaction design rules. There's been a bit of a push towards making interaction design a science lately. Maybe this has happened because so many people are trying to do it and they don't know where to start because they don't have much experience. I'm hoping people will start understanding that interaction design is a design discipline—that there are some guidelines and ways to do good practice—and creativity combined with analytical thinking are necessary to arrive at good products. And then, even more so than now, it is going to get interesting and be a really exciting time.

Chapter 2

Understanding and conceptualizing interaction

2.1   Introduction
2.2   Understanding the problem space
2.3   Conceptual models
    2.3.1   Conceptual models based on activities
    2.3.2   Conceptual models based on objects
    2.3.3   A case of mix and match?
2.4   Interface metaphors
2.5   Interaction paradigms
2.6   From conceptual models to physical design

2.1   Introduction

Imagine you have been asked to design an application to let people organize, store, and retrieve their email in a fast, efficient and enjoyable way. What would you do? How would you start? Would you begin by sketching out how the interface might look, work out how the system architecture will be structured, or even just start coding? Alternatively, would you start by asking users about their current experiences of saving email, look at existing email tools and, based on this, begin thinking about why, what, and how you were going to design the application?

Interaction designers would begin by doing the latter. It is important to realize that having a clear understanding of what, why, and how you are going to design something, before writing any code, can save enormous amounts of time and effort later on in the design process. Ill-thought-out ideas, incompatible and unusable designs can be ironed out while it is relatively easy and painless to do. Once ideas are committed to code (which typically takes considerable effort, time, and money), they become much harder to throw away—and much more painful. Such preliminary thinking through of ideas about user needs[1] and what

---

[1]User needs here are the range of possible requirements, including user wants and experiences.

kinds of designs might be appropriate is, however, a skill that needs to be learned. It is not something that can be done overnight through following a checklist, but requires practice in learning to identify, understand, and examine the issues—just like learning to write an essay or to program. In this chapter we describe what is involved. In particular, we focus on what it takes to understand and conceptualize interaction.

The main aims of this chapter are to:

- Explain what is meant by the problem space.
- Explain how to conceptualize interaction.
- Describe what a conceptual model is and explain the different kinds.
- Discuss the pros and cons of using interface metaphors as conceptual models.
- Debate the pros and cons of using realism versus abstraction at the interface.
- Outline the relationship between conceptual design and physical design.

2.2   Understanding the problem space

In the process of creating an interactive product, it can be temping to begin at the "nuts and bolts" level of the design. By this, we mean working out how to design the physical interface and what interaction styles to use (e.g., whether to use menus, forms, speech, icons, or commands). A problem with trying to solve a design problem beginning at this level is that critical usability goals and user needs may be overlooked. For example, consider the problem of providing drivers with better navigation and traffic information. How might you achieve this? One could tackle the problem by thinking straight away about a good technology or kind of interface to use. For example, one might think that augmented reality, where images are superimposed on objects in the real world (see Figure 2.1 on Color Plate 2), would be appropriate, since it can be useful for integrating additional information with an ongoing activity (e.g., overlaying X-rays on a patient during an operation). In the context of driving, it could be effective for displaying information to drivers who need to find out where they are going and what to do at certain points during their journey. In particular, images of places and directions to follow could be projected inside the car, on the dashboard or rear-view mirror. However, there is a major problem with this proposal: it is likely to be very unsafe. It could easily distract drivers, luring them to switch their attention from the road to where the images were being projected.

A problem in starting to solve a design problem at the physical level, therefore, is that usability goals can be easily overlooked. While it is certainly necessary at some point to decide on the design of physical aspects, it is better to make these kinds of design decisions *after* understanding the nature of the problem space. By this, we mean conceptualizing what you want to create and articulating why you want to do so. This requires thinking through how your design will support people in their everyday or work activities. In particular, you need to ask yourself whether the interactive product you have in mind will achieve what you hope it will. If so, how? In the above example, this involves finding out what is problematic with existing forms of navigating while driving (e.g., trying to read maps while moving the steering wheel) and how to ensure that drivers can continue to drive safely without being distracted.

Clarifying your usability and user experience goals is a central part of working out the problem space. This involves making explicit your implicit assumptions and claims. Assumptions that are found to be vague can highlight design ideas that need to be better formulated. The process of going through them can also help to determine relevant user needs for a given activity. In many situations, this involves identifying human activities and interactivities that are problematic and working out how they might be improved through being supported with a different form of interaction. In other situations it can be more speculative, requiring thinking through why a novel and innovative use of a new technology will be potentially useful.

Below is another scenario in which the problem space focuses on solving an identified problem with an existing product. Initial assumptions are presented first, followed by a further explanation of what lies behind these (assumptions are highlighted in italics):

A large software company has decided to develop an upgrade of its web browser. *They assume that there is a need for a new one, which has better and more powerful functionality.* They begin by carrying out an extensive study of people's actual use of web browsers, talking to lots of different kinds of users and observing them using their browsers. One of their main findings is that many people do not use the bookmarking feature effectively. A common finding is that it is too restrictive and underused. *In fathoming why this is the case, it was considered that the process of placing web addresses into hierarchical folders was an inadequate way of supporting the user activity* of needing to mark hundreds and sometimes thousands of websites such that any one of them could be easily returned to or forwarded onto other people. *An implication of the study was that a new way of saving and retrieving web addresses was needed.*

In working out why users find the existing feature of bookmarking cumbersome to use, a further assumption was explicated:

- *The existing way of organizing saved (favorite) web addresses into folders is inefficient because it takes too long and is prone to errors.*

A number of underlying reasons why this was assumed to be the case were further identified, including:

- It is easy to lose web addresses by placing them accidentally into the wrong folders.
- It is not easy to move web addresses between folders.
- It is not obvious how to move a number of addresses from the saved favorite list into another folder simultaneously.
- It is not obvious how to reorder web addresses once placed in folders.

Based on this analysis, a set of assumptions about the user needs for supporting this activity more effectively were then made. These included:

- If the bookmarking function was improved users would find it more useful and use it more to organize their web addresses.

- Users need a flexible way of organizing web addresses they want to keep for further reference or for sending on to other people.

A framework for explicating assumptions

Reasoning through your assumptions about why something might be a good idea enables you to see the strengths and weaknesses of your proposed design. In so doing, it enables you to be in a better position to commence the design process. We have shown you how to begin this, through operationalizing relevant usability goals. In addition, the following questions provide a useful framework with which to begin thinking through the problem space:

- Are there problems with an existing product? If so, what are they? Why do you think there are problems?

- Why do you think your proposed ideas might be useful? How do you envision people integrating your proposed design with how they currently do things in their everyday or working lives?

- How will your proposed design support people in their activities? In what way does it address an identified problem or extend current ways of doing things? Will it really help?

ACTIVITY 2.1   At the turn of the millennium, WAP-enabled (wireless application protocol) phones came into being, that enabled people to connect to the Internet using them. To begin with, the web-enabled services provided were very primitive, being text-based with limited graphics capabilities. Access was very restricted, with the downloaded information being displayed on a very small LCD screen (see Figure 2.2). Despite this major usability drawback, every telecommunication company saw this technological breakthrough as an opportunity to create innovative applications. A host of new services were explored, including text messaging, online booking of tickets, betting, shopping, viewing movies, stocks and shares, sports events and banking.

What assumptions were made about the proposed services? How reasonable are these assumptions?

Figure 2.2 An early cell phone display. Text is restricted to three or four lines at a time and scrolls line by line, making reading very cumbersome. Imagine trying to read a page from this book in this way! The newer 3G (third generation) phones have bigger displays, more akin to those provided with handheld computers.

Comment   The problem space for this scenario was very *open-ended*. There was no identifiable problem that needed to be improved or fixed. Alternatively, the new WAP technology provided opportunities to create new facilities and experiences for people. One of the main assumptions is that *people want to be kept informed* of up-to-the-minute news (e.g. sports, stocks and share prices) *wherever they are*. Other assumptions included:

- That people want to be able to decide what to do in an evening while on their way home from work (e.g., checking TV listings, movies, making restaurant reservations).
- That people *want to be able to interact with information on the move* (e.g., reading email on the train).
- That users are prepared to put up with a very small display and will be happy browsing and interacting with information using a restricted set of commands via a small number of tiny buttons.
- That people *will be happy* doing things on a mobile phone that they normally do using their PCs (e.g., reading email, surfing the web, playing video games, doing their shopping).

It is reasonable to assume that people want flexibility. They like to be able to find out about news and events wherever they are (just look at the number of people who take a radio with them to a soccer match to find out the scores of other matches being played at the same time). People also like to use their time productively when traveling, as in making phone calls. Thus it is reasonable to assume they would like to read and send email on the move. The most troublesome assumption is whether people are prepared to interact with the range of services proposed using such a restricted mode of interactivity. In particular, it is questionable whether most people are prepared to give up what they have been used to (e.g. large screen estate, ability to type messages using a normal-sized keyboard) for the flexibility of having access to very restricted Internet-based information via a cell phone they can keep in their pocket.

One of the benefits of working through your assumptions for a problem space before building anything is that it can highlight problematic concerns. In so doing, it can identify ideas that need to be reworked, before it becomes too late in the design process to make changes. Having a good understanding of the problem space can also help greatly in formulating what it is you want to design. Another key aspect of conceptualizing the problem space is to think about the overall structure of what will be built and how this will be conveyed to the users. In particular, this involves developing a conceptual model.

2.3 Conceptual models

> *"The most important thing to design is the user's conceptual model. Everything else should be subordinated to making that model clear, obvious, and substantial. That is almost exactly the opposite of how most software is designed." (David Liddle, 1996, p. 17)*

By a conceptual model is meant:

> *a description of the proposed system in terms of a set of integrated ideas and concepts about what it should do, behave and look like, that will be understandable by the users in the manner intended.*

To develop a conceptual model involves envisioning the proposed product, based on the users' needs and other requirements identified. To ensure that it is designed to be understandable in the manner intended requires doing iterative testing of the product as it is developed. A key aspect of this design process is initially to decide what the users will be doing when carrying out their tasks. For example, will they be primarily searching for information, creating documents, communicating with other users, recording events, or some other activity? At this stage, the interaction mode that would best support this needs to be considered. For example, would allowing the users to browse be appropriate, or would allowing them to ask questions directly to the system in their native language be more effective? Decisions about which kind of interaction style to use (e.g., whether to use a menu-based system, speech input, commands) should be made in relation to the interaction mode. Thus, decisions about which mode of interaction to support differ from those made about which style of interaction to have; the former being at a higher level of abstraction. The former are also concerned with determining the nature of the users' activities to support, while the latter are concerned with the selection of specific kinds of interface.

Once a set of possible ways of interacting with an interactive system has been identified, the design of the conceptual model then needs to be thought through in terms of actual concrete solutions. This entails working out the behavior of the interface, the particular interaction styles that will be used, and the "look and feel" of the interface. At this stage of "fleshing out," it is always a good idea to explore a number of possible designs and to assess the merits and problems of each one.

Another way of designing an appropriate conceptual model is to select an interface metaphor. This can provide a basic structure for the conceptual model that is couched in knowledge users are familiar with. Examples of well-known interface metaphors are the desktop and search engines (which we will cover in Section 2.4). Interaction paradigms can also be used to guide the formation of an appropriate conceptual metaphor. They provide particular ways of thinking about interaction design, such as designing for desktop applications or ubiquitous computing (these will also be covered in Section 2.5).

As with any aspect of interaction design, the process of fleshing out conceptual models should be done iteratively, using a number of methods. These include sketching out ideas, storyboarding, describing possible scenarios, and prototyping aspects of the proposed behavior of the system. All these methods will be covered in Chapter 8, which focuses on *doing* conceptual design. Here, we describe the different kinds of conceptual models, interface metaphors, and interaction paradigms to give you a good understanding of the various types prior to thinking about how to design them.

There are a number of different kinds of conceptual models. These can be broken down into two main categories: those based on activities and those based on objects.

2.3.1 Conceptual models based on activities

The most common types of activities that users are likely to be engaged in when interacting with systems are:

1. instructing
2. conversing
3. manipulating and navigating
4. exploring and browsing

A first thing to note is that the various kinds of activity are not mutually exclusive, as they can be carried out together. For example, it is possible for someone to give instructions while conversing or navigate an environment while browsing. However, each has different properties and suggests different ways of being developed at the interface. The first one is based on the idea of letting the user issue instructions to the system when performing tasks. This can be done in various interaction styles: typing in commands, selecting options from menus in a windows environment or on a touch screen, speaking aloud commands, pressing buttons, or using a combination of function keys. The second one is based on the user conversing with the system as though talking to someone else. Users speak to the system or type in questions to which the system replies via text or speech output. The third type is based on allowing users to manipulate and navigate their way through an environment of virtual objects. It assumes that the virtual environment shares some of the properties of the physical world, allowing users to use their knowledge of how physical objects behave when interacting with virtual objects. The fourth kind is based on the system providing information that is structured in such a way as to allow users to find out or learn things, without having to formulate specific questions to the system.

ACTIVITY 2.2 A company is building a wireless information system to help tourists find their way around an unfamiliar city. What would they need to find out in order to develop a conceptual model?

Comment To begin, they would need to ask: what do tourists want? Typically, they want to find out lots of things, such as how to get from A to B, where the post office is and where a good Chinese restaurant is. They then need to consider how best to support the activity of requesting information. Is it preferable to enable the tourists to ask questions of the system as if they were having a conversation with another human being? Or would it be more appropriate to allow them to ask questions as if giving instructions to a machine? Alternatively, would they prefer a system that structures information in the form of lists, maps, and recommendations that they could then explore at their leisure?

1. Instructing

This kind of conceptual model describes how users carry out their tasks through instructing the system what to do. Examples include giving instructions to a system to perform operations like tell the time, print a file, and remind the user of an appointment. A diverse range of devices has been designed based on this model, including VCRs, hi-fi systems, alarm clocks, and computers. The way in which the user issues instructions can vary from pressing buttons to typing in strings of characters. Many activities are readily supported by giving instructions.

Operating systems like Unix and DOS have been specifically designed as command-based systems, to which the user issues instructions at the prompt as a command or set of commands. In Windows and other GUI-based systems, control keys or the selection of menu options via a mouse are used. Well-known applications that are command-based include word processing, email, and CAD. Typically, a wide range of functions is provided from which users choose when they want to do something to the object they are working on. For example, a user writing a report using a word processor will want to format the document, count the numbers of words typed, and check the spelling. The user will need to instruct the system to do these operations by issuing appropriate commands. Typically, commands are carried out in a sequence, with the system responding appropriately (or not) as instructed.

One of the main benefits of an instruction-based conceptual model is that it supports quick and efficient interaction. It is particularly suited to repetitive kinds of actions performed on multiple objects. Examples include the repetitive actions of saving, deleting, and organizing email messages or files.

ACTIVITY 2.3   There are many different kinds of vending machines in the world. Each offers a range of goods, requiring the user initially to part with some money. Figure 2.3 shows photos of two different vending machines, one that provides soft drinks and the other a range of snacks. Both support the interaction style of issuing instructions. However, the way they do it is quite different.

What instructions must be issued to obtain a can of soft drink from the first machine and a bar of chocolate from the second? Why has it been necessary to design a more complex mode of interaction for the second vending machine? What problems can arise with this mode of interaction?

Comment   The first vending machine has been designed on a very simple instruction-based conceptual model. There are a small number of drinks to choose from and each is represented by a large button displaying the label of each drink. The user simply has to press one button and (hopefully) this will have the effect of returning the selected drink. The second machine is more complex, offering a wider range of snacks. The trade-off for providing more choices, however, is that the user can no longer instruct the machine by using a simple one-press action but is required to use a more complex process, involving: (i) reading off the code (e.g., C12) under the item chosen, then (ii) keying this into the number pad adjacent to the displayed items, and (iii) checking the price of the selected option and ensuring that the amount of money inserted is the same or more (depending on whether or not the machine provides change). Problems that can arise from this mode of interaction are the customer

Figure 2.3 Two vending machines, (a) one selling soft drinks, (b) the other selling a range of snacks.

misreading the code and or mistyping in the code, resulting in the machine not issuing the snack or providing the wrong sort.

A better way of designing an interface for a large number of choices of variable cost is to continue to use direct mapping, but use buttons that show miniature versions of the snacks placed in a large matrix (rather than showing actual versions). This would use the available space at the front of the vending machine more economically. The customer would need only to press the button of the object chosen and put in the correct amount of money.

Much research has been carried out on how to optimize command-based and other instruction-giving systems with respect to usabilty goals. The form of the commands (e.g., the use of abbreviations, full names, icons, and/or labels), their syntax (how best to combine different commands), and their organization (e.g., how to structure options in different menus) are examples of some of the main areas that have been investigated (Shneiderman, 1998). In addition, various cognitive issues have been investigated that we will look at in the next chapter, such as the problems people have in remembering the names of a set of commands. Less research has been carried out, however, on the best way to design the ordering and sequencing of button pressing for physical devices like cell phones, calculators, remote controls and vending machines.

ACTIVITY 2.4   Another ubiquitous vending machine is the ticket machine. Typically, a number of instructions have to be given in a sequence when using one of these. Consider ticket machines designed to issue train tickets at railway stations—how often have you (or the person in front of you) struggled to work out how to purchase a ticket and made a mistake? How many instructions have to be given? What order are they given in? Is it logical or arbitrary? Could the interaction have been designed any differently to make it more obvious to people how to issue instructions to the machine to get the desired train ticket?

Comment   Ticketing machines vary enormously from country to country and from application to application. There seems to be little attempt to standardize. Therefore, a person's knowledge of the Eurostar ticketing machine will not be very useful when buying a ticket for the Sydney Monorail or cinema tickets for the Odeon. Sometimes the interaction has been designed to get you to specify the type of ticket first (e.g. adult, child), the kind of ticket (e.g. single, return, special saver), then the destination, and finally to insert their money. Others require that the user insert a credit card first, before selecting the destination and the type of ticket.

2.   Conversing

This conceptual model is based on the idea of a person conversing with a system, where the system acts as a dialog partner. In particular, the system is designed to respond in a way another human being might when having a conversation with someone else. It differs from the previous category of instructing in being intended to reflect a more two-way communication process, where the system acts more like a partner than a machine that simply obeys orders. This kind of conceptual model has been found to be most useful for applications in which the user needs to find out specific kinds of information or wants to discuss issues. Examples include advisory systems, help facilities, and search engines. The proposed tourist application described earlier would fit into this category.

The kinds of conversation that are supported range from simple voice-recognition menu-driven systems that are interacted with via phones to more complex natural-language-based systems that involve the system parsing and responding to user queries typed in by the user. Examples of the former include banking, ticket booking, and train time inquiries, where the user talks to the system in single-word phrases (e.g., yes, no, three) in response to prompts from the system. Examples of the latter include search engines and help systems, where the user types in a specific query (e.g., how do I change the margin widths?) to which the system responds by giving various answers.

A main benefit of a conceptual model based on holding a conversation is that it allows people, especially novices, to interact with a system in a way they are already familiar with. For example, the search engine "Ask Jeeves for Kids!" allows children to ask a question in a way they would when asking their teachers or parents—rather than making them reformulate their question in terms of key words and Boolean logic. A disadvantage of this approach, however, is the misunderstandings that can arise when the search engine is unable to answer the child's question in the

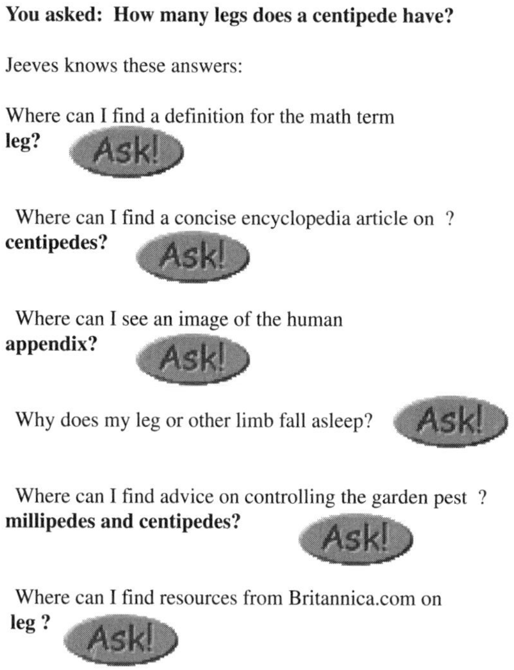

You asked: How many legs does a centipede have?

Jeeves knows these answers:

Where can I find a definition for the math term
leg? *Ask!*

Where can I find a concise encyclopedia article on ?
centipedes? *Ask!*

Where can I see an image of the human
appendix? *Ask!*

Why does my leg or other limb fall asleep? *Ask!*

Where can I find advice on controlling the garden pest ?
millipedes and centipedes? *Ask!*

Where can I find resources from Britannica.com on
leg ? *Ask!*

Figure 2.4 The response from "Ask Jeeves for Kids!" search engine when asked "how many legs does a centipede have?"

way the child expects. For example, a child might type in a seemingly simple question, like "How many legs does a centipede have?" which the search engine finds difficult to answer. Instead, the search engine replies by suggesting a number of possible websites that may be relevant but–as can be seen in Figure 2.4–can be off the mark.

Another problem that can arise from a conversational-based, conceptual model is that certain kinds of tasks are transformed into cumbersome and one-sided interactions. This is especially the case for automated phone-based systems that use auditory menus to advance the conversation. Users have to listen to a voice providing several options, then make a selection, and repeat through further layers of menus before accomplishing their goal (e.g., reaching a real human, paying a bill). Here is the beginning of a dialog between a user who wants to find out about car insurance and an insurance company's reception system:

```
<user dials an insurance company>
"Welcome to St. Paul's Insurance Company. Press 1 if new
customer, 2 if you are an existing customer".
<user presses 1>
"Thank you for calling St. Paul's Insurance Company. If you
require house insurance press 1, car insurance press 2,
travel insurance press 3, health insurance press 4, other
press 5"
<user presses 2>
"You have reached the car insurance division. If you re-
quire information about fully comprehensive insurance press
1, 3rd-party insurance press 2..."
```

"If you'd like to press 1, press 3.
If you'd like to press 3, press 8.
If you'd like to press 8, press 5..."

A recent development based on the conversing conceptual model is animated agents. Various kinds of characters, ranging from "real" people appearing at the interface (e.g., videoed personal assistants and guides) to cartoon characters (e.g., virtual and imaginary creatures), have been designed to act as the partners in the conversation with the system. In so doing, the dialog partner has become highly visible and tangible, appearing to both act and talk like a human being (or creature). The user is able to see, hear, and even touch the partner (when it is a physical toy) they are talking with, whereas with other systems based on a dialog partner (e.g., help systems) they can only hear or read what the system is saying. Many agents have also been designed to exhibit desirable human-like qualities (e.g., humorous, happy, enthusiastic, pleasant, gentle) that are conveyed through facial expressions and lifelike physical movements (head and lip movements, body movements). Others have been designed more in line with Disney-like cartoon characters, exhibiting exaggerated behaviors (funny voices, larger-than-life facial expressions).

Animated agents that exhibit human-like or creature-like physical behavior as well as "talk" can be more believable. The underlying conceptual model is conveyed much more explicitly through having the system act and talk via a visible agent. An advantage is that it can make it easier for people to work out that the interface agent (or physical toy) they are conversing with is not a human being, but a synthetic character that has been given certain human qualities. In contrast, when the dialog partner is hidden from view, it is more difficult to discern what is behind it and just how intelligent it is. The lack of visible cues can lead users into thinking it is more intelligent than it actually is. If the dialog partner then fails to understand their questions or comments, users are likely to lose patience with it. Moreover, they are likely to be less forgiving of it (having been fooled into thinking the dialog partner is more intelligent than it really is) than of a dialog partner that is represented as a cartoon character at the interface (having only assumed it was a simple partner). The flip side of imbuing dialog partners with a physical presence at the interface, however, is that they can turn out to be rather annoying (for more on this topic see Chapter 5).

3. Manipulating and navigating

This conceptual model describes the activity of manipulating objects and navigating through virtual spaces by exploiting users' knowledge of how they do this in the physical world. For example, virtual objects can be manipulated by moving, selecting, opening, closing, and zooming in and out of them. Extensions to these actions can also be included, such as manipulating objects or navigating through virtual spaces, in ways not possible in the real world. For example, some virtual worlds have been designed to allow users to teleport from place to place or to transform one object into another.

A well known instantiation of this kind of conceptual model is direct manipulation. According to Ben Shneiderman (1983), who coined the term, direct-manipulation interfaces possess three fundamental properties:

- continuous representation of the objects and actions of interest
- rapid reversible incremental actions with immediate feedback about the object of interest
- physical actions and button pressing instead of issuing commands with complex syntax

Benefits of direct manipulation interfaces include:

- helps beginners learn basic functionality rapidly
- experienced users can work rapidly on a wide range of tasks
- infrequent users can remember how to carry out operations over time
- no need for error messages, except very rarely
- users can immediately see if their actions are furthering their goals and if not do something else
- users experience less anxiety
- users gain confidence and mastery and feel in control

Apple Computer Inc. was one of the first computer companies to design an operating environment using direct manipulation as its central mode of interaction. The highly successful Macintosh desktop demonstrates the main principles of direct manipulation (see Figure 2.5). To capitalize on people's understanding of what happens to physical objects in the real world, they used a number of visual and auditory cues at the interface that were intended to emulate them. One of

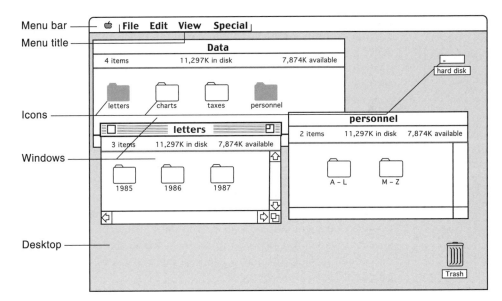

Figure 2.5 Original Macintosh desktop interface.

their assumptions was that people expect their physical actions to have physical results, so when a drawing tool is used, a corresponding line should appear and when a file is placed in the trash can a corresponding sound or visual cue showing it has been successfully thrown away is used (Apple Computer Inc., 1987). A number of specific visual and auditory cues were used to provide such feedback, including various animations and sounds (e.g. shrinking and expanding icons accompanied with '*shhhlicc*' and '*crouik*' sounds to represent opening and closing of files). Much of this interaction design was geared towards providing clues to the user to know what to do, to feel comfortable, and to enjoy exploring the interface.

Many other kinds of direct manipulation interfaces have been developed, including video games, data visualization tools and CAD systems. Virtual environments and virtual reality have similarly employed a range of interaction mechanisms that enable users to interact with and navigate through a simulated 3D physical world. For example, users can move around and explore aspects of a 3D environment (e.g., the interior of a building) while also moving objects around in the virtual environment, (e.g., rearranging the furniture in a simulated living room). Figure 2.6 on Color Plate 3 shows screen shots of some of these.

While direct manipulation and virtual environments provide a very versatile mode of interaction, they do have a number of drawbacks. At a conceptual level, some people may take the underlying conceptual model too literally and expect certain things to happen at the interface in the way they would in the physical world. A well known example of this phenomenon is of new Mac users being terrified of dragging the icon of their floppy disk to the trash can icon on the desktop to eject it from the computer for fear of deleting it in the same way files are when placed in the trash can. The conceptual confusion arises because the designers opted to use the same action (dropping) on the same object (trash can) for two completely different operations, deleting and ejecting. Another problem is that not all tasks can be described by objects and not all actions can be done directly. Some tasks are better achieved through issuing instructions and having textual descriptions rather than iconic representations. Imagine if email messages were represented as small icons in your mailbox with abbreviations of who they were from and when they were sent. Moreover, you could only move them around by dragging them with a mouse. Very quickly they would take up your desk space and you would find it impossible to keep track of them all.

4. Exploring and browsing

This conceptual model is based on the idea of allowing people to explore and browse information, exploiting their knowledge of how they do this with existing media (e.g., books, magazines, TV, radio, libraries, pamphlets, brochures). When people go to a tourist office, a bookstore, or a dentist's office, often they scan and flick through parts of the information displayed, hoping to find something interesting to read. CD-ROMs, web pages, portals and e-commerce sites are applications based on this kind of conceptual model. Much thought needs to go into structuring the information in ways that will support effective navigation, allowing people to search, browse, and find different kinds of information.

ACTIVITY 2.5   What conceptual models are the following applications based on?

(a) a 3D video game, say a car-racing game with a steering wheel and tactile, audio, and visual feedback

(b) the Windows environment

(c) a web browser

Comment

(a) A 3D video game is based on a direct manipulation/virtual environment conceptual model.

(b) The Windows environment is based on a hybrid form of conceptual model. It combines a manipulating mode of interaction where users interact with menus, scrollbars, documents, and icons, an instructing mode of interaction where users can issue commands through selecting menu options and combining various function keys, and a conversational model of interaction where agents (e.g. Clippy) are used to guide users in their actions.

(c) A web browser is also based on a hybrid form of conceptual model, allowing users to explore and browse information via hyperlinks and also to instruct the network what to search for and what results to present and save.

BOX 2.1   Which is Best—Agents, Direct Manipulations, or Commands?

An ongoing debate in interaction design concerns the pros and cons of using direct manipulation versus interface agents. Nicholas Negroponte (MIT Media Lab), a strong advocate of the agents approach, claims that they can be much more versatile than direct manipulation interfaces, allowing users to do what they want to do through delegating the boring and time-consuming tasks to an agent. He describes the analogy of a well-trained English butler, who answers the phone, tends to a person's needs, fends off callers, and tells 'white lies' if necessary on his master's behalf. Similarly, a *digital* butler is designed to read a user's email and flag the important ones, scout the web and newsgroups for interesting information, screen unwanted electronic intrusions, and so on. His vision is based on the assumption that people like to delegate work to others rather than directly manipulating computers themselves.

In opposition, Ben Shneiderman (University of Maryland) warns of the dangers of delegating tasks to agents, pointing out how difficult it is to train an agent to do all the things users want done in the way they want them done. If the agents do the tasks incorrectly or do not understand what the user wants, frustration and anger will ensue. Moreover, he argues that users do not want to be constantly monitored and told what to do by the computer. Consider the analogy of your car deciding you should be driving more slowly because it is raining. He suggests that direct manipulation has many more advantages, allowing users to enjoy mastery and being in control. He points out how people like to know what is going on, be involved in the action and have a sense of power over the computer—all of which direct manipulation interfaces support.

Another perspective on this debate is that in fact many tasks are best carried out at an abstract level, involving neither manipulation nor conversing with an agent. Issuing abstract commands based on a carefully designed set of syntax and semantics is often a very efficient and elegant way of performing operations. This is especially the case for repetitive operations, where often the same action needs to be performed on multiple objects. Examples include sorting out files, deleting accumulated email messages, opening and closing files, and installing applications comprising multiple files—which when done by direct manipulation or through delegation can be inefficient or ambiguous.

Consider how you would edit an essay using a word processor. Suppose you had referenced work by Ben Shneiderman but had spelled his name as Schneiderman, with an extra "c" throughout the essay. How could you correct this error using a direct manipulation interface? You would need to read through your essay and manually select the "c" in every "Schneiderman," highlighting and then deleting it. This is tedious and it would be easy to miss one or two. By contrast, this operation would be relatively effortless and also likely to be more accurate by issuing commands. You would simply need to instruct the word processor to *find* every "Schneiderman" and *replace* it with "Shneiderman." This could be done through either speaking the commands or typing them into a dialog box.

As a compromise, several interaction designers have recognized the need to support abstract classes of action in direct manipulation interfaces, while allowing for command-based and dialog modes of interaction in direct manipulation interfaces. However, as mentioned earlier, such redundancy can result in a more complex conceptual model that the user will have to spend more time learning.

ACTIVITY 2.6   Which conceptual model or combination of models do you think is most suited to supporting the following user activities?

        (a) downloading music off the web

        (b) programming

Comment

        (a) The activity involves selecting, saving, cataloging and retrieving large files from an external source. Users need to be able to browse and listen to samples of the music and then instruct the machine to save and catalog the files in an order that they can readily access at subsequent times. A conceptual model based on instructing and navigating would seem appropriate.

        (b) Programming involves various activities including checking, debugging, copying libraries, editing, testing, and annotating. An environment that supports this range of tasks needs to be flexible. A conceptual model that allows visualization and easy manipulation of code plus efficient instructing of the system on how to check, debug, copy, etc. is essential.

2.3.2   Conceptual models based on objects

The second category of conceptual models is based on an object or artifact, such as a tool, a book, or a vehicle. These tend to be more specific than conceptual models based on activities, focusing on the way a particular object is used in a particular context. They are often based on an analogy with something in the physical world. An example of a highly successful conceptual model based on an object is the spreadsheet (Winograd, 1996). The object this is based on is the ledger sheet.

The first spreadsheet was designed by Dan Bricklin, and called VisiCalc. It enabled people to carry out a range of tasks that previously could only be done very laboriously and with much difficulty using other software packages, a calculator, or by hand (see Figure 2.7). The main reasons why the spreadsheet has become so successful are first, that Bricklin *understood* what kind of tool would be useful to people in the financial world (like accountants) and second, he knew how to design it so that it could be used in the way that these people would find useful. Thus, at the outset, he understood (i) the kinds of activities involved in the financial side of business, and (ii) the problems people were having with existing tools when trying to achieve these activities.

A core financial activity is forecasting. This requires projecting financial results based on assumptions about a company, such as projected and actual sales, investments, infrastructure, and costs. The amount of profit or loss is calculated for different projections. For example, a company may want to determine how much loss it will incur before it will start making a profit, based on different amounts of investment, for different periods of time. Financial analysts need to see a spread of projections for different time periods. Doing this kind of multiple projecting by hand requires much effort and is subject to errors. Using a calculator can reduce the computational load of doing numerous sums, but it still requires the person to do much key pressing and writing down of partial results—again making the process vulnerable to errors.

To tackle these problems, Bricklin exploited the interactivity provided by microcomputers and developed an application that was capable of *interactive* financial

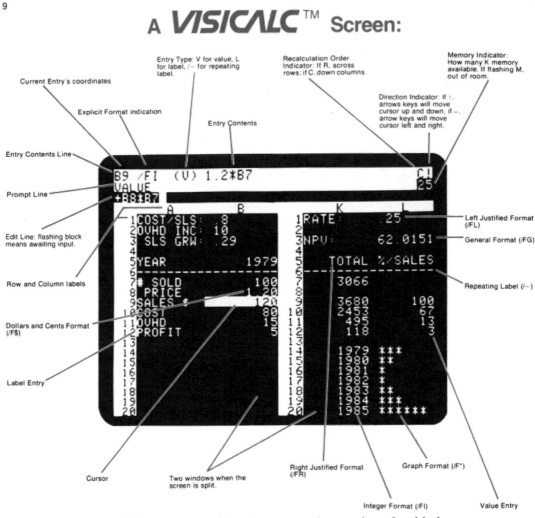

Figure 2.7 Reference card showing annotated screen dump for visicalc
(www.brinklin.com/history/refcards.htm)

modeling. Key aspects of his conceptual model were: (i) to create a spreadsheet that was *analogous* to a ledger sheet in the way it looked, with columns and rows, which allowed people to capitalize on their familiarity with how to use this kind of representation, (ii) to make the spreadsheet interactive, by allowing the user to input and change data in any of the cells in the columns or rows, and (iii) to get the computer to perform a range of different calculations and recalculations in response to user input. For example, the last column can be programmed to display the sum of all the cells in the columns preceding it. With the computer doing all the calculations, together with an easy-to-learn-and-use interface, users were provided with an *easy-to-understand* tool. Moreover, it gave them a new way of effortlessly working out any number of forecasts—*greatly extending* what they could do before with existing tools.

Another popular accounting tool intended for the home market, based on a conceptual model of an object, is Quicken. This used paper checks and registers for its basic structure. Other examples of conceptual models based on objects include most operating environments (e.g., Windows and the Mac desktop) and web portals. All provide the user with a familiar frame of reference when starting the application.

BOX 2.2 The Star Interface *(Based On Miller and Johnson, 1996 and Smith et al., 1982)*

In 1981, Xerox introduced the 8010 "Star" system, which revolutionized the way interfaces were designed for personal computing. Although not commercially successful, many of the ideas behind its design were borrowed and adapted by other companies, such as the Apple Mac and Microsoft Windows, which then became highly successful.

Star was designed as an office system, targeted at workers not interested in computing *per se*. An important design goal, therefore, was to make the computer as "invisible" to the users as possible and to design applications that were suitable for

them. The Star developers spent several person-years at the beginning of the project working out an appropriate conceptual model for such an office system. In the end they selected a conceptual model based on a physical office. They wanted the office workers to imagine the computer to be like an office environment, by acting on electronic counterparts of physical objects in the real world. Their assumption was that this would simplify and clarify the electronic world, making it seem more familiar, less alien and easier to learn, (see Figure 2.8).

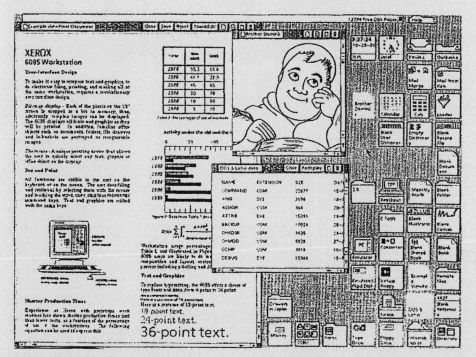

Figure 2.8 The Star interface.

2.3.3   A case of mix and match?

As we have pointed out, which kind of conceptual model is optimal for a given application obviously depends on the nature of the activity to be supported. Some are clearly suited to supporting a given activity (e.g., using manipulation and navigation for a flight simulator) while for others, it is less clear what might be best (e.g., writing and planning activities may be suited to both manipulation and giving instructions). In such situations, it is often the case that some form of hybrid conceptual model that combines different interaction styles is appropriate. For example, the tourist application in Activity 2.2 may end up being optimally designed based on a combination of conversing and exploring models. The user could ask specific questions by typing them in or alternatively browse through information. Shopping on the Internet is also often supported by a range of interaction modes. Sometimes the user may be browsing and navigating, other times communicating with an agent, at yet other times parting with credit card details via an instruction-based form fill-in. Hence, which mode of interaction is "active" depends on the stage of the activity that is being carried out.

BOX 2.3   Do Users Understand The Conceptual Model In The Way Intended?

A fundamental part of developing a conceptual model is to determine whether the ideas generated about how the system should look and behave will be understood by the users in the manner intended. Norman (1988) has provided a framework to elucidate the relationship between the design of a conceptual model and a user's understanding of it (see Figure 2.9). Essentially, there are three interacting components: the designer, the user, and the system. Behind each of these are three interlinking conceptual models:

- the design model—the model the designer has of how the system should work
- the system image—how the system actually works
- the user's model—how the user understands how the system works

In an ideal world, all three should map onto each other. Users should be able to carry out their tasks in the way intended by the designer through interacting with the system image, which makes it obvious what to do. However, if the system image does not make the design model clear to the users, it is likely that they will end up with an incorrect understanding of the system, which in turn will make them use the system ineffectively and make errors.

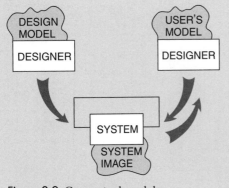

Figure 2.9 Conceptual models.

The down side of mixing interaction modes is that the underlying conceptual model can end up being more complex and ambiguous, making it more difficult for the user to understand and learn. For example, some operating and word-processing systems now make it possible for the user to carry out the same activity in a number of different ways (e.g., to delete a file the user can issue a command like CtrlD, speak to the computer by saying "delete file," or drag an icon of the file to the recycle bin). Users will have to learn the different styles to decide which they prefer. Inevitably, the learning curve will be steeper, but in the long run the benefits are that it enables users to decide how they want to interact with the system.

2.4 Interface metaphors

Another way of describing conceptual models is in terms of interface metaphors. By this is meant a conceptual model that has been developed to be similar in some way to *aspects* of a physical entity (or entities) but that also has its own behaviors and properties. Such models can be based on an activity or an object or both. As well as being categorized as conceptual models based on objects, the desktop and the spreadsheet are also examples of interface metaphors. Another example of an interface metaphor is a "search engine." The tool has been designed to invite comparison with a physical object—a mechanical engine with several parts working—together with an everyday action—searching by looking through numerous files in many different places to extract relevant information. The functions supported by a search engine also include other features besides those belonging to an engine that searches, such as listing and prioritizing the results of a search. It also does these actions in quite different ways from how a mechanical engine works or how a human being might search a library for books on a given topic. The similarities alluded to by the use of the term "search engine," therefore, are at a very general conceptual level. They are meant to conjure up the essence of the process of finding relevant information, enabling the user to leverage off this "anchor" further understanding of other aspects of the functionality provided.

Interface metaphors are based on conceptual models that combine familiar knowledge with new concepts. As mentioned in Box 2.2, the Star was based on a conceptual model of the familiar knowledge of an office. Paper, folders, filing cabinets, and mailboxes were represented as icons on the screen and were designed to possess some of the properties of their physical counterparts. Dragging a document icon across the desktop screen was seen as equivalent to picking up a piece of paper in the physical world and moving it (but of course is a very different action). Similarly, dragging an electronic document onto an electronic folder was seen as being analogous to placing a physical document into a physical cabinet. In addition, new concepts that were incorporated as part of the desktop metaphor were operations that couldn't be performed in the physical world. For example, electronic files could be placed onto an icon of a printer on the desktop, resulting in the computer printing them out.

BOX 2.4　Why Are Metaphors and Analogies So Popular?

People frequently use analogies and metaphors (here we use the terms interchangeably) as a source of inspiration to understand and explain to others what they are doing or trying to do, in terms that are familiar to them. They are an integral part of human language (Lackoff and Johnson, 1980). They are most commonly used to explain something that is unfamiliar or hard to grasp by way of comparison with something that is familiar and easy to grasp. For example, they are commonly employed in education, where teachers use them to introduce something new to students by comparing the new material to something they already understand. An example is the comparison of human evolution with a game. We are all familiar with the properties of a game: there are rules, each player has a goal to win (or lose), there are heuristics to deal with situations where there are no rules, there is the propensity to cheat when the other players are not looking, and so on. By conjuring up these properties, the analogy helps us begin to understand the more difficult concept of evolution—how it happens, what rules govern it, who cheats, and so.

It is not surprising, therefore, to see how widely metaphors and analogies have been applied in interaction design. Both have been used, in overlapping ways, to conceptualize abstract, hard to imagine, and difficult to articulate computer-based concepts and interactions in more concrete and familiar terms and as graphical visualizations at the interface. This use includes:

- as a way of conceptualizing a particular interaction style, e.g., using the system as a tool

- as a conceptual model that is instantiated as part of an interface, e.g., the desktop metaphor

- as a way of describing computers, eg., the Internet highway

- names for describing specific operations, e.g., "cut" and "paste" commands for deleting and copying objects (analogy taken from the media industry)

- as part of the training material aimed at helping learning, e.g., comparing a word processor with a typewriter.

In many instances, it is hard *not* to use metaphorical terms, as they have become so ingrained in the language we use to express ourselves. This is increasingly the case when talking about computers. Just ask yourself or someone else to describe how the Internet works. Then try doing it without using any metaphors or analogies.

ACTIVITY 2.7　Interface metaphors are often actually composites, i.e., they combine quite different pieces of familiar knowledge with the system functionality. We already mentioned the "search engine" as one such example. Can you think of any others?

Comment　Some other examples include:

　　Scrollbar—combines the concept of a scroll with a bar, as in bar chart

　　Toolbar—combines the idea of a set of tools with a bar

　　Portal website—a gateway to a particular collection of pages of networked information

---

Benefits of interface metaphors

Interface metaphors have proven to be highly successful, providing users with a familiar orienting device and helping them understand and learn how to use a system. People find it easier to learn and talk about what they are doing at the computer interface in terms familiar to them—whether they are computer-phobic or highly experienced programmers. Metaphorically based commands used in Unix, like "lint" and "pipe," have very concrete meanings in everyday language that, when used in the context of the Unix operating system, metaphorically represent some aspect of the operations they refer to. Although their meaning may appear obscure, especially to the novice, they make sense when understood in the context of programming. For example, Unix allows the programmer to send the output of one program to another by using the pipe (|) symbol. Once explained, it is easy to imagine the output from one container going to another via a pipe.

ACTIVITY 2.8   Can you think of any bizarre computing metaphors that have become common parlance whose original source of reference is (or always was) obscure?

Comment   A couple of intriguing ones are:

> Java—The programing language Java originally was called Oak, but that name had already been taken. It is not clear how the developers moved from Oak to Java. Java is a name commonly associated with coffee. Other Java-based metaphors that have been spawned include Java beans (a reusable software component) and the steaming coffee-cup icon that appears in the top left-hand corner of Java applets.

> Bluetooth—Bluetooth is used in a computing context to describe the wireless technology that is able to unite technology, communication, and consumer electronics. The name is taken from King Harald Blue Tooth, who was a 10th century legendary Viking king responsible for uniting Scandinavia and thus getting people to talk to each other.

Opposition to using interface metaphors

A mistake sometimes made by designers is to try to design an interface metaphor to look and behave literally like the physical entity it is being compared with. This misses the point about the benefit of developing interface metaphors. As stressed earlier, they are meant to be used to map familiar to unfamiliar knowledge, enabling users to understand and learn about the new domain. Designing interface metaphors only as literal models of the thing being compared with has understandably led to heavy criticism. One of the most outspoken critics is Ted Nelson (1990) who considers metaphorical interfaces as "using old half-ideas as crutches" (p. 237). Other objections to the use of metaphors in interaction design include:

Breaks the rules. Several commentators have criticized the use of interface metaphors because of the cultural and logical contradictions involved in accommodating the metaphor when instantiated as a GUI. A pet hate is the recycle bin (formerly trash can) that sits on the desktop. Logically and culturally (i.e., in the real world), it should be placed under the desk. If this same rule were followed in the virtual desktop, users would not be able to see the bin because it would be occluded by the desktop surface. A counter-argument to this objection is that it does not matter whether rules are contravened. Once people understand why the bin is on the desktop, they readily accept that the real-world rule had to be broken. Moreover, the unexpected juxtaposition of the bin on the desktop can draw to the user's attention the additional functionality that it provides.

Too constraining. Another argument against interface metaphors is that they are too constraining, restricting the kinds of computational tasks that would be useful at the interface. An example is trying to open a file that is embedded in several hundreds of files in a directory. Having to scan through hundreds of icons on a desktop or scroll through a list of files seems a very inefficient way of doing this. As discussed earlier, a better way is to allow the user to instruct the computer to open the desired file by typing in its name (assuming they can remember the name of the file).

Conflicts with design principles. By trying to design the interface metaphor to fit in with the constraints of the physical world, designers are forced into making bad design solutions that conflict with basic design principles. Ted Nelson sets up the trash can again as an example of such violation: "a hideous failure of consistency is the garbage can on the Macintosh, which means either "destroy this" or "eject it for safekeeping" (Nelson, 1990).

Not being able to understand the system functionality beyond the metaphor. It has been argued that users may get fixed in their understanding of the system based on the interface metaphor. In so doing, they may find it difficult to see what else can be done with the system beyond the actions suggested by the interface metaphor. Nelson (1990) also argues that the similarity of interface metaphors to any real objects in the world is so tenuous that it gets in the way more than it helps. We would argue the opposite: because the link is tenuous and there are only a certain number of similarities, it enables the user to see both the dissimilarities and how the metaphor has been extended.

Overly literal translation of existing bad designs. Sometimes designers fall into the trap of trying to create a virtual object to resemble a familiar physical object that is itself badly designed. A well-known example is the virtual calculator, which is designed to look and behave like a physical calculator. The interface of many physical calculators, however, has been poorly designed in the first place, based on poor conceptual models, with excessive use of modes, poor labeling of functions, and difficult-to-manipulate key sequences (Mullet and Sano, 1995). The design of the calculator in Figure 2.10(a) has even gone as far as replicating functions needing shift keys (e.g., deg, oct, and hex), which could have been re-designed as dedicated software buttons. Trying to use a virtual calculator that has been designed to emulate a poorly designed physical calculator is much harder than using the physical device itself. A better approach would have been for the designers to think about how to use the computational power of the computer to support the kinds of tasks people need to do when doing calculations (cf. the spreadsheet design). The calculator in Figure 2.10(b) has tried to do this to some extent, by moving the buttons closer to each other (minimizing the amount of mousing) and providing flexible display modes with one-to-one mappings with different functions.

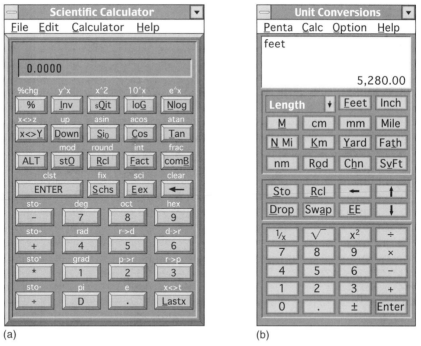

Figure 2.10 Two virtual calculators where (a) has been designed too literally and (b) more appropriately for a computer screen.

Limits the designer's imagination in conjuring up new paradigms and models. Designers may fixate on "tired" ideas, based on well known technologies, that they know people are very familiar with. Examples include travel and books for representing interaction with the web and hypermedia. One of the dangers of always looking backwards is that it restricts the designer in thinking of what new functionality to provide. For example, Gentner and Nielsen (1996) discuss how they used a book metaphor for designing the user interface to Sun Microsystems' online documentation. In hindsight they realized how it had blinkered them in organizing the online material, preventing them from introducing desirable functions such as the ability to reorder chapters according to their relevance scores after being searched.

Clearly, there are pitfalls in using interface metaphors in interaction design. Indeed, this approach has led to some badly designed conceptual models, that have resulted in confusion and frustration. However, this does not have to be the case. Provided designers are aware of the dangers and try to develop interface metaphors that effectively combine familiar knowledge with new functionality in a meaningful way, then many of the above problems can be avoided. Moreover, as we have seen with the spreadsheet example, the use of analogy as a basis for a conceptual model can be very innovative and successful, opening up the realm of computers and their applications to a greater diversity of people.

ACTIVITY 2.9   Examine a web browser interface and describe the various forms of analogy and composite interface metaphors that have been used in its design. What familiar knowledge has been combined with new functionality?

Comment   Many aspects of a web browser have been combined to create a composite interface metaphor:

- a range of toolbars, such as a button bar, navigation bar, favorite bar, history bar
- tabs, menus, organizers
- search engines, guides
- bookmarks, favorites
- icons for familiar objects like stop lights, home

These have been combined with other operations and functions, including saving, searching, downloading, listing, and navigating.

2.5 Interaction paradigms

At a more general level, another source of inspiration for informing the design of a conceptual model is an interaction paradigm. By this it is meant a particular philosophy or way of thinking about interaction design. It is intended to orient designers to the kinds of questions they need to ask. For many years the prevailing paradigm in interaction design was to develop applications for the desktop—intended to be used by single users sitting in front of a CPU, monitor, keyboard and mouse. A dominant part of this approach was to design software applications that would run using a GUI or WIMP interface (windows, icons, mouse and pull-down menus, alternatively referred to as windows, icons, menus and pointers).

As mentioned earlier, a recent trend has been to promote paradigms that move "beyond the desktop." With the advent of wireless, mobile, and handheld technologies, developers started designing applications that could be used in a diversity of ways besides running only on an individual's desktop machine. For example, in September, 2000, the clothes company Levi's, with the Dutch electronics company Philips, started selling the first commercial e-jacket—incorporating wires into the lining of the jacket to create a body-area network (BAN) for hooking up various devices, e.g., mobile phone, MP3, microphone, and headphone (see Figure 1.2(iii) in Color Plate 1). If the phone rings, the MP3 player cuts out the music automatically to let the wearer listen to the call. Another innovation was handheld interactive devices, like the PalmPilot, for which a range of applications were programmed. One was to program the PalmPilot as a multipurpose identity key, allowing guests to check in to certain hotels and enter their room without having to interact with the receptionist at the front desk.

A number of alternative interaction paradigms have been proposed by researchers intended to guide future interaction design and system development (see Figure 2.11). These include:

- ubiquitous computing (technology embedded in the environment)
- pervasive computing (seamless integration of technologies)
- wearable computing (or wearables)

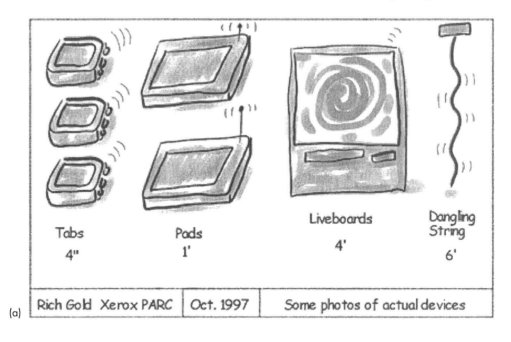

Rich Gold Xerox PARC | Oct. 1997 | Some photos of actual devices

(a)

(b)

(c)

Figure 2.11 Examples of new interaction paradigms: (a) Some of the original devices developed as part of the ubiquitous computing paradigm. Tabs are small hand-sized wireless computers which know where they are and who they are with. Pads are paper-sized devices connected to the system via radio. They know where they are and who they are with. Liveboards are large wall sized devices. The "Dangling String" created by artist Natalie Jeremijenko was attached directly to the ethernet that ran overhead in the ceiling. It spun around depending on the level of digital traffic.

(b) Ishii and Ulmer, MIT Lab (1997) Tangible bits: from GUIs of desktop PCs to Tangible User Interfaces. The paradigm is concerned with establishing a new type of HCI called "Tangible User Interfaces" (TUIs). TUIs augment the real physical world by coupling digital information to everyday physical objects and environments.

(c) Affective Computing: The project, called "BlueEyes," is creating devices with embedded technology that gather information about people. This face (with movable eyebrows, eyes and mouth) tracks your movements and facial expressions and responds accordingly.

- tangible bits, augmented reality, and physical/virtual integration
- attentive environments (computers attend to user's needs)
- the Workaday World (social aspects of technology use)

Ubiquitous computing ("ubicomp"). The late Mark Weiser (1991), an influential visionary, proposed the interaction paradigm of ubiquitous computing (Figure 2.11). His vision was for computers to disappear into the environment so that we would be no longer aware of them and would use them without thinking about them. As part of this process, they should "invisibly" enhance the world that already exists rather than create artificial ones. Existing computing technology, e.g., multimedia-based systems and virtual reality, currently do not allow us to do this. Instead, we are forced to focus our attention on the multimedia representations on the screen (e.g., buttons, menus, scrollbars) or to move around in a virtual simulated world, manipulating virtual objects.

So, how can technologies be designed to disappear into the background? Weiser did not mean ubiquity in the sense of simply making computers portable so that they can be moved from the desk into our pockets or used on trains or in bed. He meant that technology be designed to be integrated seamlessly into the physical world in ways that *extend* human capabilities. One of his prototypes was a "tabs, pads, and boards" setup whereby hundreds of computer devices equivalent in size to post-it notes, sheets of paper, and blackboards would be embedded in offices. Like the spreadsheet, such devices are assumed to be easy to use, because they capitalize on existing knowledge about how to interact and use everyday objects. Also like the spreadsheet, they provide much greater computational power. One of Weiser's ideas was that the tabs be connected to one another, enabling them to become multipurpose, including acting as a calendar, diary, identification card, and an interactive device to be used with a PC.

> *Ubiquitous computing will produce nothing fundamentally new, but by making everything faster and easier to do, with less strain and fewer mental gymnastics, it will transform what is apparently possible (Weiser, 1991, p. 940).*

Pervasive computing. Pervasive computing is a direct follow-on of ideas arising from ubiquitous computing. The idea is that people should be able to access and interact with information any place and any time, using a seamless integration of technologies. Such technologies are often referred to as smart devices or information appliances—designed to perform a particular activity. Commercial products include cell phones and handheld devices, like PalmPilots. On the domestic front, other examples currently being prototyped include intelligent fridges that signal the user when stocks are low, interactive microwave ovens that allow users to access information from the web while cooking, and smart pans that beep when the food is cooked.

Wearable computing. Many of the ideas behind ubiquitous computing have since inspired other researchers to develop technologies that are part of the environment. The MIT Media Lab has created several such innovations. One example is wearable computing (Mann, 1996). The combination of multimedia and wireless communication presented many opportunities for thinking about how to embed such technologies on people in the clothes they wear. Jewelry, head-mounted caps, glasses, shoes, and jackets have all been experimented with to provide the user with a means of interacting with digital information while on the move in the physical world. Applications that have been developed include automatic diaries that keep users up to date on what is happening and what they need to do throughout the day, and tour guides that inform users of relevant information as they walk through an exhibition and other public places (Rhodes et al., 1999).

Tangible bits, augmented reality, and physical/virtual integration. Another development that has evolved from ubiquitous computing is tangible user interfaces or tangible bits (Ishii and Ullmer, 1997). The focus of this paradigm is the "integration of computational augmentations into the physical environment", in other words, finding ways to combine digital information with physical objects and surfaces (e.g., buildings) to allow people to carry out their everyday activities. Examples include physical books embedded with digital information, greeting cards that play a digital animation when opened, and physical bricks attached to virtual objects that when grasped have a similar effect on the virtual objects. Another illustration of this approach is the one described in Chapter 1 of an *enjoyable* interface, in which a person could use a physical hammer to hit a physical key with corresponding virtual representations of the action being displayed on a screen.

Another part of this paradigm is augmented reality, where virtual representations are superimposed on physical devices and objects (as shown in Figure 2.1 on Color Plate 2). Bridging the gulf between physical and virtual worlds is also currently undergoing much research. One of the earlier precursors of this work was the Digital Desk (Wellner, 1993). Physical office tools, like books, documents and paper, were integrated with virtual representations, using projectors and video cameras. Both virtual and real documents were seamlessly combined.

Attentive environments and transparent computing. This interaction paradigm proposes that the computer attend to user's needs through anticipating what the user wants to do. Instead of users being in control, deciding what they want to do and where to go, the burden should be shifted onto the computer. In this sense the mode of interaction is much more implicit: computer interfaces respond to the user's expressions and gestures. Sensor-rich environments are used to detect the user's current state and needs. For example, cameras can detect where people are looking on a screen and decide what to display accordingly. The system should be able to determine when someone wants to make a call and which websites they want to visit at particular times. IBM's BlueEyes project is developing a range of computational devices that use non-obtrusive sensing technology, including videos and microphones, to track and identify users' actions. This information is then analyzed with respect to where users are looking, what they are doing, their gestures, and their facial expressions. In turn, this is coded in terms of the users' physical, emotional or informational state and is then used to determine what information they would like. For example, a BlueEyes-enabled computer could become active when a user first walks into a room, firing up any new email messages that have arrived. If the user shakes his or her head, it would be interpreted by the computer as "I don't want to read them," and instead show a listing of their appointments for that day.

The Workaday World. In the new paradigms mentioned above, the emphasis is on exploring how technological devices can be linked with each other and digital information in novel ways that allow people to do things they could not do before. In contrast, the Workaday World paradigm is driven primarily by conceptual and mundane concerns. It was proposed by Tom Moran and Bob Anderson (1990), when working at Xerox PARC. They were particularly concerned with the need to understand the social aspects of technology use in a way that could be useful for designers. The Workaday World paradigm focuses on the essential character of the workplace in terms of people's everyday activities, relationships, knowledge, and resources. It seeks to unravel the "set of patterns that convey the richness of the settings in which technologies live—the complex, unpredictable, multiform relationships that hold among the various aspects of working life" (p. 384).

2.6 From conceptual models to physical design

As we emphasize throughout this book, interaction design is an iterative process. It involves cycling through various design processes at different levels of detail. Primarily it involves: thinking through a design problem, understanding the user's needs, coming up with possible conceptual models, prototyping them, evaluating them with respect to usability and user experience goals, thinking about the design implications of the evaluation studies, making changes to the prototypes with respect to these, evaluating the changed prototypes, thinking through whether the changes have improved the interface and interaction, and so on. Interaction design may also require going back to the original data to gather and check the requirements. Throughout the iterations, it is important to think through and understand whether the conceptual model being developed is working in the way intended and to ensure that it is supporting the user's tasks.

Throughout this book we describe the way you should go about *doing* interaction design. Each iteration should involve *progressing through* the design in more depth. A first pass through an iteration should involve essentially thinking about the problem space and identifying some initial user requirements. A second pass should involve more extensive information gathering about users' needs and the problems they experience with the way they currently carry out their activities (see Chapter 7). A third pass should continue explicating the requirements, leading to thinking through possible conceptual models that would be appropriate (see Chapter 8). A fourth pass should begin "fleshing out" some of these using a variety of user-centered methods. A number of user-centered methods can be used to create prototypes of the potential candidates. These include using storyboarding to show how the interaction between the users and the system will take place and the laying out of cards and post-it notes to show the possible structure of and navigation through a website. Throughout the process, the various prototypes of the conceptual models should be evaluated to see if they meet users' needs. Informally asking users what they think is always a good starting point (see Chapter 12). A number of other techniques can also be used at different stages of the development of the prototypes, depending on the particular information required (see Chapters 13 and 14).

Many issues will need to be addressed when developing and testing initial prototypes of conceptual models. These include:

- the way information is to be presented and interacted with at the interface
- what combinations of media to use (e.g., whether to use sound and animations)
- the kind of feedback that will be provided
- what combinations of input and output devices to use (e.g., whether to use speech, keyboard plus mouse, handwriting recognition)
- whether to provide agents and in what format
- whether to design operations to be hardwired and activated through physical buttons or to represent them on the screen as part of the software
- what kinds of help to provide and in what format

While working through these design decisions about the nature of the interaction to be supported, issues concerning the actual physical design will need to be addressed. These will often fall out of the conceptual decisions about the way information is to be represented, the kind of media to be used, and so on. For example, these would typically include:

- *information presentation*
  –which dialogs and interaction styles to use (e.g., form fill-ins, speech input, menus)
  –how to structure items in graphical objects, like windows, dialog boxes and menus (e.g., how many items, where to place them in relation to each other)
- *feedback*
  –what navigation mechanisms to provide (e.g., forward and backward buttons)
- *media combination*
  –which kinds of icons to use

Many of these physical design decisions will be specific to the interactive product being built. For example, designing a calendar application intended to be used by business people to run on a handheld computer will have quite different constraints and concerns from designing a tool for scheduling trains to run over a large network, intended to be used by a team of operators via multiple large displays. The way the information will be structured, the kinds of graphical representations that will be appropriate, and the layout of the graphics on the screens will be quite different.

These kinds of design decisions are very practical, needing user testing to ensure that they meet with the usability goals. It is likely that numerous trade-offs will surface, so it is important to recognize that there is no right or wrong way to resolve these. Each decision has to be weighed with respect to the others. For example, if you decide that a good way of providing visibility for the calendar application on the handheld device is to have a set of "soft" navigation buttons permanently as

DILEMMA   Realism versus Abstraction?

One of the challenges facing interaction designers is whether to use realism or abstraction when designing an interface to instantiate their conceptual model. By this it is meant designing objects either (i) to give the illusion of behaving and looking like real-world counterparts or (ii) to appear as simply abstractions of the objects being represented. This concern is particularly relevant when fleshing out conceptual models that are deliberately based on an analogy to some aspect of the real world. For example, is it preferable to design a desktop to look like a real desktop, a virtual house to look like a real house, or a virtual living room to look like a real living room? Or, alternatively, is it more effective to design representations of the conceptual model as simple abstract renditions, depicting only a few salient features?

We already discussed in Chapter 1 the problems of trying to design graphical interfaces with affordances. Here, we consider more generally the dilemma of using realism at the interface. One of the main benefits of using realism is that it can enable people, especially computer phobics and novices, to feel *more comfortable* when first learning an application. The reason for this is that such representations can readily tap into people's understanding of the physical world. Hence, realistic interfaces can help users initially understand the underlying conceptual model. In contrast, overly schematic and abstract representations can appear to be too computer-like and off-putting to the newcomer. The advantage of these kinds of more abstract interfaces is that they are often more efficient to use. Furthermore, the more experienced users become, the more they may find comfortable interfaces no longer to their liking. A dilemma facing designers, therefore, is deciding between designing interfaces to make novice users feel comfortable (but more experienced users less comfortable) versus designing interfaces to be effective for more experienced users (but maybe harder to learn by novices).

One of the earliest attempts at using realism at the interface was General Magic's office system Magic Cap, which was rendered in 3D. To achieve this degree of realism required using various perceptual cues such as perspective, shadowing, and shading. The result of their efforts was a rather cute interface (see Figure 2.12). Although their intentions were well-grounded, the outcome was less successful. Many people commented on how childish and gawky it looked, having the appearance of illustrations in a children's picture book rather than a work-based application.

Mullet and Sano (1995) also point out how a 3D rendition of an object like a desk nearly always suffers from both an unnatural point of view and an awkward rendering style that ironically destroy the impression of being in a real physical space. One reason for this is that 3D depictions conflict with the effective use of display space, especially when 2D editing tasks need to be performed. As can be seen in Figure 2.12, these kinds of tasks have been represented as "flat" buttons that appear to be floating in front of the desk (e.g., mail, program manager, task manager).

In certain kinds of applications, using realism can be very effective for both novices and experienced users. Video games fall into this category, especially those where users have to react rapidly to dynamic events that happen in a virtual world in real time, say flying a plane or playing a game of virtual football. Making the characters in the game resemble humans in the way they look, move, dress, and gesture also makes them seem more convincing and lifelike, enhancing the enjoyment and fun factor (see Figure 2.13).

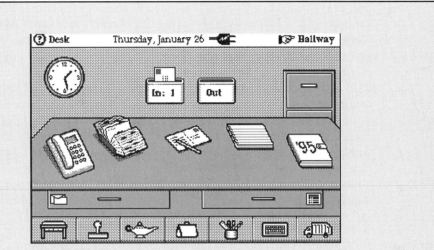

Figure 2.12 Magic Cap's 3D desktop interface.

Figure 2.13 3D avatars in computer games: A screenshot
from The Sims World.

part of the visual display, you then need to consider the consequences of doing this for the rest of the information that needs to be interacted with. Will it still be possible to structure the display to show the calendar as days in a week or a month, all on one screen?

This part of the design process is highly dependent on the context and essentially involves lots of juggling between design decisions. If you visit our website you can try out some of the interactivities provided, where you have to make such decisions when designing the physical layout for various interfaces. Here, we provide the background and rationale that can help you make appropriate choices when faced with a series of design decisions (primarily Chapters 3–5 and 8). For example, we explain why you shouldn't cram a screen full of information; why certain techniques are better than others for helping users remember how to carry out their tasks at the interface; and why certain kinds of agents appear more believable than others.

Assignment

The aim of this assignment is for you to think about the appropriateness of different kinds of conceptual model that have been designed for similar kinds of physical and electronic artifacts.

(a) Describe the conceptual model that underlie the design of:

- a personal pocket-sized calendar/diary (one week to a page)
- a wall calendar (one month to a page, usually with a picture/photo)
- a wall planner (displaying the whole year)

What is the main kind of activity and object they are based on? How do they differ for each of the three artifacts? What metaphors have been used in the design of their physical interface (think about the way time is conceptualized for each of them)? Do users understand the conceptual models these are based on in the ways intended (ask a few people to explain how they use them)? Do they match the different user needs?

(b) Now describe the conceptual models that underlie the design of:

- an electronic *personal* calendar found on a personal organizer or handheld computer
- a *shared* calendar found on the web

How do they differ from the equivalent physical artifacts? What new functionality has been provided? What interface metaphors have been used? Are the functions and interface metaphor well integrated? What problems do users have with these interactive kinds of calendars? Why do you think this is?

Summary

This chapter has explained the importance of conceptualizing interaction design before trying to build anything. It has stressed throughout the need always to be clear and explicit about the rationale and assumptions behind any design decision made. It described a taxonomy of conceptual models and the different properties of each. It also discussed interface metaphors and interaction paradigms as other ways of informing the design of conceptual models.

Key points

- It is important to have a good understanding of the problem space, specifying what it is you are doing, why and how it will support users in the way intended.

- A fundamental aspect of interaction design is to develop a conceptual model.

- There are various kinds of conceptual models that are categorized according to the activity or object they are based on.

- Interaction modes (e.g., conversing, instructing) provide a structure for thinking about which conceptual model to develop.

- Interaction styles (e.g., menus, form fill-ins) are specific kinds of interfaces that should be decided upon after the conceptual model has been chosen.

- Decisions about conceptual design also should be made before commencing any physical design (e.g., designing an icon).

- Interface metaphors are commonly used as part of a conceptual model.

- Many interactive systems are based on a hybrid conceptual model. Such models can provide more flexibility, but this can make them harder to learn.

- 3D realism is not necessarily better than 2D or other forms of representation when instantiating a conceptual model: what is most effective depends on the users' activities when interacting with a system.

- General interaction paradigms, like WIMP and ubiquitous computing, provide a particular way of thinking about how to design a conceptual model.

Further reading

LAUREL, B. (1990) (ed.) *The Art of Human Computer Design* has a number of papers on conceptual models and interface metaphors. Two that are definitely worth reading are: Tom Erickson, "Working with interface metaphors" (pp. 65–74), which is a practical hands-on guide to designing interface metaphors (covered later in this book), and Ted Nelson's polemic, "The right way to think about software design" (pp. 229–234), which is a scathing attack on the use of interface metaphors.

JOHNSON, M. AND LAKOFF, G. (1980) *Metaphors We Live By.* The University of Chicago Press. Those wanting to find out more about how metaphors are used in everyday conversations should take a look at this text.

There are many good articles on the topic of interface agents. A classic is: Lanier, J. (1995) Agents of alienation, *ACM Interactions*, 2(3), 66–72. *The Art of Human Computer Design* also provides several thought-provoking articles, including one called "Interface agents: metaphors with character" by Brenda Laurel (pp. 355–366) and another called "Guides: characterizing the interface" by Tim Oren et al. (pp. 367–382). Liam Bannon has also written a critical review of the agent approach to interface design: "Problems in human-machine interaction and communication." In *Proc HCI'97*, San Francisco, August '97.

MIT's Media Lab (www.media.mit.edu) is a good starting place to find out what is currently happening in the world of agents, wearables, and other new interaction paradigms.

INTERVIEW with Terry Winograd

Terry Winograd is a professor of computer science at Stanford University. He has done extensive research and writing on the design of human-computer interaction. His early research on natural language understanding by computers was a milestone in artificial intelligence, and he has written two books and numerous articles on that topic. His book, *Bringing Design to Software*, brings together the perspectives of a number of leading researchers and designers. See Color Plate 2 for an example of his latest research.

YR: Tell me about your background and how you moved into interaction design.

TW: I got into interaction design through a couple of intermediate steps. I started out doing research into artificial intelligence. I became interested in how people interact with computers, in particular, when using ordinary language. It became clear after years of working on that, however, that the computer was a long way off from matching human abilities. Moreover, using natural language with a computer when it doesn't really understand you can be very frustrating and in fact a very bad way to interact with it. So, rather than trying to get the computer to imitate the person, I became interested in other ways of taking advantage of what the computer can do well and what the person can do well. That led me into the general field of HCI. As I began to look at what was going on in that field and to study it, it became clear that it was not the same as other areas of computer science. The key issues were about how the technology fits with what people could do and what they wanted to do. In contrast, most of computer science is really dominated by how the mechanisms operate.

I was very attracted to thinking more in the style of design disciplines, like product design, urban design, architecture, and so on. I realized that there was an approach that you might call a design way, that puts the technical asspects into the background with respect to understanding the interaction. Through looking at these design disciplines, I realized that there was something unique about interaction design, which is that it has a dialogic temporal element. By

this I mean a human dialog not in the sense of using ordinary language, but in the sense of thinking about the sequence and the flow of interaction. So I think interaction design is about designing a space for people, where that space has to have a temporal flow. It has to have a dialog with the person.

YR: Could you tell me a bit more about what you think is involved in interaction design?

TW: One of the biggest influences is product design. I think that interaction design overlaps with it, because they both take a very strong user-oriented view. Both are concerned with finding a user group, understanding their needs, then using that understanding to come up with new ideas. They may be ones that the users don't even realize they need. It is then a matter of trying to translate who it is, what they are doing, and why they are doing it into possible innovations. In the case of product design it is products. In the case of interaction design it is the way that the computer system interacts with the person.

YR: What do you think are important inputs into the design process?

TW: One of the characteristics of design fields as opposed to traditional engineering fields is that there is much more dependence on case studies and examples than on formulas. Whereas an engineer knows how to calculate something, an architect or a designer is working in a tradition where there is a history over time of other things people have done. People have said that the secret of great design is to know what to steal and to know when some element or some way of doing things that worked before will be appropriate to your setting and then adapt it. Of course you can't apply it directly, so I think a big part of doing good design is experience and exposure. You have to have seen a lot of things in practice and understood what is good and bad about them, to then use these to inform your design.

YR: How do you see the relationship between studying interaction design and the practice of it? Is there a good dialog between research and practice?

TW: Academic study of interaction design is a tricky area because so much of it depends on a kind of tacit knowledge that comes through experience and exposure. It is not the kind of thing you can set down easily as, say, you can scientific formulas. A lot of design tends to be methodological. It is not about the design *per se* but is more about how you go about doing design, in particular, knowing what are the appropriate steps to take and how you put them together.

YR: How do you see the field of interaction design taking on board the current explosion in new technologies—for example mobile, ubiquitous, infrared, and so on? Is it different, say, from 20 years ago when it was just about designing software applications to sit on the desktop?

TW: I think a real change in people's thinking has been to move from interface design to interaction design. This has been pushed by the fact that we do have all kinds of devices nowadays. Interface design used to mean graphical interfaces, which meant designing menus and other widgets. But now when you're talking about handheld devices, gesture interfaces, telephone interfaces and so on, it is clear that you can't focus just on the widgets. The widgets may be part of any one of these devices but the design thinking as a whole has to focus on the interaction.

YR: What advice would you give to a student coming into the field on what they should be learning and looking for?

TW: I think a student who wants to learn this field should think of it as a kind of dual process, that is what Donald Schön calls "reflection in action," needing both the action and the reflection. It is important to have experience with trying to build things. That experience can be from outside work, projects, and courses where you are actually engaged in making something work. At the same time you need to be able to step back and look at it not as "What do I need to do next?" but from the perspective of what you are doing and how that fits into the larger picture.

YR: Are there any classic case studies that stand out as good exemplars of interaction design?

TW: You need to understand what has been important in the past. I still use the Xerox Star as an exemplar because so much of what we use today was there. When you go back to look at the Star you see it in the context of when it was first created. I also think some exemplars that are very interesting are ones that never actually succeeded commercially. For example, I use the PenPoint system that was developed for pen computers by Go. Again, they were thinking fresh. They set out to do something different and they were much more conscious of the design issues than somebody who was simply adapting the next version of something that already existed. PalmPilot is another good example, because they looked at the problem in a different way to make something work. Another interesting exemplar, which other people may not agree with, is Microsoft Bob—not because it was a successful program, because it wasn't, but because it was a first exploration of a certain style of interaction, using animated agents. You can see very clearly from these exemplars what design trade-offs the designers were making and why and then you can look at the consequences.

YR: Finally, what are the biggest challenges facing people working in this area?

TW: I think one of the biggest challenges is what Pelle Ehn calls the dialectic between tradition and transcendence. That is, people work and live in certain ways already, and they understand how to adapt that within a small range, but they don't have an understanding or a feel for what it would mean to make a radical change, for example, to change their way of doing business on the Internet before it was around, or to change their way of writing from pen and paper when word processors weren't around. I think what the designer is trying to do is envision things for users that the users can't yet envision. The hard part is not fixing little problems, but designing things that are both innovative and that work.

Chapter 3

Understanding users

3.1   Introduction
3.2   What is cognition?
3.3   Applying knowledge from the physical world to the digital world
3.4   Conceptual frameworks for cognition
    3.4.1   Mental models
    3.4.2   Information processing
    3.4.3   External cognition
3.5   Informing design: from theory to practice

3.1   Introduction

Imagine trying to drive a car by using just a computer keyboard. The four arrow keys are used for steering, the space bar for braking, and the return key for accelerating. To indicate left you need to press the F1 key and to indicate right the F2 key. To sound your horn you need to press the F3 key. To switch the headlights on you need to use the F4 key and, to switch the windscreen wipers on, the F5 key. Now imagine as you are driving along a road a ball is suddenly kicked in front of you. What would you do? Bash the arrow keys and the space bar madly while pressing the F4 key? How would you rate your chances of missing the ball?

Most of us would balk at the very idea of driving a car this way. Many early video games, however, were designed along these lines: the user had to press an arbitrary combination of function keys to drive or navigate through the game. There was little, if any, consideration of the user's capabilities. While some users regarded mastering an arbitrary set of keyboard controls as a challenge, many users found them very limiting, frustrating, and difficult to use. More recently, computer consoles have been designed with the user's capabilities and the demands of the activity in mind. Much better ways of controlling and interacting, such as through using joysticks and steering wheels, are provided that map much better onto the physical and cognitive aspects of driving and navigating.

In this chapter we examine some of the core cognitive aspects of interaction design. Specifically, we consider what humans are good and bad at and show how this knowledge can be used to *inform* the design of technologies that both *extend* human capabilities and *compensate* for their weaknesses. We also look at some of the influential cognitively based conceptual frameworks that have been developed for explaining the way humans interact with computers. (Other ways of conceptualizing human behavior that focus on the social and affective aspects of interaction design are presented in the following two chapters.)

The main aims of this chapter are to:

- Explain what cognition is and why it is important for interaction design.
- Describe the main ways cognition has been applied to interaction design.
- Provide a number of examples in which cognitive research has led to the design of more effective interactive products.
- Explain what mental models are.
- Give examples of conceptual frameworks that are useful for interaction design.
- Enable you to try to elicit a mental model and be able to understand what it means.

3.2   What is cognition?

Cognition is what goes on in our heads when we carry out our everyday activities. It involves cognitive processes, like thinking, remembering, learning, daydreaming, decision making, seeing, reading, writing and talking. As Figure 3.1 indicates, there are many different kinds of cognition. Norman (1993) distinguishes between two general modes: experiential and reflective cognition. The former is a state of mind in which we perceive, act, and react to events around us effectively and effortlessly. It requires reaching a certain level of expertise and engagement. Examples include driving a car, reading a book, having a conversation, and playing a video game. In contrast, reflective cognition involves thinking, comparing, and decision-making. This kind of cognition is what leads to new ideas and creativity. Examples include designing, learning, and writing a book. Norman points out that both modes are essential for everyday life but that each requires different kinds of technological support.

What goes on in the mind?

Figure 3.1 What goes on in the mind?

Cognition has also been described in terms of specific kinds of processes. These include:

- attention
- perception and recognition
- memory
- learning
- reading, speaking, and listening
- problem solving, planning, reasoning, decision making

It is important to note that many of these cognitive processes are interdependent: several may be involved for a given activity. For example, when you try to learn material for an exam, you need to attend to the material, perceive, and recognize it, read it, think about it, and try to remember it. Thus, cognition typically involves a range of processes. It is rare for one to occur in isolation. Below we describe the various kinds in more detail, followed by a summary box highlighting core design implications for each. Most relevant (and most thoroughly researched) for interaction design is memory, which we describe in greatest detail.

Attention is the process of selecting things to concentrate on, at a point in time, from the range of possibilities available. Attention involves our auditory and/or visual senses. An example of auditory attention is waiting in the dentist's waiting room for our name to be called out to know when it is our time to go in. An example of attention involving the visual senses is scanning the football results in a newspaper to attend to information about how our team has done. Attention allows us to focus on information that is relevant to what we are doing. The extent to which this process is easy or difficult depends on (i) whether we have clear goals and (ii) whether the information we need is salient in the environment:

(i) Our goals   If we know exactly what we want to find out, we try to match this with the information that is available. For example, if we have just landed at an airport after a long flight and want to find out who had won the World Cup, we might scan the headlines at the newspaper stand, check the web, call a friend, or ask someone in the street.

When we are not sure exactly what we are looking for we may browse through information, allowing it to guide our attention to interesting or salient items. For example, when we go to a restaurant we may have the general goal of eating a meal but only a vague idea of what we want to eat. We peruse the menu to find things that whet our appetite, letting our attention be drawn to the imaginative descriptions of various dishes. After scanning through the possibilities and imagining what each dish might be like (plus taking into account other factors, such as cost, who we are with, what the specials are, what the waiter recommends, whether we want a two- or three-course meal, and so on), we may then make a decision.

(ii) Information presentation   The way information is displayed can also greatly influence how easy or difficult it is to attend to appropriate pieces of information. Look at Figure 3.2 and try the activity. Here, the information-searching tasks are very precise, requiring specific answers. The information density is identical in both

South Carolina					
City	Motel/Hotel	Area code	Phone	Rates Single	Double
Charleston	Best Western	803	747-0961	$26	$30
Charleston	Days Inn	803	881-1000	$18	$24
Charleston	Holiday Inn N	803	744-1621	$36	$46
Charleston	Holiday Inn SW	803	556-7100	$33	$47
Charleston	Howard Johnsons	803	524-4148	$31	$36
Charleston	Ramada Inn	803	774-8281	$33	$40
Charleston	Sheraton Inn	803	744-2401	$34	$42
Columbia	Best Western	803	796-9400	$29	$34
Columbia	Carolina Inn	803	799-8200	$42	$48
Columbia	Days Inn	803	736-0000	$23	$27
Columbia	Holiday Inn NW	803	794-9440	$32	$39
Columbia	Howard Johnsons	803	772-7200	$25	$27
Columbia	Quality Inn	803	772-0270	$34	$41
Columbia	Ramada Inn	803	796-2700	$36	$44
Columbia	Vagabond Inn	803	796-6240	$27	$30

Pennsylvania
Bedford Motel/Hotel: Crinaline Courts
  (814) 623-9511 S: $18 D: $20
Bedford Motel/Hotel: Holiday Inn
  (814) 623-9006 S: $29 D: $36
Bedford Motel/Hotel: Midway
  (814) 623-8107 S: $21 D: $26
Bedford Motel/Hotel: Penn Manor
  (814) 623-8177 S: $19 D: $25
Bedford Motel/Hotel: Quality Inn
  (814) 623-5189 S: $23 D: $28
Bedford Motel/Hotel: Terrace
  (814) 623-5111 S: $22 D: $24
Bradley Motel/Hotel: De Soto
  (814) 362-3567 S: $20 D: $24
Bradley Motel/Hotel: Holiday House
  (814) 362-4511 S: $22 D: $25
Bradley Motel/Hotel: Holiday Inn
  (814) 362-4501 S: $32 D: $40
Breezewood Motel/Hotel: Best Western Plaza
  (814) 735-4352 S: $20 D: $27
Breezewood Motel/Hotel: Motel 70
  (814) 735-4385 S: $16 D: $18

Figure 3.2 Two different ways of structuring the same information at the interface: one makes it much easier to find information than the other. Look at the top screen and: (i) find the price for a double room at the Quality Inn in Columbia; (ii) find the phone number of the Days Inn in Charleston. Then look at the bottom screen and (i) find the price of a double room at the Holiday Inn in Bradley; (ii) find the phone number of the Holiday Inn in Bedford. Which took longer to do? In an early study Tullis found that the two screens produced quite different results: it took an average of 3.2 seconds to search the top screen and 5.5 seconds to find the same kind of information in the bottom screen. Why is this so, considering that both displays have the same density of information (31%)? The primary reason is the way the characters are grouped in the display: in the top they are grouped into vertical categories of information (e.g., place, kind of accommodation, phone number, and rates) that have columns of space between them. In the bottom screen the information is bunched up together, making it much harder to search through.

displays. However, it is much harder to find the information in the bottom screen than in the top screen. The reason for this is that the information is very poorly structured in the bottom, making it difficult to find the information. In the top the information has been ordered into meaningful categories with blank spacing between them, making it easier to select the necessary information.

Perception refers to how information is acquired from the environment, via the different sense organs (e.g., eyes, ears, fingers) and transformed into experiences of objects, events, sounds, and tastes (Roth, 1986). It is a complex process, involving other cognitive processes such as memory, attention, and language. Vision is the

DESIGN IMPLICATIONS Attention

- Make information salient when it needs attending to at a given stage of a task.
- Use techniques like animated graphics, color, underlining, ordering of items, sequencing of different information, and spacing of items to achieve this.
- Avoid cluttering the interface with too much information. This especially applies to the use of color, sound and graphics: there is a temptation to use lots of them, resulting in a mishmash of media that is distracting and annoying rather than helping the user attend to relevant information.
- Interfaces that are plain are much easier to use, like the Google search engine (see Figure 3.3). The main reason is that it is much easier for users to find where on the screen to type in their search.

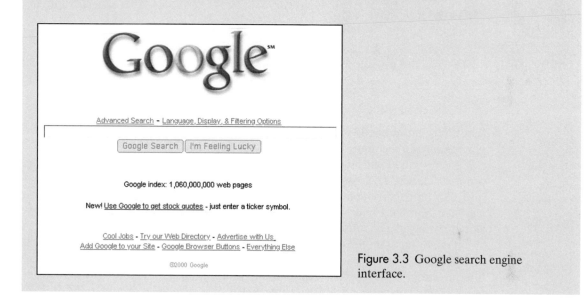

Figure 3.3 Google search engine interface.

most dominant sense for sighted individuals, followed by hearing and touch. With respect to interaction design, it is important to present information in a way that can be readily perceived in the manner intended. For example, there are many ways to design icons. The key is to make them easily distinguishable from one another and to make it simple to recognize what they are intended to represent (not like the ones in Figure 3.4).

Combinations of different media need also to be designed to allow users to recognize the composite information represented in them in the way intended. The use of sound and animation together needs to be coordinated so they happen in a logical sequence. An example of this is the design of lip-synch applications, where the animation of an avatar's or agent's face to make it appear to be talking, must be carefully synchronized with the speech that is emitted. A slight delay between the two can make it difficult and disturbing to perceive what is happening—as sometimes happens when film dubbing gets out of synch. A general design principle is

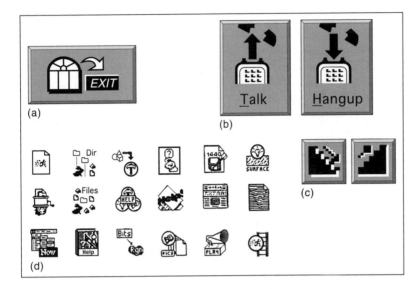

Figure 3.4 Poor icon set. What do you think the icons mean and why are they so bad?

that information needs to be represented in an appropriate form to facilitate the perception and recognition of its underlying meaning.

Memory involves recalling various kinds of knowledge that allow us to act appropriately. It is very versatile, enabling us to do many things. For example, it allows us to recognize someone's face, remember someone's name, recall when we last met them and know what we said to them last. Simply, without memory we would not be able to function.

It is not possible for us to remember everything that we see, hear, taste, smell, or touch, nor would we want to, as our brains would get completely overloaded. A filtering process is used to decide what information gets further processed and memorized. This filtering process, however, is not without its problems. Often we

---

DESIGN IMPLICATIONS   Perception

Representations of information need to be designed to be perceptible and recognizable across different media:

- Icons and other graphical representations should enable users to readily distinguish their meaning.
- Sounds should be audible and distinguishable so users understand what they represent.
- Speech output should enable users to distinguish between the set of spoken words and also be able to understand their meaning.

- Text should be legible and distinguishable from the background (e.g., it is OK to use yellow text on a black or blue background but not on a white or green background).
- Tactile feedback used in virtual environments should allow users to recognize the meaning of the various touch sensations being emulated. The feedback should be distinguishable so that, for example, the sensation of squeezing is represented in a tactile form that is different from the sensation of pushing.

forget things we would dearly love to remember and conversely remember things we would love to forget. For example, we may find it difficult to remember every-day things like people's names and phone numbers or academic knowledge like mathematical formulae. On the other hand, we may effortlessly remember trivia or tunes that cycle endlessly through our heads.

How does this filtering process work? Initially, encoding takes place, determin-ing which information is attended to in the environment and how it is interpreted. The extent to which it takes place affects our ability to recall that information later. The more attention that is paid to something and the more it is processed in terms of thinking about it and comparing it with other knowledge, the more likely it is to be remembered. For example, when learning about a topic it is much better to re-flect upon it, carry out exercises, have discussions with others about it, and write notes than just passively read a book or watch a video about it. Thus, how informa-tion is interpreted when it is encountered greatly affects how it is represented in memory and how it is used later.

Another factor that affects the extent to which information can be subse-quently retrieved is the context in which it is encoded. One outcome is that some-times it can be difficult for people to recall information that was encoded in a different context from the one they currently are in. Consider the following sce-nario:

> You are on a train and someone comes up to you and says hello. You don't recognize him for a few moments but then realize it is one of your neighbors. You are only used to seeing your neighbor in the hallway of your apartment block and seeing him out of context makes him difficult to recognize.

Another well-known memory phenomenon is that people are much better at rec-ognizing things than recalling things. Furthermore, certain kinds of information are easier to recognize than others. In particular, people are very good at recognizing thousands of pictures, even if they have only seen them briefly before.

ACTIVITY 3.1   Try to remember the dates of all the members of your family's and your closest friends' birthdays. How many can you remember? Then try to describe what is on the cover of the last DVD/CD or record you bought. Which is easiest and why?

Comment   It is likely that you remembered much better what was on the CD/DVD/record cover (the image, the colors, the title) than the birthdays of your family and friends. People are very good at remembering visual cues about things, for example the color of items, the location of objects (a book being on the top shelf), and marks on an object (e.g., a scratch on a watch, a chip on a cup). In contrast, people find other kinds of information persistently difficult to learn and remember, especially arbitrary material like birthdays and phone numbers.

Instead of requiring users to recall from memory a command name from a pos-sible set of hundreds or even thousands, GUIs provide visually based options that users can browse through until they recognize the operation they want to perform (see Figure 3.5(a) and (b)). Likewise, web browsers provide a facility of bookmarking or saving favorite URLs that have been visited, providing a visual list. This means that users need only recognize a name of a site when scanning through the saved list of URLs.

```
Microsoft Windows 2000 [Version 5.00.2195]
<c> Copyright 1985-1999 Microsoft Corp.

C:\>dir /w
 Volume in drive C has no label
 Volume Serial Number is 07D1-0109

 Directory of C:\

[BACKUP]                         [DELL]                    [DISCOVER]
[I386]                           [WINNT]                   [DRIVERS]
[Documents and settings] [Program Files]                  [temp]
[DellUtil]                       [DMI]                     [My Music]
[Downloads]                      [Palm]                    [Inetpub]
TxE8 - Backup
                    1 File(s)            1.367 bytes
                   15 Dir(s)   30.522.605.568 bytes free

C:\>cd Documents and settings

C:\Documents and settings>>dir
 Volume in drive C has no label.
 Volume Serial Number is 07D1-0109

 Directory of C:\Documents and settings

09/01/2001  11:49         <DIR>          .
09/01/2001  11:49         <DIR>          ..
09/01/2001  11:49         <DIR>          All Users
09/01/2001  12:04         <DIR>          Administrator
              0 File<s>                  0 bytes
              4Dir<s>     30,522,605,568 bytes free

C:\Documents and settings>cd Administrator

C:\Documents and Settings\Administrator>dir
 Volume in drive C has no label..
 Volume Serial Number is 07D1-0109

 Directory of C:\Documents and settings\Administrator

09/01/2001  12:04         <DIR>          .
09/01/2001  12:04         <DIR>          ..
09/01/2001  11:49         <DIR>          Start Menu
09/01/2001  11:49         <DIR>          My Documents
09/01/2001  11:49         <DIR>          Favorites
09/01/2001  11:49         <DIR>          Desktop
24/01/2001  17:16         <DIR>          Abisuite
              0 File<s>                  0 bytes
              7Dir<s>     30,522,605,568 bytes free

C:\Documents and settings\Administrator>cd My Documents

C:\Documents and settings\Administrator\My Documents>
```

Figure 3.5(a) A DOS-based interface, requiring the user to type in commands.

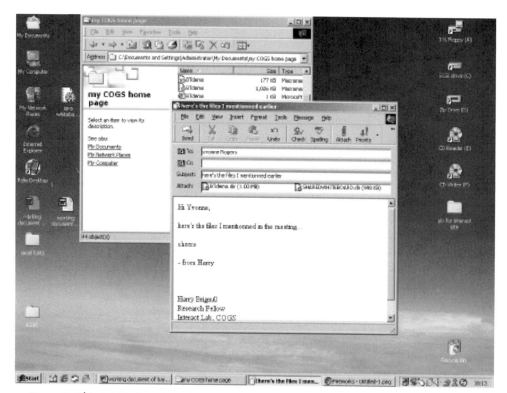

Figure 3.5(b) A Windows-based interface, with menus, icons, and buttons.

ACTIVITY 3.2 What strategies do you use to help you remember things?

Comment People often write down what they need to remember on a piece of paper. They also ask others to remind them. Another approach is to use various mental strategies, like mnemonics. A mnemonic involves taking the first letters of a set of words in a phrase or set of concepts and using them to make a more memorable phrase, often using bizarre and idiosyncratic connections. For example, some people have problems working out where east is in relation to west and vice versa (i.e., is it to the left or right). A mnemonic to help figure this out is to take the first letters of the four main points of the compass and then use them in the phrase "Never Eat Shredded Wheat" mentally recited in a clockwise sequence.

A growing problem for computer users is file management. The number of documents created, images and videoclips downloaded, emails and attachments saved, URLs bookmarked, and so on increases every day. A major problem is finding them again. Naming is the most common means of encoding them, but trying to remember a name of a file you created some time back can be very difficult, especially if there are tens of thousands of named files. How might such a process be facilitated, bearing in mind people's memory abilities? Mark Lansdale, a British psychologist, has been researching this problem of information retrieval for many

BOX 3.1   The Problem with the Magical Number 7 Plus or Minus 2

Perhaps the best known finding in psychology (certainly the one that nearly all students remember many years after they have finished their studies) is George Miller's (1956) theory that 7±2 chunks of information can be held in short-term memory at any one time. By short-term memory he meant a memory store in which information was assumed to be processed when first perceived. By chunks he meant a range of items like numbers, letters, or words. According to Miller's theory, therefore, people's immediate memory capacity is very limited. They are able to remember only a few words or numbers that they have heard or seen. If you are not familiar with this phenomenon, try out the following exercise: read the first list below (or get someone to read it to you), cover it up, and then try to recall as many of the items as possible. Repeat this for the other lists.

- 3, 12, 6, 20, 9, 4, 0, 1, 19, 8, 97, 13, 84
- cat, house, paper, laugh, people, red, yes, number, shadow, broom, rain, plant, lamp, chocolate, radio, one, coin, jet
- t, k, s, y, r, q, x, p, a, z, l, b, m, e

How many did you correctly remember for each list? Between 5 and 9, as suggested by Miller's theory?

Chunks can also be combined items that are meaningful. For example, it is possible to remember the same number of two-word phrases like hot chocolate, banana split, cream cracker, rock music, cheddar cheese, leather belt, laser printer, tree fern, fluffy duckling, cold rain. When these are all muddled up (e.g., split belt, fern crackers, banana laser, printer cream, cheddar tree, rain duckling, hot rock), however, it is much harder to remember as many chunks. This is mainly because the first set contains all meaningful two-word phrases that have been heard before and require less time to be processed in short-term memory, whereas the second set are completely novel phrases that don't exist in the real world. You need to spend time linking the two parts of the phrase together while trying to memorize them. This takes more time and effort to achieve. Of course it is possible to do if you have time to spend rehearsing them, but if you are asked to do it having

heard them only once in quick succession, it is most likely you will remember only a few.

You may be thinking by now, "Ok, this is interesting, but what has it got to do with interaction design?" Well not only does this 50-year-old theory have a special place in psychology, it has also made a big impression in HCI. Unfortunately, however, for the wrong reasons. Many designers have heard or read about this phenomenon and think, ah, here is a bit of psychology I can usefully apply to interface design. Would you agree with them? If so, how might people's ability to only remember 7±2 chunks that they have just read or heard be usefully applied to interaction design?

According to a survey by Bob Bailey (2000), several designers have been led to believe the following guidelines and have even created interfaces based on them:

- Have only seven options on a menu.
- Display only seven icons on a menu bar.
- Never have more than seven bullets in a list.
- Place only seven tabs at the top of a website page.
- Place only seven items on a pull-down menu.

All of these are wrong. Why? The simple reason is that these are all items that can be scanned and rescanned visually and hence do *not* have to be recalled from short-term memory. They don't just flash up on the screen and disappear, requiring the user to remember them before deciding which one to select. If you were asked to find an item of food most people crave in the set of single words listed above, would you have any problem? No, you would just scan the list until you recognized the one (chocolate) that matched the task and then select it—just as people do when interacting with menus, lists, and tabs—regardless of whether they comprise three or 30 items. What the users are required to do here is not remember as many items as possible, having only heard or seen them once in a sequence, but instead *scan* through a set of items until they *recognize* the one they want. Quite a different task. Furthermore, there is much more useful psychological research that can be profitably applied to interaction design.

years. He suggests that it is profitable to view this process as involving two memory processes: recall-directed, followed by recognition-based scanning. The first refers to using memorized information about the required file to get as close to it as possible. The more exact this is, the more success the user will have in tracking down the desired file. The second happens when recall has failed to produce what a user wants and so requires reading through directories of files.

To illustrate the difference between these two processes, consider the following scenario: a user is trying to access a couple of websites visited the day before that compared the selling price of cars offered by different dealers. The user is able to recall the name of one website: "alwaysthecheapest.com". She types this in and the website appears. This is an example of successful recall-directed memory. However, the user is unable to remember the name of the second one. She vaguely remembers it was something like 'autobargains.com'; but typing this in proves unsuccessful. Instead, she switches to scanning her bookmarks/favorites, going to the list of most recent ones saved. She notices two or three URLs that could be the one desired, and on the second attempt she finds the website she is looking for. In this situation, the user initially tries recall-directed memory and when this fails, adopts the second strategy of recognition-based scanning—which takes longer but eventually results in success.

Lansdale proposes that file management systems should be designed to optimize both kinds of memory processes. In particular, systems should be developed that let users use whatever memory they have to limit the area being searched and then represent the information in this area of the interface so as to maximally assist them in finding what they need. Based on this theory, he has developed a prototype system called MEMOIRS that aims at improving users' recall of information they had encoded so as to make it easier to recall later (Lansdale and Edmunds, 1992). The system was designed to be flexible, providing the user with a range of ways of encoding documents mnemonically, including time stamping (see Figure 3.6), flagging, and attribution (e.g., color, text, icon, sound or image).

More flexible ways of helping users track down the files they want are now beginning to be introduced as part of commercial applications. For example, various search and find tools, like Apple's Sherlock, have been designed to enable the user to type a full or partial name or phrase that the system then tries to match by listing all the files it identifies containing the requested name/phrase. This method, however, is still quite limited, in that it allows users to encode and retrieve files using only alphanumericals.

DESIGN IMPLICATIONS Memory

- Do not overload users' memories with complicated procedures for carrying out tasks.
- Design interfaces that promote *recognition* rather than *recall* by using menus, icons, and consistently placed objects.
- Provide users with a variety of ways of encoding electronic information (e.g., files, emails, images) to help them remember where they have stored them, through the use of color, flagging, time stamping, icons, etc.

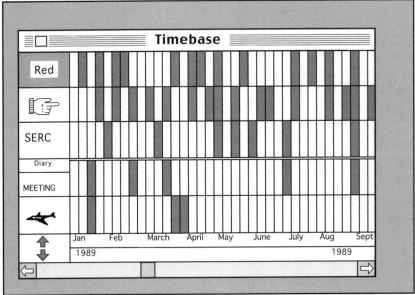

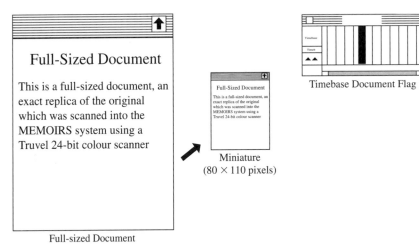

Figure 3.6 Memoirs tool.

BOX 3.2 A Case of Too Much Memory Load?

Phone banking has become increasingly popular in the last few years. It allows customers to carry out financial transactions, such as paying bills and checking the balance of their accounts, at their convenience. One of the problems confronting banks that provide this facility, however, is how to manage security concerns. Anyone can phone up a bank and pretend to be someone else. How do the banks prevent fraudulent transactions?

One solution has been to develop rigorous security measures whereby customers must provide various pieces of information before gaining access to their accounts. Typically, these include providing the answers to a combination of the following:

- their zip code or post code
- their mother's maiden name
- their birthplace
- the last school they attended
- the first school they attended
- a password of between 5 and 10 letters
- a *memorable* address (not their home)
- a *memorable* date (not their birthday)

Many of these are relatively easy to remember and recall as they are very familiar. But consider the last two. How easy is it for someone to come up with such *memorable* information and then be able to recall it readily? Perhaps the customer can give the address and birthday of another member of their family as a memorable address and date. But what about the request for a password? Sup-

pose a customer selects the word "interaction" as a password—fairly easy to remember. The problem is that the bank operators do not ask for the full password, because of the danger that someone in the vicinity might overhear and write it down. Instead they are instructed to ask the customer to provide specific letters from it, like the 7th followed by the 5th. However, such information does not spring readily to mind. Instead, it requires mentally counting each letter of the password until the desired one is reached. How long does it take you to determine the 7th letter of the password "interaction"? How did you do it?

To make things harder, banks also randomize the questions they ask. Again, this is to prevent someone who might be overhearing from memorizing the sequence of information. However, it also means that the customers themselves cannot learn the sequence of information required, meaning they have to generate different information every time they call up the bank.

This requirement to remember and recall such information puts a big memory load on customers. Some people find such a procedure quite nerve-wracking and are prone to forget certain pieces of information. As a coping strategy they write down their details on a sheet of paper. Having such an external representation at hand makes it much easier for them to read off the necessary information rather than having to recall it from memory. However, it also makes them vulnerable to the very fraud the banks were trying to prevent, should anyone else get hold of that piece of paper!

ACTIVITY 3.3   How else might banks solve the problem of providing a secure system while making the memory load relatively easy for people wanting to use phone banking? How does phone banking compare with online banking?

Comment   An alternative approach is to provide the customers with a PIN number (it could be the same as that of their ATM card) and ask them to key this in on their phone keypad, followed by asking one or two questions like their zip or post code, as a backup. Online banking has similar security risks to phone banking and hence this requires a number of security measures to be enforced. These include that the user sets up a nickname and a password. For example, some banks require typing in three randomly selected letters from a password each time the user logs on. This is harder to do online than when asked over the phone, mainly because it interferes with the normally highly automated process of typing in a password. You really have to think about what letters and numbers are in your password; for example, has it got two letter f's after the number 6, or just one?

Learning can be considered in terms of (i) how to use a computer-based application or (ii) using a computer-based application to understand a given topic. Jack Carroll (1990) and his colleagues have written extensively about how to design interfaces to help learners develop computer-based skills. A main observation is that people find it very hard to learn by following sets of instructions in a manual. Instead, they much prefer to "learn through doing." GUIs and direct manipulation interfaces are good environments for supporting this kind of learning by supporting exploratory interaction and importantly allowing users to "undo" their actions, i.e., return to a previous state if they make a mistake by clicking on the wrong option. Carroll has also suggested that another way of helping learners is by using a "training-wheels" approach. This involves restricting the possible functions that can be carried out by a novice to the basics and then extending these as the novice becomes more experienced. The underlying rationale is to make initial learning more tractable, helping the learner focus on simple operations before moving on to more complex ones.

There have also been numerous attempts to harness the capabilities of different technologies to help learners understand topics. One of the main benefits of interactive technologies, such as web-based, multimedia, and virtual reality, is that they provide alternative ways of representing and interacting with information that are not possible with traditional technologies (e.g., books, video). In so doing, they have the potential of offering learners the ability to explore ideas and concepts in different ways.

| ACTIVITY 3.4 | Ask a grandparent, child, or other person who has not used a cell phone before to make and answer a call using it. What is striking about their behavior? |

Comment   First-time users often try to apply their understanding of a land-line phone to operating a cell phone. However, there are marked differences in the way the two phones operate, even for the simplest of tasks, like making a call. First, the power has to be switched on when using a cell phone, by pressing a button (but not so with land-line phones), then the number has to be keyed in, including at all times the area code (in the UK), even if the callee is in the same area (but not so with land-lines), and finally the "make a call" button must be pressed (but not so with land-line phones). First-time users may intuitively know how to switch the phone on but not know which key to hit, or that it has to be held down for a couple of seconds. They may also forget to key in the area code if they are in the same area as the person they are calling, and to press the "make a call" key. They may also forget to press the "end a call" button (this is achieved through putting the receiver down with a land-line phone). Likewise, when answering a call, the first-time user may forget to press the "accept a call" button or not know which one to press. These additional actions are quick to learn, once the user understands the need to explicitly instruct the cell phone when they want to make, accept, or end a call.

Reading, speaking and listening: these three forms of language processing have both similar and different properties. One similarity is that the meaning of

---

BOX 3.3   Learning the "Difficult Stuff" through Interactive Multimedia: the Role of Dynalinking

Children (and adults) often have problems learning the difficult stuff—by this we mean mathematical formulae, notations, laws of physics, and other abstract concepts. One of the main reasons is that they find it difficult to relate their concrete experiences of the physical world with these higher-level abstractions. Research has shown, however, that it is possible to facilitate this kind of learning through the use of interactive multimedia. In particular, different representations of the same process (e.g., a graph, a formula, a sound, a simulation) can be displayed and interacted with in ways that make their relationship with each other more explicit to the learner. This process of linking and manipulating multimedia representations at the interface is called dynalinking (Rogers and Scaife, 1998).

An example where we have found dynalinking beneficial is in helping children and students learn ecological concepts (e.g., food webs, carbon cycles, and energy). In one of our projects, we built a simple ecosystem of a pond using multimedia. The concrete simulation showed various organisms swimming and moving around and occasionally an event where one would eat another (e.g., a snail eating the weed). This was annotated and accompanied by various eating sounds (e.g., chomping) to attract the children's attention. The children could also interact with the simulation. When an organism was clicked on, it would say what it was and what it ate (e.g., "I'm a weed. I make my own food").

The simulation was dynalinked with other abstract representations of the pond ecosystem. One of these was a food web diagram (See Figure 3.7 in Color Plate 4). The children were encouraged to interact with the interlinked diagrams in various ways and to observe what happened in the concrete simulation when something was changed in the diagram and vice versa. Our study showed that children enjoyed interacting with the simulation and diagrams and, importantly, that they understood much better the purpose of the abstract diagrams and how to use them to reason about the ecosystem.

Dynalinking is a powerful form of interaction and can be used in a range of domains to explicitly show relationships among multiple dimensions, especially when the information to be understood or learned is complex. For example, it can be useful for domains like economic forecasting, molecular modeling, and statistical analyses.

---

sentences or phrases is the same regardless of the mode in which it is conveyed. For example, the sentence "Computers are a wonderful invention" essentially has the same meaning whether one reads it, speaks it, or hears it. However, the ease with which people can read, listen, or speak differs depending on the person, task, and context. For example, many people find listening much easier than reading. Specific differences between the three modes include:

- Written language is permanent while listening is transient. It is possible to reread information if not understood the first time round. This is not possible with spoken information that is being broadcast.

DESIGN IMPLICATIONS   Learning

- Design interfaces that encourage exploration.
- Design interfaces that constrain and guide users to select appropriate actions.
- Dynamically link representations and abstractions that need to be learned.

- Reading can be quicker than speaking or listening, as written text can be rapidly scanned in ways not possible when listening to serially presented spoken words.

- Listening requires less cognitive effort than reading or speaking. Children, especially, often prefer to listen to narratives provided in multimedia or web-based learning material than to read the equivalent text online.

- Written language tends to be grammatical while spoken language is often ungrammatical. For example, people often start a sentence and stop in mid-sentence, letting someone else start speaking.

- There are marked differences between people in their ability to use language. Some people prefer reading to listening, while others prefer listening. Likewise, some people prefer speaking to writing and vice versa.

- Dyslexics have difficulties understanding and recognizing written words, making it hard for them to write grammatical sentences and spell correctly.

- People who are hard of hearing or hard of seeing are also restricted in the way they can process language.

Many applications have been developed either to capitalize on people's reading, writing and listening skills, or to support or replace them where they lack or have difficulty with them. These include:

- interactive books and web-based material that help people to read or learn foreign languages

- speech-recognition systems that allow users to provide instructions via spoken commands (e.g., word-processing dictation, home control devices that respond to vocalized requests)

- speech-output systems that use artificially generated speech (e.g., written-text-to-speech systems for the blind)

- natural-language systems that enable users to type in questions and give text-based responses (e.g., Ask Jeeves search engine)

- cognitive aids that help people who find it difficult to read, write, and speak. A number of special interfaces have been developed for people who have problems with reading, writing, and speaking (e.g., see Edwards, 1992).

- various input and output devices that allow people with various disabilities to have access to the web and use word processors and other software packages

Helen Petrie and her team at the Sensory Disabilities Research Lab in the UK have been developing various interaction techniques to allow blind people to access the web and other graphical representations, through the use of auditory navigation and tactile diagrams.

Problem-solving, planning, reasoning and decision-making are all cognitive processes involving reflective cognition. They include thinking about what to do, what the options are, and what the consequences might be of carrying out a given action. They often involve conscious processes (being aware of what one is thinking

DESIGN IMPLICATIONS Reading, Speaking and Listening

- Keep the length of speech-based menus and instructions to a minimum. Research has shown that people find it hard to follow spoken menus with more than three or four options. Likewise, they are bad at remembering sets of instructions and directions that have more than a few parts.

- Accentuate the intonation of artificially generated speech voices, as they are harder to understand than human voices.
- Provide opportunities for making text large on a screen, without affecting the formatting, for people who find it hard to read small text.

about), discussion with others (or oneself), and the use of various kinds of artifacts, (e.g., maps, books, and pen and paper). For example, when planning the best route to get somewhere, say a foreign city, we may ask others, use a map, get instructions from the web, or a combination of these. Reasoning also involves working through different scenarios and deciding which is the best option or solution to a given problem. In the route-planning activity we may be aware of alternative routes and reason through the advantages and disadvantages of each route before deciding on the best one. Many a family argument has come about because one member thinks he or she knows the best route while another thinks otherwise.

Comparing different sources of information is also common practice when seeking information on the web. For example, just as people will phone around for a range of quotes, so too, will they use different search engines to find sites that give the best deal or best information. If people have knowledge of the pros and cons of different search engines, they may also select different ones for different kinds of queries. For example, a student may use a more academically oriented one when looking for information for writing an essay, and a more commercially based one when trying to find out what's happening in town.

The extent to which people engage in the various forms of reflective cognition depends on their level of experience with a domain, application, or skill. Novices tend to have limited knowledge and will often make assumptions about what to do using other knowledge about similar situations. They tend to act by trial and error, exploring and experimenting with ways of doing things. As a result they may start off being slow, making errors and generally being inefficient. They may also act irrationally, following their superstitions and not thinking ahead to the consequences of their actions. In contrast, experts have much more knowledge and experience and are able to select optimal strategies for carrying out their tasks. They are likely to be able to think ahead more, considering what the consequences might be of opting for a particular move or solution (as do expert chess players).

DESIGN IMPLICATION Problem-Solving, Planning, Reasoning and Decision-Making

- Provide additional hidden information that is easy to access for users who wish to understand more about how to carry out an activity more effectively (e.g., web searching).

3.3 Applying knowledge from the physical world to the digital world

As well as understanding the various cognitive processes that users engage in when interacting with systems, it is also useful to understand the way people cope with the demands of everyday life. A well known approach to applying knowledge about everyday psychology to interaction design is to *emulate*, in the digital world, the strategies and methods people commonly use in the physical world. An assumption is that if these work well in the physical world, why shouldn't they also work well in the digital world? In certain situations, this approach seems like a good idea. Examples of applications that have been built following this approach include electronic post-it notes in the form of "stickies," electronic "to-do" lists, and email reminders of meetings and other events about to take place. The stickies application displays different colored notes on the desktop in which text can be inserted, deleted, annotated, and shuffled around, enabling people to use them to remind themselves of what they need to do—analogous to the kinds of externalizing they do when using paper stickies. Moreover, a benefit is that electronic stickies are more durable than paper ones—they don't get lost or fall off the objects they are stuck to, but stay on the desktop until explicitly deleted.

In other situations, however, the simple emulation approach can turn out to be counter-productive, forcing users to do things in bizarre, inefficient, or inappropriate ways. This can happen when the activity being emulated is more complex than is assumed, resulting in much of it being oversimplified and not supported effectively. Designers may notice something salient that people do in the physical world and then fall into the trap of trying to copy it in the electronic world without thinking through how and whether it will work in the new context (remember the poor design of the virtual calculator based on the physical calculator described in the previous chapter).

Consider the following classic study of real-world behavior. Ask yourself, first, whether it is useful to emulate at the interface, and second, how it could be extended as an interactive application.

Tom Malone (1983) carried out a study of the "natural history" of physical offices. He interviewed people and studied their offices, paying particular attention to their filing methods and how they organized their papers. One of his findings was that whether people have messy offices or tidy offices may be more significant than people realize. Messy offices were seen as being chaotic with piles of papers everywhere and little organization. Tidy offices, on the other hand, were seen as being well organized with good use of a filing system. In analyzing these two types of offices, Malone suggested what they reveal in terms of the underlying cognitive behaviors of the occupants. One of his observations was that messy offices may appear chaotic but in reality often reflect a coping strategy by the person: documents are left lying around in obvious places to act as reminders that something has to be done with them. This observation suggests that using piles is a fundamental strategy, regardless of whether you are a chaotic or orderly person.

Such observations about people's coping strategies in the physical world bring to mind an immediate design implication about how to support electronic file management: to capitalize on the "pile" phenomenon by trying to emulate it in the electronic world. Why not let people arrange their electronic files into piles as they do with paper files? The danger of doing this is that it could heavily constrain the way people manage their files, when in fact there may be far more effective and flexible ways of filing in the electronic world. Mark Lansdale (1988) points out how introducing unstructured piles of electronic documents on a desktop would be counterproductive, in the same way as building planes to flap their wings in the way birds do (someone seriously thought of doing this).

But there may be benefits of emulating the pile phenomenon by using it as a kind of interface metaphor that is extended to offer other functionality. How might this be achieved? A group of interface designers at Apple Computer (Mandler et al., 1992) tackled this problem by adopting the philosophy that they were going to build an application that went beyond physical-world capabilities, providing new functionality that only the computer could provide and that enhanced the interface. To begin their design, they carried out a detailed study of office behavior and analyzed the many ways piles are created and used. They also examined how people use the default hierarchical file-management systems that computer operating systems provide. Having a detailed understanding of both enabled them to create a conceptual model for the new functionality—which was to provide various interactive organizational elements based around the notion of using piles. These included providing the user with the means of creating, ordering, and visualizing piles of files. Files could also be encoded using various external cues, including date and color. New functionality that could not be achieved with physical files included the provision of a scripting facility, enabling files in piles to be ordered in relation to these cues (see Figure 3.8).

Emulating real-world activity at the interface can be a powerful design strategy, provided that new functionality is incorporated that extends or supports the users in their tasks in ways not possible in the physical world. The key is really to understand the nature of the problem being addressed in the electronic world in relation to the various coping and externalizing strategies people have developed to deal with the physical world.

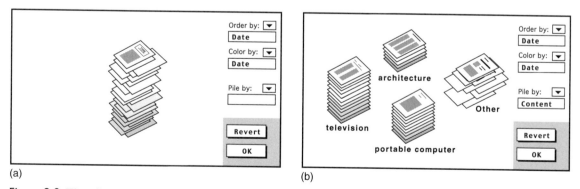

(a)   (b)

Figure 3.8 The pile metaphor as it appears at the interface.

3.4 Conceptual frameworks for cognition

In the previous section we described the pros and cons of applying knowledge of people's coping strategies in the physical world to the digital world. Another approach is to apply theories and conceptual frameworks to interaction design. In this section we examine three of these approaches, which each have a different perspective on cognition:

- mental models
- information processing
- external cognition

3.4.1 Mental models

In Chapter 2 we pointed out that a successful system is one based on a conceptual model that enables users to readily learn a system and use it effectively. What happens when people are learning and using a system is that they develop knowledge of how to use the system and, to a lesser extent, how the system works. These two kinds of knowledge are often referred to as a user's mental model.

Having developed a mental model of an interactive product, it is assumed that people will use it to make inferences about how to carry out tasks when using the interactive product. Mental models are also used to fathom what to do when something unexpected happens with a system and when encountering unfamiliar systems. The more someone learns about a system and how it functions, the more their mental model develops. For example, TV engineers have a "deep" mental model of how TVs work that allows them to work out how to fix them. In contrast,

an average citizen is likely to have a reasonably good mental model of how to operate a TV but a "shallow" mental model of how it works.

Within cognitive psychology, mental models have been postulated as internal constructions of some aspect of the external world that are manipulated enabling predictions and inferences to be made (Craik, 1943). This process is thought to involve the "fleshing out" and the "running" of a mental model (Johnson-Laird, 1983). This can involve both unconscious and conscious mental processes, where images and analogies are activated.

ACTIVITY 3.5 To illustrate how we use mental models in our everyday reasoning, imagine the following two scenarios:

(a) You arrive home from a holiday on a cold winter's night to a cold house. You have a small baby and you need to get the house warm as quickly as possible. Your house is centrally heated. Do you set the thermostat as high as possible or turn it to the desired temperature (e.g. 70°F)?

(b) You arrive home from being out all night, starving hungry. You look in the fridge and find all that is left is an uncooked pizza. The instructions on the packet say heat the oven to 375°F and then place the pizza in the oven for 20 minutes. Your oven is electric. How do you heat it up? Do you turn it to the specified temperature or higher?

Comment Most people when asked the first question imagine the scenario in terms of what they would do in their own house and choose the first option. When asked why, a typical explanation that is given is that setting the temperature to be as high as possible increases the rate at which the room warms up. While many people may believe this, it is incorrect. Thermostats work by switching on the heat and keeping it going at a constant speed until the desired temperature set is reached, at which point they cut out. They cannot control the rate at which heat is given out from a heating system. Left at a given setting, thermostats will turn the heat on and off as necessary to maintain the desired temperature.

When asked the second question, most people say they would turn the oven to the specified temperature and put the pizza in when they think it is at the desired temperature. Some people answer that they would turn the oven to a higher temperature in order to warm it up more quickly. Electric ovens work on the same principle as central heating and so turning the heat up higher will not warm it up any quicker. There is also the problem of the pizza burning if the oven is too hot!

Why do people use erroneous mental models? It seems that in the above scenarios, they are running a mental model based on a general valve theory of the way something works (Kempton, 1986). This assumes the underlying principle of "more is more": the more you turn or push something, the more it causes the desired effect. This principle holds for a range of physical devices, such as taps and radio controls, where the more you turn them, the more water or volume is given. However, it does not hold for thermostats, which instead function based on the principle of an on-off switch. What seems to happen is that in everyday life people develop a core set of abstractions about how things work, and apply these to a range of devices, irrespective of whether they are appropriate.

Using incorrect mental models to guide behavior is surprisingly common. Just watch people at a pedestrian crossing or waiting for an elevator (lift). How many times do they press the button? A lot of people will press it at least twice. When asked why, a common reason given is that they think it will make the lights change faster or ensure the elevator arrives. This seems to be another example of following the "more is more" philosophy: it is believed that the more times you press the button, the more likely it is to result in the desired effect.

Another common example of an erroneous mental model is what people do when the cursor freezes on their computer screen. Most people will bash away at all manner of keys in the vain hope that this will make it work again. However, ask them how this will help and their explanations are rather vague. The same is true when the TV starts acting up: a typical response is to hit the top of the box repeatedly with a bare hand or a rolled-up newspaper. Again, ask people why and their reasoning about how this behavior will help solve the problem is rather lacking.

The more one observes the way people interact with and behave towards interactive devices, the more one realizes just how strange their behavior can get—especially when the device doesn't work properly and they don't know what to do. Indeed, research has shown that people's mental models of the way interactive devices work is poor, often being incomplete, easily confusable, based on inappropriate analogies, and superstition (Norman, 1983). Not having appropriate mental models available to guide their behavior is what causes people to become very frustrated—often resulting in stereotypical "venting" behavior like those described above.

On the other hand, if people could develop better mental models of interactive systems, they would be in a better position to know how to carry out their tasks efficiently and what to do if the system started acting up. Ideally, they should be able to develop a mental model that matches the conceptual model developed by the designer. But how can you help users to accomplish this? One suggestion is to educate them better. However, many people are resistant to spending much time learning about how things work, especially if it involves reading manuals and other documentation. An alternative proposal is to design systems to be more transparent, so that they are easier to understand. This doesn't mean literally revealing the guts of the system (cf. the way some phone handsets—see Figure 3.9 on Color Plate 4—and iMacs are made of transparent plastic to reveal the colorful electronic circuitry inside), but requires developing an easy-to-understand system image (see Chapter 2 for explanation of this term in relation to conceptual models). Specifically, this involves providing:

- useful feedback in response to user input
- easy-to-understand and intuitive ways of interacting with the system

In addition, it requires providing the right kind and level of information, in the form of:

- clear and easy-to-follow instructions
- appropriate online help and tutorials
- context-sensitive guidance for users, set at their level of experience, explaining how to proceed when they are not sure what to do at a given stage of a task.

DILEMMA | How Much Transparency?

How much and what kind of transparency do you think a designer should provide in an interactive product? This is not a straightforward question to answer and depends a lot on the requirements of the targeted user groups. Some users simply want to get on with their tasks and don't want to have to learn about how the thing they are using works. In this situation, the system should be designed to make it obvious what to do and how to use it. For example, most cell-phone users want a simple "plug-and-play" type interface, where it is straightforward how to carry out functions like saving an address, text messaging, and making a call. Functions that require too much learning can be off-putting. Users simply won't bother to make the extra effort, meaning that many of the functions provided are never used. Other users like to understand how the device they are using works, in order to make informed decisions about how to carry out their tasks, especially if there are numerous ways of doing something. Some search engines have been designed with this in mind: they provide background information on how they work and how to improve one's searching techniques (see Figure 3.10).

Thus, the extent to which designers should provide extensive information about how to use a system and how it works, as part of the system image, needs to be appraised in terms of what different people want to know and how much they are prepared to learn.

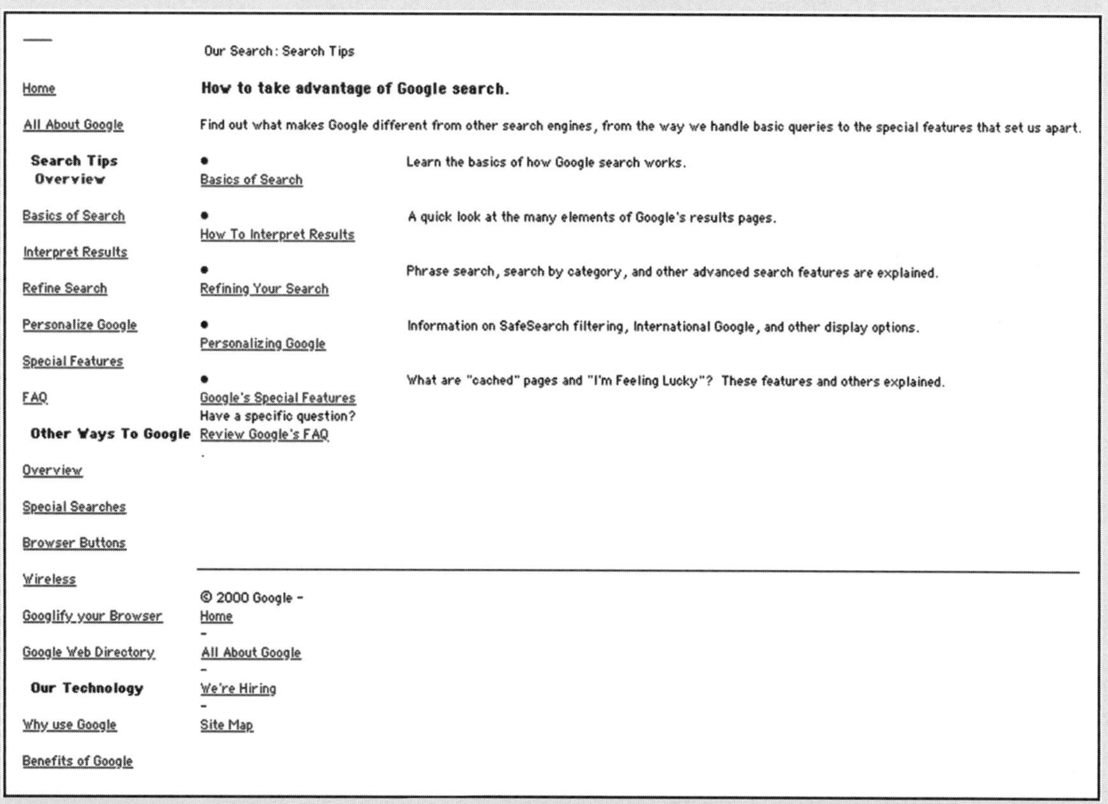

Our Search: Search Tips

How to take advantage of Google search.

Find out what makes Google different from other search engines, from the way we handle basic queries to the special features that set us apart.

Home
All About Google
Search Tips Overview
Basics of Search
Interpret Results
Refine Search
Personalize Google
Special Features
FAQ
Other Ways To Google
Overview
Special Searches
Browser Buttons
Wireless
Googlify your Browser
Google Web Directory
Our Technology
Why use Google
Benefits of Google

• Basics of Search — Learn the basics of how Google search works.

• How To Interpret Results — A quick look at the many elements of Google's results pages.

• Refining Your Search — Phrase search, search by category, and other advanced search features are explained.

• Personalizing Google — Information on SafeSearch filtering, International Google, and other display options.

• Google's Special Features — What are "cached" pages and "I'm Feeling Lucky"? These features and others explained.

Have a specific question? Review Google's FAQ

© 2000 Google –
Home
-
All About Google
-
We're Hiring
-
Site Map

Figure 3.10 The Google search engine, which provides extensive information about how to make your searching strategy more effective.

3.4.2   Information processing

Another approach to conceptualizing how the mind works has been to use metaphors and analogies (see also Chapter 2). A number of comparisons have been made, including conceptualizing the mind as a reservoir, a telephone network, and a digital computer. One prevalent metaphor from cognitive psychology is the idea that the mind is an information processor. Information is thought to enter and exit the mind through a series of ordered processing stages (see Figure 3.11). Within these stages, various processes are assumed to act upon mental representations. Processes include comparing and matching. Mental representations are assumed to comprise images, mental models, rules, and other forms of knowledge.

The information processing model provides a basis from which to make predictions about human performance. Hypotheses can be made about how long someone will take to perceive and respond to a stimulus (also known as reaction time) and what bottlenecks occur if a person is overloaded with too much information. The best known approach is the human processor model, which models the cognitive processes of a user interacting with a computer (Card et al., 1983). Based on the information processing model, cognition is conceptualized as a series of processing stages, where perceptual, cognitive, and motor processors are organized in relation to one another (see Figure 3.12). The model predicts which cognitive processes are involved when a user interacts with a computer, enabling calculations to be made of how long a user will take to carry out various tasks. This can be very useful when comparing different interfaces. For example, it has been used to compare how well different word processors support a range of editing tasks.

The information processing approach is based on modeling mental activities that happen *exclusively* inside the head. However, most cognitive activities involve people interacting with external kinds of representations, like books, documents, and computers—not to mention one another. For example, when we go home from wherever we have been we do not need to remember the details of the route because we rely on cues in the environment (e.g., we know to turn left at the red house, right when the road comes to a T-junction, and so on). Similarly, when we are at home we do not have to remember where everything is because information is "out there." We decide what to eat and drink by scanning the items in the fridge, find out whether any messages have been left by glancing at the answering machine to see if there is a flashing light, and so on. To what extent, therefore, can we say that information processing models are truly representative of everyday cognitive activities? Do they adequately account for cognition as it happens in the real world and, specifically, how people interact with computers and other interactive devices?

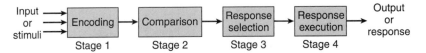

Figure 3.11 Human information processing model.

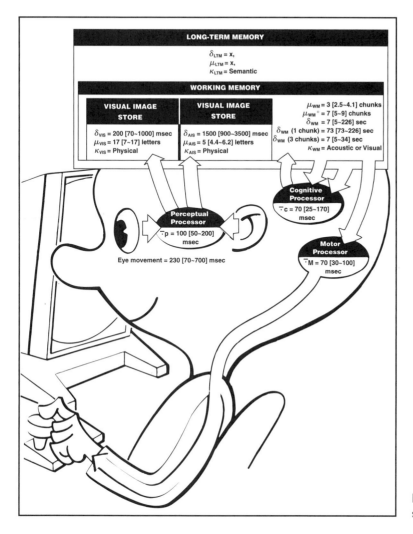
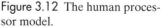

Figure 3.12 The human processor model.

Several researchers have argued that existing information processing approaches are too impoverished:

> *The traditional approach to the study of cognition is to look at the pure intellect, isolated from distractions and from artificial aids. Experiments are performed in closed, isolated rooms, with a minimum of distracting lights or sounds, no other people to assist with the task, and no aids to memory or thought. The tasks are arbitrary ones, invented by the researcher. Model builders build simulations and descriptions of these isolated situations. The theoretical analyses are self-contained little structures, isolated from the world, isolated from any other knowledge or abilities of the person. (Norman, 1990, p. 5)*

Instead, there has been an increasing trend to study cognitive activities in the context in which they occur, analyzing cognition as it happens "in the wild"

(Hutchins, 1995). A central goal has been to look at how structures in the environment can both aid human cognition and reduce cognitive load. A number of alternative frameworks have been proposed, including external cognition and distributed cognition. In this chapter, we look at the ideas behind external cognition—which has focused most on how to inform interaction design (distributed cognition is described in the next chapter).

3.4.3 External cognition

People interact with or create information through using a variety of external representations, e.g., books, multimedia, newspapers, web pages, maps, diagrams, notes, drawings, and so on. Furthermore, an impressive range of tools has been developed throughout history to aid cognition, including pens, calculators, and computer-based technologies. The *combination* of external representations and physical tools have greatly extended and supported people's ability to carry out cognitive activities (Norman, 1993). Indeed, they are such an integral part that it is difficult to imagine how we would go about much of our everyday life without them.

External cognition is concerned with explaining the cognitive processes involved when we interact with different external representations (Scaife and Rogers, 1996). A main goal is to explicate the cognitive benefits of using different representations for different cognitive activities and the processes involved. The main ones include:

1. externalizing to reduce memory load
2. computational offloading
3. annotating and cognitive tracing

1. Externalizing to reduce memory load

A number of strategies have been developed for transforming knowledge into external representations to reduce memory load. One such strategy is externalizing things we find difficult to remember, such as birthdays, appointments, and addresses. Diaries, personal reminders and calendars are examples of cognitive artifacts that are commonly used for this purpose, acting as external reminders of what we need to do at a given time (e.g., buy a card for a relative's birthday).

Other kinds of external representations that people frequently employ are notes, like "stickies," shopping lists, and to-do lists. Where these are placed in the environment can also be crucial. For example, people often place post-it notes in prominent positions, such as on walls, on the side of computer monitors, by the front door and sometimes even on their hands, in a deliberate attempt to ensure they do remind them of what needs to be done or remembered. People also place things in piles in their offices and by the front door, indicating what needs to be done urgently and what can wait for a while.

Externalizing, therefore, can help reduce people's memory burden by:

- reminding them to do something (e.g., to get something for their mother's birthday)

- reminding them of what to do (e.g., to buy a card)
- reminding them of when to do something (send it by a certain date)

2. Computational offloading

Computational offloading occurs when we use a tool or device in conjunction with an external representation to help us carry out a computation. An example is using pen and paper to solve a math problem.

ACTIVITY 3.6

(a) Multiply 2 by 3 in your head. Easy. Now try multiplying 234 by 456 in your head. Not as easy. Try doing the sum using a pen and paper. Then try again with a calculator. Why is it easier to do the calculation with pen and paper and even easier with a calculator?

(b) Try doing the same two sums using Roman numerals.

Comment

(a) Carrying out the sum using pen and the paper is easier than doing it in your head because you "offload" some of the computation by writing down partial results and using them to continue with the calculation. Doing the same sum with a calculator is even easier, because it requires only eight simple key presses. Even more of the computation has been offloaded onto the tool. You need only follow a simple internalized procedure (key in first number, then the multiplier sign, then next number and finally the equals sign) and then read of the result from the external display.

(b) Using roman numerals to do the same sum is much harder. 2 by 3 becomes ll × lll, and 234 by 456 becomes CCXXXlllI × CCCCXXXXXVI. The first calculation may be possible to do in your head or on a bit of paper, but the second is incredibly difficult to do in your head or even on a piece of paper (unless you are an expert in using Roman numerals or you "cheat" and transform it into Arabic numerals). Calculators do not have Roman numerals so it would be impossible to do on a calculator.

Hence, it is much harder to perform the calculations using Roman numerals than algebraic numerals—even though the problem is equivalent in both conditions. The reason for this is the two kinds of *representation* transform the task into one that is easy and more difficult, respectively. The kind of tool used also can change the nature of the task to being more or less easy.

3. Annotating and cognitive tracing

Another way in which we externalize our cognition is by modifying representations to reflect changes that are taking place that we wish to mark. For example, people often cross things off in a to-do list to show that they have been completed. They may also reorder objects in the environment, say by creating different piles as the nature of the work to be done changes. These two kinds of modification are called annotating and cognitive tracing:

- *Annotating* involves modifying external representations, such as crossing off or underlining items.

- *Cognitive tracing* involves externally manipulating items into different orders or structures.

Annotating is often used when people go shopping. People usually begin their shopping by planning what they are going to buy. This often involves looking in their cupboards and fridge to see what needs stocking up. However, many people are aware that they won't remember all this in their heads and so often externalize it as a written shopping list. The act of writing may also remind them of other items that they need to buy that they may not have noticed when looking through the cupboards. When they actually go shopping at the store, they may cross off items on the shopping list as they are placed in the shopping basket or cart. This provides them with an annotated externalization, allowing them to see at a glance what items are still left on the list that need to be bought.

Cognitive tracing is useful in situations where the current state of play is in a state of flux and the person is trying to optimize their current position. This typically happens when playing games, such as:

- in a card game, the continued rearrangement of a hand of cards into suits, ascending order, or same numbers to help determine what cards to keep and which to play, as the game progresses and tactics change
- in Scrabble, where shuffling around letters in the tray helps a person work out the best word given the set of letters (Maglio et al., 1999)

It is also a useful strategy for letting users know what they have studied in an online learning package. An interactive diagram can be used to highlight all the nodes visited, exercises completed, and units still to study.

A general cognitive principle for interaction design based on the external cognition approach is to provide external representations at the interface that reduce memory load and facilitate computational offloading. Different kinds of information visualizations can be developed that reduce the amount of effort required to make inferences about a given topic (e.g., financial forecasting, identifying pro-

---

BOX 3.4   Context-Sensitive Information: Shopping Reminders on the Move

A number of researchers have begun developing wireless communication systems that use GPRS technology to provide people on the move with context-sensitive information. This involves providing people with information (such as reminders and to-do lists) whenever it is appropriate to their location. For example, one such system called comMotion, which is being developed at MIT (Marmasse and Schmandt, 2000) uses a speech-output system to inform people when they are driving past a store that sells the groceries they need to buy, such as milk.

How useful is this kind of externalization? Are people really that bad at remembering things? In what way will it be an improvement over other reminder techniques, like shopping lists written on paper or lists stored on PalmPilots or other pocket computers? Sure, there are certain people who have debilitating memory problems (e.g. Alzheimer's) and who may greatly benefit from having such prosthetic memory devices. But what about those who aren't afflicted? What would happen to them if they started relying more and more on spoken reminders popping up all over the place to tell them what they should be doing when and where? They may well be reminded to buy the milk, but at what price? Losing their own ability to remember?

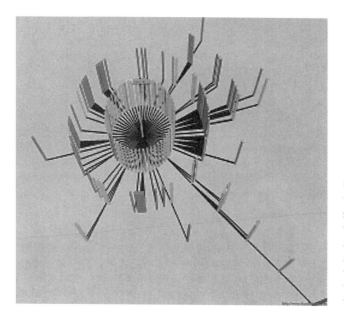

Figure 3.13 Information visualization. Visual Insights' site map showing web page use. Each page appears as a 3D color rod and is positioned radially, with the position showing the location of the page in the site.

gramming bugs). In so doing, they can extend or amplify cognition, allowing people to perceive and do activities that they couldn't do otherwise. For example, a number of information visualizations have been developed that present masses of data in a form that makes it possible to make cross comparisons between dimensions at a glance (see Figure 3.13). GUIs can also be designed to reduce memory load significantly, enabling users to rely more on external representations to guide them through their interactions.

3.5 Informing design: from theory to practice

Theories, models, and conceptual frameworks provide abstractions for thinking about phenomena. In particular, they enable generalizations to be made about cognition across different situations. For example, the concept of mental models provides a means of explaining why and how people interact with interactive products in the way they do across a range of situations. The information processing model has been used to predict the usability of a range of different interfaces.

Theory in its pure form, however, can be difficult to digest. The arcane terminology and jargon used can be quite off-putting to those not familiar with it. It also requires much time to become familiar with it—something that designers and engineers can't afford when working to meet deadlines. Researchers have tried to help out by making theory more accessible and practical. This has included translating it into:

- design principles and concepts
- design rules
- analytic methods
- design and evaluation methods

A main emphasis has been on transforming theoretical knowledge into tools that can be used by designers. For example, Card et al's (1983) psychological model of the human processor, mentioned earlier, was simplified into another model called GOMS (an acronym standing for goals, operators, methods, and selection rules). The four components of the GOMS model describe how a user performs a computer-based task in terms of goals (e.g., save a file) and the selection of methods and operations from memory that are needed to achieve them. This model has also been transformed into the keystroke level method that essentially provides a formula for determining the amount of time each of the methods and operations takes. One of the main attractions of the GOMS approach is that it allows quantitative predictions to be made (see Chapter 14 for more on this).

Another approach has been to produce various kinds of design principles, such as the ones we discussed in Chapter 1. More specific ones have also been proposed for designing multimedia and virtual reality applications (Rogers and Scaife, 1998). Thomas Green (1990) has also proposed a framework of cognitive dimensions. His overarching goal is to develop a set of high-level concepts that are both valuable and easy to use for evaluating the designs of informational artifacts, such as software applications. An example dimension from the framework is "viscosity," which simply refers to resistance to local change. The analogy of stirring a spoon in syrup (high viscosity) versus milk (low viscosity) quickly gives the idea. Having understood the concept in a familiar context, Green then shows how the dimension can be further explored to describe the various aspects of interacting with the information structure of a software application. In a nutshell, the concept is used to examine "how much extra work you have to do if you change your mind." Different kinds of viscosity are described, such as knock-on viscosity, where performing one goal-related action makes necessary the performance of a whole train of extraneous actions. The reason for this is constraint density: the new structure that results from performing the first action violates some constraint that must be rectified by the second action, which in turn leads to a different violation, and so on. An example is editing a document using a word processor without widow control. The action of inserting a sentence at the beginning of the document means that the user must then go through the rest of the document to check that all the headers and bodies of text still lie on the same page.

DILEMMA   Evolutionary versus Revolutionary Upgrading

A constant dilemma facing designers involved in upgrading software is where and how to place new functions. Decisions have to be made on how to incorporate them with the existing interface design. Do they try to keep the same structure and add more buttons/menu options, or do they design a new model of interaction that is better suited to organizing and categorizing the increased set of functions? If the former strategy is followed, users do not have to learn a new conceptual model every time they upgrade a piece of software. The down side of trying to keep the same interface structure, however, is that it can easily get overloaded.

A problem when upgrading software, therefore, is working out how to redesign the interaction so that the amount of relearning, relative to the gains from the new functionality, is acceptable by users.

Assignment

The aim of this assignment is for you to elicit mental models from people. In particular, the goal is for you to understand the nature of people's knowledge about an interactive product in terms of how to use it and how it works.

(a) First, elicit your own mental model. Write down how you think a cash machine (ATM) works. Then answer the following questions (abbreviated from Payne, 1991):

How much money are you allowed to take out?

If you took this out and then went to another machine and tried to withdraw the same amount, what would happen?

What is on your card?

How is the information used?

What happens if you enter the wrong number?

Why are there pauses between the steps of a transaction?

How long are they? What happens if you type ahead during the pauses?

What happens to the card in the machine?

Why does it stay inside the machine?

Do you count the money? Why?

Next, ask two other people the same set of questions.

(b) Now analyze your answers. Do you get the same or different explanations? What do the findings indicate? How accurate are people's mental models of the way ATMs work? How transparent are the ATM systems they are talking about?

(c) Next, try to interpret your findings with respect to the design of the system. Are any interface features revealed as being particularly problematic? What design recommendations do these suggest?

(d) Finally, how might you design a better conceptual model that would allow users to develop a better mental model of ATMs (assuming this is a desirable goal)?

This exercise is based on an extensive study carried out by Steve Payne on people's mental models of ATMs. He found that people do have mental models of ATMs, frequently resorting to analogies to explain how they work. Moreover, he found that people's explanations were highly variable and based on ad hoc reasoning.

Summary

This chapter has explained the importance of understanding users, especially their cognitive aspects. It has described relevant findings and theories about how people carry out their everyday activities and how to learn from these when designing interactive products. It has provided illustrations of what happens when you design systems with the user in mind and what happens when you don't. It has also presented a number of conceptual frameworks that allow ideas about cognition to be generalized across different situations.

Key points

- Cognition comprises many processes, including thinking, attention, learning, memory, perception, decision-making, planning, reading, speaking, and listening.

- The way an interface is designed can greatly affect how well people can perceive, attend, learn, and remember how to carry out their tasks.

- The main benefits of conceptual frameworks and cognitive theories are that they can explain user interaction and predict user performance.

- The conceptual framework of mental models provides a way of conceptualizing the user's understanding of the system.

- Research findings and theories from cognitive psychology need to be carefully reinterpreted in the context of interaction design to avoid oversimplification and misapplication.

Further reading

MULLET, K., AND SANO, D. (1995) *Designing Visual Interfaces*. New Jersey: SunSoft Press. This is an excellent book on the do's and don'ts of interactive graphical design. It includes many concrete examples that have followed (or not) design principles based on cognitive issues.

CARROLL, J. (1991) (ed.) *Designing Interaction*. Cambridge: Cambridge University Press. This edited volume provides a good collection of papers on cognitive aspects of interaction design.

NORMAN, D. (1988) *The Psychology of Everyday Things*. New York: Basic Books.

NORMAN, D. (1993) *Things that Make Us Smart*. Reading, MA: Addison-Wesley. These two early books by Don Norman provide many key findings and observations about people's behavior and their use of artifacts. They are written in a stimulating and thought-provoking way, using many examples from everyday life to illustrate conceptual issues. He also presents a number of psychological theories, including external cognition, in an easily digestible form.

ROGERS, Y., RUTHERFORD, A., AND BIBBY, P. (1992) (eds.) *Models in the Mind*. Orlando: Academic Press. This volume provides a good collection of papers on eliciting, interpreting, and theorizing about mental models in HCI and other domains.

For more on dynalinking and interactivity see www.cogs.susx.ac.uk/ECOi

Chapter 4

Designing for collaboration and communication

4.1 Introduction
4.2 Social mechanisms in communication and collaboration
    4.2.1   Conversational mechanisms
    4.2.2   Designing collaborative technologies to support conversation
    4.2.3   Coordination mechanisms
    4.2.4   Designing collaborative technologies to support coordination
    4.2.5   Awareness mechanisms
    4.2.6   Designing collaborative technologies to support awareness
4.3   Ethnographic studies of collaboration and communication
4.4   Conceptual frameworks
    4.4.1   The language/action framework
    4.4.2   Distributed cognition

4.1   Introduction

Imagine going into school or work each day and sitting in a room all by yourself with no distractions. At first, it might seem blissful. You'd be able to get on with your work. But what if you discovered you had no access to email, phones, the Internet and other people? On top of that there is nowhere to get coffee. How long would you last? Probably not very long. Humans are inherently social: they live together, work together, learn together, play together, interact and talk with each other, and socialize. It seems only natural, therefore, to develop interactive systems that support and extend these different kinds of sociality.

There are many kinds of sociality and many ways of studying it. In this chapter our focus is on how people communicate and collaborate in their working and everyday lives. We examine how collaborative technologies (also called groupware) have been designed to support and extend communication and collaboration. We also look at the social factors that influence the success or failure of user adoption of such technologies. Finally, we examine the role played by ethnographic studies and theoretical frameworks for informing system design.

The main aims of this chapter are to:

- Explain what is meant by communication and collaboration.
- Describe the main kinds of social mechanisms that are used by people to communicate and collaborate.
- Outline the range of collaborative systems that have been developed to support this kind of social behavior.
- Consider how field studies and socially-based theories can inform the design of collaborative systems.

4.2   Social mechanisms in communication and collaboration

A fundamental aspect of everyday life is talking, during which we pass on knowledge to each other. We continuously update each other about news, changes, and developments on a given project, activity, person, or event. For example, friends and families keep each other posted on what's happening at work, school, at the pub, at the club, next door, in soap operas, and in the news. Similarly, people who work together keep each other informed about their social lives and everyday happenings—as well as what is happening at work, for instance when a project is about to be completed, plans for a new project, problems with meeting deadlines, rumors about closures, and so on.

The kinds of knowledge that are circulated in different social circles are diverse, varying among social groups and across cultures. The frequency with which knowledge is disseminated is also highly variable. It can happen continuously throughout the day, once a day, weekly or infrequently. The means by which communication happens is also flexible—it can take place via face to face conversations, telephone, videophone, messaging, email, fax, and letters. Non-verbal communication also plays an important role in augmenting face to face conversation, involving the use of facial expressions, back channeling (the "aha's" and "umms"), voice intonation, gesturing, and other kinds of body language.

All this may appear self-evident, especially when one reflects on how we interact with one another. Less obvious is the range of social mechanisms and practices that have evolved in society to enable us to be social and maintain social order. Various rules, procedures, and etiquette have been established whose function is to let people know how they should behave in social groups. Below we describe three main categories of social mechanisms and explore how technological systems have been and can be designed to facilitate these:

- the use of conversational mechanisms to facilitate the flow of talk and help overcome breakdowns during it
- the use of coordination mechanisms to allow people to work and interact together
- the use of awareness mechanisms to find out what is happening, what others are doing and, conversely, to let others know what is happening

4.2.1 Conversational mechanisms

Talking is something that is effortless and comes naturally to most people. And yet holding a conversation is a highly skilled collaborative achievement, having many of the qualities of a musical ensemble. Below we examine what makes up a conversation. We begin by examining what happens at the beginning:

A: Hi there.

B: Hi!

C: Hi.

A: All right?

C: Good. How's it going?

A: Fine, how are you?

C: Good.

B: OK. How's life treating you?

Such mutual greetings are typical. A dialog may then ensue in which the participants take turns asking questions, giving replies, and making statements. Then when one or more of the participants wants to draw the conversation to a close, they do so by using either implicit or explicit cues. An example of an implicit cue is when a participant looks at his watch, signaling indirectly to the other participants that he wants the conversation to draw to a close. The other participants may choose to acknowledge this cue or carry on and ignore it. Either way, the first participant may then offer an explicit signal, by saying, "Well, I must be off now. Got work to do," or, "Oh dear, look at the time. Must dash. Have to meet someone." Following the acknowledgment by the other participants of such implicit and explicit signals, the conversation draws to a close, with a farewell ritual. The different participants take turns saying, "Bye," "Bye then," "See you," repeating themselves several times, until they finally separate.

Such conversational mechanisms enable people to coordinate their "talk" with one another, allowing them to know how to start and stop. Throughout a conversation further "turn-taking" rules are followed, enabling people to know when to listen, when it is their cue to speak, and when it is time for them to stop again to allow the others to speak. Sacks, Schegloff and Jefferson (1978)—who are famous for their work on conversation analysis—describe these in terms of three basic rules:

- rule 1—the current speaker chooses the next speaker by asking an opinion, question, or request

- rule 2—another person decides to start speaking

- rule 3—the current speaker continues talking

The rules are assumed to be applied in the above order, so that whenever there is an opportunity for a change of speaker to occur (e.g., someone comes to the end of a sentence), rule 1 is applied. If the listener to whom the question or opinion is addressed does not accept the offer to take the floor, the second rule is applied and someone else taking part in the conversation may take up the opportunity and offer a view on the matter. If this does not happen then the third rule is applied and the current speaker continues talking. The rules are cycled through recursively until someone speaks again.

To facilitate rule following, people use various ways of indicating how long they are going to talk and on what topic. For example, a speaker might say right at the beginning of their turn in the conversation that he has three things to say. A speaker may also explicitly request a change in speaker by saying, "OK, that's all I want to say on that matter. So, what do you think?" to a listener. More subtle cues to let others know that their turn in the conversation is coming to an end include the lowering or raising of the voice to indicate the end of a question or the use of phrases like, "You know what I mean?" or simply, "OK?" Back channeling (uh-huh, mmm), body orientation (e.g., moving away from or closer to someone), gaze (staring straight at someone or glancing away), and gesture (e.g. raising of arms) are also used in different combinations when talking, to signal to others when someone wants to hand over or take up a turn in the conversation.

Another way in which conversations are coordinated and given coherence is through the use of adjacency pairs (Shegloff and Sacks, 1973). Utterances are assumed to come in pairs in which the first part sets up an expectation of what is to come next and directs the way in which what does come next is heard. For example, A may ask a question to which B responds appropriately:

A: So shall we meet at 8:00?

B: Um, can we make it a bit later, say 8:30?

Sometimes adjacency pairs get embedded in each other, so it may take some time for a person to get a reply to their initial request or statement:

A: So shall we meet at 8:00?

B: Wow, look at him.

A: Yes, what a funny hairdo!

B: Um, can we make it a bit later, say 8:30?

For the most part people are not aware of following conversational mechanisms, and would be hard pressed to articulate how they can carry on a conversation. Furthermore, people don't necessarily abide by the rules all the time. They may interrupt each other or talk over each other, even when the current speaker has clearly indicated a desire to hold the floor for the next two minutes to finish an argument. Alternatively, a listener may not take up a cue from a speaker to answer a question or take over the conversation, but instead continue to say nothing even though the speaker may be making it glaringly obvious it is the listener's turn to say something. Many a time a teacher will try to hand over the conversation to a student in a seminar, by staring at her and asking a specific question, only to see the student look at the floor, and say nothing. The outcome is an embarrassing silence, followed by either the teacher or another student picking up the conversation again.

Other kinds of breakdowns in conversation arise when someone says something that is ambiguous and the other person misinterprets it to mean something else. In such situations the participants will collaborate to overcome the misunderstanding by using repair mechanisms. Consider the following snippet of conversation between two people:

> A: Can you tell me the way to get to the Multiplex Ranger cinema?

> B: Yes, you go down here for two blocks and then take a right (pointing to the right), go on till you get to the lights and then it is on the left.

> A: Oh, so I go along here for a couple of blocks and then take a right and the cinema is at the lights (pointing ahead of him)?

> A: No, you go on *this* street for a couple of blocks (gesturing more vigorously than before to the street to the right of him while emphasizing the word "this").

> B: Ahhhh! I thought you meant *that* one: so it's *this* one (pointing in same direction as the other person).

> A: Uh-hum, yes that's right, *this* one.

Detecting breakdowns in conversation requires the speaker and listener to be attending to what the other says (or does not say). Once they have understood the nature of the failure, they can then go about repairing it. As shown in the above example, when the listener misunderstands what has been communicated, the speaker repeats what she said earlier, using a stronger voice intonation and more exaggerated gestures. This allows the speaker to repair the mistake and be more explicit to the listener, allowing her to understand and follow better what they are saying. Listeners may also signal when they don't understand something or want further clarification by using various tokens, like "Huh?", "Quoi?" or "What?" (Schegloff, 1982) together with giving a puzzled look (usually frowning). This is especially the case when the speaker says something that is vague. For example, they might say "I want it" to their partner, without saying what *it* is they want. The partner may reply using a token or, alternatively, explicitly ask, "What do you mean by *it*?"

Taking turns also provides opportunities for the listener to initiate repair or request clarification, or for the speaker to detect that there is a problem and to initiate repair. The listener will usually wait for the next turn in the conversation before interrupting the speaker, to give the speaker the chance to clarify what is being said by completing the utterance (Suchman, 1987).

ACTIVITY 4.1   How do people repair breakdowns in conversations when using the phone or email?

Comment   In these settings people cannot see each other and so have to rely on other means of repairing their conversations. Furthermore, there are more opportunities for breakdowns to occur and fewer mechanisms available for repair. When a breakdown occurs over the phone, people will often shout louder, repeating what they said several times, and use stronger intonation. When a breakdown occurs via email, people may literally spell out what they meant, making things much more explicit in a subsequent email. If the message is beyond repair they may resort to another mode of communication that allows greater flexibility of expression, either telephoning or speaking to the recipient face to face.

Kinds of conversations

Conversations can take a variety of forms, such as an argument, a discussion, a heated debate, a chat, a tête-à-tête, or giving someone a "telling off." A well-known distinction in conversation types is between formal and informal communication. Formal communication involves assigning certain roles to people and prescribing *a priori* the types of turns that people are allowed to take in a conversation. For example, at a board meeting, it is decided who is allowed to speak, who speaks when, who manages the turn-taking, and what the participants are allowed to talk about.

In contrast, informal communication is the chat that goes on when people socialize. It also commonly happens when people bump into each other and talk briefly. This can occur in corridors, at the coffee machine, when waiting in line, and walking down the street. Informal conversations include talking about impersonal things like the weather (a favorite) and the price of living, or more personal things, like how someone is getting on with a new roommate. It also provides an opportunity to pass on gossip, such as who is going out to dinner with whom. In office settings, such chance conversations have been found to serve a number of functions, including coordinating group work, transmitting knowledge about office culture, establishing trust, and general team building (Kraut et al, 1990). It is also the case that people who are in physical proximity, such as those whose offices or desks are close to one another, engage much more frequently in these kinds of informal chats than those who are in different corridors or buildings. Most companies and organizations are well aware of this and often try to design their office space so that people who need to work closely together are placed close to one another in the same physical space.

4.2.2   Designing collaborative technologies to support conversation

As we have seen, "talk" and the way it is managed is integral to coordinating social activities. One of the challenges confronting designers is to consider how the different kinds of communication can be facilitated and supported in settings where there may be obstacles preventing it from happening "naturally." A central concern has been to develop systems that allow people to communicate with each other when they are in *physically different locations* and thus not able to communicate in the usual face to face manner. In particular, a key issue has been to determine how to allow people to carry on communicating as if they were in the same place, even though they are geographically separated—sometimes many thousands of miles apart.

Email, videoconferencing, videophones, computer conferencing, chatrooms and messaging are well-known examples of some of the collaborative technologies that have been developed to enable this to happen. Other less familiar systems are collaborative virtual environments (CVEs) and media spaces. CVEs are virtual worlds where people meet and chat. These can be 3D graphical worlds where users explore rooms and other spaces by teleporting themselves around in the guise of avatars (See Figure 4.1 on Color Plate 5), or text and graphical "spaces" (often called MUDs and MOOs) where users communicate with each other via some form of messaging. Media spaces are distributed systems comprising audio, video, and computer systems that "extend the world of desks, chairs, walls and ceilings" (Harrison et al., 1997), enabling people distributed over space and time to communicate and interact with one another as if they were physically present. The various collaborative technologies have been designed to support different kinds of communication, from informal to formal and from one-to-one to many-to-many conversations. Collectively, such technologies are often referred to as computer-mediated communication (CMC).

ACTIVITY 4.2   Do you think it is better to develop technologies that will allow people to talk at a distance as if they were face to face, or to develop technologies that will support new ways of conversing?

Comment   On the one hand, it seems a good idea to develop technologies supporting people communicating at a distance that emulate the way they hold conversations in face to face situations. After all, this means of communicating is so well established and second nature to people. Phones and videoconferencing have been developed to essentially support face to face conversations. It is important to note, however, that conversations held in this way are not the same as when face to face. People have adapted the way they hold conversations to fit in with the constraints of the respective technologies. As noted earlier, they tend to shout more when misunderstood over the phone. They also tend to speak more loudly when talking on the phone, since they can't monitor how well the person can hear them at the other end of the phone. Likewise, people tend to project themselves more when videoconferencing. Turn-taking appears to be much more explicit, and greetings and farewells more ritualized.

On the other hand, it is interesting to look at how the new communication technologies have been extending the way people talk and socialize. For example, SMS text messaging has provided people with quite different ways of having a conversation at a distance. People (especially teenagers) have evolved a new form of fragmentary conversation (called "texting") that they continue over long periods. The conversation comprises short phrases that are typed in, using the key pad, commenting on what each is doing or thinking, allowing the other to keep posted on current developments. These kinds of "streamlined" conversations are coordinated simply by taking turns sending and receiving messages. Online chatting has also enabled effectively hundreds and even thousands of people to take part in the same conversations, which is not possible in face to face settings.

The range of systems that support computer-mediated communication is quite diverse. A summary table of the different types is shown in Table 4.1, highlighting how they support, extend and differ from face to face communication. A conventionally accepted classification system of CMC is to categorize them in terms of either synchronous or asynchronous communication. We have also included a third category: systems that support CMC in combination with other collaborative activities, such as meetings, decision-making, learning, and collaborative authoring of documents. Although some communication technologies are not strictly speaking computer-based (e.g., phones, video-conferencing) we have included these in the classification of CMC, as most now are display-based and interacted with or controlled via an interface. (For more detailed overviews of CMC, see Dix et al. (Chapter 13, 1998) and Baecker et al. (Part III and IV, 1993).

Table 4.1 Classification of computer-mediated communication (CMC) into three types: (I) Synchronous communication, (ii) Asynchronous communication and (iii) CMC combined with other activity

i. Synchronous communication

Where conversations in real time are supported by letting people talk with each other either using their voices or through typing. Both modes seek to support non-verbal communication to varying degrees.

Examples:

- Talking with voice: video phones, video conferencing (desktop or wall), media spaces.
- Talking via typing: text messaging (typing in messages using cell phones), instant messaging (real-time interaction via PCs) chatrooms, collaborative virtual environments (CVEs).

New kinds of functionality:

- CVEs allow communication to take place via a combination of graphical representations of self (in the form of avatars) with a separate chatbox or overlaying speech bubbles.
- CVEs allow people to represent themselves as virtual characters, taking on new personas (e.g., opposite gender), and expressing themselves in ways not possible in face-to-face settings.
- CVEs, MUDs and chatrooms have enabled new forms of conversation mechanisms, such as multi-turn-taking, where a number of people can contribute and keep track of a multi-streaming text-based conversation.
- Instant messaging allows users to multitask by holding numerous conversations at once.

Benefits:

- Not having to physically face people may increase shy people's confidence and self-esteem to converse more in "virtual" public.
- It allows people to keep abreast of the goings-on in an organization without having to move from their office.
- It enables users to send text and images instantly between people using instant messaging.
- In offices, instant messaging allows users to fire off quick questions and answers without the time lag of email or phone-tag.

Problems:

- Lack of adequate bandwidth has plagued video communication, resulting in poor-quality images that frequently break up, judder, have shadows, and appear as unnatural images.
- It is difficult to establish eye contact (normally an integral and subconscious part of face-to-face conversations) in CVEs, video conferencing, and videophones.
- Having the possibility of hiding behind a persona, a name, or an avatar in a chatroom gives people the opportunity to behave differently. Sometimes this can result in people becoming aggressive or intrusive.

ii. Asynchronous communication

Where communication between participants takes place remotely and at different times. It relies not on time-dependent turn-taking but on participants initiating communication and responding to others when they want or are able to do so.

Examples:

- email, bulletin boards, newsgroups, computer conferencing

New kinds of functionality:

- Attachments of different sorts (including annotations, images, music) for email and computer conferencing can be sent.
- Messages can be archived and accessed using various search facilities.

Benefits:

- Ubiquity: Can read any place, any time.
- Flexibility: Greater autonomy and control of when and how to respond, so can attend to it in own time rather than having to take a turn in a conversation at a particular cue.
- Powerful: Can send the same message to many people.
- Makes some things easier to say: Do not have to interact with person so can be easier to say things than when face to face (e.g., announcing sudden death of colleague, providing feedback on someone's performance).

(Continued)

Table 4.1   *(Continued)*

Problems:

- Flaming: When a user writes incensed angry email expressed in uninhibited language that is much stronger than normally used when interacting with the same person face to face. This includes the use of impolite statements, exclamation marks, capitalized sentences or words, swearing, and superlatives. Such "charged" communication can lead to misunderstandings and bad feelings among the recipients.

- Overload: Many people experience message overload, receiving over 30 emails or other messages a day. They find it difficult to cope and may overlook an important message while working through their ever increasing pile of email—especially if they have not read it for a few days. Various interface mechanisms have been designed to help people manage their email better, including filtering, threading, and the use of signaling to indicate the level of importance of a message (via the sender or recipient), through color coding, bold font, or exclamation marks placed beside a message.

- False expectations: An assumption has evolved that people will read their messages several times a day and reply to them there and then. However, many people have now reverted to treating email more like postal mail, replying when they have the time to do so.

iii. CMC combined with other activity

People often talk with each other while carrying out other activities. For example, designing requires people to brainstorm together in meetings, drawing on whiteboards, making notes, and using existing designs. Teaching involves talking with students as well as writing on the board and getting students to solve problems collaboratively. Various meeting- and decision- support systems have been developed to help people work or learn while talking together.

Examples:

- Customized electronic meeting rooms have been built that support people in face-to-face meetings, via the use of networked workstations, large public displays, and shared software tools, together with various techniques to help decision-making. One of the earliest systems was the University of Arizona's GroupSystem (see Figure 4.2).

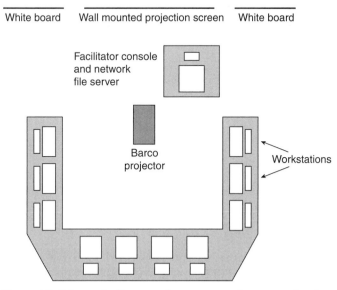

Figure 4.2 Schematic diagram of a group meeting room, showing relationship of work-station, whiteboards and video projector.

(Continued)

Table 4.1   *(Continued)*

Figure 4.3 An ACTIVBoard whiteboard developed by Promethean (U.K. company) that allows children to take control of the front-of-class display. This allows them to add comments and type in queries, rather than having to raise their hands and hope the teacher sees them.

- Networked classrooms: Recently schools and universities have realized the potential of using combinations of technologies to support learning. For example, wireless communication, portable devices and interactive whiteboards are being integrated in classroom settings to allow the teacher and students to learn and communicate with one another in novel interactive ways (see Figure 4.3).
- Argumentation tools which record the design rationale and other arguments used in a discussion that lead to decisions in a design (e.g. gIBIS, Conklin and Begeman, 1989). These are mainly designed for people working in the same physical location.
- Shared authoring and drawing tools that allow people to work on the same document at the same time. This can be remotely over the web (e.g., shared authoring tools like Shredit) or on the same drawing surface in the same room using multiple mouse cursors (e.g., KidPad, Benford et al., 2000).

New kinds of functionality:
- Allows new ways of collaboratively creating and editing documents.
- Supports new forms of collaborative learning.
- Integrates of different kinds of tools.

Benefits:
- Supports talking while carrying out other activities at the same time, allowing multi-tasking—which is what happens in face-to-face settings.
- Speed and efficiency: allows multiple people to be working on same document at same time.
- Greater awareness: allows users to see how one another are progressing in real time.

Problems:
- WYSIWIS (what you see is what I see)—It can be difficult to see what other people are referring to when in remote locations, especially if the document is large and different users have different parts of the document on their screens.
- Floor control—users may want to work on the same piece of text or design, potentially resulting in file conflicts. These can be overcome by developing various social and technological floor-control policies.

ACTIVITY 4.3   One of the earliest technological innovations (besides the telephone and telegraph) developed for supporting conversations at a distance was the videophone. Despite numerous attempts by the various phone companies to introduce them over the last 50 years (see Figure 4.4), they have failed each time. Why do you think this is so?

Comment   One of the biggest problems with commercial videophones is that the bandwidth is too low, resulting in poor resolution and slow refresh rate. The net effect is the display of unacceptable images: the person in the picture appears to move in sudden jerks; shadows are left behind when a speaker moves, and it is difficult to read lips or establish eye contact. There is also the social acceptability issue of whether people want to look at pocket-sized images of each other when talking. Sometimes you don't want people to see what state you are in or where you are.

Another innovation has been to develop systems that allow people to communicate and interact with each other in ways not possible in the physical world. Rather than try to imitate or facilitate face to face communication (like the above systems), designers have tried to develop new kinds of interactions. For example, ClearBoard was developed to enable facial expressions of participants to be made visible to others by using a transparent board that showed their face to the others (Ishii et al., 1993). HyperMirror was designed to provide an environment in which the participants could feel they were in the same virtual place even

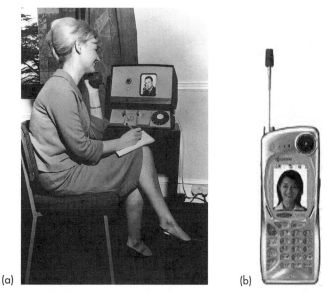

(a)   (b)

Figure 4.4 (a) One of British Telecom's early videophones and (b) a recent mobile "visualphone" developed in Japan.

BOX 4.1   Supporting Informal Chatting through Audio-Video Links

A number of researchers have tried to capitalize on the social phenomenon of informal communication and the important role it plays at work. In particular, they have been interested in finding ways of using audio-video links to *mimic* physical settings that are conducive to informal communication for people who are geographically separated. One of the first systems to be built, at Bellcore in 1989, was the *VideoWindow System* (see Figure 4.5). The goal was to design a shared space that would allow people in different locations to carry on a conversation as they would do if sitting in the same room drinking coffee together. Two lounge areas that were 50 miles apart were connected with high-bandwidth video channels and full-duplex four-channel audio. Connecting them was a 3 × 8 foot "picture-window" onto which video images were projected. The large size was meant to allow viewers to see a room of people roughly the same size as themselves. The system was designed to be active 24 hours a day, so that anyone entering one room could speak to whoever happened to be in the other room.

A study by Kraut et al. (1990) of how effective the system was showed that in general, many of the interactions that took place between the re-

mote conversants were indeed indistinguishable from similar face-to-face interactions—with the exception that they spoke a bit louder and constantly talked about the video system. However, they also found that people who were in the same room tended to talk more with each other than with those in the video-linked room. Various usability problems were identified as contributing to this reluctance to talk with video images of other people. One of these was the tendency for people to move closer to the picture window to strike up a conversation with someone (which is what one would normally do in a face-to-face setting); this had the opposite effect to what the person intended, as it moved his or her head out of the picture and also out of microphone range, meaning he or she could not be seen or heard. Thus rather than getting nearer to the other person, this behavior had the counter-intuitive effect of removing him or her from the "picture." Moreover, there was no way for participants to know whether they were being seen and heard by the others in the other room. This inability to monitor how others are or are not "receiving" you caused numerous problems. Another problem was that the system allowed only public conversations, meaning that

Figure 4.5 Diagram of VideoWindow system in use.

they could be heard by everyone in the rooms. Such public broadcasting contrasts with how people normally engage in informal face-to-face conversations, where they will often whisper and conspire with each other when a topic becomes more private or secret. Such private conversations clearly could not be supported by the VideoWindow system.

A number of other synchronous audio-video systems have since been developed that have tried to incorporate different kinds of conversational mechanisms to facilitate informal communication in video-linked locations. For example, Cruiser was designed to support informal communication by placing separate audio and video equipment on the desktop of each person who was connected to the system (Fish, 1989). This set-up differed from the VideoWindow system in that it enabled both public and private interactions to take place. It also provided additional functionality that allowed people to initiate conversations by typing in a cruise command followed by a question like, "I'm bored. Anyone want a chat?" or "Can someone help me?"—the aim here being explicitly to encourage people to engage in the kind of talk that they normally do when they bump into each other, but this time over the computer network. A further conversation mechanism built into Cruiser was a "glance" feature that allowed users

to check whether the person they wanted to talk to was in fact available before trying to initiate a conversation.

Commercial systems are now available that support multiple connections among sites. These can be very useful for virtual centers that have multiple groups working at a number of different sites. For example, the Distributed Systems Technology Center (a research collaboration between Australian universities and industries) has been using a commercial videoconferencing system (Polycom Viewstation 128) that enables people at its main sites (Brisbane, Sydney, and Melbourne) to keep in contact via regular informal and formal meetings. Formal meetings among the sites involve all the teams (over a hundred people) getting together and reporting on their projects. Each site has a camera which it controls, projecting different images to the other sites of what is happening at the site. This can be an image of the person who is doing the talking, a person being talked about, or a person who is being funny. The images from the different sites are displayed side by side on a large screen at each site. Weekly informal project meetings are also held among small groups. Informal chatting is also supported by having a continuous virtual presence via a video wall of the kitchen of one Queensland University site in another, and vice versa (see Figure 4.6).

Figure 4.6 A commercial videoconferencing system being used to support informal chatting among researchers across sites at Queensland University. In contrast to the VideoWindow system, a window of each site is shown in the top left-hand corner of their display to let the participants monitor their own behavior.

(a)   (b)   (c)

Figure 4.7 Hypermirror in action, showing perception of virtual personal space. (a) A woman is in one room (indicated by arrow on screen) (b) the man and woman in the other room chat to each other and move apart when they notice they are "overlapping" her (c) virtual personal space is established.

though they were physically in different places (Morikawa and Maesako, 1998). Mirror reflections of people in different places were synthesized and projected onto a single screen, so that they appeared side by side in the same virtual space. In this way, the participants could see both themselves and others in the same seamless virtual space. Observations of people using the system showed how quickly they adapted to perceiving themselves and others in this way. For example, participants quickly became sensitized to the importance of virtual personal space, moving out of the way if they perceived they were overlapping someone else on the screen (see Figure 4.7).

4.2.3   Coordination mechanisms

Coordination takes place when a group of people act or interact together to achieve something. For example, consider what is involved in playing a game of basketball. Teams have to work out how to play with each other and to plan a set of tactics that they think will outwit the other team. For the game to proceed both teams need to follow (and sometimes contravene) the rules of the game. An incredible amount of coordination is required within a team and between the competing teams in order to play.

In general, collaborative activities require us to coordinate with each other, whether playing a team game, moving a piano, navigating a ship, working on a large software project, taking orders and serving meals in a restaurant, constructing a bridge or playing tennis. In particular, we need to figure out how to interact with one another to progress with our various activities. To help us we use a number of coordinating mechanisms. Primarily, these include:

- verbal and non-verbal communication
- schedules, rules and conventions
- shared external representations

Verbal and non-verbal communication

When people are working closely together they talk to each other, issuing commands and letting others know how they are progressing with their part. For example, when two or more people are collaborating together, as in moving a piano, they shout to each other commands like "Down a bit, left a bit, now straight forward" to coordinate their actions with each other. As in a conversation, nods, shakes, winks, glances, and hand-raising are also used in combination with such coordination "talk" to emphasize and sometimes replace it.

In formal settings, like meetings, explicit structures such as agendas, memos, and minutes are employed to coordinate the activity. Meetings are chaired, with secretaries taking minutes to record what is said and plans of actions agreed upon. Such minutes are subsequently distributed to members to remind them of what was agreed in the meeting and for those responsible to act upon what was agreed.

For time-critical and routinized collaborative activities, especially where it is difficult to hear others because of the physical conditions, gestures are frequently used (radio-controlled communication systems may also be used). Various kinds of hand signals have evolved, with their own set of standardized syntax and semantics. For example, the arm and baton movements of a conductor coordinate the different players in an orchestra, while the arm and baton movements of a ground marshal at an airport signal to a pilot how to bring the plane into its allocated gate.

ACTIVITY 4.4   How much communication is non-verbal? Watch a soap opera on the TV and turn down the volume and look at the kinds and frequency of gestures that are used. Are you able to understand what is going on? How do radio soaps compensate for not being able to use non-verbal gestures? How do people compensate when chatting online?

Comment   Soaps are good to watch for observing non-verbal behavior as they tend to be overcharged, with actors exaggerating their gestures and facial expressions to convey their emotions. It is often easy to work out what kind of scene is happening from their posture, body movement, gestures, and facial expressions. In contrast, actors on the radio use their voice a lot more, relying on intonation and surrounding sound effects to help convey emotions. When chatting online, people use emoticons and other specially evolved verbal codes.

Schedules, rules, and conventions

A common practice in organizations is to use various kinds of schedules to organize the people who are part of it. For example, consider how a university manages to coordinate the people within it with its available resources. A core task is allocating the thousands of lectures and seminars that need to be run each week with the substantially smaller number of rooms available. A schedule has to be devised that allows students to attend the lectures and seminars for their given courses, taking into account numerous rules and constraints. These include:

- A student cannot attend more than one lecture at a given time.
- A professor cannot give more than one lecture or seminar at a given time.
- A room cannot be allocated to more than one seminar or lecture at a given time.
- Only a certain number of students can be placed in a room, depending on its size.

BOX 4.2 Shared Calendars—My Time Or Yours?

Trying to schedule meetings for different people in an organization to attend can be a nightmare. Typically, a secretary will send out email or try to call those who need to be assembled for a given meeting. Some of those people may not be there at the time they are contacted and so the secretary must wait for them to reply before being able to set up a meeting. In the meantime, the others who have replied (but have not heard back about the meeting) may start to fill up the times they said they would be free with other engagements. When the secretary eventually gets back to them with a proposed date, it is often too late. The consequence is that the secretary will have to start again, coming up with a new set of times. The more people to be scheduled, the more difficult it becomes to keep track of people's calendars in order to find mutually free time slots to arrange a meeting. All in all, it is a very time-consuming and laborious activity, one in which it would seem a computer-based scheduling tool would be very useful.

Indeed, a number of shared calendars have been developed. Some of the most recent ones have been developed as web applications, allowing individuals to use them both as *personal calendars*, reminding them of events they have to attend, and as *public calendars*, broadcasting to anyone accessing their web pages, when they are busy and when they are free.

A number of studies of the implementation of shared calendars in various organizations have found them to be successful computer-supported coordination tools. For example, a study of one system, called MeetingMaker, showed that the users found it had greatly simplified meeting scheduling and was much faster than manual scheduling (Mosier and Tammaro, 1997). The shared tool provided a range of facilities, including supporting group scheduling, resource and room scheduling, to-do lists and various permissions to permit others to schedule meetings. The system could also inform individual users of upcoming events and meetings via a pop-up dialog box.

Other shared calendars, however, have not been so successful. These are usually ones that have been designed to allow people to look at a person's schedule, and upon finding a free slot, book a meeting with them without any form of negotiation. The normal etiquette when arranging a meeting is to ask someone when they are free to meet and suggest a number of dates and times. When a meeting is simply imposed, people can feel that their privacy has been invaded, especially if they had planned to use the allocated slot to work on something else. A typical response is to simply stop using the shared calendar. The problem with one person doing this, however, is that the rest of the group cannot continue using the shared calendar as a coordinating tool, since that person is now excluded.

The more successful scheduling tools, like MeetingMaker, have overcome this sensitive privacy issue by providing users with a proxy function. This allows users to mark off parts of their calendars as "private" while giving permissions to others to read and/or write to their calendars. Providing this more flexible control allows users to decide *a priori* which times of the week they will make themselves available for meetings and which times they want to keep private to get on with their own work, without revealing to others what they are up to.

Other coordinating mechanisms that are employed by groups working together are rules and conventions. These can be formal or informal. Formal rules, like the compulsory attendance of seminars, writing monthly reports, and filling in of timesheets, enable organizations to maintain order and keep track of what its members are doing. Conventions, like keeping quiet in a library or removing meal trays after finishing eating in a cafeteria, are a form of courtesy to others.

Shared external representations

Shared external representations are commonly used to coordinate people. We have already mentioned one example, that of shared calendars that appear on user's monitors as graphical charts, email reminders, and dialog boxes. Other kinds that are commonly used include forms, checklists, and tables. These are presented on public noticeboards or as part of other shared spaces. They can also be attached to documents and folders. They function by providing external information of who is working on what, when, where, when a piece of work is supposed to be finished, and who it goes to next. For example, a shared table of who has completed the checking of files for a design project (see Figure 4.8), provides the necessary information from which other members of the group can at a glance update their model of the current progress of that project. Importantly, such external representations can be readily updated by annotating. If a project is going to take longer than planned, this can be indicated on a chart or table by extending the line representing it, allowing others to see the change when they pass by and glance up at the whiteboard.

Shared externalizations allow people to make various inferences about the changes or delays with respect to their effect on their current activities. Accordingly,

Sheet no	Gary copied in	Kate & Gary plot file created	Mark checked by Phil	Kate plot sent	Mark plot file created	Mark plot sent mylar
596S6	✓	✓				
S7	✓	✓				
S8	✓					
S9	✓					
S10	✓					

Figure 4.8 An external representation used to coordinate collaborative work in the form of a print-out table showing who has completed the checking of files and who is down to do what.

they may need to reschedule their work and annotate the shared workplan. In so doing, these kinds of coordination mechanisms are considered to be *tangible*, providing important representations of work and responsibility that can be changed and updated as and when needed.

4.2.4   Designing collaborative technologies to support coordination

Shared calendars, electronic schedulers, project management tools, and workflow tools that provide interactive forms of scheduling and planning are some of the main kinds of collaborative technologies that have been developed to support coordination. A specific mechanism that has been implemented is the use of conventions. For example, a shared workspace system (called POLITeam) that supported email and document sharing to allow politicians to work together at different sites introduced a range of conventions. These included how folders and files should be organized in the shared workspace. Interestingly, when the system was used in practice, it was found that the conventions were often violated (Mark, et al., 1997). For example, one convention that was set up was that users should always type in the code of a file when they were using it. In practice, very few people did this, as pointed out by an administrator: "They don't type in the right code. I must correct them. I must sort the documents into the right archive. And that's annoying".

The tendency of people not to follow conventions can be due to a number of reasons. If following conventions requires additional work that is extraneous to the users' ongoing work, they may find it gets in the way. They may also perceive the convention as an unnecessary burden and "forget" to follow it all the time. Such "productive laziness" (Rogers, 1993) is quite common. A simple analogy to everyday life is forgetting to put the top back on the toothpaste tube: it is a very simple convention to follow and yet we are all guilty sometimes (or even all the time) of not doing this. While such actions may only take a tiny bit of effort, people often don't do them because they perceive them as tedious and unnecessary. However, the consequence of not doing them can cause grief to others.

When designing coordination mechanisms it is important to consider how socially acceptable they are to people. Failure to do so can result in the users not using the system in the way intended or simply abandoning it. A key part is getting the right balance between human coordination and system coordination. Too much system control and the users will rebel. Too little control and the system breaks down. Consider the example of file locking, which is a form of concurrency control. This is used by most shared applications (e.g., shared authoring tools, file-sharing systems) to prevent users from clashing when trying to work on the same part of a shared document or file at the same time. With file locking, whenever someone is working on a file or part of it, it becomes inaccessible to others. Information about who is using the file and for how long may be made available to the other users, to show why they can't work on a particular file. When file-locking mechanisms are used in this way, however, they are often considered too rigid as a form of coordination, primarily because they don't let other users negotiate with the first user about when they can have access to the locked file.

A more flexible form of coordination is to include a social policy of floor control. Whenever a user wants to work on a shared document or file, he must initially request "the floor." If no one else is using the specified section or file at that time, then he is given the floor. That part of the document or file then becomes locked, preventing others from having access to it. If other users want access to the file, they likewise make a request for the floor. The current user is then notified and can then let the requester know how long the file will be in use. If not acceptable, the requester can try to negotiate a time for access to the file. This kind of coordination mechanism, therefore, provides more scope for negotiation between users on how to collaborate, rather than simply receiving a point-blank "permission denied" response from the system when a file is being used by someone else.

BOX 4.3 Turning Technology Inside Out:
Online versus Physical Coordination Mechanisms

Many software applications now exist to support coordination, notably project management systems. From the project manager's perspective, they provide a flexible means of scheduling, distributing, and monitoring collaborative work and enabling them continuously to remind people of deadlines and milestones via the use of email and other kinds of representations. From the perspective of the individuals working in the company, they give them a means of letting others know when they are available for meetings and where they will be.

In practice, however, project management systems that rely exclusively on computer-mediated coordination mechanisms have not been found to be as effective as hoped. This tends to happen when the system is used to coordinate a large number of events or projects. People begin not to take notice of the numerous internal reminders and messages that are sent to them by the system, finding them to be too intrusive, overwhelming, or annoying. This can then lead to missing important meetings and deadlines. A work-around in some organizations has been to print out the schedules and events that have been entered into the project management online database and display them as paper-based external representations (see Whittaker and Schwartz, 1995). A study that looked at the creation and use of shared external representations in collaborative work (Bellotti and Rogers, 1997) found that in several cases, information that is represented online becomes re-represented as a physical entity because the online version often gets lost, forgotten, or overlooked. This was particularly prevalent at new media companies producing web content that needed to be updated regularly. The various groups had to be coordinated across a number of parallel-running, time-critical projects.

At one site, a project coordinator would write up on a physical whiteboard every morning the main projects, schedules, and deadlines relevant for that day extracted from the online project management software. When asked why she laboriously wrote down by hand information that could be readily accessed by everyone over the computer network, she replied that, owing to the multiplication of projects and people working on them, it had become very difficult to keep track of everything that was going on. Moreover, people had become de-sensitized to the many email reminders that the software application provided, so they often forgot their significance immediately after having acknowledged them. Consequently, everyone (including herself) needed to be reminded of what was urgent and what needed dealing with that day. Placing this critical information on a physical whiteboard in a prominent public place that was clearly distinct from the continuous stream of other online information and messages provided a more effective public reminder of what was urgent and needed doing that day. In essence, the company had resorted to *turning the technology inside out*."

| ACTIVITY 4.5 | Why are whiteboards so useful for coordinating projects? How might electronic whiteboards be designed to extend this practice? |

Comment   Physical whiteboards are very good as coordinating tools as they display information that is external and public, making it highly visible for everyone to see. Furthermore, the information can be easily annotated to show up-to-date modifications to a schedule. Whiteboards also have a gravitational force, drawing people to them. They provide a meeting place for people to discuss and catch up with latest developments.

Electronic whiteboards have the added advantage that important information can be animated to make it stand out. Important information can also be displayed on multiple displays throughout a building and can be extracted from existing databases and software, thereby making the project coordinator's work much easier. The boards could also be used to support on-the-fly meetings in which individuals could use electronic pens to sketch out ideas that could then be stored electronically. In such settings they could also be interacted with via wireless handheld computers, allowing information to be "scraped" off or "squirted" onto the whiteboard.

4.2.5 Awareness mechanisms

Awareness involves knowing who is around, what is happening, and who is talking with whom (Dourish and Bly, 1992). For example, when we are at a party, we move around the physical space, observing what is going on and who is talking to whom, eavesdropping on others' conversations and passing on gossip to others. A specific kind of awareness is peripheral awareness. This refers to a person's ability to maintain and constantly update a sense of what is going on in the physical and social context, through keeping an eye on what is happening in the periphery of their vision. This might include noting whether people are in a good or bad mood by the way they are talking, how fast the drink and food is being consumed, who has entered or left the room, how long someone has been absent, and whether the lonely guy in the corner is finally talking to someone—all while we are having a conversation with someone else. The combination of direct observations and peripheral monitoring keeps people informed and updated of what is happening in the world.

Similar ways of becoming aware and keeping aware take place in other contexts, such as a place of study or work. Importantly, this requires fathoming when is an appropriate time to interact with others to get and pass information on. Seeing a professor slam the office door signals to students that this is definitely not a good time to ask for an extension on an assignment deadline. Conversely, seeing teachers with beaming faces, chatting openly to other students suggests they are in a good mood and therefore this would be a good time to ask them if it would be all right to miss next week's seminar because of an important family engagement. The knowledge that someone is amenable or not rapidly spreads through a company, school, or other institution. People are very eager to pass on both good and bad news to others and will go out of their way to gossip, loitering in corridors, hanging around at the photocopier and coffee machine "spreading the word."

Figure 4.9 An external representation used to signal to others a person's availability.

In addition to monitoring the behaviors of others, people will organize their work and physical environment to enable it to be successfully monitored by others. This ranges from the use of subtle cues to more blatant ones. An example of a subtle cue is when someone leaves their dorm or office door slightly ajar to indicate that they can be approached. A more blatant one is the complete closing of their door together with a "do not disturb" notice prominently on it, signaling to everyone that under no circumstances should they be disturbed (see Figure 4.9).

Overhearing and overseeing

People who work closely together also develop various strategies for coordinating their work, based on an up-to-date awareness of what the others are doing. This is especially so for interdependent tasks, where the outcome of one person's activity is needed for others to be able to carry out their tasks. For example, when putting on a show, the performers will constantly monitor what one another is doing in order to coordinate their performance efficiently.

The metaphorical expression "closely-knit teams" exemplifies this way of collaborating. People become highly skilled in reading and tracking what others are doing and the information they are attending to. A well-known study of this phenomenon is described by Christian Heath and Paul Luff (1992), who looked at how two controllers worked together in a control room in the London Underground. An overriding observation was that the actions of one controller were tied very closely to what the other was doing. One of the controllers was responsible for the movement of trains on the line (controller A), while the other was responsible for providing information to passengers about the current service (controller B). In many instances, it was found that controller B overheard what controller A was doing and saying, and acted accordingly—even though controller A had not said anything explicitly to him. For example, on overhearing controller A discussing a problem with a train driver over the in-cab intercom system, controller B inferred from the ensuing conversation that there was going to be a disruption to the service and so started announcing this to the passengers on the platform before controller A had even finished talking with the train driver. At other times, the two controllers keep a lookout for each other, monitoring the environment for actions and events which they might have not noticed but may be important for them to know about so that they can act appropriately.

ACTIVITY 4.6   What do you think happens when one person of a closely knit team does not see or hear something or misunderstands what has been said, while the others in the group assume they have seen, heard, or understood what has been said?

Comment   In such circumstances, the person is likely to carry on as normal. In some cases this will result in inappropriate behavior. Repair mechanisms will then need to be set in motion. The knowledgeable participants may notice that the other person has not acted in the manner expected. They may then use one of a number of subtle repair mechanisms, say coughing or glancing at something that needs attending to. If this doesn't work, they may then resort to explicitly stating aloud what had previously been signaled implicitly. Conversely, the unaware participant may wonder why the event hasn't happened and, likewise, look over at the other people, cough to get their attention or explicitly ask them a question. The kind of repair mechanism employed at a given moment will depend on a number of factors, including the relationship among the participants (e.g., whether one is more senior than the others—this determines who can ask what), perceived fault or responsibility for the breakdown and the severity of the outcome of not acting there and then on the new information.

4.2.6   Designing collaborative technologies to support awareness

The various observations about awareness have led system developers to consider how best to provide awareness information for people who need to work together but who are not in the same physical space. Various technologies have been employed along with the design of specific applications to convey information about what people are doing and the progress of their ongoing work. As mentioned previously, audio-video links have been developed to enable remote colleagues to keep in touch with one another. Some of these systems have also been developed to provide awareness information about remote partners, allowing them to find out what one another is doing. One of the earliest systems was Portholes, developed at Xerox PARC research labs (Dourish and Bly, 1992). The system presented regularly-updated digitized video images of people in their offices from a number of different locations (in the US and UK). These were shown in a matrix display on people's workstations. Clicking on one of the images had the effect of bringing up a dialog box providing further information about that individual (e.g., name, phone number) together with a set of lightweight action buttons (e.g., email the person, listen to a pre-recorded audio snippet). The system provided changing images of people throughout the day and night in their offices, letting others see at a glance whether they were in their offices, what they were working on, and who was around (see Figure 4.10). Informal evaluation of the

Figure 4.10 A screen dump of Portholes, showing low resolution monochrome images from offices in the US and UK PARC sites. (Permission from Xerox Research Centre, Europe)

set-up suggested that having access to such information led to a shared sense of community.

The emphasis in the design of these early awareness systems was largely on supporting peripheral monitoring, allowing people to see each other and their progress. Dourish and Bellotti (1992) refer to this as shared feedback. More recent distributed awareness systems provide a different kind of awareness information. Rather than place the onus on participants to find out about each other, they have been designed to allow users to notify each other about specific kinds of events. Thus, there is less emphasis on monitoring and being monitored and more on explicitly letting others know about things. Notification mechanisms are also used to provide information about the status of shared objects and the progress of collaborative tasks.

Hence, there has been a shift towards supporting a collective "stream of consciousness" that people can attend to when they want to, and likewise provide information for when they want to. An example of a distributed awareness system is Elvin, developed at the University of Queensland (Segall and Arnold, 1997), which provides a range of client services. A highly successful client is Tickertape, which is a lightweight instant messaging system, showing small color-coded messages that scroll from right to left across the screen (Fitzpatrick et al., 1999). It has been most useful as a "chat" and local organizing tool, allowing people in different locations to effortlessly send brief messages and requests to the public tickertape display (see Figure 4.11). It has been used for a range of functions, including organizing shared

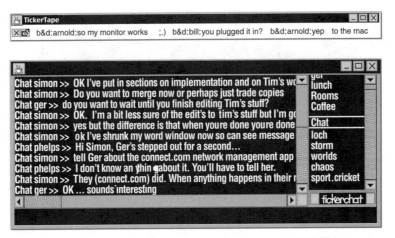

Figure 4.11 The Tickertape and Tickerchat interface for ELVIN awareness service.

events (e.g. lunch dates), making announcements, and as an "always-on" communication tool for people working together on projects but who are not physically co-located. It is also often used as a means of mediating help between people. For example, when I was visiting the University of Queensland, I asked for help over Tickertape. Within minutes, I was inundated with replies from people logged onto the system who did not even know me. At the time, I was having problems working out the key mappings between the PC that I was using in Australia and a Unix editor I couldn't find a way of quitting from on a remote machine in the UK. The suggestions that appeared on Tickertape quickly led to a discussion among the participants, and within five minutes someone had come over to my desk and sorted the problem out for me!

In addition to presenting awareness information as streaming text messages, more abstract forms of representation have been used. For example, a communication tool called Babble, developed at IBM (Erickson et al., 1999), provides a dynamic visualization of the participants in an ongoing chat-like conversation. A large 2D circle is depicted with colored marbles on each user's monitor. Marbles inside the circle convey those individuals active in the current conversation. Marbles outside the circle convey users involved in other conversations. The more active a participant is in the conversation, the more the corresponding marble is moved towards the center of the circle. Conversely, the less engaged a person is in the ongoing conversation, the more the marble moves towards the periphery of the circle (see Figure 4.12).

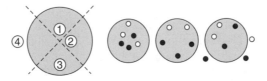

Figure 4.12 The Babble interface, with dynamic visualization of participants in ongoing conversation.

4.3   Ethnographic studies of collaboration and communication

One of the main approaches to informing the design of collaborative technologies that takes into account social concerns is carrying out an ethnographic study (a type of field study). Observations of the setting, be it home, work, school, public place, or other setting, are made, examining the current work and other collaborative practices people engage in. The way existing technologies and everyday artifacts are used is also analyzed. The outcome of such studies can be very illuminating, revealing how people currently manage in their work and everyday environments. They also provide a basis from which to consider how such existing settings might be improved or enhanced through the introduction of new technologies, and can also expose problematic assumptions about how collaborative technologies will or should be used in a setting (for more on *how to use* ethnography to inform design, see Chapter 9; *how to do* ethnography is covered in Chapter 12).

Many studies have analyzed in detail how people carry out their work in different settings (Plowman et al., 1995). The findings of these studies are used both to inform the design of a *specific* system, intended for a particular workplace, and more generally, to provide input into the design of new technologies. They can also highlight problems with existing system design methods. For example, an early study by Lucy Suchman (1983) looked at the way existing office technologies were being designed in relation to how people actually worked. She observed what really happened in a number of offices and found that there was a big mismatch between the way work was actually accomplished and the way people were supposed to work using the office technology provided. She argued that designers would be much better positioned to develop systems that could match the way people behave and use technology, if they began by considering the *actual details* of work practice.

In her later, much-cited study of how pairs of users interacted with an interactive help system—intended as a facility for using with a photocopier—Suchman (1987) again stressed the point that the design of interactive systems would greatly benefit from analyses that focused on the *unique details* of the user's particular situation—rather than being based on preconceived models of how people ought to (and will) follow instructions and procedures. Her detailed analysis of how the help system was unable to help users in many situations, highlighted the inadequacy of basing the design of an interactive system purely on an abstract user model.

Since Suchman's seminal work, a large number of ethnographic studies have examined how work gets done in a range of companies (e.g., fashion, design, multimedia, newspapers) and local government. Other settings have also recently come under scrutiny to see how technologies are used and what people do at home, in public places, in schools, and even cyberspace. Here, the objective has been to understand better the social aspects of each setting and then to come up with implications for the design of future technologies that will support and extend these. For more on the way user studies can inform future technologies, see the interview at the end of this chapter with Abigail Sellen.

4.4 Conceptual frameworks

A number of conceptual frameworks of the "social" have been adapted from other disciplines, like sociology and anthropology. As with the conceptual frameworks derived from cognitive approaches, the aim has been to provide analytic frameworks and concepts that are more amenable to design concerns. Below, we briefly describe two well known approaches, that have quite distinct origins and ways of informing interaction design. These are:

- Language/action framework
- Distributed cognition

The first describes how a model of the way people communicate was used to inform the design of a collaborative technology. The second describes a theory that is used primarily to analyze how people carry out their work, using a variety of technologies.

4.4.1 The language/action framework

The basic premise of the language/action framework is that people act through language (Winograd and Flores, 1986). It was developed to inform the design of systems to help people work more effectively through improving the way they communicate with one another. It is based on various theories of how people use language in their everyday activities, most notably speech act theory.

Speech act theory is concerned with the functions utterances have in conversations (Austin, 1962; Searle, 1969). A common function is a request that is asked indirectly (known as an indirect speech act). For example, when someone says, "It's hot in here" they may really be asking if it would be OK to open the window because they need some fresh air. Speech acts range from formalized statements (e.g., I hereby declare you man and wife) to everyday utterances (e.g., how about dinner?).

There are five categories of speech acts:

- Assertives—commit the speaker to something being the case
- Commissives—commit the speaker to some future action
- Declarations—pronounce something has happened
- Directives—get the listener to do something
- Expressives—express a state of affairs, such as apologizing or praising someone

Each utterance can vary in its force. For example, a command to do something has quite a different force from a polite comment about the state of affairs.

The language/action approach was developed further into a framework called conversations for action (CfA). Essentially, this framework describes the sequence of actions that can follow from a speaker making a request of someone else. It depicts a conversation as a kind of "dance" (see Figure 4.13) involving a series of steps that are seen as following the various speech acts. Different dance steps ensue depending on the speech acts followed. The most straightforward kind of dance involves progressing from state 1 through to state 5 of the conversation,

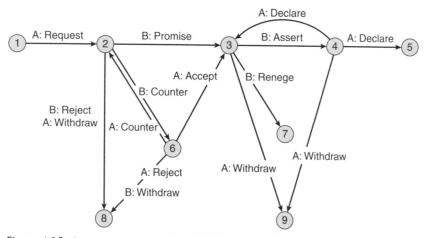

Figure 4.13 Conversation for action (CfA) diagram (from Winograd and Flores, 1986, p. 65).

in a linear order. For example, A (state 1) may request B to do homework (state 2), B may promise to do it after she has watched a TV program (state 3), B may then report back to A that the homework is done (state 4) and A, having looked at it, declares that this is the case (state 5). In reality, conversation dances tend to be more complex. For example, A may look at the homework and see that it is very shoddy and request that B complete it properly. The conversation is thus moved back a step. B may promise to do the homework but may in fact not do it at all, thereby canceling their promise (state 7), or A may say that B doesn't need to do it any more (state 9). B may also suggest an alternative, like cooking dinner (moving to state 6).

The CfA framework was used as the basis of a conceptual model for a commercial software product called the Coordinator. The goal was to develop a system to facilitate communication in a variety of work settings, like sales, finance, general management, and planning. The Coordinator was designed to enable electronic messages to be sent between people in the form of explicit speech acts. When sending someone a request, say "Could you get the report to me", the sender was also required to select the menu option "request." This was placed in the subject header of the message, thereby explicitly specifying the nature of the speech act. Other speech-act options included offer, promise, inform, and question (see Figure 4.14). The system also asked the user to fill in the dates by which the request should be completed. Another user receiving such a message had the option of responding with another labeled speech act. These included:

- acknowledge
- promise
- counter-offer
- decline
- free form

Table A:	*Menu items for initiating a new conversation.*
Request	Sender wants receiver to do something.
Offer	Sender offers to do something, pending acceptance.
Promise	Sender promises to do something (request is implicit).
What if	Opens a joint exploration of a space of possibilities.
Inform	Sender provides information.
Question	A request for information.
Note	A simple exchange of messages (as in ordinary E-mail).

Figure 4.14 Menu items for initiating a conversation.

Thus, the Coordinator was designed to provide a straightforward conversational structure, allowing users to make clear the status of their work and, likewise, to be clear about the status of others' work in terms of various commitments. To reiterate, a core rationale for developing this system was to try to improve people's ability to communicate more effectively. Earlier research had shown how communication could be improved if participants were able to distinguish among the kinds of commitments people make in conversation and also the time scales for achieving them. These findings suggested to Winograd and Flores that they might achieve their goal by designing a communication system that enabled users to develop a better awareness of the value of using "speech acts." Users would do this by being explicit about their intentions in their email messages to one another.

Normally, the application of a theory backed up with empirical research is regarded as a fairly innocuous and systematic way of informing system design. However, in this instance it opened up a very large can of worms. Much of the research community at the time was incensed by the assumptions made by Winograd and Flores in applying speech act theory to the design of the Coordinator System. Many heated debates ensued, often politically charged. A major concern was the extent to which the system *prescribed* how people should communicate. It was pointed out that asking users to specify *explicitly* the nature of their implicit speech acts was contrary to what they normally do in conversations. Forcing people to communicate in such an artificial way was regarded as highly undesirable. While some people may be very blatant about what they want doing, when they want it done by, and what they are prepared to do, most people tend to use more subtle and indirect forms of communication to advance their collaborations with others. The problem that Winograd and Flores came up against was people's resistance to radically change their way of communicating.

Indeed, many of the people who tried using the Coordinator System in their work organizations either abandoned it or resorted to using only the free-form message facility, which had no explicit demands associated with it. In these contexts, the system failed because it was asking too much of the users to change the way they communicated and worked. However, it should be noted that the Coordinator was successful in other kinds of organizations, namely those that are highly structured and need a highly structured system to support them. In particular, the most successful use of the Coordinator and its successors has been in organizations, like large manufacturing divisions of companies, where there is a great need for considerable management of orders and where previous support has been mainly in the form of a hodgepodge of paper forms and inflexible task-specific data processing applications (Winograd, 1994).

4.4.2 Distributed cognition

In the previous chapter we described how traditional approaches to modeling cognition have focussed on what goes on inside one person's head. We also mentioned that there has been considerable dissatisfaction with this approach, as it ignores how people interact with one another and their use of artifacts and external representations in their everyday and working activities. To redress this situation, Ed Hutchins and his colleagues developed the distributed cognition approach as a new paradigm for conceptualizing human work activities (e.g., Hutchins, 1995) (see Figure 4.15).

The distributed cognition approach describes what happens in a cognitive system. Typically, this involves explaining the interactions among people, the artifacts

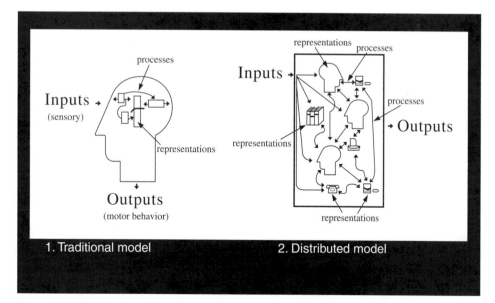

Figure 4.15 Comparison of traditional and distributed cognition approaches.

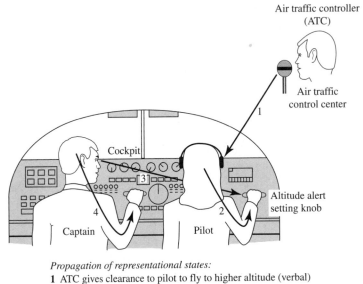

Air traffic controller
(ATC)

Air traffic
control center

Cockpit

Altitude alert
setting knob

Captain   Pilot

Propagation of representational states:
1 ATC gives clearance to pilot to fly to higher altitude (verbal)
2 Pilot changes altitude meter (mental and physical)
3 Captain observes pilot (visual)
4 Captain flies to higher altitude (mental and physical)

Figure 4.16 A cognitive system in which information is propagated through different media.

they use, and the environment they are working in. An example of a cognitive system is an airline cockpit, where a top-level goal is to fly the plane. This involves:

- the pilot, co-pilot and air traffic controller *interacting* with one another
- the pilot and co-pilot *interacting* with the instruments in the cockpit
- the pilot and co-pilot *interacting* with the environment in which the plane is flying (e.g., sky, runway).

A primary objective of the distributed cognition approach is to describe these interactions in terms of how information is propagated through different media. By this is meant how information is represented and re-represented as it moves across individuals and through the array of artifacts that are used (e.g., maps, instrument readings, scribbles, spoken word) during activities. These transformations of information are referred to as changes in *representational state*.

This way of describing and analyzing a cognitive activity contrasts with other cognitive approaches (e.g., the information processing model) in that it focuses not on what is happening inside the heads of *each* individual but on what is happening across individuals and artifacts. For example, in the cognitive system of the cockpit, a number of people and artifacts are involved in the activity of "flying to a higher altitude." The air traffic controller initially tells the co-pilot when it is safe to fly to a higher altitude. The co-pilot then alerts the pilot, who is flying the plane, by moving a knob on the instrument panel in front of them, indicating that it is now safe to fly (see Figure 4.16). Hence, the information concerning this activity is transformed through different media (over the radio, through the co-pilot, and via a change in the position of an instrument).

A distributed cognition analysis typically involves examining:

- the distributed problem solving that takes place (including the way people work together to solve a problem)
- the role of verbal and non-verbal behavior (including what is said, what is implied by glances, winks, etc., and what is not said)
- the various coordinating mechanisms that are used (e.g., rules, procedures)
- the various communicative pathways that take place as a collaborative activity progresses
- how knowledge is shared and accessed

In addition, an important part of a distributed cognition analysis is to identify the problems, breakdowns, and concomitant problem-solving processes that emerge to deal with them. The analysis can be used to predict what would happen to the way information is propagated through a cognitive system, using a different arrangement of technologies and artifacts and what the consequences of this would be for the current work setting. This is especially useful when designing and evaluating new collaborative technologies.

DILEMMA Who Should Have Control?

A core design dilemma facing those involved in developing collaborative technologies is how much control to implement and how much to leave to the users to configure for themselves. Should coordinating mechanisms like rules, procedures, and conventions be designed as part of the system architecture, or should the system be designed to be more free and open, allowing all users to do the same things? For example, when designing a file-sharing system is it more desirable to allow all users free access to all files, or is it preferable to implement some kind of social protocol that allows different user privileges and permissions? Likewise, when designing shared applications such as shared workspaces and collaborative authoring tools, to what extent should system-mediated control mechanisms be implemented that prescribe (and make quite clear) to users how they should share and collaborate? What happens when it is left up to the users to decide upon their own social protocols on how they should coordinate and collaborate with one another? Does it result in anarchy, or do they succeed in creating a shared environment that supports a harmonious way of working?

When working through these design concerns, it is important to consider what happens if too much or too little control is implemented in the collaborative technology. If there is too much "social engineering," then it could result in the users refusing to use it in the way intended. For example, the Coordinator system was found to be unusable in a number of organizations because it required people to radically change their way of communicating in ways considered unacceptable. Similarly, many of the conventions implemented in the POLITeam workspace system (e.g., always typing in the code of a file when using it) were not followed because they required the users to do extra work that was seen by them as being tedious and unnecessary.

Conversely, if not enough consideration is given to the way control is handled, the resulting system could end up being unusable and unacceptable. For example, some of the early shared calendar systems that had a free-for-all policy (anyone could look at anyone else's calendar and arrange a meeting in a free slot) were found to be too much of an invasion of people's privacy.

There are several other well known conceptual frameworks that are used to analyze how people collaborate and communicate, including activity theory, ethnomethodology, situated action and common ground theory.

Assignment

The aim of this design activity is for you to analyze the design of a collaborative virtual environment (CVE) with respect to how it is designed to support collaboration and communication.

Visit an existing CVE (many are freely downloadable) such as V-Chat (vchat.microsoft.com), one of the many Worlds Away environments (www.worlds.net), or the Palace (www.communities.com). Try to work out how they have been designed to take into account the following:

(a) *General social issues*

What is the purpose of the CVE?

What kinds of conversation mechanisms are supported?

What kinds of coordination mechanisms are provided?

What kinds of social protocols and conventions are used?

What kinds of awareness information is provided?

Does the mode of communication and interaction seem natural or awkward?

(b) *Specific interaction design issues*

What form of interaction and communication is supported (e.g., text/audio/video)?

What other visualizations are included? What information do they convey?

How do users switch between different modes of interaction (e.g., exploring and chatting)? Is the switch seamless?

Are there any social phenomena that occur specific to the context of the CVE that wouldn't happen in face to face settings (e.g., flaming)?

(c) *Design issues*

What other features might you include in the CVE to improve communication and collaboration?

Summary

In this chapter we have looked at some core aspects of sociality, namely communication and collaboration. We examined the main social mechanisms that people use in different settings in order to collaborate. A number of collaborative technologies have been designed to support and extend these mechanisms. We looked at representative examples of these, highlighting core interaction design concerns. A particular concern is social acceptability that is critical for the success or failure of technologies intended to be used by groups of people working or communicating together. We also discussed how ethnographic studies and theoretical frameworks can play a valuable role when designing new technologies for work and other settings.

Key points

- Social aspects are the actions and interactions that people engage in at home, work, school, and in public.

- The three main kinds of social mechanism used to coordinate and facilitate social aspects are conversation, coordination, and awareness.

- Talk and the way it is managed is integral to coordinating social activities.

- Many kinds of computer-mediated communication systems have been developed to enable people to communicate with one another when in physically different locations.

- External representations, rules, conventions, verbal and non-verbal communication are all used to coordinate activities among people.

- It is important to take into account the social protocols people use in face to face collaboration when designing collaborative technologies.

- Keeping aware of what others are doing and letting others know what you are doing are important aspects of collaborative working and socializing.

- Ethnographic studies and conceptual frameworks play an important role in understanding the social issues to be taken into account in designing collaborative systems.

- Getting the right level of control between users and system is critical when designing collaborative systems.

Further reading

DIX, A., FINLAY, J., ABOWD, G., AND BEALE, R. (1998) *Human-Computer Interaction*. Upper Saddle River, NJ: Prentice Hall. This textbook provides a comprehensive overview of groupware systems and the field of CSCW in Chapters 13 and 14.

ENGESTROM, Y AND MIDDLETON, D. (1996) (eds.) *Cognition and Communication at Work*. Cambridge: Cambridge University Press. A good collection of classic ethnographic studies that examine the relationship between different theoretical perspectives and field studies of work practices.

PREECE, J. (2000) *Online Communities: Designing Usability, Supporting Sociability*. New York: John Wiley and Sons. This book combines usability and sociability issues to do with designing online communities.

BAECKER, R. M., GRUDIN, J., BUXTON, W. A. S., AND GREENBERG, S. (eds.) (1995) *Readings in Human-Computer Interaction: Toward the Year 2000,* (second edition) San Francisco, Ca.: Morgan Kaufmann, 1995.

BAECKER, R. M. (ed.) (1993) *Readings in Groupware and Computer-Supported Cooperative Work: Assisting Human-Human Collaboration,* San Mateo, Ca.: Morgan Kaufmann. These two collections of readings include a number of representative papers from the field of CSCW, ranging from social to system architecture issues.

MUNRO, A.J., HOOK, K. AND BENYON, D. (eds.) (1999) *Social Navigation of Information Space*. New York: Springer Verlag. Provides a number of illuminating papers that explore how people navigate information spaces in real and virtual worlds and how people interact with one another in them.

INTERVIEW with Abigail Sellen

Abigail Sellen is a senior researcher at Hewlett Packard Labs in Bristol, UK. Her work involves carrying out user studies to inform the development of future products, including appliances and web-based services. She has a background in cognitive science and human factors engineering, having obtained her doctorate at the University of California, San Diego. Prior to this Abigail worked at Xerox Research Labs in Cambridge, UK, and Apple Computer Inc. She has also worked as an academic researcher at the Computer Systems Research Institute at the University of Toronto, Canada and the Applied Psychology Unit in Cambridge, UK. She has written widely on the social and cognitive aspects of paper use, video conferencing, input devices, human memory, and human error, all with an eye to the design of new technologies.

YR: Could you tell me what you do at Hewlett Packard Research Labs?

AS: Sure, I've been at HP Labs for about three years now as a member of its User Studies and Design Group. This is a smallish group consisting of five social scientists and three designers. Our work can best be described as doing three things: we do projects that are group-led around particular themes, like for example, how people use digital music or how people capture documents using scanning technology. We do consulting work for development teams at HP, and thirdly, we do a little bit of our own individual work, like writing papers and books, and giving talks.

YR: Right. Could you tell me about user studies, what they are and why you consider them important?

AS: OK. User studies essentially involve looking at how people behave either in their natural habitats or in the laboratory, both with old technologies and with new ones. I think there are many different questions that these kinds of studies can help you answer. Let me name a few. One question is: who is going to be the potential user for a particular device or service that you are thinking of developing? A second question—which I think is key—is, what is the potential

value of a particular product for a user? Once we know this, we can then ask, for a particular situation or task, what features do we want to deliver and how best should we deliver those features? This includes, for example, what would the interface look like? Finally, I think user studies are important to understand how users' lives may change and how they will be affected by introducing a new technology. This has to take into account the social, physical, and technological context into which it will be introduced.

YR: So it sounds like you have a set of general questions you have in mind when you do a user study. Could you now describe how you would do a user study and what kinds of things you would be looking for?

AS: Well, I think there are two different classes of user studies and both are quite different in the ways you go about them. There are evaluation studies, where we take a concept, a prototype or even a developed technology and look at how it is used and then try to modify or improve it based on what we find. The second class of user studies is more about discovering what people's unmet needs may be. This means trying to develop new concepts and ideas for things that people may never have thought of before. This is difficult because you can't necessarily just ask people what they would like or what they would use. Instead, you have to make inferences from studying people in different situations and try to understand from this what they might need or value.

YR: In the book we mention the importance of taking into account social aspects, such as awareness of others, how people communicate with each other and so on. Do you think these issues are important when you are doing these two kinds of user studies?

AS: Well, yes, and in particular I think social aspects really are playing to that second class of user study I mentioned where you are trying to discover what people's unmet needs or requirements may be. Here you are trying to get rich descriptions about what people do in the context of their everyday lives—whether this is in their working lives, their home lives, or lives on the move. I'd say getting the social aspects understood is often very important in trying to understand what value new products and services might bring to people's day-to-day activities, and also how they would fit into those existing activities.

YR: And what about cognitive aspects, such as how people carry out their tasks, what they remember, what they are bad at remembering? Is that also important to look into when you are doing these kinds of studies?

AS: Yes, if you think about evaluation studies, then cognitive aspects are extremely important. Looking at cognitive aspects can help you understand the nature of the user interaction, in particular what processes are going on in their heads. This includes issues like learning how users perceive a device and how they form a mental model of how something works. Cognitive issues are especially important to consider when we want to contrast one device with another or think about new and better ways in which we might design an interface.

YR: I wonder if you could describe to me briefly one of your recent studies where you have looked at cognitive and social aspects.

AS: How about a recent study we did to do with building devices for reading digital documents? When we first set out on this study, before we could begin to think about how to build such devices, we had to begin by asking, "What do we mean by reading?" It turned out there was not a lot written about the different ways people read in their day-to-day lives. So the first thing we did was a very broad study looking at how people read in work situations. The technique we used here was a combination of asking people to fill out a diary about their reading activities during the course of a day and interviewing them at the end of each day. The interviews were based around what was written in the diaries, which turned out to be a good way of unpacking more details about what people had been doing.

That initial study allowed us to categorize all the different ways people were reading. What we found out is that actually you can't talk about reading in a generic sense but that it falls into at least 10 different categories. For example, sometimes people skim read, sometimes they read for the purpose of writing something, and sometimes they read very reflectively and deeply, marking up their documents as they go. What quickly emerged from this first study was that if you're designing a device for reading it might look very different depending on the kind of reading the users are doing. So, for example, if they're reading by themselves, the screen size and viewing angle may not be as important as if they're reading with others. If they're skim reading, the ability to quickly flick through pages is important. And if they're reading and writing, then this points to the need for a pen-based interface. All of these issues become important design considerations.

This study then led to the development of some design concepts and ideas for new kinds of reading devices. At this stage we involved designers to develop different "props" to get feedback and reactions from potential users. A prop could be anything from a quick sketch to an animation to a styrofoam 3D mockup. Once you have this initial design work, you can then begin to develop working prototypes and test them with realistic tasks in both laboratory and natural settings. Some of this work we have already completed, but the project has had an impact on several different research and development efforts.

YR: Would you say that user studies are going to become an increasingly important part of the interaction design process, especially as new technologies like ubiquitous computing and handheld devices come into being—and where no one really knows what applications to develop?

AS: Yes. I think the main contribution of user studies, say, 15 years ago was in the area of evaluation and usability testing. I think that role is changing now in that user studies researchers are not only those who evaluate devices and interfaces but also those who develop new concepts. Also, another important development is a change in the way the research is carried out. More and more I am finding that teams are drawing together people from other disciplines, such as sociologists, marketing people, designers, and people from business and technology development.

YR: So they are essentially working as a multidisciplinary team. Finally, what is it like to work in a large organization like HP, with so many different departments?

AS: One thing about working for a large organization is that you get a lot of variety in what you can do. You can pick and choose to some extent and, depending on the organization, don't have to be tied to a particular product. If, on the other hand, you work for a smaller organization such as a start-up company, inevitably there is lots of pressure to get things out the door quickly. Things are often very focused. Whether large or small, however, I think one of the hardest things I have found in working for corporate research is learning to work with the development teams. They put huge pressures on you because they have huge pressures on them. You really have to work at effectively incorporating user studies findings into the development process. This can be incredibly challenging, but it's also satisfying to have an impact on real products.

Chapter 5

Understanding how interfaces affect users

5.1 Introduction
5.2 What are affective aspects?
5.3 Expressive interfaces
5.4 User frustration
5.5 A debate: the application of anthropomorphism to interaction design
5.6 Virtual characters: agents
    5.6.1 Kinds of agents
    5.6.2 General design concerns: believability of virtual characters

5.1 Introduction

An overarching goal of interaction design is to develop interactive systems that elicit positive responses from users, such as feeling at ease, being comfortable, and enjoying the experience of using them. More recently, designers have become interested in how to design interactive products that elicit specific kinds of emotional responses in users, motivating them to learn, play, be creative, and be social. There is also a growing concern with how to design websites that people can trust, that make them feel comfortable about divulging personal information or making a purchase.

We refer to this newly emerging area of interaction design as affective aspects. In this chapter we look at how and why the design of computer systems cause certain kinds of emotional responses in users. We begin by looking in general at expressive interfaces, examining the role of an interface's appearance on users and how it affects usability. We then examine how computer systems elicit negative responses, e.g., user frustration. Following this, we present a debate on the controversial topic of anthropomorphism and its implications for designing applications to have human-like qualities. Finally, we examine the range of virtual characters designed to motivate people to learn, buy, listen, etc., and consider how useful and appropriate they are.

The main aims of this chapter are to:

- Explain what expressive interfaces are and the affects they can have on people.
- Outline the problems of user frustration and how to reduce them.
- Debate the pros and cons of applying anthropomorphism in interaction design.
- Assess the believability of different kinds of agents and virtual characters.
- Enable you to critique the persuasive impact of e-commerce agents on customers.

5.2 What are affective aspects?

In general, the term "affective" refers to producing an emotional response. For example, when people are happy they smile. Affective behavior can also cause an emotional response in others. So, for example, when someone smiles it can cause others to feel good and smile back. Emotional skills, especially the ability to express and recognize emotions, are central to human communication. Most of us are highly skilled at detecting when someone is angry, happy, sad, or bored by recognizing their facial expressions, way of speaking, and other body signals. We are also very good at knowing what emotions to express in given situations. For example, when someone has just heard they have failed an exam we know it is not a good time to smile and be happy. Instead we try to empathize.

It has been suggested that computers be designed to recognize and express emotions in the same way humans do (Picard, 1998). The term coined for this approach is "affective computing". A long-standing area of research in artificial intelligence and artificial life has been to create intelligent robots and other computer-based systems that behave like humans and other creatures. One well-known project is MIT's COG, in which a number of researchers are attempting to build an artificial two-year-old. One of the offsprings of COG is Kismet (Breazeal, 1999), which has been designed to engage in meaningful social interactions with humans (see Figure 5.1). Our concern in this chapter takes a different approach: how can interactive systems be designed (both deliberately and inadvertently) to make people respond in certain ways?

Figure 5.1 Kismet the robot expressing surprise, anger, and happiness.

5.3 Expressive interfaces

A well-known approach to designing affective interfaces is to use expressive icons and other graphical elements to convey emotional states. These are typically used to indicate the current state of a computer. For example, a hallmark of the Apple computer is the icon of a smiling Mac that appears on the screen when the machine is first started (see Figure 5.2(a)). The smiling icon conveys a sense of friendliness, inviting the user to feel at ease and even smile back. The appearance of the icon on the screen can also be very reassuring to users, indicating that their computer is working fine. This is especially useful when they have just rebooted the computer after it has crashed and where previous attempts to reboot have failed (usually indicated by a sad icon face—see Figure 5.2(b)). Other ways of conveying the status of a system are through the use of:

- dynamic icons, e.g., a recycle bin expanding when a file is placed into it
- animations, e.g., a bee flying across the screen indicating that the computer is doing something, like checking files
- spoken messages, using various kinds of voices, telling the user what needs to be done
- various sounds indicating actions and events (e.g. window closing, files being dragged, new email arriving)

One of the benefits of these kinds of expressive embellishments is that they provide reassuring feedback to the user that can be both informative and fun.

The style of an interface, in terms of the shapes, fonts, colors, and graphical elements that are used and the way they are combined, influences how pleasurable it is to interact with. The more effective the use of imagery at the interface, the more engaging and enjoyable it can be (Mullet and Sano, 1995). Conversely, if little thought is given to the appearance of an interface, it can turn out looking like a dog's dinner. Until recently, HCI has focused primarily on getting the usability right, with little attention being paid to how to design aesthetically pleasing interfaces. Interestingly, recent research suggests that the aesthetics of an interface can

Figure 5.2 (a) Smiling and (b) sad Apple Macs.

have a positive effect on people's perception of the system's usability (Tractin-sky, 1997). Moreover, when the "look and feel" of an interface is pleasing (e.g., beautiful graphics, nice feel to the way the elements have been put together, well-designed fonts, elegant use of images and color) users are likely to be more tolerant of its usability (e.g., they may be prepared to wait a few more seconds for a website to download). As we have argued before, interaction design should not just be about usability *per se*, but should also include aesthetic design, such as how pleasurable an interface is to look at (or listen to). The key is to get the right balance between usability and other design concerns, like aesthetics (See Figure 5.3 on Color Plate 6).

ACTIVITY 5.1   A question of style or stereotype? Figure 5.4 shows two differently designed dialog boxes. Describe how they differ in terms of style. Of the two, which one do you prefer? Why? Which one do you think (i) Europeans would like the most and (ii) Americans would like the most?

Comment   Aaron Marcus, a graphic designer, created the two designs in an attempt to provide appealing interfaces. Dialog box A was designed for white American females while dialog box B was designed for European adult male intellectuals. The rationale behind Marcus's ideas was that European adult male intellectuals like "suave prose, a restrained treatment of information density, and a classical approach to font selection (e.g., the use of serif type in axial symmetric layouts similar to those found in elegant bronze European building identification signs)." In contrast, white American females "prefer a more detailed presentation, curvilinear shapes and the absence of some of the more brutal terms . . . favored by male software engineers."

When the different interfaces were empirically tested by Teasley et al. (1994), their results did not concur with Marcus's assumptions. In particular, they found that the European dialog box was liked the best by all people and was considered most appropriate for all users. Moreover, the round dialog box designed for women was strongly disliked by everyone. The assumption that women like curvilinear features clearly was not true in this context. At the very least, displaying the font labels in a circular plane makes them more difficult to read than when presented in the conventionally accepted horizontal plane.

Another popular kind of expressive interface is the friendly interface agent. A general assumption is that novices will feel more at ease with this kind of "companion" and will be encouraged to try things out, after listening, watching, following, and interacting with them. For example, Microsoft pioneered a new class of agent-based software, called Bob, aimed at new computer users (many of whom were seen as computer-phobic). The agents were presented as friendly characters, including a friendly dog and a cute bunny. An underlying assumption was that having these kinds of agents on the screen would make the users feel more comfortable and at ease with using the software. An interface metaphor of a warm, cozy living room, replete with fire, furnishings, and furniture was provided (see Figure 5.5)—again intended to convey a comfortable feeling.

Since the creation of Bob, Microsoft has developed other kinds of agents, including the infamous "Clippy" (a paper clip that has human-like qualities), as part

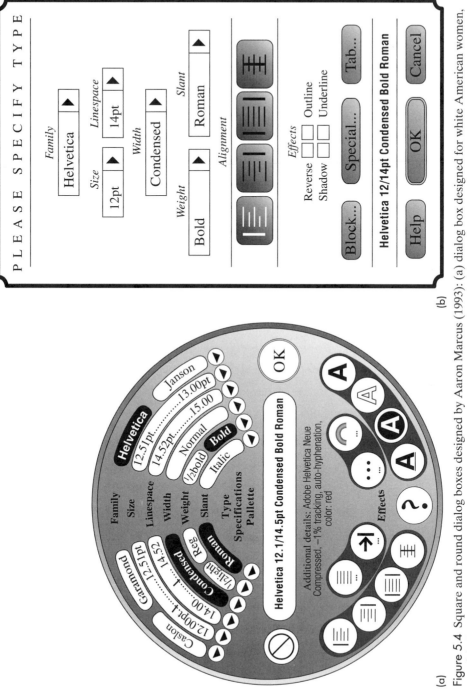

(a)

(b)

Figure 5.4 Square and round dialog boxes designed by Aaron Marcus (1993): (a) dialog box designed for white American women, and (b) dialog box designed for European adult male intellectuals.

Figure 5.5 "At home with Bob" software.

of their Windows operating environment.[1] The agents typically appear at the bottom of the screen whenever the system "thinks" the user needs help carrying out a particular task. They, too, are depicted as cartoon characters, with friendly warm personalities. As mentioned in Chapter 2, one of the problems of using agents in this more general context is that some users do not like them. More experienced users who have developed a reasonably good mental model of the system often find such agent helpers very trying and quickly find them annoying intrusions, especially when they distract them from their work. (We return to anthropomorphism and the design of interface agents later in Section 5.5).

Users themselves have also been inventive in expressing their emotions at the computer interface. One well-known method is the use of emoticons. These are keyboard symbols that are combined in various ways to convey feelings and emotions by simulating facial expressions like smiling, winking, and frowning on the screen. The meaning of an emoticon depends on the content of the message and where it is placed in the message. For example, a smiley face placed at the end of a message can mean that the sender is happy about a piece of news she has just written about. Alternatively, if it is placed at the end of a comment in the body of the message, it usually indicates that this comment is not intended to be taken seriously. Most emoticons are designed to be interpreted with the viewer's head tilted over to the left (a result of the way the symbols are represented on the screen). Some of the best known ones are presented in Table 5.1. A recently created shorthand language, used primarily by teenagers when online chatting or texting is the use of abbreviated words. These are formed by keying in various numbers and let-

---

[1]On the Mac version of Microsoft's Office 2001, Clippy is replaced by an anthropomorphized Mac computer with big feet and a hand that conveys various gestures and moods.

Table 5.1   Some commonly used emoticons.

Emotion	Expression	Emoticon	Possible meanings
Happy	Smile	:) or :D	(i) Happiness, or (ii) previous comment not to be taken seriously
Sad	Mouth down	:(or :-<	Disappointed, unhappy
Cheeky	Wink	;) or ;-)	Previous comment meant as tongue-in-cheek
Mad	Brows raised	>:	Mad about something
Very angry	Angry face	>:-(	Very angry, cross
Embarrassed	Mouth open	:O	Embarrassed, shocked
Sick	Looking sick	:x	Feeling ill
Naïve	Schoolboyish look	<:-)	Smiley wearing a dunce's cap to convey that the sender is about to ask a stupid question.

ters in place of words, e.g., "I 1 2 CU 2nite". As well as being creative, the shorthand can convey emotional connotations.

Expressive forms like emoticons, sounds, icons, and interface agents have been primarily used to (i) convey emotional states and/or (ii) elicit certain kinds of emotional responses in users, such as feeling at ease, comfort, and happiness. However, in many situations computer interfaces *inadvertently* elicit negative emotional responses. By far the most common is user frustration, to which we now turn our attention.

5.4   User frustration

Everyone at some time or other gets frustrated when using a computer. The effect ranges from feeling mildly amused to extremely angry. There are myriads of reasons why such emotional responses occur:

- when an application doesn't work properly or crashes
- when a system doesn't do what the user wants it to do
- when a user's expectations are not met
- when a system does not provide sufficient information to let the user know what to do
- when error messages pop up that are vague, obtuse, or condemning
- when the appearance of an interface is too noisy, garish, gimmicky, or patronizing
- when a system requires users to carry out many steps to perform a task, only to discover a mistake was made somewhere along the line and they need to start all over again

ACTIVITY 5.2   Provide specific examples for each of the above categories from your own experience, when you have become frustrated with an interactive device (e.g., telephone, VCR, vending machine, PDA, computer). In doing this, write down any further types of frustration that come to mind. Then prioritize them in terms of how annoying they are. What are the worst types?

Comment   In the text below we provide examples of common frustrations experienced when using computer systems. The worst include unhelpful error messages and excessive housekeeping tasks. You no doubt came up with many more.

Often user frustration is caused by bad design, no design, inadvertent design, or ill-thought-out design. It is rarely caused deliberately. However, its impact on users can be quite drastic and make them abandon the application or tool. Here, we present some examples of classic user-frustration provokers that could be avoided or reduced by putting more thought into the design of the conceptual model.

1.   Gimmicks

Cause: When a users' expectations are not met and they are instead presented with a gimmicky display.
Level of frustration: Mild
This can happen when clicking on a link to a website only to discover that it is still "under construction." It can be still more annoying when the website displays a road-sign icon of "men at work" (see Figure 5.6). Although the website owner may think such signs amusing, it serves to underscore the viewer's frustration at having made the effort to go to the website only to be told that it is incomplete (or not even started in some cases). Clicking on links that don't work is also frustrating.
How to avoid or help reduce the frustration:
By far the best strategy is to avoid using gimmicks to cover up the real crime. In this example it is much better to put material live on the web only when it is complete and working properly. People very rarely return to sites when they see icons like the one in Figure 5.6.

2.   Error Messages

Cause: When a system or application crashes and provides an "unexpected" error message.
Level of frustration: High
Error messages have a long history in computer interface design, and are notorious for their incomprehensibility. For example, Nielsen (1993) describes an early system that was developed that allowed only for one line of error messages. Whenever the

Figure 5.6 Men at work icon sign indicating "website under construction." According to AltaVista, there were over 12 million websites containing the phrase "under construction" in January 2001.

error message was too long, the system truncated it to fit on the line, which the users would spend ages trying to decipher. The full message was available only by pressing the PF1 (help key) function key. While this may have seemed like a natural design solution to the developers, it was not at all obvious to the users. A much better design solution would have been to use the one line of the screen to indicate how to find more information about the current error ("press the PF1 key for explanation").

The use of cryptic language and developer's jargon in error messages is a major contributing factor in user frustration. It is one thing to have to cope when something goes wrong but it is another to have to try to understand an obscure message that pops up by way of explanation. One of my favorites, which sometimes appears on the screen when I'm trying to do something perfectly reasonable like paste some text into a document, using a word processor, is: "The application Word Wonder has unexpectedly quit due to a Type 2 error."

It is very clear from what the system has just done (closed the application very rapidly) that it has just crashed, so such feedback is not very helpful. Letting the user know that the error is of a Type 2 kind is also not very useful. How is the average user meant to understand this? Is there a list of error types ready at hand to tell the user how to solve the problem for each error? Moreover, such a reference invites the user to worry about how many more error types there might be. The tone of the message is also annoying. The adjective "unexpectedly" seems condescending, implying almost that it is the fault of the user rather than the computer. Why include such a word at all? After all, how else could the application have quit? One could never imagine the opposite situation: an error message pops up saying, "The application has *expectedly* quit, due to poor coding in the operating system."
How to avoid or help reduce the frustration:

Ideally, error messages should be treated as how-to-fix-it messages. Instead of explicating what has happened, they should state the cause of the problem and what the user needs to do to fix it. Shneiderman (1998) has developed a detailed set of guidelines on how to develop helpful messages that are easy to read and understand. Box 5.1 summarizes the main recommendations.

BOX 5.1   Main Guidelines on How to Design Good Error Messages *(Adapted from Shneiderman, 1998)*

- Rather than condemn users, messages should be courteous, indicating what users need to do to set things right.
- Avoid using terms like FATAL, ERROR, INVALID, BAD, and ILLEGAL.
- Avoid long code numbers and uppercase letters.
- Audio warnings should be under the user's control, since they can cause much embarrassment.

- Messages should be precise rather than vague.
- Messages should provide a help icon or command to allow users to get context-sensitive help.
- Messages should be provided at multiple levels, so that short messages can be supplemented with longer explanations.

ACTIVITY 5.3 Below are some common error messages expressed in harsh computer jargon that can be quite threatening and offensive. Rewrite them in more usable, useful, and friendly language that would help users to understand the cause of the problem and how to fix it. For each message, imagine a specific context where such a problem might occur.

> SYNTAX ERROR
>
> INVALID FILENAME
>
> INVALID DATA
>
> APPLICATION ZETA HAS UNEXPECTEDLY QUIT DUE TO A TYPE 4 ERROR
>
> DRIVE ERROR: ABORT, RETRY OR FAIL?

Comment How specific the given advice can be will depend on the kind of system it is. Here are suggestions for hypothetical systems.

> SYNTAX ERROR—There is a problem with the way you have written the command. Check for typos.
>
> INVALID FILENAME—Choose another file name that uses only 20 characters or less and is lower case without any spaces.
>
> INVALID DATA—There is a problem with the data you have entered. Try again, checking that no decimal points are used.
>
> APPLICATION ZETA HAS UNEXPECTEDLY QUIT DUE TO A TYPE 4 ERROR—The application you were working on crashed because of an internal memory problem. Try rebooting and increasing the amount of allocated memory to the application.
>
> DRIVE ERROR: ABORT, RETRY OR FAIL?—There is a problem with reading your disk. Try inserting it again.

3. Overburdening the user

Cause: Upgrading software so that users are required to carry out excessive housekeeping tasks
Level of frustration: Medium to high
Another pervasive frustrating user experience is upgrading a piece of software. It is now common for users to have to go through this housekeeping task on a regular basis, especially if they run a number of applications. More often than not it tends to be a real chore, being very time-consuming and requiring the user to do a whole range of things, like resetting preferences, sorting out extensions, checking other configurations, and learning new ways of doing things. Often, problems can develop that are not detected till some time later, when a user tries an operation that worked fine before but mysteriously now fails. A common problem is that settings get lost or do not copy over properly during the upgrade. As the number of options for customizing an application or operating system increases for each new upgrade, so, too, does the headache of having to reset all the relevant preferences. Wading through myriads of dialog boxes and menus and figuring out which checkbox to

> "You do not have the plug-in needed to view the audio/x-pn-real-audio plug-in-type information on this page. To get plug-in now, view plug-in directory"

Figure 5.7a Typical message in dialog box that appears when trying to run an applet on a website that needs a plug-in the user does not have.

click on, can be a very arduous task. To add to the frustration, users may also discover that several of their well-learned procedures for carrying out tasks have been substantially changed in the upgrade.

A pet frustration of mine over the years has been trying to run various websites that require me to install a new plug-in. Achieving this is never straightforward. I have spent huge amounts of time trying to install what I assume to be the correct plug-in—only to discover that it is not yet available or incompatible with the operating system or machine I am using.

What typically happens is I'll visit a tempting new website, only to discover that my browser is not suitably equipped to view it. When my browser fails to run the applet, a helpful dialog box will pop up saying that a plug-in of X type is required. It also usually directs me to another website from where the plug-in can be downloaded (see Figure 5.7a). Websites that offer such plug-ins, however, are not organized around my specific needs but are designed more like hardware stores (a bad conceptual model), offering hundreds (maybe even thousands) of plug-ins covering all manner of applications and systems. Getting the right kind of plug-in from the vast array available requires knowing a number of things about your machine and the kind of network you are using. In going through the various options

---

WEB PLUG-IN DIRECTORY
Here is where you find the links to all of the plug-ins available on the net. Simply find a plug-in you're interested in, view what platforms it currently (or will 'soon') support and click on its link. If you know of a plug-in not listed on this page please take a moment and tell us about it with our all new reporting system!

Plug-ins by Category

The Full List	This is the whole list, but I gotta warn ya its getting big
MultiMedia	Multi-Media Plug-Ins, AVI, QuickTime, ShockWave...
Graphics	Graphic Plug-Ins, PNG, CMX, DWG...
Sound	Sound & MIDI Plug-Ins, MIDI, ReadAudio, TrueSpeech...
Document	Document Viewer Plug-Ins, Acrobat, Envoy, MS Word...
Productivity	Productivity Plug-Ins, Map Viewers, Spell Checkers...
VRML/3-D	VRML & QD3D Plug-Ins

Plug-ins by platform

Macintosh	Macintosh Plug-Ins
OS/2	IBM OS/2 Plug-Ins
Unix	Unix Plug-Ins
Windows	Windows Plug-Ins

Figure 5.7b Directory of plug-ins available on a plug-in site directed to from Netscape.

to narrow down which plug-in is required, it is easy to overlook something and end up with an inappropriate plug-in. Even when the right plug-in has been downloaded and placed in the appropriate system folder, it may not work. A number of other things usually need to be done, like specifying mime-type and suffix. The whole process can end up taking huge amounts of time, rather than the couple of minutes most users would assume.

How to avoid or help reduce the frustration:

Users should not have to spend large amounts of time on housekeeping tasks. Upgrading should be an effortless and largely automatic process. Designers need to think carefully about the trade-offs incurred when introducing upgrades, especially the amount of relearning required. Plug-ins that users have to search for, download, and set up themselves should be phased out and replaced with more powerful browsers that automatically download the right plug-ins and place them in the appropriate desktop folder reliably, or, better still, interpret the different file types themselves.

4. Appearance

Cause: When the appearance of an interface is unpleasant
Level of frustration: Medium
As mentioned earlier, the appearance of an interface can affect its usability. Users get annoyed by:

- websites that are overloaded with text and graphics, making it difficult to find the information desired and slow to access
- flashing animations, especially banner ads, which are very distracting
- the copious use of sound effects and Muzak, especially when selecting options, carrying out actions, starting up CD-ROMS, running tutorials, or watching website demos
- featuritis—an excessive number of operations, represented at the interface as banks of icons or cascading menus
- childish designs that keep popping up on the screen, such as certain kinds of helper agents
- poorly laid out keyboards, pads, control panels, and other input devices that cause the user to press the wrong keys or buttons when trying to do something else

How to avoid or help reduce the frustration:

Interfaces should be designed to be simple, perceptually salient, and elegant and to adhere to usability design principles, well-thought-out graphic design principles, and ergonomic guidelines (e.g. Mullet and Sano, 1996).

5.3.1 Dealing with user frustration

One way of coping with computer-induced frustration is to vent and take it out on the computer or other users. As mentioned in Chapter 3, a typical response to seeing the cursor freeze on the screen is repeatedly to bash every key on the keyboard.

Another way of venting anger is through flaming. When upset or annoyed by a piece of news or something in an email message, people may overreact and respond by writing things in email that they wouldn't dream of saying face to face. They often use keyboard symbols to emphasize their anger or frustration, e.g., exclamation marks (!!!!), capital letters (WHY DID YOU DO THAT?) and repeated question marks (??????) that can be quite offensive to those on the receiving end. While such venting behavior can make the user feel temporarily less frustrated, it can be very unproductive and can annoy the recipients. Anyone who has received a flame knows just how unpleasant it is.

In the previous section, we provided some suggestions on how systems could be improved to help reduce commonly caused frustrations. Many of the ideas discussed throughout the book are also concerned with designing technologies and interfaces that are usable, useful, and enjoyable. There will always be situations, however, in which systems do not function in the way users expect them to, or in which the user misunderstands something and makes a mistake. In these circumstances, error messages (phrased as "how-to-fix-it" advice) should be provided that explain what the user needs to do.

Another way of providing information is through online help, such as tips, handy hints, and contextualized advice. Like error messages, these need to be designed to guide users on what to do next when they get stuck and it is not obvious from the interface what to do. The signaling used at the interface to indicate that such online help is available needs careful consideration. A cartoon-based agent with a catchy tune may seem friendly and helpful the first time round but can quickly become annoying. A help icon or command that is activated by the users themselves when they want help is often preferable.

BOX 5.2 Should Computers Say They're Sorry?

A provocative idea is that computers should apologize when they make a mistake. Reeves and Naas (1996), for example, argue that they should be polite and courteous in the same way as people are to one another. While apologizing is normal social etiquette in human behavior, especially when someone makes a mistake, would you agree that computers should be made to behave in the same way? Would users be as forgiving of computers as they are of one another? For example, what would most users think if, after a system had crashed, it came up with a spoken or written apology such as, "I'm really sorry I crashed. I'll try not to do it again"? Would they think that the computer was being sincere? Would the apology make them forgive the computer in the way they forgive other people, after receiving such an apology? Or would it have no effect at all? Worse still, would users perceive such messages as vacuous statements and regard them simply as condescending, thereby increasing their level of frustration? How else might systems communicate with users when they have committed an error?

5.5 A debate: the application of anthropomorphism to interaction design

In this section we present a debate. Read through the arguments for and against the motion and then the evidence provided. Afterwards decide for yourself whether you support the motion.

The motion

The use of anthropomorphism in interaction design is an effective technique and should be exploited further.

Background

A controversial debate in interaction design is whether to exploit the phenomenon of anthropomorphism (the propensity people have to attribute human qualities to objects). It is something that people do naturally in their everyday lives and is commonly exploited in the design of technologies (e.g., the creation of humanlike animals and plants in cartoon films, the design of toys that have human qualities). The approach is also becoming more widespread in interaction design, through the introduction of agents in a range of domains.

What is anthropomorphism? It is well known that people readily attribute human qualities to their pets and their cars, and, conversely, are willing to accept human attributes that have been assigned by others to cartoon characters, robots, toys, and other inanimate objects. Advertisers are well aware of this phenomenon and often create humanlike characters out of inanimate objects to promote their products. For example, breakfast cereals, butter, and fruit drinks have all been transmogrified into characters with human qualities (they move, talk, have personalities, and show emotions), enticing the viewer to buy them. Children are especially susceptible to this kind of "magic," as witnessed in their love of cartoons, where all manner of inanimate objects are brought to life with humanlike qualities.

Examples of its application to system design

The finding that people, especially children, have a propensity to accepting and enjoying objects that have been given humanlike qualities has led many designers into capitalizing on it, most prevalently in the design of human-computer dialogs modeled on how humans talk to each other. A range of animated screen characters, such as agents, friends, advisors and virtual pets, have also been developed.

Anthropomorphism has also been used in the development of cuddly toys that are embedded with computer systems. Commercial products like ActiMates™ have been designed to try to encourage children to learn through playing with the cuddly toys. For example, Barney attempts to motivate play in children by using human-based speech and movement (Strommen, 1998). The toys are programmed to react to the child and make comments while watching TV together or working together on a computer-based task (see Figure 1.2 in Color Plate 1). In particular, Barney is programmed to congratulate the child whenever he or she gets a right answer and also to react to the content on screen with appropriate emotions (e.g., cheering at good news and expressing concern at bad news).

Arguments for exploiting this behavior

An underlying argument in favor of the anthropomorphic approach is that furnishing interactive systems with personalities and other humanlike attributes makes them more enjoyable and fun to interact with. It is also assumed that they can motivate people to carry out the tasks suggested (e.g., learning material, purchasing goods) more strongly than if they are presented in cold, abstract computer language. Being addressed in first person (e.g., "Hello Chris! Nice to see you again. Welcome back. Now what were we doing last time? Oh yes, exercise 5. Let's start again.") is much more endearing than being addressed in the impersonal third person ("User 24, commence exercise 5"), especially for children. It can make them feel more at ease and reduce their anxiety. Similarly, interacting with screen characters like tutors and wizards can be much pleasanter than interacting with a cold dialog box or blinking cursor on a blank screen. Typing a question in plain English, using a search engine like Ask Jeeves (which impersonates the well-known fictitious butler), is more natural and personable than thinking up a set of keywords, as required by other search engines. At the very least, anthropomorphic interfaces are a harmless bit of fun.

Arguments against exploring this behavior

There have been many criticisms of the anthropomorphic approach. Shneiderman (1998), one of the best known critics, has written at length about the problems of attributing human qualities to computer systems. His central argument is that anthropomorphic interfaces, especially those that use first-person dialog and screen characters, are downright deceptive. An unpleasant side effect is that they can make people feel anxious, resulting in them feeling inferior or stupid. A screen tutor that wags its finger at the user and says, "Now, Chris, that's not right! Try again. You can do better." is likely to feel more humiliating than a system dialog box saying, "Incorrect. Try again."

Anthropomorphism can also lead people into a false sense of belief, enticing them to confide in agents called "software bots" that reside in chatrooms and other electronic spaces, pretending to be conversant human beings. By far the most common complaint against computers pretending to have human qualities, however, is that people find them very annoying and frustrating. Once users discover that the system cannot really converse like a human or does not possess real human qualities (like having a personality or being sincere), they become quickly disillusioned and subsequently distrust it. E-commerce sites that pretend to be caring by presenting an assortment of virtual assistants, receptionists, and other such helpers are seen for what they really are—artificial and flaky. Children and adults alike also are quickly bored and annoyed with applications that are fronted by artificial screen characters (e.g., tutor wizards) and simply ignore whatever they might suggest.

Evidence for the motion

A number of studies have investigated people's reactions and responses to computers that have been designed to be more humanlike. A body of work reported by Reeves and Nass (1996) has identified several benefits of the anthropomorphic approach. They found that computers that were designed to flatter and praise users when they did something right had a positive impact on how they felt about themselves. For example, an educational program was designed to say, "Your question makes an interesting and useful distinction. Great job!" after a user had contributed a new question to it. Students enjoyed the experience and were more willing to continue working with the computer than were other students who were not praised by the computer for doing the same things. In another study, Walker et al. (1994) compared people's responses to a talking-face display and an equivalent text-only one and found that people spent more time with the talking-face display than the text-only one. When given a questionnaire to fill in, the face-display group made fewer mistakes and wrote down more comments. In a follow-up study, Sproull et al. (1996) again found that users reacted quite differently to the two interfaces, with users presenting themselves in a more positive light to the talking-face display and generally interacting with it more.

Evidence against the motion

Sproull et al.'s studies also revealed, however, that the talking-face display made some users feel somewhat disconcerted and displeased. The choice of a stern talking face may have been a large contributing factor. Perhaps a different kind of response would have been elicited if a friendlier smiling face had been used. Nevertheless, a number of other studies have shown that increasing the "humanness" of an interface is counterproductive. People can be misled into believing that a computer is like a human, with human levels of intelligence. For example, one study investigating user's responses to interacting with agents at the interface represented as human guides found that the users expected the agents to be more humanlike than they actually were. In particular, they expected the agents to have personality, emotion, and motivation—even though the guides were portrayed on the screen as simple black and white static icons (see Figure 5.8). Furthermore, the users became disappointed when they discovered the agents did not have any of these characteristics (Oren et al., 1990). In another study comparing an anthropomorphic interface that spoke in the first person and was highly personable (HI THERE, JOHN! IT'S NICE TO MEET YOU, I SEE YOU ARE READY NOW) with a mechanistic one that spoke in third person (PRESS THE ENTER KEY TO

Figure 5.8 Guides of historical characters.

BEGIN SESSION), the former was rated by college students as less honest and it made them feel less responsible for their actions (Quintanar et al., 1982).

Casting your vote: On the basis of this debate and any other articles on the topic (see Section 5.6 and the recommended readings at the end of this chapter) together with your experiences with anthropomorphic interfaces, make up your mind whether you are for or against the motion.

5.6 Virtual characters: agents

As mentioned in the debate above, a whole new genre of cartoon and life-like characters has begun appearing on our computer screens—as agents to help us search the web, as e-commerce assistants that give us information about products, as characters in video games, as learning companions or instructors in educational programs, and many more. The best known are videogame stars like Lara Croft and Super Mario. Other kinds include virtual pop stars (See Figure 5.9 on Color Plate 6), virtual talk-show hosts, virtual bartenders, virtual shop assistants, and virtual newscasters. Interactive pets (e.g., Aibo) and other artificial anthropomorphized characters (e.g., Pokemon, Creatures) that are intended to be cared for and played with by their owners have also proved highly popular.

5.6.1 Kinds of agents

Below we categorize the different kinds of agents in terms of the degree to which they anthropomorphize and the kind of human or animal qualities they emulate. These are (1) synthetic characters, (2) animated agents, (3) emotional agents, and (4) embodied conversational interface agents.

1. Synthetic characters

These are commonly designed as 3D characters in video games or other forms of entertainment, and can appear as a first-person avatar or a third-person agent. Much effort goes into designing them to be lifelike, exhibiting realistic human movements, like walking and running, and having distinct personalities and traits. The design of the characters' appearance, their facial expressions, and how their lips move when talking are also considered important interface design concerns.

Bruce Blumberg and his group at MIT are developing autonomous animated creatures that live in virtual 3D environments. The creatures are autonomous in that they decide what to do, based on what they can sense of the 3D world, and how they feel, based on their internal states. One of the earliest creatures to be developed was Silas T. Dog (Blumberg, 1996). The 3D dog looks like a cartoon creature (colored bright yellow) but is designed to behave like a real dog (see Figure 5.10). For example, he can walk, run, sit, wag his tail, bark, cock his leg, chase sticks, and rub his head on people when he is happy. He navigates through his world by using his "nose" and synthetic vision. He also has been programmed with various internal goals and needs that he tries to satisfy, including wanting to play

Figure 5.10 User interacting with Silas the dog in (a) physical world (b) virtual world, and (c) close-up of Silas.

and have company. He responds to events in the environment; for example, he becomes aggressive if a hamster enters his patch.

A person can interact with Silas by making various gestures that are detected by a computer-vision system. For example, the person can pretend to throw a stick, which is recognized as an action that Silas responds to. An image of the person is also projected onto a large screen so that he can be seen in relation to Silas (see Figure 5.10). Depending on his mood, Silas will run after the stick and return it (e.g., when he is happy and playful) or cower and refuse to fetch it (e.g., when he is hungry or sad).

2.   Animated agents

These are similar to synthetic characters except they tend to be designed to play a collaborating role at the interface. Typically, they appear at the side of the screen as tutors, wizards and helpers intended to help users perform a task. This might be designing a presentation, writing an essay or learning about a topic. Most of the characters are designed to be cartoon-like rather than resemble human beings.

An example of an animated agent is Herman the Bug, who was developed by Intellimedia at North Carolina State University to teach children from kindergarten to high school about biology (Lester et al., 1997). Herman is a talkative, quirky insect that flies around the screen and dives into plant structures as it provides problem-solving advice to students (See Figure 5.11 on Color Plate 7). When providing its explanations it performs a range of activities including walking, flying, shrinking, expanding, swimming, bungee jumping, acrobatics, and teleporting. Its behavior includes 30 animated segments, 160 canned audio clips, and a number of songs. Herman offers advice on how to perform tasks and also tries to motivate students to do them.

3.   Emotional agents

These are designed with a predefined personality and set of emotions that are manipulated by users. The aim is to allow people to change the moods or emotions of agents and see what effect it has on their behavior. Various mood changers are provided at the interface in the form of sliders and icons. The effect of requesting an animated agent to become very happy, sad, or grumpy is seen through changes to their behavior. For example, if a user moves a slider to a "scared" position on an emotional scale, the agent starts behaving scared, hiding behind objects and making frightened facial expressions.

The Woggles are one of the earliest forms of emotional agents (Bates, 1994). A group of agents was designed to appear on the screen that played games with one another, such as hide and seek (See Figure 5.12 on Color Plate 7). They were designed as different colored bouncy balls with cute facial expressions. Users could change their moods (e.g., from happy to sad) by moving various sliders, which in turn changed their movement (e.g., they bounced less), facial expression (e.g., they no longer smiled), and how willing they were to play with the other Woggles.

4. Embodied conversational interface agents

Much of the research on embodied conversational interface agents has been concerned with how to emulate human conversation. This has included modeling various conversational mechanisms such as:

- recognizing and responding to verbal and non-verbal input
- generating verbal and non-verbal output
- coping with breakdowns, turn-taking and other conversational mechanisms
- giving signals that indicate the state of the conversation as well as contributing new suggestions for the dialog (Cassell, 2000, p.72)

In many ways, this approach is the most anthropomorphic in its aims of all the agent research and development.

Rea is an embodied real-estate agent with a humanlike body that she uses in humanlike ways during a conversation (Cassell, 2000). In particular, she uses eye gaze, body posture, hand gestures, and facial expressions while talking (See Figure 5.13 on Color Plate 8). Although the dialog appears relatively simple, it involves a sophisticated underlying set of conversational mechanisms and gesture-recognition techniques. An example of an actual interaction with Rea is:

Mike approaches the screen and Rea turns to face him and says:

"Hello. How can I help you?"

Mike: "I'm looking to buy a place near MIT."

Rea nods, indicating she is following.

Rea: "I have a house to show you" (picture of a house appears on the screen). "It is in Somerville."

Mike: "Tell me about it."

Rea looks up and away while she plans what to say.

Rea: "It's big."

Rea makes an expansive gesture with her hands.

> Mike brings his hands up as if to speak, so Rea does not continue, waiting for him to speak.
>
> Mike: "Tell me more about it."
>
> Rea: "Sure thing. It has a nice garden . . ."

ACTIVITY 5.4   Which of the various kinds of agents described above do you think are the most convincing? Is it those that try to be as humanlike as possible or those that are designed to be simple, cartoon-based animated characters?

Comment   We argue that the agents that are the most successful are ironically those that are least like humans. The reasons for this include that they appear less phony and are not trying to pretend they are more intelligent or human than they really are. However, others would argue that the more humanlike they are, the more believable they are and hence the more convincing.

5.6.2 General design concerns

Believability of virtual characters

One of the major concerns when designing agents and virtual characters is how to make them believable. By believability is meant "the extent to which users interacting with an agent come to believe that it has its own beliefs, desires and personality" (Lester and Stone, 1997, p 17). In other words, a virtual character that a person can believe in is taken as one that allows users to suspend their disbelief. A key aspect is to match the personality and mood of the character to its actions. This requires deciding what are appropriate behaviors (e.g., jumping, smiling, sitting, raising arms) for different kinds of emotions and moods. How should the emotion "very happy" be expressed? Through a character jumping up and down with a big grin on its face? What about moderately happy—through a character jumping up and down with a small grin on its face? How easy is it for the user to distinguish between these two and other emotions that are expressed by the agents? How many emotions are optimal for an agent to express?

Appearance

The appearance of an agent is very important in making it believable. Parsimony and simplicity are key. Research findings suggest that people tend to prefer simple cartoon-based screen characters to detailed images that try to resemble the human form as much as possible (Scaife and Rogers, 2001). Other research has also found that simple cartoon-like figures are preferable to real people pretending to be artificial agents. A project carried out by researchers at Apple Computer Inc. in the 80s found that people reacted quite differently to different representations of the same interface agent. The agent in question, called Phil, was created as part of a promotional

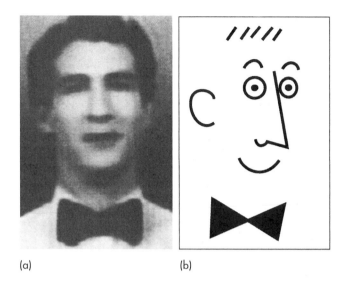

(a) (b)

Figure 5.14 Two versions of Phil, the agent assistant that appeared in Apple's promotional video called the Knowledge Navigator (a) as a real actor pretending to be a computer agent and (b) as a cartoon being an agent. Phil was created by Doris Mitsch and the actor Phil was Scott Freeman.

video called "The Knowledge Navigator." He was designed to respond and behave just like a well-trained human assistant. In one version, he was played by a real actor that appeared on a university professor's computer screen. Thus, he was portrayed as an artificial agent but was played by a real human. The actor was a smartly dressed assistant wearing a white shirt and bow tie. He was also extremely polite. He performed a number of simple tasks at the computer interface, such as reminding the professor of his appointments for that day and alerting him to phone calls waiting. Many people found this version of Phil unrealistic. After viewing the promotional video, people complained about him, saying that he seemed too stupid. In another version, Phil was designed as a simple line-drawn cartoon with limited animation (see Figure 5.14) and was found to be much more likeable (see Laurel, 1993).

Behavior

Another important consideration in making virtual characters believable is how convincing their behavior is when performing actions. In particular, how good are they at pointing out relevant objects on the screen to the user, so that the user knows what they are referring to? One way of achieving this is for the virtual character to "lead" with its eyes. For example, Silas the dog turns to look at an object or a person before he actually walks over to it (e.g., to pick the object up or to invite the person to play). A character that does not lead with its eyes looks very mechanical and as such not very life-like (Maes, 1995).

As mentioned previously, an agent's actions need also to match their underlying emotional state. If the agent is meant to be angry, then its body posture, movements, and facial expression all need to be integrated to show this. How this can be achieved effectively can be learned from animators, who have a long tradition in this field. For example, one of their techniques is to greatly exaggerate expressions and movements as a way of conveying and drawing attention to an emotional state of a character.

Mode of interaction

The way the character communicates with the user is also important. One approach has been towards emulating human conversations as much as possible to make the character's way of talking more convincing. However, as mentioned in the debate above, a drawback of this kind of masquerading is that people can get annoyed easily and feel cheated. Paradoxically, a more believable and acceptable dialog with a virtual character may prove to be one that is based on a simple *artificial* mode of interaction, in which prerecorded speech is played at certain choice points in the interaction and the user's responses are limited to selecting menu options. The reason why this mode of interaction may ultimately prove more effective is because the user is in a better position to understand what the agent is capable of doing. There is no pretence of a stupid agent pretending to be a smart human.

Assignment

This assignment requires you to write a critique of the persuasive impact of virtual sales agents on customers. Consider what it would take for a virtual sales agent to be believable, trustworthy, and convincing, so that customers would be reassured and happy to buy something based on its recommendations.

(a) Look at some e-commerce sites that use virtual sales agents (use a search engine to find sites or start with Miss Boo at boo.com, which was working at time of printing) and answer the following:

- What do the virtual agents do?
- What type of agent are they?
- Do they elicit an emotional response from you? If so, what is it?
- What kind of personality do they have?
- How is this expressed?
- What kinds of behavior do they exhibit?
- What are their facial expressions like?
- What is their appearance like? Is it realistic or cartoon-like?
- Where do they appear on the screen?
- How do they communicate with the user (text or speech)?
- Is the level of discourse patronizing or at the right level?
- Are the agents helpful in guiding the customer towards making a purchase?
- Are they too pushy?
- What gender are they? Do you think this makes a difference?
- Would you trust the agents to the extent that you would be happy to buy a product from them? If not, why not?
- What else would it take to make the agents persuasive?

(b) Next, look at an e-commerce website that does not include virtual sales agents but is based on a conceptual model of browsing (e.g., Amazon.com). How does it compare with the agent-based sites you have just looked at?

- Is it easy to find information about products?
- What kind of mechanism does the site use to make recommendations and guide the user in making a purchase?
- Is any kind of personalization used at the interface to make the user feel welcome or special?
- Would the site be improved by having an agent? Explain your reasons either way.

(c) Finally, discuss which site you would trust most and give your reasons for this.

Summary

This chapter has described the different ways interactive products can be designed (both deliberately and inadvertently) to make people respond in certain ways. The extent to which users will learn, buy a product online, chat with others, and so on depends on how comfortable they feel when using a product and how well they can trust it. If the interactive product is frustrating to use, annoying, or patronizing, users easily get angry and despondent, and often stop using it. If, on the other hand, the system is a pleasure, enjoyable to use, and makes the users feel comfortable and at ease, then they are likely to continue to use it, make a purchase, return to the website, continue to learn, etc. This chapter has described various interface mechanisms that can be used to elicit positive emotional responses in users and ways of avoiding negative ones.

Key points

- Affective aspects of interaction design are concerned with the way interactive systems make people respond in emotional ways.
- Well-designed interfaces can elicit good feelings in people.
- Aesthetically pleasing interfaces can be a pleasure to use.
- Expressive interfaces can provide reassuring feedback to users as well as be informative and fun.
- Badly designed interfaces often make people frustrated and angry.
- Anthropomorphism is the attribution of human qualities to objects.
- An increasingly popular form of anthropomorphism is to create agents and other virtual characters as part of an interface.
- People are more accepting of believable interface agents.
- People often prefer simple cartoon-like agents to those that attempt to be humanlike.

Further reading

TURKLE, S. (1995) *Life on the Screen*. New York: Simon and Schuster. This classic covers a range of social impact and affective aspects of how users interact with a variety of computer-based applications. Sherry Turkle discusses at length how computers, the Internet, software, and the design of interfaces affect our identities.

Two very good papers on interface agents can be found in Brenda Laurel's (ed.) *The Art of Human-Computer Interface Design* (1990) Reading, MA.: Addison Wesley:

LAUREL, B. (1990) Interface agents: metaphor with character, 355–366

OREN. T., SALOMON, G., KREITMAN, K., AND DON. A. (1990) Guides: characterizing the interface, 367–381

MAES, P. (1995) Artificial life meets entertainment: lifelike autonomous agents. *Communications of the ACM,* 38. (11), 108–114. Pattie Maes has written extensively about the role and design of intelligent agents at the interface. This paper provides a good review of some of her work in this field.

Excerpts from a lively debate between Pattie Maes and Ben Shneiderman on "Direct manipulation vs. interface agents" can be found *ACM Interactions Magazine,* 4 (6) (1997), 42–61.

Chapter 6

The process of interaction design

6.1 Introduction
6.2 What is interaction design about?
    6.2.1 Four basic activities of interaction design
    6.2.2 Three key characteristics of the interaction design process
6.3 Some practical issues
    6.3.1 Who are the users?
    6.3.2 What do we mean by "needs"?
    6.3.3 How do you generate alternative designs?
    6.3.4 How do you choose among alternative designs?
6.4 Lifecycle models: showing how the activities are related
    6.4.1 A simple lifecycle model for interaction design
    6.4.2 Lifecycle models in software engineering
    6.4.3 Lifecycle models in HCI

6.1. Introduction

Design is a practical and creative activity, the ultimate intent of which is to develop a product that helps its users achieve their goals. In previous chapters, we looked at different kinds of interactive products, issues you need to take into account when doing interaction design and some of the theoretical basis for the field. This chapter is the first of four that will explore *how* we can design and build interactive products.

Chapter 1 defined interaction design as being concerned with "designing interactive products to support people in their everyday and working lives." But how do you go about doing this?

Developing a product must begin with gaining some understanding of what is required of it, but where do these requirements come from? Whom do you ask about them? Underlying good interaction design is the philosophy of user-centered design, i.e., involving users throughout development, but who are the users? Will they know what they want or need even if we can find them to ask? For an innovative product, users are unlikely to be able to envision what is possible, so where do these ideas come from?

In this chapter, we raise and answer these kinds of questions and discuss the four basic activities and key characteristics of the interaction design process that were introduced in Chapter 1. We also introduce a lifecycle model of interaction design that captures these activities and characteristics.

The main aims of this chapter are to:

- Consider what 'doing' interaction design involves.
- Ask and provide answers for some important questions about the interaction design process.
- Introduce the idea of a lifecycle model to represent a set of activities and how they are related.
- Describe some lifecycle models from software engineering and HCI and discuss how they relate to the process of interaction design.
- Present a lifecycle model of interaction design.

6.2 What is interaction design about?

There are many fields of design, for example graphic design, architectural design, industrial and software design. Each discipline has its own interpretation of "designing." We are not going to debate these different interpretations here, as we are focussing on interaction design, but a general definition of "design" is informative in beginning to understand what it's about. The definition of design from the *Oxford English Dictionary* captures the essence of design very well: "(design is) a plan or scheme conceived in the mind and intended for subsequent execution." The act of designing therefore involves the development of such a plan or scheme. For the plan or scheme to have a hope of ultimate execution, it has to be informed with knowledge about its use and the target domain, together with practical constraints such as materials, cost, and feasibility. For example, if we conceived of a plan for building multi-level roads in order to overcome traffic congestion, before the plan could be executed we would have to consider drivers' attitudes to using such a construction, the viability of the structure, engineering constraints affecting its feasibility, and cost concerns.

In interaction design, we investigate the artifact's use and target domain by taking a user-centered approach to development. This means that users' concerns direct the development rather than technical concerns.

Design is also about trade-offs, about balancing conflicting requirements. If we take the roads plan again, there may be very strong environmental arguments for stacking roads higher (less countryside would be destroyed), but these must be balanced against engineering and financial limitations that make the proposition less attractive. Getting the balance right requires experience, but it also requires the development and evaluation of alternative solutions. Generating alternatives is a key principle in most design disciplines, and one that should be encouraged in interaction design. As Marc Rettig suggested: "To get a good idea, get lots of ideas" (Rettig, 1994). However, this is not necessarily easy, and unlike many design disciplines, interaction designers are not generally trained to generate alternative designs. However, the ability to brainstorm and contribute alternative ideas can be learned, and techniques from other design disciplines can be successfully used in interaction design. For example, Danis and Boies (2000) found that using techniques from graphic design that encouraged the generation of alternative designs stimulated innovative interactive systems design. See also the interview with Gillian Crampton Smith at the end of this chapter for her views on how other aspects of traditional design can help produce good interaction design.

Although possible, it is unlikely that just one person will be involved in developing and using a system and therefore the plan must be communicated. This requires it to be captured and expressed in some suitable form that allows review, revision, and improvement. There are many ways of doing this, one of the simplest being to produce a series of sketches. Other common approaches are to write a description in natural language, to draw a series of diagrams, and to build prototypes. A combination of these techniques is likely to be the most effective. When users are involved, capturing and expressing a design in a suitable format is especially important since they are unlikely to understand jargon or specialist notations. In fact, a form that users can interact with is most effective, and building prototypes of one form or another (see Chapter 8) is an extremely powerful approach.

So interaction design involves developing a plan which is informed by the product's intended use, target domain, and relevant practical considerations. Alternative designs need to be generated, captured, and evaluated by users. For the evaluation to be successful, the design must be expressed in a form suitable for users to interact with.

ACTIVITY 6.1

Imagine that you want to design an electronic calendar or diary for yourself. You might use this system to plan your time, record meetings and appointments, mark down people's birthdays, and so on, basically the kinds of things you might do with a paper-based calendar. Draw a sketch of the system outlining its functionality and its general look and feel. Spend about five minutes on this.

Having produced an outline, now spend five minutes reflecting on how you went about tackling this activity. What did you do first? Did you have any particular artifacts or experience to base your design upon? What process did you go through?

Comment

The sketch I produced is shown in Figure 6.1. As you can see, I was quite heavily influenced by the paper-based books I currently use! I had in mind that this calendar should allow me to record meetings and appointments, so I need a section representing the days and months. But I also need a section to take notes. I am a prolific note-taker, and so for me this was a key requirement. Then I began to wonder about how I could best use hyperlinks. I certainly want to keep addresses and telephone numbers in my calendar, so maybe there could be a link between, say, someone's name in the calendar and their entry in my address book that will give me their contact details when I need them? But I still want the ability to be able to turn page by page, for when I'm scanning or thinking about how to organize my time. A search facility would be useful too.

The first thing that came into my head when I started doing this was my own paper-based book where I keep appointments, maps, telephone numbers, and other small notes. I also thought about my notebook and how convenient it would be to have the two combined. Then I sat and sketched different ideas about how it might look (although I'm not very good at sketching). The sketch in Figure 6.1 is the version I'm happiest with. Note that my sketch

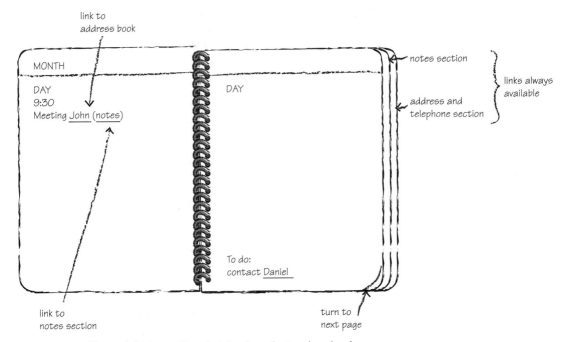

Figure 6.1 An outline sketch of an electronic calendar.

has a strong resemblance to a paper-based book, yet I've also tried to incorporate electronic capabilities. Maybe once I have evaluated this design and ensured that the tasks I want to perform are supported, then I will be more receptive to changing the look away from a paper-based "look and feel."

The exact steps taken to produce a product will vary from designer to designer, from product to product, and from organization to organization. In this activity, you may have started by thinking about what you'd like such a system to do for you, or you may have been thinking about an existing paper calendar. You may have mixed together features of different systems or other record-keeping support. Having got or arrived at an idea of what you wanted, maybe you then imagined what it might look like, either through sketching with paper and pencil or in your mind.

6.2.1 Four basic activities of interaction design

Four basic activities for interaction design were introduced in Chapter 1, some of which you will have engaged in when doing Activity 6.1. These are: identifying needs and establishing requirements, developing alternative designs that meet those requirements, building interactive versions so that they can be communicated and assessed, and evaluating them, i.e., measuring their acceptability. They are fairly generic activities and can be found in other designs disciplines too. For example, in architectural design (RIBA, 1988) basic requirements are established in a work stage called "inception", alternative design options are considered in a "feasibility" stage and "the brief" is developed through outline proposals and scheme design. During this time, prototypes may be built or perspectives may be drawn to give clients a better indication of the design being developed. Detail design specifies all components, and working drawings are produced. Finally, the job arrives on site and building commences.

We will be expanding on each of the basic activities of interaction design in the next two chapters. Here we give only a brief introduction to each.

Identifying needs and establishing requirements

In order to design something to support people, we must know who our target users are and what kind of support an interactive product could usefully provide. These needs form the basis of the product's requirements and underpin subsequent design and development. This activity is fundamental to a user-centered approach, and is very important in interaction design; it is discussed further in Chapter 7.

Developing alternative designs

This is the core activity of designing: actually suggesting ideas for meeting the requirements. This activity can be broken up into two sub-activities: conceptual design and physical design. Conceptual design involves producing the conceptual model for the product, and a conceptual model describes what the product should do, behave and look like. Physical design considers the detail of the product including the colors, sounds, and images to use, menu design, and icon design. Alternatives are considered at every point. You met some of the ideas for conceptual design in Chapter 2; we go into more detail about conceptual and physical design in Chapter 8.

Building interactive versions of the designs

Interaction design involves designing interactive products. The most sensible way for users to evaluate such designs, then, is to interact with them. This requires an interactive version of the designs to be built, but that does not mean that a software version is required. There are different techniques for achieving "interaction," not all of which require a working piece of software. For example, paper-based prototypes are very quick and cheap to build and are very effective for identifying problems in the early stages of design, and through role-playing users can get a real sense of what it will be like to interact with the product. This aspect is also covered in Chapter 8.

Evaluating designs

Evaluation is the process of determining the usability and acceptability of the product or design that is measured in terms of a variety of criteria including the number of errors users make using it, how appealing it is, how well it matches the requirements, and so on. Interaction design requires a high level of user involvement throughout development, and this enhances the chances of an acceptable product being delivered. In most design situations you will find a number of activities concerned with quality assurance and testing to make sure that the final product is "fit-for-purpose." Evaluation does not replace these activities, but complements and enhances them. We devote Chapters 10 through 14 to the important subject of evaluation.

The activities of developing alternative designs, building interactive versions of the design, and evaluation are intertwined: alternatives are evaluated through the interactive versions of the designs and the results are fed back into further design. This iteration is one of the key characteristics of the interaction design process, which we introduced in Chapter 1.

6.2.2   Three key characteristics of the interaction design process

There are three characteristics that we believe should form a key part of the interaction design process. These are: a user focus, specific usability criteria, and iteration.

The need to *focus on users* has been emphasized throughout this book, so you will not be surprised to see that it forms a central plank of our view on the interaction design process. While a process cannot, in itself, guarantee that a development will involve users, it can encourage focus on such issues and provide opportunities for evaluation and user feedback.

Specific usability and user experience goals should be identified, clearly documented, and agreed upon at the beginning of the project. They help designers to choose between different alternative designs and to check on progress as the product is developed.

Iteration allows designs to be refined based on feedback. As users and designers engage with the domain and start to discuss requirements, needs, hopes and aspirations, then different insights into what is needed, what will help, and what is feasible will emerge. This leads to a need for iteration, for the activities to inform each other and to be repeated. However good the designers are and however clear the users may think their vision is of the required artifact, it will be necessary to revise ideas in light of feedback, several times. This is particularly true if you are trying to innovate. Innovation rarely emerges whole and ready to go. It takes time, evolution, trial and error, and a great deal of patience. Iteration is inevitable because designers never get the solution right the first time (Gould and Lewis, 1985).

We shall return to these issues and expand upon them in Chapter 9.

6.3   Some practical issues

Before we consider how the activities and key characteristics of interaction design can be pulled together into a coherent process, we want to consider some questions highlighted by the discussion so far. These questions must be answered if we are going to be able to "do" interaction design in practice. These are:

- Who are the users?
- What do we mean by needs?
- How do you generate alternative designs?
- How do you choose among alternatives?

6.3.1 Who are the users?

In Chapter 1, we said that an overarching objective of interaction design is to optimize the interactions people have with computer-based products, and that this requires us to support needs, match wants, and extend capabilities. We also stated above that the activity of identifying these needs and establishing requirements was fundamental to interaction design. However, we can't hope to get very far with this intent until we know who the users are and what they want to achieve. As a starting point, therefore, we need to know who we consult to find out the users' requirements and needs.

Identifying the users may seem like a straightforward question, but in fact there are many interpretations of "user." The most obvious definition is those people who interact directly with the product to achieve a task. Most people would agree with this definition; however, there are others who can also be thought of as users. For example, Holtzblatt and Jones (1993) include in their definition of "users" those who manage direct users, those who receive products from the system, those who test the system, those who make the purchasing decision, and those who use competitive products. Eason (1987) identifies three categories of user: primary, secondary and tertiary. Primary users are those likely to be frequent hands-on users of the system; secondary users are occasional users or those who use the system through an intermediary; and tertiary users are those affected by the introduction of the system or who will influence its purchase.

The trouble is that there is a surprisingly wide collection of people who all have a stake in the development of a successful product. These people are called *stakeholders*. Stakeholders are "people or organizations who will be affected by the system and who have a direct or indirect influence on the system requirements" (Kotonya and Sommerville, 1998). Dix et al. (1993) make an observation that is very pertinent to a user-centered view of development, that "It will frequently be the case that the formal 'client' who orders the system falls very low on the list of those affected. Be very wary of changes which take power, influence or control from some stakeholders without returning something tangible in its place."

Generally speaking, the group of stakeholders for a particular product is going to be larger than the group of people you'd normally think of as users, although it will of course include users. Based on the definition above, we can see that the group of stakeholders includes the development team itself as well as its managers, the direct users and their managers, recipients of the product's output, people who may lose their jobs because of the introduction of the new product, and so on.

For example, consider again the calendar system in Activity 6.1. According to the description we gave you, the user group for the system has just one member: you. However, the stakeholders for the system would also include people you make appointments with, people whose birthdays you remember, and even companies that produce paper-based calendars, since the introduction of an electronic calendar may increase competition and force them to operate differently.

This last point may seem a little exaggerated for just one system, but if you think of others also migrating to an electronic version, and abandoning their paper calendars, then you can see how the companies may be affected by the introduction of the system.

The net of stakeholders is really quite wide! We do not suggest that you need to involve all of the stakeholders in your user-centered approach, but it is important to be aware of the wider impact of any product you are developing. Identifying the stakeholders for your project means that you can make an informed decision about who should be involved and to what degree.

ACTIVITY 6.2   Who do you think are the stakeholders for the check-out system of a large supermarket?

Comment   First, there are the check-out operators. These are the people who sit in front of the machine and pass the customers' purchases over the bar code reader, receive payment, hand over receipts, etc. Their stake in the success and usability of the system is fairly clear and direct. Then you have the customers, who want the system to work properly so that they are charged the right amount for the goods, receive the correct receipt, are served quickly and efficiently. Also, the customers want the check-out operators to be satisfied and happy in their work so that they don't have to deal with a grumpy assistant. Outside of this group, you then have supermarket managers and supermarket owners, who also want the assistants to be happy and efficient and the customers to be satisfied and not complaining. They also don't want to lose money because the system can't handle the payments correctly. Other people who will be affected by the success of the system include other supermarket employees such as warehouse staff, supermarket suppliers, supermarket owners' families, and local shop owners whose business would be affected by the success or failure of the system. We wouldn't suggest that you should ask the local shop owner about requirements for the supermarket check-out system. However, you might want to talk to warehouse staff, especially if the system links in with stock control or other functions.

6.3.2   What do we mean by "needs"?

If you had asked someone in the street in the late 1990s what she 'needed', I doubt that the answer would have included interactive television, or a jacket which was wired for communication, or a smart fridge. If you presented the same person with these possibilities and asked whether she would buy them if they were available, then the answer would have been different. When we talk about identifying needs, therefore, it's not simply a question of asking people, "What do you need?" and then supplying it, because people don't necessarily know what is possible (see Suzanne Robertson's interview at the end of Chapter 7 for "un-dreamed-of" requirements). Instead, we have to approach it by understanding the characteristics and capabilities of the users, what they are trying to achieve, how they achieve it currently, and whether they would achieve their goals more effectively if they were supported differently.

There are many dimensions along which a user's capabilities and characteristics may vary, and that will have an impact on the product's design. You have met some of these in Chapter 3. For example, a person's physical characteristics may affect the design: size of hands may affect the size and positioning of input buttons, and motor abilities may affect the suitability of certain input and output devices; height is relevant in designing a physical kiosk, for example; and strength in designing a child's toy—a toy should not require too much strength to operate, but may require strength greater than expected for the target age group to change batteries or perform other operations suitable only for an adult. Cultural diversity and experience may affect the terminology the intended user group is used to, or how nervous about technology a set of users may be.

If a product is a new invention, then it can be difficult to identify the users and representative tasks for them; e.g., before microwave ovens were invented, there were no users to consult about requirements and there were no representative tasks to identify. Those developing the oven had to imagine who might want to use such an oven and what they might want to do with it.

It may be tempting for designers simply to design what they would like, but their ideas would not necessarily coincide with those of the target user group. It is *imperative* that representative users from the real target group be consulted. For example, a company called Netpliance was developing a new "Internet appliance," i.e., a product that would seamlessly integrate all the services necessary for the user to achieve a specific task on the Internet (Isensee et al., 2000). They took a user-centered approach and employed focus group studies and surveys to understand their customers' needs. The marketing department led these efforts, but developers observed the focus groups to learn more about their intended user group. Isensee et al. (p. 60) observe that "It is always tempting for developers to create products they would want to use or similar to what they have done before. However, in the Internet appliance space, it was essential to develop for a new audience that desires a simpler product than the computer industry has previously provided."

In these circumstances, a good indication of future behavior is current or past behavior. So it is always useful to start by understanding similar behavior that is already established. Apart from anything else, introducing something new into people's lives, especially a new "everyday" item such as a microwave oven, requires a culture change in the target user population, and it takes a long time to effect a culture change. For example, before cell phones were so widely available there were no users and no representative tasks available for study, *per se.* But there were standard telephones and so understanding the tasks people perform with, and in connection with, standard telephones was a useful place to start. Apart from making a telephone call, users also look up people's numbers, take messages for others not currently available, and find out the number of the last person to ring them. These kinds of behavior have been translated into memories for the telephone, answering machines, and messaging services for mobiles. In order to maximize the benefit of e-commerce sites, traders have found that referring back to customers' non-electronic habits and behaviors can be a good basis for enhancing e-commerce activity (CHI panel, 2000; Lee et al., 2000).

6.3.3   How do you generate alternative designs?

A common human tendency is to stick with something that we know works. We probably recognize that a better solution may exist out there somewhere, but it's very easy to accept this one because we know it works—it's "good enough." Settling for a solution that is good enough is not, in itself, necessarily "bad," but it may be undesirable because good alternatives may never be considered, and considering alternative solutions is a crucial step in the process of design. But where do these alternative ideas come from?

One answer to this question is that they come from the individual designer's flair and creativity. While it is certainly true that some people are able to produce wonderfully inspired designs while others struggle to come up with any ideas at all, very little in this world is completely new. Normally, innovations arise through cross-fertilization of ideas from different applications, the evolution of an existing product through use and observation, or straightforward copying of other, similar products. For example, if you think of something commonly believed to be an "invention," such as the steam engine, this was in fact inspired by the observation that the steam from a kettle boiling on the stove lifted the lid. Clearly there was an amount of creativity and engineering involved in making the jump from a boiling kettle to a steam engine, but the kettle provided the inspiration to translate experience gained in one context into a set of principles that could be applied in another. As an example of evolution, consider the word processor. The capabilities of suites of office software have gradually increased from the time they first appeared. Initially, a word processor was just an electronic version of a typewriter, but gradually other capabilities, including the spell-checker, thesaurus, style sheets, graphical capabilities, etc., were added.

So although creativity and invention are often wrapped in mystique, we do understand something of the process and of how creativity can be enhanced or inspired. We know, for instance, that browsing a collection of designs will inspire designers to consider alternative perspectives, and hence alternative solutions. The field of case-based reasoning (Maher and Pu, 1997) emerged from the observation that designers solve new problems by drawing on knowledge gained from solving previous similar problems. As Schank (1982; p. 22) puts it, "An expert is someone who gets reminded of just the right prior experience to help him in processing his current experiences." And while those experiences may be the designer's own, they can equally well be others'.

A more pragmatic answer to this question, then, is that alternatives come from looking at other, similar designs, and the process of inspiration and creativity can be enhanced by prompting a designer's own experience and by looking at others' ideas and solutions. Deliberately seeking out suitable sources of inspiration is a valuable step in any design process. These sources may be very close to the intended new product, such as competitors' products, or they may be earlier versions of similar systems, or something completely different.

ACTIVITY 6.3   Consider again the calendar system introduced at the beginning of the chapter. Reflecting on the process again, what do you think inspired your outline design? See if you can identify any elements within it that you believe are truly innovative.

Comment   For my design, I haven't seen an electronic calendar, although I have seen plenty of other software-based systems. My main sources of inspiration were my current paper-based books.

Some of the things you might have been thinking of include your existing paper-based calendar, and other pieces of software you commonly use and find helpful or easy to use in some way. Maybe you already have access to an electronic calendar, which will have given you some ideas, too. However, there are probably other aspects that make the design somehow unique to you and may be innovative to a greater or lesser degree.

All this having been said, under some circumstances the scope to consider alternative designs may be limited. Design is a process of balancing constraints and constantly trading off one set of requirements with another, and the constraints may be such that there are very few viable alternatives available. As another example, if you are designing a software system to run under the Windows operating system, then elements of the design will be prescribed because you must conform to the Windows "look and feel," and to other constraints intended to make Windows programs consistent for the user. We shall return to style guides and standards in Chapter 8.

If you are producing an upgrade to an existing system, then you may face other constraints, such as wanting to keep the familiar elements of it and retain the same "look and feel." However, this is not necessarily a rigid rule. Kent Sullivan reports that when designing the Windows 95 operating system to replace the Windows 3.1 and Windows for Workgroups 3.11 operating systems, they initially focused too much on consistency with the earlier versions (Sullivan, 1996).

BOX 6.1   A Box Full of Ideas

The innovative product design company IDEO was introduced in Chapter 1. It has been involved in the development of many artifacts including the first commercial computer mouse and the PalmPilot V. Underlying some of their creative flair is a collection of weird and wonderful engineering housed in a large flatbed filing cabinet called the TechBox (see Figure 6.2). The TechBox holds around 200 gizmos and interesting materials, divided into categories: "Amazing Materials," "Cool Mechanisms," "Interesting Manufacturing Processes," "Electronic Technologies," and "Thermal and Optical." Each item has been placed in the box because it represents a neat idea or a new process. Staff at IDEO take along a selection of items from the TechBox to brainstorming meetings. The items may be chosen because they provide useful visual props or possible solutions to a particular issue, or simply to provide some light relief.

Each item is clearly labeled with its name and category, but further information can be found by accessing the TechBox's online catalog. Each item has its own page detailing what the item is, why it's interesting, where it came from, and who has used it or knows more about it. For example, the page in Figure 6.3 relates to a metal injection-molding technique.

Other items in the box include an example of metal-coated wood, materials with and without holes that stretch, bend, and change shape or color at different temperatures.

Each TechBox has its own curator who is responsible for maintaining and cataloging the items and for promoting its use within the office. Anyone can submit a new item for consideration and

Figure 6.2 The TechBox at IDEO.

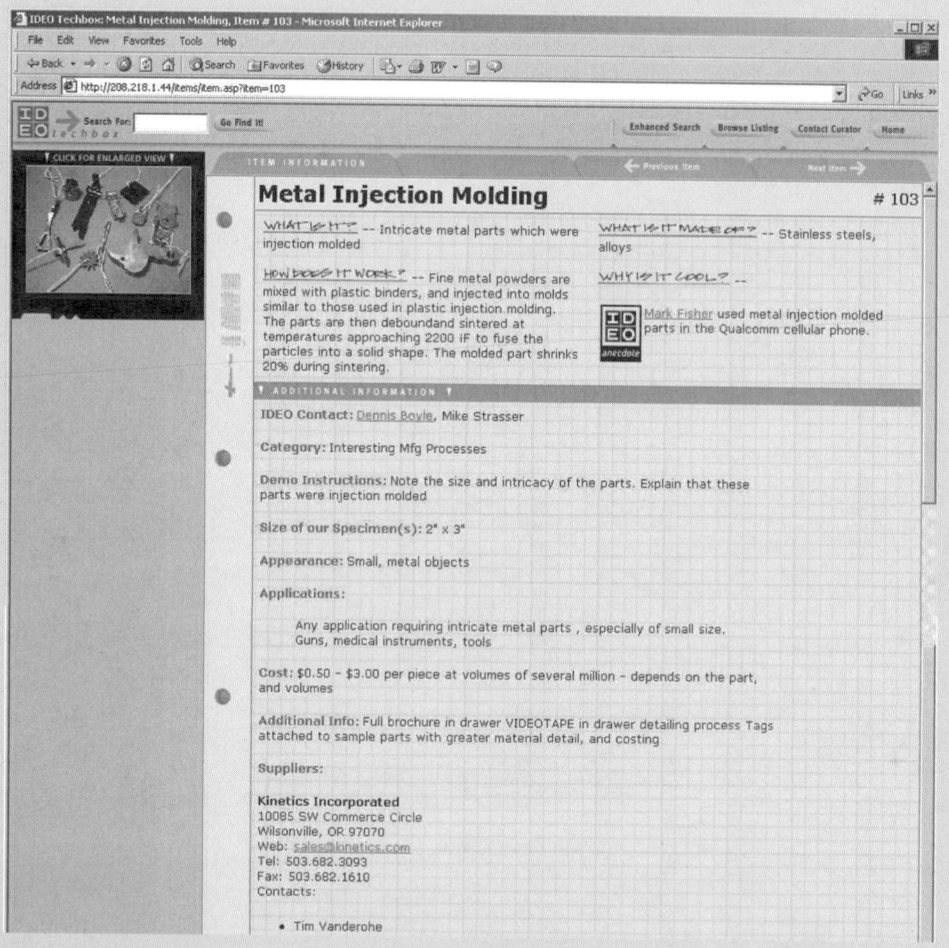

Figure 6.3 The web page for the metal injection molding.

as items become common place, they are removed from the TechBox to make way for the next generation of fascinating contraptions.

How are these things used? Well here is one example from Patrick Hall at the London IDEO office (see Figure 6.4):

> IDEO was asked to review the design of a mass-produced hand-held medical product that was deemed to be too big.

As well as brainstorming and other conventional idea-generation methods, I was able to immediately pick out items which I knew about from having used the TechBox in the past: Deep Draw; Fibre-Optic magnifier; Metal Injection molding; Flexy Battery. Further browsing and searching using the keywords search engine highlighted in-mold assembly and light-intensifying film. The

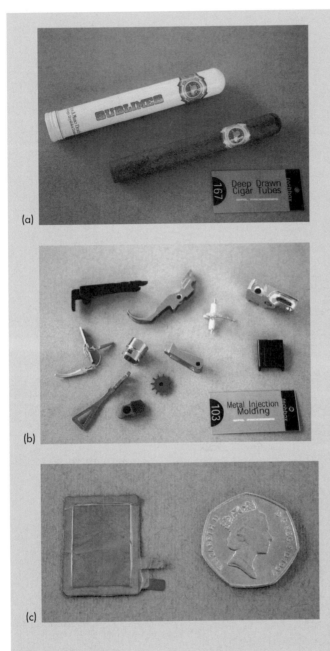

(a)

(b)

(c)

Figure 6.4 Items from the TechBox used in the design of a medical product. (a) Deep Draw—A metal-forming process to generate close-ended cylindrical parts; (b) Metal Injection Molding—A molding and sintering process to produce complex metal parts in high numbers; (c) Flexy Battery—a lithium polymer cell from Varta that is very thin (intended for Smart Cards) and can be formed into cylindrical shapes.

associated web pages for these items enabled me to learn more about these items immediately and indicated who to talk to in IDEO to find out more, and the details of vendors to approach.

The project ended at the feasibility phase, with the client pursuing the technologies I had suggested. Only the fiber-optic magnifier proved (immediately) not to be worth pursuing (because of cost).

DILEMMA   Copying for Inspiration: Is It Legal?

Designers draw on their experience of design when approaching a new project. This includes the use of previous designs that they know work, both designs they have created themselves and those that others have created. Others' creations often spark inspiration that also leads to new ideas and innovation. This is well known and understood. However, the expression of an idea is protected by copyright, and people who infringe that copyright can be taken to court and prosecuted. Note that copyright covers the expression of an idea and not the idea itself. This means, for example, that while there are numerous word processors all with similar functionality, this does not represent an infringement of copyright as the idea has been expressed in different ways, and it's the expression that's been copyrighted. Copyright is free and is automatically invested in the author of something, e.g., the writer of a book or a programmer who develops a program, unless he signs the copyright over to someone else. Authors writing for academic journals often are asked to sign over their copyright to the publisher of the journal. Various limitations and special conditions can apply, but basically, the copyright is no longer theirs. People who produce something through their employment, such as programs or products, may have in their employment contract a statement saying that

the copyright relating to anything produced in the course of that employment is automatically assigned to the employer and does not remain with the employee.

On the other hand, patenting is an alternative to copyright that does protect the idea rather than the expression. There are various forms of patenting, each of which is designed to allow the inventor the chance to capitalize on an idea. It is unusual for software to be patented, since it is a long, slow, and expensive process, although there is a recent trend towards patenting business processes. For example, Amazon, the on-line bookstore, has patented its "one-click" purchasing process, which allows regular users simply to choose a book and buy it with one mouse click (US Patent No. 5960411, September 29, 1999). This is possible because the system stores its customers' details and "recognizes" them when they access the site again.

So the dilemma comes in knowing when it's OK to use someone else's work as a source of inspiration and when you are infringing copyright or patent law. The issues around this question are complex and detailed, and well beyond the scope of this book, but more information and examples of law cases that have been brought successfully and unsuccessfully can be found in Bainbridge (1999).

6.3.4   How do you choose among alternative designs?

Choosing among alternatives is about making design decisions: Will the device use keyboard entry or a touch screen? Will the device provide an automatic memory function or not? These decisions will be informed by the information gathered about users and their tasks, and by the technical feasibility of an idea. Broadly speaking, though, the decisions fall into two categories: those that are about externally visible and measurable features, and those that are about characteristics internal to the system that cannot be observed or measured without dissecting it. For example, externally visible and measurable factors for a building design include the ease of access to the building, the amount of natural light in rooms, the width of corridors, and the number of power outlets. In a photocopier, externally visible and measurable factors include the physical size of the machine, the speed and quality of copying, the different sizes of paper it can use, and so on. Underlying each of these factors are other considerations that cannot be observed or studied without dissecting the building or the machine. For example, the number of power outlets will be dependent on how the wiring within the building is designed and the capacity of the main power supply; the choice of materials used in a photocopier may depend on its friction rating and how much it deforms under certain conditions.

In an interactive product there are similar factors that are externally visible and measurable and those that are hidden from the users' view. For example, exactly why the response time for a query to a database (or a web page) is, say, 4 seconds will almost certainly depend on technical decisions made when the database was constructed, but from the users' viewpoint the important observation is the fact that it does take 4 seconds to respond.

In interaction design, the way in which the users interact with the product is considered the driving force behind the design and so we concentrate on the externally visible and measurable behavior. Detailed internal workings are important only to the extent that they affect the external behavior. This does not mean that design decisions concerning a system's internal behavior are any less important: however, the tasks that the user will perform should influence design decisions no less than technical issues.

So, one answer to the question posed above is that we choose between alternative designs by letting users and stakeholders interact with them and by discussing their experiences, preferences and suggestions for improvement. This is fundamental to a user-centered approach to development. This in turn means that the designs must be available in a form that can be reasonably evaluated with users, not in technical jargon or notation that seems impenetrable to them.

One form traditionally used for communicating a design is documentation, e.g., a description of how something will work or a diagram showing its components. The trouble is that a static description cannot capture the dynamics of behavior, and for an interaction device we need to communicate to the users what it will be like to actually operate it.

In many design disciplines, *prototyping* is used to overcome potential client misunderstandings and to test the technical feasibility of a suggested design and its production. Prototyping involves producing a limited version of the product with the purpose of answering specific questions about the design's feasibility or appropriateness. Prototypes give a better impression of the user experience than simple descriptions can ever do, and there are different kinds of prototyping that are suitable for different stages of development and for eliciting different kinds of information. One experience illustrating the benefits of prototyping is described in Box 6.2. So one important aspect of choosing among alternatives is that prototypes should be built and evaluated by users. We'll revisit the issue of prototyping in Chapter 8.

Another basis on which to choose between alternatives is "quality," but this requires a clear understanding of what "quality" means. People's views of what is a quality product vary, and we don't always write it down. Whenever we use anything we have some notion of the level of quality we are expecting, wanting, or needing. Whether this level of quality is expressed formally or informally does not matter. The point is that it exists and we use it consciously or subconsciously to evaluate alternative items. For example, if you have to wait too long to download

BOX 6.2 The Value of Prototyping

I learned the value of a prototype through a very effective role-playing exercise. I was on a course designed to introduce new graduates to different possible careers in industry. One of the themes was production and manufacturing and the aim of one group exercise was to produce a notebook. Each group was told that it had 30 minutes to deliver 10 books to the person in charge. Groups were given various pieces of paper, scissors, sticky tape, staples, etc., and told to organize ourselves as best we could. So my group set to work organizing ourselves into a production line, with one of us cutting up the paper, another stapling the pages together, another sealing the binding with the sticky tape, and so on. One person was even in charge of quality assurance. It took us less than 10 minutes to produce the 10 books, and we rushed off with our delivery. When we showed the person in

charge, he replied, "That's not what I wanted, I need it bigger than that." Of course, the size of the notebook wasn't specified in the description of the task, so we found out how big he wanted it, got some more materials, and scooted back to produce 10 more books. Again, we set up our production line and produced 10 books to the correct size. On delivery we were again told that it was not what was required: he wanted the binding to work the other way around. This time we got as many of the requirements as we could and went back, developed one book, and took that back for further feedback and refinement before producing the 10 required.

If we had used prototyping as a way of exploring our ideas and checking requirements in the first place, we could have saved so much effort and resource!

a web page, then you are likely to give up and try a different site—you are applying a certain measure of quality associated with the time taken to download the web page. If one cell phone makes it easy to perform a critical function while another involves several complicated key sequences, then you are likely to buy the former rather than the latter. You are applying a quality criterion concerned with efficiency.

Now, if you are the only user of a product, then you don't necessarily have to express your definition of "quality" since you don't have to communicate it to anyone else. However, as we have seen, most projects involve many different stakeholder groups, and you will find that each of them has a different definition of quality and different acceptable limits for it. For example, although all stakeholders may agree on targets such as "response time will be fast" or "the menu structure will be easy to use," exactly what each of them means by this is likely to vary. Disputes are inevitable when, later in development, it transpires that "fast" to one set of stakeholders meant "under a second," while to another it meant "between 2 and 3 seconds." Capturing these different views in clear unambiguous language early in development takes you halfway to producing a product that will be regarded as "good" by all your stakeholders. It helps to clarify expectations, provides a benchmark against which products of the development process can be measured, and gives you a basis on which to choose among alternatives.

The process of writing down formal, verifiable–and hence measurable–usability criteria is a key characteristic of an approach to interaction design called *usability engineering* that has emerged over many years and with various proponents (Whiteside et al., 1988; Nielsen, 1993). Usability engineering involves specifying quantifiable measures of product performance, documenting them in a usability specification, and assessing the product against them. One way in which this approach is used is to make changes to subsequent versions of a system based on feedback from carefully documented results of usability tests for the earlier version. We shall return to this idea later when we discuss evaluation.

ACTIVITY 6.4   Consider the calendar system that you designed in Activity 6.1. Suggest some usability criteria that you could use to determine the calendar's quality. You will find it helpful to think in terms of the usability goals introduced in Chapter 1: effectiveness, efficiency, safety, utility, learnability, and memorability. Be as specific as possible. Check your criteria by considering exactly what you would measure and how you would measure its performance.

Having done that, try to do the same thing for the user experience goals introduced in Chapter 1; these relate to whether a system is satisfying, enjoyable, motivating, rewarding, and so on.

Comment   Finding measurable characteristics for some of these is not easy. Here are some suggestions, but you may have found others. Note that the criteria must be measurable and very specific.

- *Effectiveness*: Identifying measurable criteria for this goal is particularly difficult since it is a combination of the other goals. For example, does the system support you in keeping appointments, taking notes, and so on. In other words, is the calendar used?
- *Efficiency*: Assuming that there is a search facility in the calendar, what is the response time for finding a specific day or a specific appointment?
- *Safety*: How often does data get lost or does the user press the wrong button? This may be measured, for example, as the number of times this happens per hour of use.
- *Utility*: How many functions offered by the calendar are used every day, how many every week, how many every month? How many tasks are difficult to complete in a reasonable time because functionality is missing or the calendar doesn't support the right subtasks?
- *Learnability*: How long does it take for a novice user to be able to do a series of set tasks, e.g., make an entry into the calendar for the current date, delete an entry from the current date, edit an entry in the following day?
- *Memorability*: If the calendar isn't used for a week, how many functions can you remember how to perform? How long does it take you to remember how to perform your most frequent task?

Finding measurable characteristics for the user experience criteria is even harder, though. How do you measure satisfaction, fun, motivation or aesthetics? What is entertaining to one person may be boring to another; these kinds of criteria are subjective, and so cannot be measured objectively.

6.4   Lifecycle models: showing how the activities are related

Understanding what activities are involved in interaction design is the first step to being able to do it, but it is also important to consider how the activities are related to one another so that the full development process can be seen. The term *lifecycle model*[1] is used to represent a model that captures a set of activities and how they are related. Sophisticated models also incorporate a description of when and how to move from one activity to the next and a description of the deliverables for each activity. The reason such models are popular is that they allow developers, and particularly managers, to get an overall view of the development effort so that progress can be tracked, deliverables specified, resources allocated, targets set, and so on.

Existing models have varying levels of sophistication and complexity. For projects involving only a few experienced developers, a simple process would probably be adequate. However, for larger systems involving tens or hundreds of developers with hundreds or thousands of users, a simple process just isn't enough to provide the management structure and discipline necessary to engineer a usable product. So something is needed that will provide more formality and more discipline. Note that this does not mean that innovation is lost or that creativity is stifled. It just means that a structured process is used to provide a more stable framework for creativity.

However simple or complex it appears, any lifecycle model is a simplified version of reality. It is intended as an abstraction and, as with any good abstraction, only the amount of detail required for the task at hand should be included. Any organization wishing to put a lifecycle model into practice will need to add detail specific to its particular circumstances and culture. For example, Microsoft wanted to maintain a small-team culture while also making possible the development of very large pieces of software. To this end, they have evolved a process that has been called "synch and stabilize," as described in Box 6.3.

In the next subsection, we introduce our view of what a lifecycle model for interaction design might look like that incorporates the four activities and the three key characteristics of the interaction design process discussed above. This will form the basis of our discussion in Chapters 7 and 8. Depending on the kind of system being developed, it may not be possible or appropriate to follow this model for every element of the system, and it is certainly true that more detail would be required to put the lifecycle into practice in a real project.

Many other lifecycle models have been developed in fields related to interaction design, such as software engineering and HCI, and our model is evolved from these ideas. To put our interaction design model into context we include here a description of five lifecycle models, three from software engineering and two from HCI, and consider how they relate to it.

---

[1]Sommerville (2001) uses the term process model to mean what we mean by lifecycle model, and refers to the waterfall model as the software lifecycle. Pressman (1992) talks about paradigms. In HCI the term "lifecycle model" is used more widely. For this reason, and because others use "process model" to represent something that is more detailed than a lifecycle model (e.g., Comer, 1997) we have chosen to use lifecycle model.

BOX 6.3   How Microsoft Builds Software *(Cusumano and Selby, 1997)*

Microsoft is one of the largest software companies in the world and builds some very complex software; for example, Windows 95 contains more than 11 million lines of code and required more than 200 programmers. Over a two-and-a-half-year period from the beginning of 1993, two researchers, Michael Cusumano and Richard Selby, were given access to Microsoft project documents and key personnel for study and interview. Their aim was to build up an understanding of how Microsoft produces software. Rather than adopt the structured software engineering practices others have followed, Microsoft's strategy has been to cultivate entrepreneurial flexibility throughout its software teams. In essence, it has tried to scale up the culture of a loosely-structured, small software team. "The objective is to get many small teams (three to eight developers each) or individual programmers to work together as a single relatively large team in order to build large products relatively quickly while still allowing

individual programmers and teams freedom to evolve their designs and operate nearly autonomously" (p. 54).

In order to maintain consistency and to ensure that products are eventually shipped, the teams synchronize their activities daily and periodically stabilize the whole product. Cusumano and Selby have therefore labeled Microsoft's unique process "synch and stabilize." Figure 6.5 shows an overview of this process, which is divided into three phases: the planning phase, the development phase and the stabilization phase.

The planning phase begins with a vision statement that defines the goals of the new product and the user activities to be supported by the product. (Microsoft uses a method called activity-based planning to identify and prioritize the features to be built; we return to this in Chapter 9.) The program managers together with the developers then write a functional specification in enough detail to describe features and to develop schedules and al-

---

Planning Phase Define product vision, specifications, and schedule

• **Vision Statement** Product and program management use extensive customer input to identify and priority-order product features.

• **Specification Document** Based on vision statement, program management and development group define feature functionality, architectural issues, and component interdependencies.

• **Schedule and Feature Team Formation** Based on specification document, program management coordinates schedule and arranges feature teams that each contain approximately 1 program manager, 3–8 developers, and 3–8 testers (who work in parallel 1:1 with developers).

Development Phase Feature development in 3 or 4 sequential subprojects that each results in a milestone release

Program managers coordinate evolution of specification. Developers design, code, and debug. Testers pair with developers for continuous testing.

• **Subproject I** First 1/3 of features (Most critical features and shared components)

• **Subproject II** Second 1/3 of features

• **Subproject III** Final 1/3 of features (Least critical features)

Stabilization Phase Comprehensive internal and external testing, final product stabilization, and ship

Program managers coordinate OEMs and ISVs and monitor customer feedback. Developers perform final debugging and code stabilization. Testers recreate and isolate errors.

• **Internal Testing** Thorough testing of complete product within the company.

• **External Testing** Thorough testing of complete product outside the company by "beta" sites, such as OEMs, ISVs, and end users.

• **Release preparation** Prepare final release of "golden master" disks and documentation for manufacturing.

Figure 6.5 Overview of the synch and stabilize development approach.

locate staff. The feature list in this document will change by about 30% during the course of development, so the list is not fixed at this time. In the next phase, the development phase, the feature list is divided into three or four parts, each with its own small development team, and the schedule is divided into sequential subprojects, each with its own deadline (milestone). The teams work in parallel on a set of features and synchronize their work by putting together their code and finding errors on a daily and weekly basis. This is necessary because many programmers may be working on the same code at once. For example, during the peak development of Excel 3.0, 34 developers were actively changing the same source code on a daily basis.

At the end of a subproject, i.e., on reaching a milestone, all errors are found and fixed, thus stabilizing the product, before moving on to the next subproject and eventually to the final milestone, which represents the release date. Figure 6.6 shows an overview of the milestone structure for a project with three subprojects. This synch-and-stabilize approach has been used to develop Excel, Office, Publisher, Windows 95, Windows NT, Word, and Works, among others.

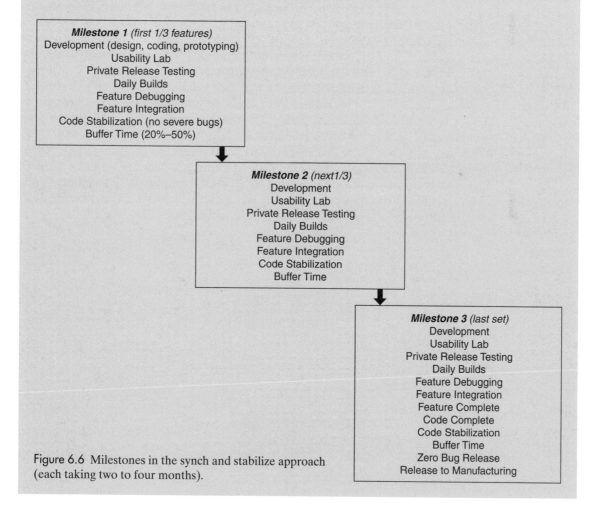

Figure 6.6 Milestones in the synch and stabilize approach (each taking two to four months).

6.4.1   A simple lifecycle model for interaction design

We see the activities of interaction design as being related as shown in Figure 6.7. This model incorporates iteration and encourages a user focus. While the outputs from each activity are not specified in the model, you will see in Chapter 7 that our description of establishing requirements includes the need to identify specific usability criteria.

The model is not intended to be prescriptive; that is, we are not suggesting that this is how all interactive products are or should be developed. It is based on our observations of interaction design and on information we have gleaned in the research for this book. It has its roots in the software engineering and HCI lifecycle models described below, and it represents what we believe is practiced in the field.

Most projects start with identifying needs and requirements. The project may have arisen because of some evaluation that has been done, but the lifecycle of the new (or modified) product can be thought of as starting at this point. From this activity, some alternative designs are generated in an attempt to meet the needs and requirements that have been identified. Then interactive versions of the designs are developed and evaluated. Based on the feedback from the evaluations, the team may need to return to identifying needs or refining requirements, or it may go straight into redesigning. It may be that more than one alternative design follows this iterative cycle in parallel with others, or it may be that one alternative at a time is considered. Implicit in this cycle is that the final product will emerge in an evolutionary fashion from a rough initial idea through to the finished product. Exactly how this evolution happens may vary from project to project, and we return to this issue in Chapter 8. The only factor limiting the number of times through the cycle is the resources available, but whatever the number is, development ends with an evaluation activity that ensures the final product meets the prescribed usability criteria.

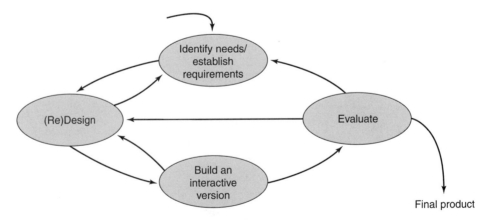

Figure 6.7 A simple interaction design model.

6.4.2 Lifecycle models in software engineering

Software engineering has spawned many lifecycle models, including the waterfall, the spiral, and rapid applications development (RAD). Before the waterfall was first proposed in 1970, there was no generally agreed approach to software development, but over the years since then, many models have been devised, reflecting in part the wide variety of approaches that can be taken to developing software. We choose to include these specific lifecycle models for two reasons: First, because they are representative of the models used in industry and they have all proved to be successful, and second, because they show how the emphasis in software development has gradually changed to include a more iterative, user-centered view.

The waterfall lifecycle model

The waterfall lifecycle was the first model generally known in software engineering and forms the basis of many lifecycles in use today. This is basically a linear model in which each step must be completed before the next step can be started (see Figure 6.8). For example, requirements analysis has to be completed before

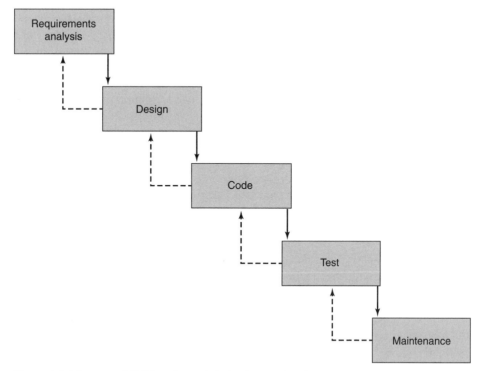

Figure 6.8 The waterfall lifecycle model of software development.

design can begin. The names given to these steps varies, as does the precise definition of each one, but basically, the lifecycle starts with some requirements analysis, moves into design, then coding, then implementation, testing, and finally maintenance. One of the main flaws with this approach is that requirements change over time, as businesses and the environment in which they operate change rapidly. This means that it does not make sense to freeze requirements for months, or maybe years, while the design and implementation are completed.

Some feedback to earlier stages was acknowledged as desirable and indeed practical soon after this lifecycle became widely used (Figure 6.8 does show some limited feedback between phases). But the idea of iteration was not embedded in the waterfall's philosophy. Some level of iteration is now incorporated in most versions of the waterfall, and review sessions among developers are commonplace. However, the opportunity to review and evaluate with *users* was not built into this model.

The spiral lifecycle model

For many years, the waterfall formed the basis of most software developments, but in 1988 Barry Boehm (1988) suggested the spiral model of software development (see Figure 6.9). Two features of the spiral model are immediately clear from Figure 6.9: risk analysis and prototyping. The spiral model incorporates them in an iterative framework that allows ideas and progress to be repeatedly checked and evaluated. Each iteration around the spiral may be based on a different lifecycle model and may have different activities.

In the spiral's case, it was not the need for user involvement that inspired the introduction of iteration but the need to identify and control risks. In Boehm's approach, development plans and specifications that are focused on the risks involved in developing the system drive development rather than the intended functionality, as was the case with the waterfall. Unlike the waterfall, the spiral explicitly encourages alternatives to be considered, and steps in which problems or potential problems are encountered to be re-addressed.

The spiral idea has been used by others for interactive devices (see Box 6.4). A more recent version of the spiral, called the WinWin spiral model (Boehm et al., 1998), explicitly incorporates the identification of key stakeholders and their respective "win" conditions, i.e., what will be regarded as a satisfactory outcome for each stakeholder group. A period of stakeholder negotiation to ensure a "win-win" result is included.

Rapid Applications Development (RAD)

During the 1990s the drive to focus upon users became stronger and resulted in a number of new approaches to development. The Rapid Applications Development (RAD) approach attempts to take a user-centered view and to minimize the risk caused by requirements changing during the course of the project. The ideas be-

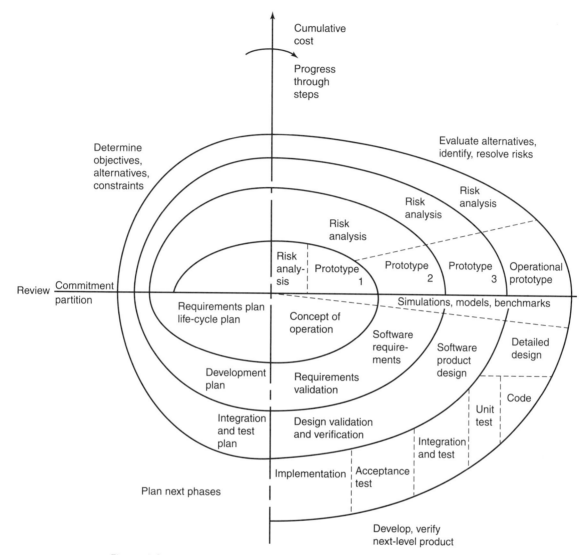

Figure 6.9 The spiral lifecycle model of software development.

hind RAD began to emerge in the early 1990s, also in response to the inappropriate nature of the linear lifecycle models based on the waterfall. Two key features of a RAD project are:

- Time-limited cycles of approximately six months, at the end of which a system or partial system must be delivered. This is called time-boxing. In effect, this breaks down a large project into many smaller projects that can deliver products incrementally, and enhances flexibility in terms of the development techniques used and the maintainability of the final system.

- JAD (Joint Application Development) workshops in which users and developers come together to thrash out the requirements of the system (Wood and Silver, 1995). These are intensive requirements-gathering sessions in which difficult issues are faced and decisions are made. Representatives from each identified stakeholder group should be involved in each workshop so that all the relevant views can be heard.

A basic RAD lifecycle has five phases (see Figure 6.10): project set-up, JAD workshops, iterative design and build, engineer and test final prototype, implementation review. The popularity of RAD has led to the emergence of an industry-standard RAD-based method called DSDM (Dynamic Systems Development Method) (Millington and Stapleton, 1995). This was developed by a non-profit-making DSDM consortium made up of a group of companies that recognized the need for some standardization in the field. The first of nine principles stated as underlying DSDM is that "active user involvement is imperative." The DSDM lifecycle is more complicated than the one we've shown here. It involves five phases: feasibility study, business study, functional model iteration, design and build iteration, and implementation. This is only a generic process and must be tailored for a particular organization.

ACTIVITY 6.5   How closely do you think the RAD lifecycle model relates to the interaction design model described in Section 6.4.1?

Comment   RAD and DSDM explicitly incorporate user involvement, evaluation and iteration. User involvement, however, appears to be limited to the JAD workshop, and iteration appears to be limited to the design and build phase. The philosophy underlying the interaction design model is present, but the flexibility appears not to be. Our interaction design process would be appropriately used within the design and build stage.

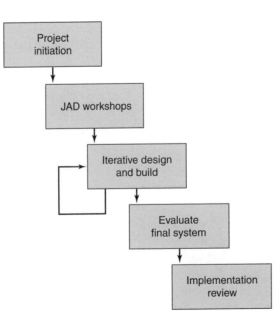

Figure 6.10 A basic RAD lifecycle model of software development.

BOX 6.4   A Product Design Process for Internet Appliances

Netpliance, which has moved into the market of providing Internet appliances, i.e. one-stop products that allow a user to achieve a specific Internet-based task, have adopted a user-centered approach to development based on RAD (Isensee et al., 2000). They attribute their ability to develop systems from concept to delivery in seven months to this strong iterative approach: the architecture was revised and iterated over several days; the code was developed with weekly feedback sessions from users; components were typically revised four times, but some went through 12 cycles. Their simple spiral model is shown in Figure 6.11.

The target audience for this appliance, called the i-opener, were people who did not use or own a PC and who may have been uncomfortable around computers. The designers were therefore looking to design something that would be as far away from the "traditional" PC model as possible in terms of both hardware and software. In designing the software, they abandoned the desktop metaphor of the Windows operating system and concentrated on an interface that provided good support for the user's task. For the hardware design, they needed to get away from the image of a large heavy box with lots of wires and plugs, any one of which may be faulty and cause the user problems.

The device provides three functions: sending and receiving email, categorical content, and web accessibility. That is it. There are no additional features, no complicated menus and options. The device is streamlined to perform these tasks and no more. This choice of functions was based on user studies and testing that served to identify the most frequently used functions, i.e., those that most appropriately supported the users. An example screen showing the news channel for i-opener is shown in Figure 6.12.

Identifying requirements for a new device is difficult. There is no direct experience of using a similar product, and so it is difficult to know what will be used, what will be needed, what will be frustrating, and what will be ignored. The Netpliance team started to gather information for their device by focusing on existing data about PC users: demographics, usability studies, areas of dissatisfaction, etc. They employed marketing research, focus groups, and user surveys to identify the key features of the appliance, and concentrated on delivering these fundamentals well.

The team was multidisciplinary and included hardware engineers, user interface designers, marketing specialists, test specialists, industrial designers, and visual designers. Users were involved throughout development and the whole team took an active part in the design. The interface was designed first, to meet user requirements, and then the hardware and software were developed to fit the interface. In all of this, the emphasis was on a lean development process with a minimum of documentation, early prototyping, and frequent iterations for each component. For example, the design of the hardware proceeded from sketches through pictures to physical prototypes that the users could touch, pick up, move around, and so on. To complement prototyping, the team also used usage scenarios, which are basically descriptions of the appliance's use to achieve a task. These helped developers to understand how the product could be used from a user's perspective. We will return to similar techniques in Chapter 7.

Implementation was achieved through rapid cycles of implement and test. Small usability tests were conducted throughout implementation to

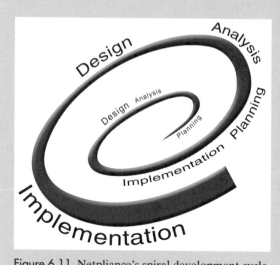

Figure 6.11 Netpliance's spiral development cycle.

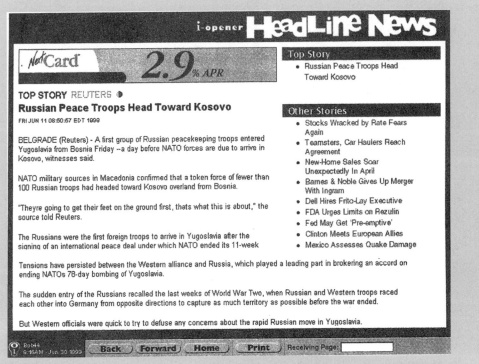

Figure 6.12 The news channel as part of the categorical content.

find and fix usability problems. Developers and their families or friends were encouraged to use the appliance so that designers could enjoy the same experience as the users (called "eating your own dogfood"!). For these field tests, the product was instrumented so that the team could monitor how often each function was used. This data helped to prioritize the development of features as the product release deadline approached.

6.4.3   Lifecycle models in HCI

Another of the traditions from which interaction design has emerged is the field of HCI (human–computer interaction). Fewer lifecycle models have arisen from this field than from software engineering and, as you would expect, they have a stronger tradition of user focus. We describe two of these here. The first one, the Star, was derived from empirical work on understanding how designers tackled HCI design problems. This represents a very flexible process with evaluation at its core. In contrast, the second one, the usability engineering lifecycle, shows a more structured approach and hails from the usability engineering tradition.

The Star Lifecycle Model

About the same time that those involved in software engineering were looking for alternatives to the waterfall lifecycle, so too were people involved in HCI looking for alternative ways to support the design of interfaces. In 1989, the Star lifecycle

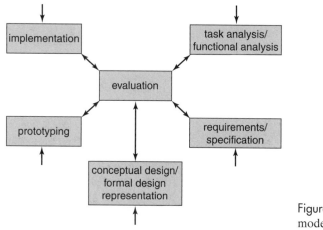

Figure 6.13 The Star lifecycle model.

model was proposed by Hartson and Hix (1989) (see Figure 6.13). This emerged from some empirical work they did looking at how interface designers went about their work. They identified two different modes of activity: analytic mode and synthetic mode. The former is characterized by such notions as top-down, organizing, judicial, and formal, working from the systems view towards the user's view; the latter is characterized by such notions as bottom-up, free-thinking, creative and *ad hoc,* working from the user's view towards the systems view. Interface designers move from one mode to another when designing. A similar behavior has been observed in software designers (Guindon, 1990).

Unlike the lifecycle models introduced above, the Star lifecycle does not specify any ordering of activities. In fact, the activities are highly interconnected: you can move from any activity to any other, provided you first go through the evaluation activity. This reflects the findings of the empirical studies. Evaluation is central to this model, and whenever an activity is completed, its result(s) must be evaluated. So a project may start with requirements gathering, or it may start with evaluating an existing situation, or by analyzing existing tasks, and so on.

ACTIVITY 6.6 The Star lifecycle model has not been used widely and successfully for large projects in industry. Consider the benefits of lifecycle models introduced above and suggest why this may be.

Comment One reason may be that the Star lifecycle model is extremely flexible. This may be how designers work in practice, but as we commented above, lifecycle models are popular because "they allow developers, and particularly managers, to get an overall view of the development effort so that progress can be tracked, deliverables specified, resources allocated, targets set, and so on." With a model as flexible as the Star lifecycle, it is difficult to control these issues without substantially changing the model itself.

The Usability Engineering Lifecycle

The Usability Engineering Lifecycle was proposed by Deborah Mayhew in 1999 (Mayhew, 1999). Many people have written about usability engineering, and as

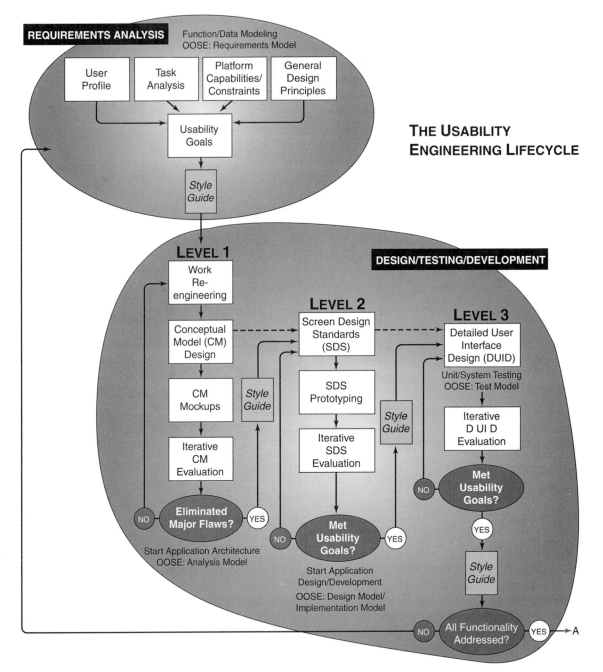

Figure 6.14 The Usability Engineering Lifecycle.

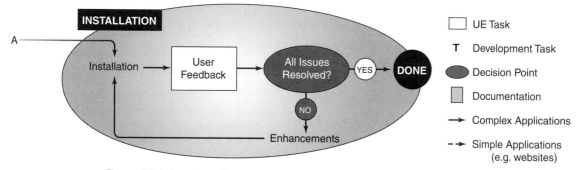

Figure 6.14 (*continued*).

Mayhew herself says, "I did not invent the concept of a Usability Engineering Lifecycle. Nor did I invent any of the Usability Engineering tasks included in the lifecycle . . . .". However, what her lifecycle does provide is a holistic view of usability engineering and a detailed description of how to perform usability tasks, and it specifies how usability tasks can be integrated into traditional software development lifecycles. It is therefore particularly helpful for those with little or no expertise in usability to see how the tasks may be performed alongside more traditional software engineering activities. For example, Mayhew has linked the stages with a general development approach (rapid prototyping) and a specific method (object-oriented software engineering (OOSE, Jacobson et al, 1992)) that have arisen from software engineering.

The lifecycle itself has essentially three tasks: requirements analysis, design/testing/development, and installation, with the middle stage being the largest and involving many subtasks (see Figure 6.14). Note the production of a set of usability goals in the first task. Mayhew suggests that these goals be captured in a style guide that is then used throughout the project to help ensure that the usability goals are adhered to.

This lifecycle follows a similar thread to our interaction design model but includes considerably more detail. It includes stages of identifying requirements, designing, evaluating, and building prototypes. It also explicitly includes the style guide as a mechanism for capturing and disseminating the usability goals of the project. Recognizing that some projects will not require the level of structure presented in the full lifecycle, Mayhew suggests that some substeps can be skipped if they are unnecessarily complex for the system being developed.

ACTIVITY 6.7 Study the usability engineering lifecycle and identify how this model differs from our interaction design model described in Section 6.4.1, in terms of the iterations it supports.

Comment One of the main differences between Mayhew's model and ours is that in the former the iteration between design and evaluation is contained within the second phase. Iteration between the design/test/development phase and the requirements analysis phase occurs only after the conceptual model and the detailed designs have been developed, prototyped, and evaluated one at a time. Our version models a return to the activity of identifying needs and establishing requirements after evaluating any element of the design.

Assignment

Nowadays, timepieces (such as clocks, wristwatches etc) have a variety of functions. They not only tell the time and date but they can speak to you, remind you when it's time to do something, and provide a light in the dark, among other things. Mostly, the interface for these devices, however, shows the time in one of two basic ways: as a digital number such as 23:40 or through an analog display with two or three hands—one to represent the hour, one for the minutes, and one for the seconds.

In this assignment, we want you to design an innovative timepiece for your own use. This could be in the form of a wristwatch, a mantelpiece clock, an electronic clock, or any other kind of clock you fancy. Your goal is to be inventive and exploratory. We have broken this assignment down into the following steps to make it clearer:

(a) Think about the interactive product you are designing: what do you want it to do for you? Find 3–5 potential users and ask them what they would want. Write a list of requirements for the clock, together with some usability criteria based on the definition of usability used in Chapter 1.

(b) Look around for similar devices and seek out other sources of inspiration that you might find helpful. Make a note of any findings that are interesting, useful or insightful.

(c) Sketch out some initial designs for the clock. Try to develop at least two distinct alternatives that both meet your set of requirements.

(d) Evaluate the two designs, using your usability criteria and by role playing an interaction with your sketches. Involve potential users in the evaluation, if possible. Does it do what you want? Is the time or other information being displayed always clear?

Design is iterative, so you may want to return to earlier elements of the process before you choose one of your alternatives.

Once you have a design with which you are satisfied, you can send it to us and we shall post a representative sample of those we receive to our website. Details of how to format your submission are available from our website.

Summary

In this chapter, we have looked at the process of interaction design, i.e., what activities are required in order to design an interactive product, and how lifecycle models show the relationships between these activities. A simple interaction design model consisting of four activities was introduced and issues surrounding the identification of users, generating alternative designs, and evaluating designs were discussed. Some lifecycle models from software engineering and HCI were introduced.

Key points

- The interaction design process consists of four basic activities: identifying needs and establishing requirements, developing alternative designs that meet those requirements, building interactive versions of the designs so that they can be communicated and assessed, and evaluating them.

- Key characteristics of the interaction design process are explicit incorporation of user involvement, iteration, and specific usability criteria.

- Before you can begin to establish requirements, you must understand who the users are and what their goals are in using the device.

- Looking at others' designs provides useful inspiration and encourages designers to consider alternative design solutions, which is key to effective design.

- Usability criteria, technical feasibility, and users' feedback on prototypes can all be used to choose among alternatives.

- Prototyping is a useful technique for facilitating user feedback on designs at all stages.

- Lifecycle models show how development activities relate to one another.

- The interaction design process is complementary to lifecycle models from other fields.

Further reading

RUDISILL, M., LEWIS, C., POLSON, P. B., AND MCKAY, T. D. (1995) (eds.) *Human-Computer Interface Design: Success Stories, Emerging Methods, Real-World Context*. San Francisco: Morgan Kaufmann. This collection of papers describes the application of different approaches to interface design. Included here is an account of the Xerox Star development, some advice on how to choose among methods, and some practical examples of real-world developments.

BERGMAN, ERIC (2000) (ed.) *Information Appliances and Beyond*. San Francisco: Morgan Kaufmann. This book is an edited collection of papers which report on the experience of designing and building a variety of 'information appliances', i.e., purpose-built computer-based products which perform a specific task. For example, the Palm Pilot, mobile telephones, a vehicle navigation system, and interactive toys for children.

MAYHEW, DEBORAH J. (1999) *The Usability Engineering Lifecycle*. San Francisco: Morgan Kaufmann. This is a very practical book about product user interface design. It explains how to perform usability tasks throughout development and provides useful examples along the way to illustrate the techniques. It links in with two software development based methods: rapid prototyping and object-oriented software engineering.

SOMMERVILLE, IAN (2001) *Software Engineering* (6th edition). Harlow, UK: Addison-Wesley. If you are interested in pursuing the software engineering aspects of the lifecycle models section, then this book provides a useful overview of the main models and their purpose.

NIELSEN, JAKOB (1993) *Usability Engineering*. San Francisco: Morgan Kaufmann. This is a seminal book on usability engineering. If you want to find out more about the philosophy, intent, history, or pragmatics of usability engineering, then this is a good place to start.

INTERVIEW with Gillian Crampton Smith

Gillian Crampton Smith is Director of the Interaction Design Institute Ivrea near Milan, Italy.

Prior to this, she was at the Royal College of Art where she started and directed the Computer Related Design Department, developing a program to enable artist-designers to develop and apply their traditional skills and knowledge to the design of all kinds of interactive products and systems.

GC: I believe that things should work but they should also delight. In the past, when it was really difficult to make things work, that was what people concentrated on. But now it's much easier to make software and much easier to make hardware. We've got a load of technologies but they're still often not designed for people—and they're certainly not very enjoyable to use. If we think about other things in our life, our clothes, our furniture, the things we eat with, we choose what we use because they have a meaning beyond their practical use. Good design is partly about working really well, but it's also about what something looks like, what it reminds us of, what it refers to in our broader cultural environment. It's this side that interactive systems haven't really addressed yet. They're only just beginning to become part of culture. They are not just a tool for professionals any more, but an environment in which we live.

HS: How do you think we can improve things?

GC: The parallel with architecture is quite an interesting one. In architecture, a great deal of time and expense is put into the initial design; I don't think very much money or time is put into the initial design of software. If you think of the big software engineering companies, how many people work in the design side rather than on the implementation side?

HS: When you say design do you mean conceptual design, or task design, or something else?

GC: I mean all phases of design. Firstly there's research—finding out about people. This is not necessarily limited to finding out about what they want necessarily, because if we're designing new things, they are probably things people don't even know they

could have. At the Royal College of Art we tried to work with users, but to be inspired by them, and not constrained by what they know is possible.

The second stage is thinking, "What should this thing we are designing do?" You could call that conceptual design. Then a third stage is thinking how do you represent it, how do you give it form? And then the fourth stage is actually crafting the interface—exactly what color is this pixel? Is this type the right size, or do you need a size bigger? How much can you get on a screen?—all those things about the details.

One of the problems companies have is that the feedback they get is. "I wish it did x." Software looks as if it's designed, not with a basic model of how it works that is then expressed on the interface, but as a load of different functions that are strung together. The desktop interface, although it has great advantages, encourages the idea that you have a menu and you can just add a few more bits when people want more things. In today's word processors, for instance, there isn't a clear conceptual model about how it works, or an underlying theory people can use to reason about why it is not working in the way they expect.

HS: So in trying to put more effort into the design aspect of things, do you think we need different people in the team?

GC: Yes. People in the software field tend to think that designers are people who know how to give the product form, which of course is one of the things they do. But a graphic designer, for instance, is somebody who also thinks at a more strategic level, "What is the message that these people want to get over and to whom?" and then, "What is the best way to give form to a message like that?" The part you see is the beautiful design, the lovely poster or record sleeve, or elegant book, but behind that is a lot of thinking about how to communicate ideas via a particular medium.

HS: If you've got people from different disciplines, have you experienced difficulties in communication?

GC: Absolutely. I think that people from different disciplines have different values, so different results and different approaches are valued. People have different temperaments, too, that have led them to the different fields in the first place, and they've been trained in different ways. In my view the big difference between the way engineers are trained and the way designers are trained is that engineers are trained to focus in on a solution from the beginning whereas designers are trained to focus out to begin with and then focus in. They focus out and try lots of different alternatives, and they pick some and try them out to see how they go. Then they refine down. This is very hard for both the engineers and the designers because the designers are thinking the engineers are trying to hone in much too quickly and the engineers can't bear the designers faffing about. They are trained to get their results in a completely different way.

HS: Is your idea to make each more tolerant of the other?

GC: Yes, my idea is not to try to make renaissance people, as I don't think it's feasible. Very few people can do everything well. I think the ideal team is made up of people who are really confident and good at what they do and open-mined enough to realize there are very different approaches. There's the scientific approach, the engineering approach, the design approach. All three are different and that's their value—you don't want everybody to be the same. The best combination is where you have engineers who understand design and designers who understand engineering.

It's important that people know their limitations too. If you realize that you need an ergonomist, then you go and find one and you hire them to consult for you. So you need to know what you don't know as well as what you do.

HS: What other aspects of traditional design do you think help with interaction design?

GC: I think the ability to visualize things. It allows people to make quick prototypes or models or sketches so that a group of people can talk about something concrete. I think that's invaluable in the process. I think also making things that people like is just one of the things that good designers have a feel for.

HS: Do you mean aesthetically like or like in its whole sense?

GC: In its whole sense. Obviously there's the aesthetic of what something looks like or feels like but

there's also the aesthetic of how it works as well. You can talk about an elegant way of doing something as well as an elegant look.

HS: Another trait I've seen in designers is being protective of their design.

GC: I think that is both a vice and a virtue. In order to keep a design coherent you need to keep a grip on the whole and to push it through as a whole. Otherwise it can happen that people try to make this a bit smaller and cut bits out of that, and so on, and before you know where you are the coherence of the design is lost. It is quite difficult for a team to hold a coherent vision of a design. If you think of other design fields, like film-making, for instance, there is one director and everybody accepts that it's the director's vision. One of the things that's wrong with products like Microsoft Word, for instance, is that there's no coherent idea in it that makes you think, "Oh yes, I understand how this fits with that."

Design is always a balance between things that work well and things that look good, and the ideal design satisfies everything, but in most designs you have to make trade-offs. If you're making a game it's more important that people enjoy it and that it looks good than to worry if some of it's a bit difficult. If you're making a fighter cockpit then the most important thing is that pilots don't fall out of the sky, and so this informs the trade-offs you make. The question is, who decides how to decide the criteria for the tradeoffs that inevitably need to be made. This is not a matter of engineering: it's a matter of values—cultural, emotional, aesthetic.

HS: I know this is a controversial issue for some designers. Do you think users should be part of the design team?

GC: No, I don't. I think it's an abdication of responsibility. Users should definitely be involved as a source of inspiration, suggesting ideas, evaluating proposals—saying, "Yes, we think this would be great" or "No, we think this is an appalling idea." But in the end, if designers aren't better than the general public at designing things, what are they doing as designers?

Chapter 7

Identifying needs and establishing requirements

7.1 Introduction
7.2 What, how, and why?
    7.2.1 What are we trying to achieve in this design activity?
    7.2.2 How can we achieve this?
    7.2.3 Why bother? The importance of getting it right
    7.2.4 Why *establish* requirements?
7.3 What are requirements?
    7.3.1 Different kinds of requirements
7.4 Data gathering
    7.4.1 Data-gathering techniques
    7.4.2 Choosing between techniques
    7.4.3 Some basic data-gathering guidelines
7.5 Data interpretation and analysis
7.6 Task description
    7.6.1 Scenarios
    7.6.2 Use cases
    7.6.3 Essential use cases
7.7 Task analysis
    7.7.1 Hierarchical Task Analysis (HTA)

7.1 Introduction

An interaction design project may aim to replace or update an established system, or it may aim to develop a totally innovative product with no obvious precedent. There may be an initial set of requirements, or the project may have to begin by producing a set of requirements from scratch. Whatever the initial situation and whatever the aim of the project, the users' needs, requirements, aspirations, and expectations have to be discussed, refined, clarified, and probably re-scoped. This requires an understanding of, among other things, the users and their capabilities, their current tasks and goals, the conditions under which the product will be used, and constraints on the product's performance.

As we discussed in Chapter 6, identifying users' needs is not as straightforward as it sounds. Establishing requirements is also not simply writing a wish list of features. Given the iterative nature of interaction design, isolating requirements activities from design activities and from evaluation activities is a little artificial, since in practice they are all intertwined: some design will take place while requirements are being established, and the design will evolve through a series of evaluation–re-design cycles. However, each of these activities can be distinguished by its own emphasis and its own techniques.

This chapter provides a more detailed overview of identifying needs and establishing requirements. We introduce different kinds of requirements and explain some useful techniques.

The main aims of this chapter are to:

- Describe different kinds of requirements.
- Enable you to identify examples of different kinds of requirements from a simple description.
- Explain how different data-gathering techniques may be used, and enable you to choose among them for a simple description.
- Enable you to develop a "scenario," a "use case," and an "essential use case" from a simple description.
- Enable you to perform hierarchical task analysis on a simple description.

7.2　What, how, and why?

7.2.1　What are we trying to achieve in this design activity?

There are two aims. One aim is to understand as much as possible about the users, their work, and the context of that work, so that the system under development can support them in achieving their goals; this we call "identifying needs." Building on this, our second aim is to produce, from the needs identified, a set of stable requirements that form a sound basis to move forward into thinking about design. This is not necessarily a major document nor a set of rigid prescriptions, but you need to be sure that it will not change radically in the time it takes to do some design and get feedback on the ideas. Because the end goal is to produce this set of requirements, we shall sometimes refer to this as the requirements activity.

7.2.2　How can we achieve this?

The whole chapter is devoted to explaining how to achieve these aims, but first we give an overview of where we're heading.

At the beginning of the requirements activity, we know that we have a lot to find out and to clarify. At the end of the activity we will have a set of stable requirements that can be moved forward into the design activity. In the middle, there are activities concerned with gathering data, interpreting or analyzing[1] the data, and

---

[1]We use *interpretation* to mean the initial investigation of the data, while *analysis* is a more detailed study, using a particular frame of reference and notation.

capturing the findings in a form that can be expressed as requirements. Broadly speaking, these activities progress in a sequential manner: first gather some data, then interpret it, then extract some requirements from it, but it gets a lot messier than this, and the activities influence one another as the process iterates. One of the reasons for this is that once you start to analyze data, you may find that you need to gather some more data to clarify or confirm some ideas you have. Another reason is that the way in which you document your requirements may affect your analysis, since it will enable you to identify and express some aspects more easily than others. For example, using a notation which emphasizes the data-flow characteristics of a situation will lead the analysis to focus on this aspect rather than, for example, on data structure. Analysis requires some kind of framework, theory or hypothesis to provide a frame of reference, however informal, and this will inevitably affect the requirements you extract. To overcome this, it is important to use a complementary set of data-gathering techniques and data-interpretation techniques, and to constantly revise and refine the requirements. As we discuss below, there are different kinds of requirements, and each can be emphasized or de-emphasized by the different techniques.

Identifying needs and establishing requirements is itself an iterative activity in which the subactivities inform and refine one another. It does not last for a set number of weeks or months and then finish. In practice, requirements evolve and develop as the stakeholders interact with designs and see what is possible and how certain facilities can help them. And as shown in the lifecycle model in Chapter 6, the activity itself will be repeatedly revisited.

7.2.3 Why bother? The importance of getting it right

An article published in January 2000 (Taylor, 2000) investigated the causes of IT project failure. The article admits that "there is no single cause of IT project failure," but requirements issues figured highly in the findings. The research involved detailed questioning of 38 IT professionals in the UK. When asked about which project stages caused failure, respondents mentioned "requirements definition" more than any other phase. When asked about cause of failure, "unclear objectives and requirements" was mentioned more than anything else, and for critical success factors, "clear, detailed requirements" was mentioned most often.

As stressed in previous chapters, understanding what the product under development should do and ensuring that it supports stakeholders' needs are critically important activities in any product development. If the requirements are wrong then the product will at best be ignored and at worst be despised by the users, and will cause grief and lost productivity. In either case, the implications for both producer and customer are serious: anxiety and frustration, lost revenue, loss of customer confidence, and so on. However we look at it, getting the requirements of the product wrong is a very bad move and something to be avoided at all costs.

Taking a user-centered approach to development is one way to address this. If users' voices and needs are clearly heard and taken into account, then it is more likely that the end result will meet users' needs and expectations. Involving users isn't always easy, however, and we explore in more detail how to do this effectively in Chapter 9. Here we focus on establishing the requirements, while keeping the emphasis clearly on users' needs.

7.2.4 Why *establish* requirements?

The activity of understanding what a product should do has been given various labels—for example, requirements gathering, requirements capture, requirements elicitation, requirements analysis, and requirements engineering. The first two imply that requirements exist out there and we simply need to pick them up or catch them. "Elicitation" implies that "others" (presumably the clients or users) know the requirements and we have to get them to tell us. Requirements, however, are not that easy to identify. You might argue that, in some cases, customers must know what the requirements are because they know the tasks that need to be performed, and may have asked for a system to be built in the first place. However, they may not have articulated requirements as yet, and even if they have an initial set of requirements, they probably have not explored them in sufficient detail for development to begin.

The term "requirements analysis" is normally used to describe the activity of investigating and analyzing an initial set of requirements that have been gathered, elicited, or captured. Analyzing the information gathered is an important step, since it is this interpretation of the facts, rather than the facts themselves, that inspires the design. Requirements engineering is a better term than the others because it recognizes that developing a set of requirements is an iterative process of evolution and negotiation, and one that needs to be carefully managed and controlled.

We chose the term *establishing* requirements to represent the fact that requirements arise from the data-gathering and interpretation activities and have been established from a sound understanding of the users' needs. This also implies that requirements can be justified by and related back to the data collected.

7.3 What are requirements?

Before we go any further, we need to explain what we mean by a requirement. Intuitively, you probably have some understanding of what a requirement is, but we should be clear. A requirement is a statement about an intended product that specifies what it should do or how it should perform. One of the aims of the requirements activity is to make the requirements as specific, unambiguous, and clear as possible. For example, a requirement for a website might be that the time to download any complete page is less than 5 seconds. Another less precise example might be that teenage girls should find the site appealing. In the case of this latter example, further investigation would be necessary to explore exactly what teenage girls would find appealing. Requirements come in many different forms and at many different levels of abstraction, but we need to make sure that the requirements are as clear as possible and that we understand how to tell when they have been fulfilled. The example requirement shown in Figure 7.1 is expressed using a template from the Volere process (Robertson and Robertson, 1999), which you'll hear more about later in this chapter and in Suzanne Roberston's interview at the end of this

Requirement #: **75**   Requirement Type: **9**   Event/use case #: **6**

Description: **The product shall issue an alert if a weather station fails to transmit readings.**

Rationale: **Failure to transmit readings might indicate that the weather station is faulty and needs maintenance, and that the data used to predict freezing roads may be incomplete.**

Source: **Road Engineers**
Fit Criterion: **For each weather station the product shall communicate to the user when the recorded number of each type of reading per hour is not within the manufacturer's specified range of the expected number of readings per hour.**

Customer Satisfaction: **3**   Customer Dissatisfaction: **5**
Dependencies: **None**   Conflicts: **None**
Supporting Materials: **Specification of Rosa Weather Station**
History: **Raised by GBS, 28 July 99**

Volere
Copyright © Atlantic Systems Guild

Figure 7.1 An example requirement using the Volere template.*

chapter. This template requires quite a bit of information about the requirement itself, including something called a "fit criterion," which is a way of measuring when the solution meets the requirement. In Chapter 6 we emphasized the need to establish specific usability criteria for a product early on in development, and this part of the template encourages this.

7.3.1 Different kinds of requirements

In software engineering, two different kinds of requirements have traditionally been identified: functional requirements, which say what the system should do, and non-functional requirements, which say what constraints there are on the system and its development. For example, a functional requirement for a word processor may be that it should support a variety of formatting styles. This requirement might then be decomposed into more specific requirements detailing the kind of formatting required such as formatting by paragraph, by character, and by document, down to a very specific level such as that character formatting must include 20 typefaces, each with bold, italic, and standard options. A non-functional requirement for a word processor might be that it must be able to run on a variety of platforms such as PCs, Macs and Unix machines. Another might be that it must be able to function on a computer with 64 MB RAM. A different kind of non-functional requirement would be that it must be delivered in six months' time. This represents a constraint on the development activity itself rather than on the product being developed.

If we consider interaction devices in general, other kinds of non-functional requirements become relevant such as physical size, weight, color, and production

---

*See Figure 7.5 for an explanation of these fields.

feasibility. For example, when the PalmPilot was developed (Bergman and Haitani, 2000), an overriding requirement was that it should be physically as small as possible, allowing for the fact that it needed to incorporate batteries and an LCD display. In addition, there were extremely tight constraints on the size of the screen, and that had implications for the number of pixels available to display information. For example, formatting lines or certain typefaces may become infeasible to use if they take up even one extra pixel. Figure 7.2 shows two screen shots from the PalmPilot development. As you can see, removing the line at the left-hand side of the display in the top window released sufficient pixels to display the missing "s" in the bottom window.

Interaction design requires us to understand the functionality required and the constraints under which the product must operate or be developed. However, instead of referring to all requirements that are not functional as simply "non-functional" requirements, we prefer to refine this into further categories. The following is not an exhaustive list of the different requirements we need to be looking out for (see the figure in Suzanne Robertson's interview at the end of this chapter for a more detailed list), nor is it a tight categorization, however, it does illustrate the variety of requirements that need to be captured.

Functional requirements capture what the product should do. For example, a functional requirement for a smart fridge might be that it should be able to tell when the butter tray is empty. Understanding the functional requirements for an interactive product is very important.

Data requirements capture the type, volatility, size/amount, persistence, accuracy, and value of the amounts of the required data. All interactive devices have to handle greater or lesser amounts of data. For example, if the system under consid-

Figure 7.2   Every pixel counts.

eration is to operate in the share-dealing application domain, then the data must be up-to-date and accurate, and is likely to change many times a day. In the personal banking domain, data must be accurate, must persist over many months and probably years, is very valuable, and there is likely to be a lot of it.

Environmental requirements or *context of use* refer to the circumstances in which the interactive product will be expected to operate. Four aspects of the environment must be considered when establishing requirements. First is the physical environment such as how much lighting, noise, and dust is expected in the operational environment. Will users need to wear protective clothing, such as large gloves or headgear, that might affect the choice of interaction paradigm? How crowded is the environment? For example, an ATM operates in a very public physical environment. Using speech to interact with the customer is therefore likely to be problematic.

The second aspect of the environment is the social environment. The issues raised in Chapter 4 regarding the social aspects of interaction design, such as collaboration and coordination, need to be explored in the context of the current development. For example, will data need to be shared? If so, does the sharing have to be synchronous, e.g., does everyone need to be viewing the data at once, or asynchronous, e.g., two people authoring a report take turns in editing and adding to it? Other factors include the physical location of fellow team members, e.g., do collaborators have to communicate across great distances?

The third aspect is the organizational environment, e.g., how good is user support likely to be, how easily can it be obtained, and are there facilities or resources for training? How efficient or stable is the communications infrastructure? How hierarchical is the management? and so on.

Finally, the technical environment will need to be established: for example, what technologies will the product run on or need to be compatible with, and what technological limitations might be relevant?

User requirements capture the characteristics of the intended user group. In Chapter 6 we mentioned the relevance of a user's abilities and skills, and these are an important aspect of user requirements. But in addition to these, a user may be a novice, an expert, a casual, or a frequent user. This affects the ways in which interaction is designed. For example, a novice user will require step-by-step instructions, probably with prompting, and a constrained interaction backed up with clear information. An expert, on the other hand, will require a flexible interaction with more wide-ranging powers of control. If the user is a frequent user, then it would be important to provide short cuts such as function keys rather than expecting them to type long commands or to have to navigate through a menu structure. A casual or infrequent user, rather like a novice, will require clear instructions and easily understood prompts and commands, such as a series of menus. The collection of attributes for a "typical user" is called a *user profile*. Any one device may have a number of different user profiles.

Note that user requirements are not the same as usability requirements. We discuss the latter below.

Usability requirements capture the usability goals and associated measures for a particular product. In Chapter 6 we introduced the idea of usability engineering,

BOX 7.1   Underwater PCs

Developing a PC for undersea divers to take underwater has one major environmental factor: it's surrounded by water! However, waterproofing is not the main issue for the designers at WetPC, a company who have produced such a system. The interface has proved to be more of a problem. Divers typically have only one hand free to operate the computer, and are likely to be swimming and moving up and down in the water at the same time. So a traditional interface design is no good. Early prototypes of the computer used voice recognition, but the bubbles made too much noise and distorted the sound. Tracker balls were also inappropriate because the divers are not working on a flat surface. So the main developer at WetPC, Bruce Macdonald, devised a "keyboard" called a KordGrip that has five keys (see Figure 7.3(a)). Combinations of keys represent different symbols, so that divers can choose items from menus. They can perform operations such as controlling a camera and sending messages. The system is also linked to a GPS system that tells the divers where they are. This makes it much easier to mark the location of mines and other underwater discoveries (See Figure 7.3(b) on Color Plate 8).

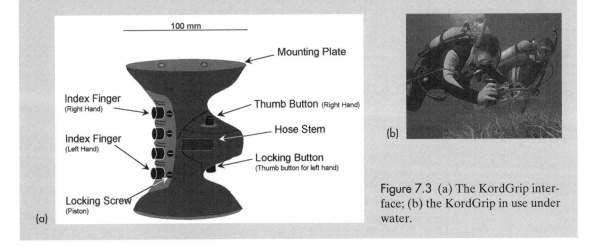

Figure 7.3 (a) The KordGrip interface; (b) the KordGrip in use under water.

an approach in which specific measures for the usability goals of the product are established and agreed upon early in the development process and are then revisited, and used to track progress as development proceeds. This both ensures that usability is given due priority and facilitates progress tracking. In Chapter 1 we described a number of usability goals: effectiveness, efficiency, safety, utility, learnability, and memorability. If we are to follow the philosophy of usability engineering and meet these usability goals, then we must identify the appropriate requirements. Chapter 1 also described some user experience goals, such as making products that are fun, enjoyable, pleasurable, aesthetically pleasing, and motivating. As we observed in Chapter 6, it is harder to identify quantifiable measures that allow us to track these qualities, but an understanding of how important each of these is to the current development should emerge as we learn more about the intended product.

Usability requirements are related to other kinds of requirement we must establish, such as the kinds of users expected to interact with the product.

ACTIVITY 7.1   Suggest one key functional, data, environmental, user and usability requirement for each of the following scenarios:

     (a) A system for use in a university's self-service cafeteria that allows users to pay for their food using a credit system.

     (b) A system to control the functioning of a nuclear power plant.

     (c) A system to support distributed design teams, e.g., for car design.

Comment   You may have come up with alternative suggestions; these are indicative of the kinds of answer we might expect.

     (a) *Functional*: The system will calculate the total cost of purchases.

        Data: The system must have access to the price of products in the cafeteria.

        Environmental: Cafeteria users will be carrying a tray and will most likely be in a reasonable rush. The physical environment will be noisy and busy, and users may be talking with friends and colleagues while using the system.

        User: The majority of users are likely to be under 25 and comfortable dealing with technology.

        Usability: The system needs to be simple so that new users can use the system immediately, and memorable for more frequent users. Users won't want to wait around for the system to finish processing, so it needs to be efficient and to be able to deal easily with user errors.

     (b) *Functional*: The system will be able to monitor the temperature of the reactors.

        Data: The system will need access to temperature readings.

        Environmental: The physical environment is likely to be uncluttered and to impose few restrictions on the console itself unless there is a need to wear protective clothing (depending on where the console is to be located).

        User: The user is likely to be a well-trained engineer or scientist who is competent to handle technology.

        Usability: Outputs from the system, especially warning signals and gauges, must be clear and unambiguous.

     (c) *Functional*: The system will be able to communicate information between remote sites.

        Data: The system must have access to design information that will be captured in a common file format (such as AutoCAD).

        Environmental: Physically distributed over a wide area. Files and other electronic media need to be shared. The system must comply with available communication protocols and be compatible with network technologies.

        User: Professional designers, who may be worried about technology but who are likely to be prepared to spend time learning a system that will help them perform their jobs better. The design team is likely to be multi-lingual.

        Usability: Keeping transmission error rate low is likely to be of high priority.

7.4 Data gathering

So how do we go about determining requirements? Data gathering is an important part of the requirements activity and also of evaluation. In this chapter, we concentrate on data gathering for the requirements activity. Further information about the techniques we present here and how to apply them in evaluation is in Chapters 12 through 14.

The purpose of data gathering is to collect sufficient, relevant, and appropriate data so that a set of stable requirements can be produced. Even if a set of initial requirements exists, data gathering will be required to expand, clarify, and confirm those initial requirements. Data gathering needs to cover a wide spectrum of issues because the different kinds of requirement we need to establish are quite varied, as we saw above. We need to find out about the tasks that users currently perform and their associated goals, the context in which the tasks are performed, and the rationale for why things are the way they are.

There is essentially a small number of basic techniques for data gathering, but they are flexible and can be combined and extended in many ways; this makes the possibilities for data gathering very varied, to give full leverage on understanding the variety of requirements we seek. These techniques are questionnaires, interviews, focus groups and workshops, naturalistic observation, and studying documentation. Some of them, such as the interview, require active participation from stakeholders, while others, such as studying documentation, require no involvement at all. In addition, various props can be used in data-gathering sessions, such as descriptions of common tasks and prototypes of possible new functionality. See Section 7.6 and Chapter 8 for further information on how to develop these props. Box 7.2 gives an

---

BOX 7.2 Combining Data-Gathering Techniques and Props to Understand Different Requirements

Rudman and Engelbeck (1996) describe how they used different techniques to establish the requirements for a complex graphical user interface for a telephone company, and how different methods resulted in understanding different requirements. They used five different techniques:

- On-site observation allowed them to understand the nature of the current business.
- Participatory prototyping, i.e., active involvement of the stakeholders in designing a prototype, allowed them to take advantage of the employees' knowledge.
- Interviews aimed at understanding the background business of the company allowed them to understand the complex nature of the wider domain.

- Interviews aimed at understanding the decision sequences of employees allowed them to create dialogs to support two-party negotiations.
- Role-playing prototype walkthroughs using simulated scenarios also helped to create dialogs to support two-party negotiations.

The difference between the third and fourth techniques lies in the focus of the questioning and in the notation used to capture data. In the third technique, interviewers focused on understanding the domain of the application and captured information using semantic nets, which are specifically designed to represent such information. In the fourth technique, decision trees were used to understand the goals, decision points, and options considered by employees when dealing with a customer.

example of how different methods and props can be combined to gain maximum advantage, while Box 7.3 describes a very different approach aimed at prompting inspiration rather than simple data gathering.

7.4.1 Data-gathering techniques

In addition to the most common forms of data-gathering techniques listed above, if a system is currently operational then data logging may be used. This involves instrumenting the software to record users' activity in a log that can be examined later. Each of the techniques will yield different kinds of data and are useful in different circumstances. In most cases, they are also used in evaluation, and how to implement them is described in Chapters 12 through 14. Here we describe what each technique involves and explain the circumstances for which they are most suitable, in the context of the requirements activity. The discussion is summarized in Table 7.1 on page 214.

Questionnaires. Most of us are familiar with questionnaires. They are a series of questions designed to elicit specific information from us. The questions may require different kinds of answers: some require a simple YES/NO, others ask us to choose from a set of pre-supplied answers, and others ask for a longer response or comment. Sometimes questionnaires are sent in electronic form and arrive via email or are posted on a website, and sometimes they are given to us on paper. In most cases the questionnaire is administered at a distance, i.e., no one is there to help you answer the questions or to explain what they mean.

Well-designed questionnaires are good at getting answers to specific questions from a large group of people, and especially if that group of people is spread across a wide geographical area, making it infeasible to visit them all. Questionnaires are often used in conjunction with other techniques. For example, information obtained through interviews might be corroborated by sending a questionnaire to a wider group of stakeholders to confirm the conclusions.

Interviews. Interviews involve asking someone a set of questions. Often interviews are face-to-face, but they don't have to be. Companies spend large amounts of money conducting telephone interviews with their customers finding out what they like or don't like about their service. If interviewed in their own work or home setting, people may find it easier to talk about their activities by showing the interviewer what they do and what systems and other artifacts they use. The context can also trigger them to remember certain things, for example a problem they have downloading email, which they would not have recalled had the interview taken place elsewhere.

Interviews can be broadly classified as structured, unstructured or semi-structured, depending on how rigorously the interviewer sticks to a prepared set of questions.

In the requirements activity, interviews are good at getting people to explore issues and unstructured interviews are often used early on to elicit scenarios (see Section 7.6 below). Interacting with a human rather than a sterile, impersonal piece of paper or electronic questionnaire encourages people to respond, and can make the exercise more pleasurable. In the context of establishing requirements, it is equally important for development team members to meet stakeholders and for users to feel involved. This on its own may be sufficient motivation to arrange interviews.

BOX 7.3   An Artist-designer Approach to Users

An alternative approach to understanding users and their needs was taken in a European Union-funded project called the Presence Project (Gaver et al., 1999). This work arose from research looking at novel interaction techniques to increase the presence of elderly people in their local community. Three different groups were studied: one in Oslo, Norway, one near Amsterdam, The Netherlands, and one near Pisa, Italy. One of the problems with designing for an unknown culture is that it can be difficult to understand or appreciate the needs of that culture. Rather than take a more traditional approach of questionnaires, interviews or ethnographic studies, this project used "cultural probes." These probes consisted of a wallet containing a variety of items: 8 to 10 postcards, about seven maps, a disposable camera, a photo album, and a media diary (see Figure 7.4). The intent was for the recipients to look through the wallet and answer questions associated with certain probes

that they contained, then to return the items directly to the researchers when they had finished with them.

The postcards had pictures on the front and questions on the back, and were pre-addressed and stamped so that they could be easily returned. Questions included "Please tell us a piece of advice or insight that has been important to you," "What place does art have in your life?" and "Tell us about your favorite device." The maps and associated inquiries were designed to find out about the participants' attitudes towards their environment. They were printed on various textured papers and were in the form of folding envelopes, also to facilitate their return. On local maps, participants were asked to mark sites where they would go to meet people, to be alone, to daydream and where they would like to go, but can't. On a map of the world, they were asked to mark places where they had been.

Participants were asked to use the camera to take pictures of their home, what they will wear today (whenever "today" was), the first person you see today, something desirable, and something boring. In the photo album they were asked to tell the researchers their story in pictures. The media diary was to record their use of television and radio.

The approach taken by these researchers was not to identify specific user needs but to seek inspiration that would lead to new opportunities, new pleasures, new forms of sociability, and new cultural forms. Hence they were seeking inspiration rather than requirements.

The probes were returned over a period of a month or so, at different rates and in different quantities for each group. The data were not analyzed *per se*, but the resulting designs reflect what the designers learned.

For the Dutch site, they proposed building a network of computer displays with which the elderly could help inhabitants communicate their values and attitudes about the culture.

For the Norwegians, they proposed that the elders should lead a community-wide conversation about social issues, publishing questions

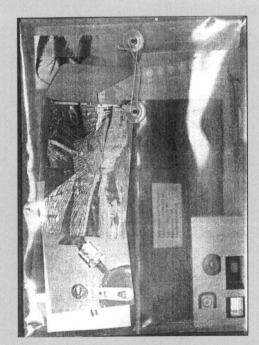

Figure 7.4 A cultural probe package.

from the library that would be sent for public response to electronic systems in cafes, trams, or public spaces.

For the Italian village near Pisa, they plan to create social and pastoral radioscapes allowing them to create flexible communications networks and to listen to the surrounding countryside.

"What we learned about the elders is only half the story, however. The other half is what the elders learned from the probes. They provoked the groups to think about the roles they play and the pleasures they experience, hinting to them that our designs might suggest new roles and new experiences." (Gaver et al., 1999, p. 29)

However, interviews are time consuming and it may not be feasible to visit all the people you'd like to see.

Focus groups and workshops. Interviews tend to be one on one, and elicit only one person's perspective. As an alternative or as corroboration, it can be very revealing to get a group of stakeholders together to discuss issues and requirements. These sessions can be very structured with set topics for discussion, or can be unstructured. In this latter case, a facilitator is required who can keep the discussion on track and can provide the necessary focus or redirection when appropriate. In some development methods, workshops have become very formalized. For example, the workshops used in Joint Application Development (Wood and Silver, 1995) are very structured, and their contents and participants are all prescribed.

In the requirements activity, focus groups and workshops are good at gaining a consensus view and/or highlighting areas of conflict and disagreement. On a social level it also helps for stakeholders to meet designers and each other, and to express their views in public. It is not uncommon for one set of stakeholders to be unaware that their views are different from another's even though they are in the same organization. On the other hand, these sessions need to be structured carefully and the participants need to be chosen carefully. It is easy for one or a few people to dominate discussions, especially if they have control, higher status, or influence over the other participants.

Naturalistic observation. It can be very difficult for humans to explain what they do or to even describe accurately how they achieve a task. So it is very unlikely that a designer will get a full and true story from stakeholders by using any of the techniques listed above. The scenarios and other props used in interviews and workshops will help prompt people to be more accurate in their descriptions, but observation provides a richer view. Observation involves spending some time with the stakeholders as they go about their day-to-day tasks, observing work as it happens, in its natural setting. A member of the design team shadows a stakeholder, making notes, asking questions (but not too many), and observing what is being done in the natural context of the activity. This is an invaluable way to gain insights into the tasks of the stakeholders that can complement other investigations. The level of involvement of the observer in the work being observed is variable along a spectrum with no involvement (outside observation) at one end and full involvement (participant observation) at the other.

Table 7.1   Overview of data-gathering techniques used in the requirements activity

Technique	Good for	Kind of data	Advantages	Disadvantages	Detail for designing in
Questionnaires	Answering specific questions	Quantitative and qualitative data	Can reach many people with low resource	The design is crucial. Response rate may be low. Responses may not be what you want	Chapter 13
Interviews	Exploring issues	Some quantitative but mostly qualitative data	Interviewer can guide interviewee if necessary. Encourages contact between developers and users	Time consuming. Artificial environment may intimidate interviewee	Chapter 13
Focus groups and workshops	Collecting multiple viewpoints	Some quantitative but mostly qualitative data	Highlights areas of consensus and conflict. Encourages contact between developers and users	Possibility of dominant characters	Chapter 13
Naturalistic observation	Understanding context of user activity	Qualitative	Observing actual work gives insights that other techniques can't give	Very time consuming. Huge amounts of data	Chapter 12
Studying documentation	Learning about procedures, regulations and standards	Quantitative	No time commitment from users required	Day-to-day working will differ from documented procedures	N/A

Not only can naturalistic observation help fill in details and nuances that simply did not come out of the other investigations, it also provides context for tasks. Contextualizing the work or behavior that a device is to support provides data that other techniques cannot, and from which we can evolve requirements.

In the requirements activity, observation is good for understanding the nature of the tasks and the context in which they are performed. However, it requires more time and commitment from a member of the design team, and it can result in a huge amount of data.

Studying documentation. Procedures and rules are often written down in manuals and these are a good source of data about the steps involved in an activity and any regulations governing a task. Such documentation should not be used as the only source, however, as everyday practices may augment them and may have been devised by those concerned to make the procedures work in a practical setting. Taking a user-centered view of development means that we are interested in the everyday practices rather than an idealized account.

Other documentation that might be studied includes diaries or job logs that are written by the stakeholders during the course of their work.

In the requirements activity, studying documentation is good for understanding legislation and getting some background information on the work. It also doesn't involve stakeholder time, which is a limiting factor on the other techniques.

7.4.2 Choosing between techniques

Table 7.1 provides some information to help you choose a set of techniques for a specific project. It tells you the kind of information you can get, e.g., answers to specific questions, and the kind of data it yields, e.g., qualitative or quantitative. It also includes some advantages and disadvantages for each technique. The kind of information you want will probably be determined by where you are in the cycle of iterations. For example, at the beginning of the project you may not have any specific questions that need answering, so it's better to spend time exploring issues through interviews rather than sending out questionnaires. Whether you want qualitative or quantitative data may also be affected by the point in development you have reached, but is also influenced by the kind of analysis you need to do.

The resources available will influence your choice, too. For example, sending out questionnaires nationwide requires sufficient time, money, and people to do a good design, try it out (i.e., pilot it), issue it, collate the results and analyze them. If you only have three weeks and no one on the team has designed a survey before, then this is unlikely to be a success.

Finally, the location and accessibility of the stakeholders need to be considered. It may be attractive to run a workshop for a large group of stakeholders, but if they are spread across a wide geographical area, it is unlikely to be practical.

Olson and Moran (1996) suggest that choosing between data-gathering techniques rests on two issues: the nature of the data gathering technique itself and the task to be studied.

Data-gathering techniques differ in two main respects:

1. The amount of time they take and the level of detail and risk associated with the findings. For example, they claim that a naturalistic observation will take two days of effort and three months of training, while interviews take one day of effort and one month of training (p. 276).

2. The knowledge the analyst must have about basic cognitive processes.

Tasks can be classified along three scales:

1. Is the task a set of sequential steps or is it a rapidly overlapping series of subtasks?

2. Does the task involve high information content with complex visual displays to be interpreted, or low information content where simple signals are sufficient to alert the user?

3. Is the task intended to be performed by a layman without much training or by a practitioner skilled in the task domain?

Box 7.4 summarizes two examples to show how techniques can be chosen using these dimensions.

So, when choosing between techniques for data gathering in the requirements activity, you need to consider the nature of the technique, the knowledge required of the analyst, the nature of the task to be studied, the availability of stakeholders and other resources, and the kind of information you need.

7.4.3   Some basic data-gathering guidelines

Organizing your first data-gathering session may seem daunting, but if you plan the sessions well, and know what your objectives are then this will increase your confidence and make the whole exercise a lot more comfortable. Below we list some data-gathering guidelines to support the requirements activity.

- Focus on identifying the stakeholders' needs. This may be achieved by studying their existing behavior and support tools, or by looking at other products,

---

BOX 7.4   Coordinated Methods *(Olson and Moran, 1996)*

For a walk-up-and-use system. An ATM is an example of a system with a simple task flow and relatively low information content that is targeted for the layman. Because of the user base, the emphasis will be on the ease with which the user can learn to operate the device. An understanding of the user's mental model may also yield insights, as evidenced by the assignment set at the end of Chapter 3.

To establish the components of the task, simple questionnaires might suffice, supplemented with naturalistic observation, i.e., observing current users at existing machines. The initial design guided by guidelines and checklists could be documented as a storyboard. A mockup of the entire system using a rapid prototyping system such as Visual Basic can be used to observe users' difficulties. After a series of such prototyping sessions, the system could be installed in a friendly site and logging data could be gathered.

For a high-performance system. The example used here is a system to support back-room workers at a bank who are reconciling the machine register with the information written on the back of the deposit slip by the customer. The task requires overlapping activation of physical actions and mental capabilities, is relatively high in information content, and is targeted for a skilled user.

This task is less obvious to the designer and we need to employ several techniques to understand it. If there is an existing system in place, then naturalistic observation and interview can be used. More detailed discovery of the objects, actions, and kinds of thinking can come from using interviews. Task analysis will help to understand the details of the task, and once understood, a series of design and evaluation steps follow, including prototyping, detailed analysis of the visual display and usability tests. The design would then iterate until it meets preset target criteria.

such as a competitor's product or an earlier release of your product under development.

- Involve all the stakeholder groups. It is very important to make sure that you get all the views of the right people. This may seem an obvious comment, but it is easy to overlook certain sections of the stakeholder population if you're not careful. We were told about one case where a large distribution and logistics company reimplemented their software systems and were very careful to involve all the clerical, managerial, and warehouse staff in their development process, but on the day the system went live, the productivity of the operation fell by 50%. On investigation it was found that the bottleneck was not in their own company, but in the suppliers' warehouses that had to interact with the new system. No one had asked them how they worked, and the new system was incompatible with their working routines.

- Involving only one representative from each stakeholder group is not enough, especially if the group is large. Everyone you involve in data gathering will have their own perspective on the situation, the task, their job and how others interact with them. If you only involve one representative stakeholder then you will only get a narrow view.

- Use a combination of data gathering techniques. Each technique will yield a certain kind of information, from a certain perspective. Using different techniques is one way of making sure that you get different perspectives (called triangulation, see Chapter 10), and corroboration of findings. For example, use observation to understand the context of task performance, interviews to target specific user groups, questionnaires to reach a wider population, and focus groups to build a consensus view.

- Support the data-gathering sessions with suitable props, such as task descriptions and prototypes if available. Since the requirements activity is iterative, prototypes or descriptions generated during one session may be reused or revisited in another with the same or a different set of stakeholders. Using props will help to jog people's memories and act as a focus for discussions.

- Run a pilot session if possible to ensure that your data-gathering session is likely to go as planned. This is particularly important for questionnaires where there is no one to help the users with ambiguities or other difficulties, but also applies to interview questions, workshop formats, and props. Any data collected during pilot sessions cannot be treated equally with other data, so don't mix them up. After running the pilot it is likely that some changes will be needed before running the session "for real."

- In an ideal world, you would understand what you are looking for and what kinds of analysis you want to do, and design the data-capture exercise to collect the data you want. However, data gathering is an expensive and time-consuming activity that is often tightly constrained on resources. Sometimes pragmatic constraints mean that you have to make compromises on the ideal situation, but before you can make sensible compromises, you need to know what you'd *really* like.

- How you record the data during a face-to-face data-gathering session is just as important as the technique(s) you use. Video recording, audio recording, and note taking are the main options. Video and audio recording provide the most accurate record of the session, but they can generate huge amounts of data. You also need to decide on practical issues that can have profound effects on the data collected, such as where to position the camera. Note taking can be harder unless this is the person's only role in the session, but note taking always involves an element of interpretation. Taking impartial, accurate notes is difficult but can be improved with practice.

ACTIVITY 7.2   For each of the situations below, consider what kinds of data gathering would be appropriate and how you might use the different techniques introduced above. You should assume that you are at the beginning of the development and that you have sufficient time and resources to use any of the techniques.

(a) You are developing a new software system to support a small accountant's office. There is a system running already with which the users are reasonably happy, but it is looking dated and needs upgrading.

(b) You are looking to develop an innovative device for diabetes sufferers to help them record and monitor their blood sugar levels. There are some products already on the market, but they tend to be large and unwieldy. Many diabetes sufferers rely on manual recording and monitoring methods involving a ritual with a needle, some chemicals, and a written scale.

(c) You are developing a website for a young person's fashion e-commerce site.

Comment

(a) As this is a small office, there are likely to be few stakeholders. Some period of observation is always important to understand the context of the new and the old system. Interviewing the staff rather than giving them questionnaires is likely to be appropriate because there aren't very many of them, and this will yield richer data and give the developers a chance to meet the users. Accountancy is regulated by a variety of laws and it would also pay to look at documentation to understand some of the constraints from this direction. So we would suggest a series of interviews with the main users to understand the positive and negative features of the existing system, a short observation session to understand the context of the system, and a study of documentation surrounding the regulations.

(b) In this case, your user group is spread about, so talking to all of them is infeasible. However, it is important to interview some, possibly at a local diabetic clinic, making sure that you have a representative sample. And you would need to observe the existing manual operation to understand what is required. A further group of stakeholders would be those who use or have used the other products on the market. These stakeholders can be questioned to find out the problems with the existing devices so that the new device can improve on them. A questionnaire sent to a wider group in order to back up the findings from the interviews would be appropriate, as might a focus group where possible.

(c) Again, you are not going to be able to interview all your users. In fact, the user group may not be very well defined. Interviews backed up by questionnaires and focus groups would be appropriate. Also, in this case, identifying similar or competing sites and evaluating them will help provide information for producing an improved product.

The problems of choosing among data-gathering techniques for the requirements activity have been recognized in requirements engineering. For example ACRE (ACquisition REquirements) is a quite extensive set of guidance to help requirements engineers choose between a variety of techniques for data gathering, including interviews and observation. The framework also includes other techniques from software engineering, knowledge engineering, and the social sciences. For more information on this framework, see Maiden and Rugg (1996).

7.5 Data interpretation and analysis

Once the first data-gathering session has been conducted, interpretation and analysis can begin. It's a good idea to start interpretation as soon after the gathering session as possible. The experience will be fresh in the minds of the participants and this can help overcome any bias caused by the recording approach. It is also a good idea to discuss the findings with others to get a variety of perspectives on the data.

The aim of the interpretation is to begin structuring and recording descriptions of requirements. Using a template such as the one suggested in Volere (Figure 7.5) highlights the kinds of information you should be looking for and guides the data interpretation and analysis. Note that many of the entries are concerned with trace-

Requirement #: **Unique Id**   Requirement Type: **Template section**   Event/use case #: **Origin of the requirement**

Description: **A one-sentence statement of the intention of the requirement**

Rationale: **Why is the requirement considered important or necessary?**

Source: **Who raised this requirement?**

Fit Criterion: **A quantification of the requirement used to determine whether the solution meets the requirement.**

Customer Satisfaction: **Measures the desire to have the requirement implemented**   Customer Dissatisfaction: **Unhappiness if it is not implemented**

Dependencies: **Other requirements with a change effect**   Conflicts: **Requirements that contradict this one**

Supporting Materials: **Pointer to supporting information**

History: **Origin and changes to the requirement**

Volere

Copyright © Atlantic Systems Guild

Figure 7.5 The Volere shell for requirements.

ability. For example, who raised the requirement and where can more information about it be found. This information may be captured in documents or in diagrams drawn during analysis. Providing links with raw data as captured on video or audio recordings can be harder, although just as important. Haumer et al. (2000) have developed a tool that records concrete scenarios using video, speech, and graphic media, and relates these recorded observations to elements of a corresponding design. This helps designers to keep track of context and usage information while analyzing and designing for the system.

More focused analysis of the data will follow initial interpretation. Different techniques and notations exist for investigating different aspects of the system that will in turn give rise to the different requirements. For example, functional requirements have traditionally been analyzed and documented using data-flow diagrams,

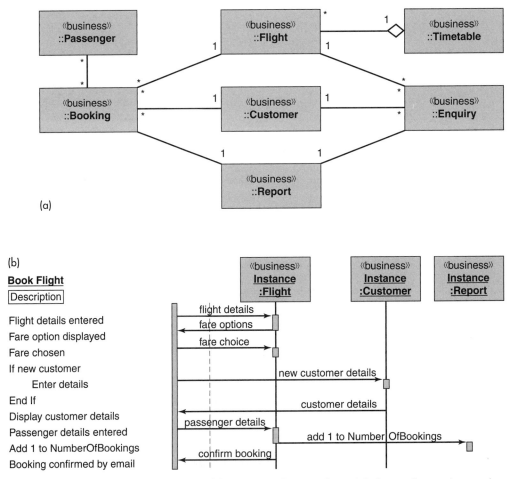

Figure 7.6 (a) Class diagram and (b) sequence diagram that might be used to analyze and capture static structure and dynamic behavior (respectively) if the system is being developed using an object-oriented approach.

state charts, work-flow charts, etc. (e.g., Sommerville, 2001). Data requirements can be expressed using entity-relationship diagrams, for example. If the development is to take an object-oriented approach, then functional and data requirements are combined in class diagrams, with behavior being expressed in state charts and sequence diagrams, among others. Examples of two such diagrams representing a portion of a holiday booking system are given in Figure 7.6. These diagrams can be linked to the requirements through the "Event/use case" field in the template in Figure 7.5.

We don't go into the detail of how diagrams such as these might be developed, as whole books are dedicated to them. Instead, we describe four techniques that have a user-centered focus and are used to understand users' goals and tasks: scenarios, use cases, essential use cases, and task analysis. All of them may be used during data-gathering sessions, and their output used as props in subsequent data-gathering sessions.

The requirements activity iterates a number of times before a set of stable requirements evolves. As more interpretation and analysis techniques are applied, a deeper understanding of requirements will emerge and the requirements descriptions will expand and clarify.

The 5th Wave   **By Rich Tennant**

"Okay, well, I think we all get the gist of where Jerry was going with the site map."

DILEMMA   How formal should your notation be?

Many forms of notation are used in design activities. Each discipline has its own set of symbols, graphs, and mnemonics that communicate precisely and clearly among people from the same discipline. But, in the early stages of design, designers are well known for their "back-of-an-envelope" sketches that capture the essence of an idea. At what stage should these sketches be transformed into more formal notations?

When we have identified needs and established requirements, they must be documented somehow. Whether this is in a purely textual form, or in prototypical form, or more formal box and line notations, our findings must be documented. When Verplank (1994) speaks about producing software-based prototypes, he talks emphatically about the importance of allowing ideas to flourish before they are formalized in the computer medium. Once cast in this way, we "get sucked into thinking that the design already works and all we have to do is fix it." The same could be said of formal paper-based notations. In interaction design, we have many notations to choose from, arising from the various disciplines that underpin our field (see

Figure 1.3). How quickly should we formalize our ideas in structured notation, and for how long should we leave the ideas fluid and flexible?

A counterargument to Verplank's position is that trying to write our findings in a more structured fashion also helps us to understand better what we've found and what's missing. The problem is that any notation has its strengths and weaknesses, and we must be aware of these when we commit our ideas to a specific notation so that our thinking and our ideas don't become too influenced by the foibles of the notation itself.

Yet again, there is also a question of who the requirements are being documented for. If users are to read and understand them, then the notation shouldn't contain technical jargon or symbols. On the other hand, if it's for communicating precise meaning within a team of developers, a more formal, specialized notation may be more appropriate.

Choosing the medium for the message can affect how the message is received and hence the meaning that is communicated, so it's important to get the medium right.

7.6 Task description

Descriptions of business tasks have been used within software development for many years. During the 1970s and 1980s, "business scenarios" were commonly used as the basis for acceptance testing, i.e., the last testing stage before the customer paid the final fee installment and "accepted" the system. In more recent years, due to the emphasis on involving users earlier in the development lifecycle and the large number of new interaction devices now being developed, task descriptions are used throughout development, from early requirements activities through prototyping, evaluation, and testing. Consequently, more time and effort has been put into understanding how best to structure and use them.

There are different flavors of task descriptions, and we shall introduce three of them here: scenarios, use cases, and essential use cases. Each of these may be used to describe either existing tasks or envisioned tasks with a new device. They are not mutually exclusive and are often used in combination to capture different perspectives or to document different stages during the development lifecycle.

In this section and the next, we use two main examples to illustrate the application of techniques. These are a library catalog service and a shared diary or calendar system. The library catalog is similar to any you might find in a public or university library, and allows you to access the details of books held in the library: for example, to search for books by a particular author, or by subject, to identify the location of a book you want to borrow, and to check on a member's current loans and status.

The shared calendar application is to support a university department. Members of the department currently keep their own calendars and communicate their whereabouts to the department's administrator, who keeps the information in a central paper calendar. Unfortunately, the central calendar and the individuals' calendars easily become out of step as members of the department arrange their own engagements. It is hoped that having a shared calendar in which individuals can enter their own engagements will help overcome the confusion that often ensues due to this mismatch. Shared calendars raise some interesting aspects of collaboration and coordination, as discussed in Chapter 4, Box 4.2. In particular, people don't usually like to have their time filled with appointments without their consent, and so a mechanism is needed for people to protect some time from being booked by others.

7.6.1 Scenarios

A scenario is an "informal narrative description" (Carroll, 2000). It describes human activities or tasks in a story that allows exploration and discussion of contexts, needs, and requirements. It does not explicitly describe the use of software or other technological support to achieve a task. Using the vocabulary and phrasing of users means that the scenarios can be understood by the stakeholders, and they are able to participate fully in the development process. In fact, the construction of scenarios by stakeholders is often the first step in establishing requirements.

Imagine that you have just been invited along to talk to a group of users who perform data entry for a university admissions office. You walk in, and are greeted by Sandy, the supervisor, who starts by saying something like:

> *Well, this is where the admissions forms arrive. We receive about 50 a day during the peak application period. Brian here opens the forms and checks that they are complete, that is, that all the documentation has been included. You see, we require copies of relevant school exam results or evidence of work experience before we can process the application. Depending on the result of this initial inspection, the forms get passed to . . . .*

Telling stories is a natural way for people to explain what they are doing or how to achieve something. It is therefore something that stakeholders can easily relate to. The focus of such stories is also naturally likely to be about what the users are trying to achieve, i.e., their goals. Understanding why people do things as they do and what they are trying to achieve in the process allows us to concentrate on the human activity rather than interaction with technology.

This is not to say that the human activity should be preserved and reflected in any new device we are trying to develop, but understanding what people do now is a good starting point for exploring the constraints, contexts, irritations, facilitators and so on under which the humans operate. It also allows us to identify the stakeholders and the products involved in the activity. Repeated reference to a particular form, book, behavior, or location indicates that this is somehow central to the activity being performed and that we should take care to understand what it is and the role it plays.

A scenario that might be generated by potential users of a library catalog service is given below:

> *Say I want to find a book by George Jeffries. I don't remember the title but I know it was published before 1995. I go to the catalog and enter my user password. I don't understand why I have to do this, since I can't get into the library to use the catalog without passing through security gates. However, once my password has been confirmed, I am given a choice of searching by author or by date, but not the combination of author and date. I tend to choose the author option because the date search usually identifies too many entries. After about 30 seconds the catalog returns saying that there are no entries for George Jeffries and showing me the list of entries closest to the one I've sought. When I see the list, I realize that in fact I got the author's first name wrong and it's Gregory, not George. I choose the entry I want and the system displays the location to tell me where to find the book.*

In this limited scenario of existing system use, there are some things of note: the importance of getting the author's name right, the annoyance concerning the need to enter a password, the lack of flexible search possibilities, and the usefulness of showing a list of similar entries when an exact match isn't clear. These are all indicators of potential design choices for the new catalog system. The scenario also tells us one (possibly common) use of the catalog system: to search for a book by an author when we don't know the title.

The level of detail present in a scenario varies, and there is no particular guidance about how much or how little should be included. Often scenarios are generated during workshop or interview sessions to help explain or discuss some aspect of the user's goals. They can be used to imagine potential uses of a device as well as to capture existing behavior. They are not intended to capture a full set of requirements, but are a very personalized account, offering only one perspective.

A simple scenario for the shared-calendar system that was elicited in an informal interview describes how one function of the calendar might work: to arrange a meeting between several people.

> *The user types in all the names of the meeting participants together with some constraints such as the length of the meeting, roughly when the meeting needs to take place, and possibly where it needs to take place. The system then checks against the individuals' calendars and the central departmental calendar and presents the user with a series of dates on which everyone is free all at the same time. Then the meeting could be confirmed and written into peoples' calendars. Some people, though, will want to be asked before the calendar entry is made. Perhaps the system could email them automatically and ask that it be confirmed before it is written in."*

An example of a futuristic scenario, devised by Symbian, showing one vision of how wireless devices might be used in the future is shown in Figure 7.7.

In this chapter, we refer to scenarios only in their role of helping to establish requirements. They have a continuing role in the design process that we shall return to in Chapter 8.

A businesswoman traveling to Paris from the US

A businesswoman is traveling from San Francisco to Paris on a business trip. On her way to the airport she narrowly misses a traffic delay. She avoids the traffic jam because her Smartphone beeps, then sends her a text message warning her of the traffic accident on her normal route from her office to the airport.

Upon arrival at the airport, the location-sensitive Smartphone notifies the airline that she will be checking in shortly, and an airline employee immediately finds her and takes her baggage. Her on-screen display shows that her flight is on time and provides a map to her gate. On her way to the gate she downloads tourist information such as maps and events occurring in Paris during her stay.

Once she finds her seat on the plane, she begins to review all the information she has downloaded. She notices than an opera is playing in Paris that she has been wanting to see, and she books her ticket. Her Smartphone can make the booking using her credit card number, which it has stored in its memory. This means that she does not need to re-enter the credit card number each time she uses wCommerce (i.e., wireless commerce), facilities. The security written into the software of the Smartphone protects her against fraud.

The Smartphone stores the opera booking along with several emails that she writes on the plane. As soon as she steps off the plane, the Smartphone makes the calls and automatically sends the emails.

As she leaves the airport, a map appears on her Smartphone's display, guiding her to her hotel.

Figure 7.7 A scenario showing how two technologies, a Smartphone and wCommerce (wireless commerce), might be used.

Capturing scenarios of existing behavior and goals helps in determining new scenarios and hence in gathering data useful for establishing the new requirements. The next activity is intended to help you appreciate how a scenario of existing activity can help identify the requirements for a future application to support the same user goal.

ACTIVITY 7.3   Write a scenario of how you would currently go about choosing a new car. This should be a brand new car, not a second-hand car. Having written it, think about the important aspects of the task, your priorities and preferences. Then imagine a new interactive product that supports you in your goal and takes account of these issues. Write a futuristic scenario showing how this product would support you.

Comment   The following example is a fairly generic view of this process. Yours will be different, but you may have identified similar concerns and priorities.

The first thing I would do is to observe cars on the road and identify ones that I like the look of. This may take some weeks. I would also try to identify any consumer reports that will include an assessment of car performance. Hopefully, these initial activities will result in me identifying a likely car to buy. The next stage will be to visit a car showroom and see at first hand what the car looks like, and how comfortable it is to sit in. If I still feel positive about the car, then I'll ask for a test drive. Even a short test drive helps me to understand how well the car handles, how noisy is the engine, how smooth are the gear changes, and so on. Once I've driven the car myself, I can usually tell whether I would like to own it or not.

From this scenario, it seems that there are broadly two stages involved in the task: researching the different cars available, and gaining first-hand experience of potential purchases. In the former, observing cars on the road and getting actual and maybe critical information about them has been highlighted. In the latter, the test drive seems to be quite significant.

For many people buying a new car, the smell and touch of the car's exterior and interior, and the driving experience itself are often the most influential factors in choosing a particular model. Other more factual attributes such as fuel consumption, amount of room inside, colors available, and price may rule out certain makes and models, but at the end of the day, cars are often chosen according to how easy they are to handle and how comfortable they are inside. This makes the test drive a vital part of the process of choosing a new car.

Taking these comments into account, we've come up with the following scenario describing how a new "one-stop" shop for new cars might operate. This product makes use of immersive virtual reality technology that is already used for other applications such as designing buildings and training bomb disposal experts.

I want to buy a new car, so I go down the street to the local "one-stop car shop." The shop has a number of booths in it, and when I go in I'm directed to an empty booth. Inside there's a large seat that reminds me of a racing car seat, and in front of that a large display screen, keyboard and printer. As I sit down, the display jumps into life. It offers me the options of browsing through video clips of new cars which have been released in the last two years, or of searching through video clips of cars by make, by model, or by year. I can choose as many of these as I like. I also have the option of searching through and reading or printing consumer reports that have been produced about the cars I'm interested in. I spend about an hour looking through materials and deciding that I'd like to experience a couple that look promising. I can of course go away and come back later, but I'd like to have a go with some of those I've found. By flicking a switch in my armrest, I can call up the options for virtual reality simulations for any of the cars I'm interested in. These are really great as they allow me to take the car for a test drive, simulating everything about the driving experience in this car, from road holding, to windscreen display, and front pedal pressure to dash board layout. It even re-creates the atmosphere of being inside the car.

Note that the product includes support for the two research activities mentioned in the original scenario, as well as the important test drive facility. This would be only a first cut scenario which would then be refined through discussion and further investigation.

7.6.2   Use cases

Use cases also focus on user goals, but the emphasis here is on a user–system interaction rather than the user's task itself. They were originally introduced through the object-oriented community in the book *Object-Oriented Software Engineering* (Jacobson et al., 1992). Although their focus is specifically on the interaction between the user (called an "actor") and a software system, the stress is still very much on the user's perspective, not the system's. The term "scenario" is also used in the context of use cases. In this context, it represents one path through the use case, i.e., one particular set of conditions. This meaning is consistent with the definition given above in that they both represent one specific example of behavior.

A use case is associated with an actor, and it is the actor's goal in using the system that the use case wants to capture. In this technique, the main use case describes what is called the "normal course" through the use case, i.e., the set of actions that the analyst believes to be most commonly performed. So, for example, if through data gathering we have found that most users of the library go to the catalog to check the location of a book before going to the shelves, then the normal course for the use case would include this sequence of events. Other possible sequences, called alternative courses, are then listed at the bottom of the use case.

A use case for arranging a meeting using the shared calendar application, with the normal course being that the meeting is written into the calendar automatically, might be:

1. The user chooses the option to arrange a meeting.
2. The system prompts user for the names of attendees.
3. The user types in a list of names.
4. The system checks that the list is valid.
5. The system prompts the user for meeting constraints.
6. The user types in meeting constraints.
7. The system searches the calendars for a date that satisfies the constraints.
8. The system displays a list of potential dates.
9. The user chooses one of the dates.
10. The system writes the meeting into the calendar.
11. The system emails all the meeting participants informing them of the appointment.

Alternative courses:

5. If the list of people is invalid,
   5.1 The system displays an error message.
   5.2 The system returns to step 2.
8. If no potential dates are found,
   8.1 The system displays a suitable message.
   8.2 The system returns to step 5.

Note that the number associated with the alternative course indicates the step in the normal course that is replaced by this action or set of actions. Also note how specific the use case is about how the user and the system will interact.

Use cases may be described graphically. Figure 7.8 shows the use case diagram for the above calendar example. The actor "Administrator" is associated with the use case "Arrange a meeting." Another actor we might identify for the calendar system is the "Departmental member" who updates his own calendar entries, also shown in Figure 7.8. Actors may be associated with more than one use case, so for

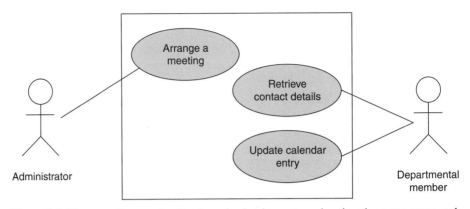

Figure 7.8 Use case diagram for the shared calendar system showing three use cases and two actors.

example the actor "Departmental member" can be associated with a use case "Retrieve contact details" as well as the "Update calendar entry" use case. Each use case may also be associated with more than one actor.

This kind of description has a different style and a different focus from the scenarios described above. The layout is more formal, and the structure of "good" use cases has been discussed by many (e.g., Cockburn, 1995; Gough et al., 1995; Ben Achour, 1999). The description also focuses on the user–system interaction rather than on the user's activities; thus a use case presupposes that technology is being used. This kind of detail is more useful at conceptual design stage than during requirements or data gathering, but use cases have been found to help some stakeholders express their views on how existing systems are used and how a new system might work.

To develop a use case, first identify the actors, i.e., the people or other systems that will be interacting with the system under development. Then examine these actors and identify their goal or goals in using the system. Each of these will be a use case.

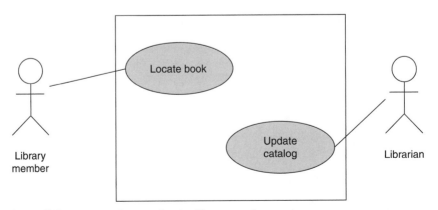

Figure 7.9 Use case diagram for the library catalog service.

ACTIVITY 7.4   Consider the example of the library catalog service again. One use case is "Locate book," and this would be associated with the "Library member" actor. Identify one other main actor and an associated use case, and draw a use case diagram.

Write out the use case for "Locate book" including the normal and some alternative courses. You may assume that the normal course is for users to go to the catalog to find the location, and that the most common path to find this is through a search by author.

Comment   One other main actor is the "Librarian." A use case for the "Librarian" would be "Update catalog." Figure 7.9 is the associated use case diagram. There are other use cases you may have identified.

The use case for "Locate book" might be something like this:

1. The system prompts for user name and password.
2. The user enters his or her user name and password into the catalog system.
3. The system verifies the user's password.
4. The system displays a menu of choices.
5. The user chooses the search option.
6. The system displays the search menu.
7. The user chooses to search by author.
8. The system displays the search author screen.
9. The user enters the author's name.
10. The system displays search results.
11. The user chooses the required book.
12. The system displays details of chosen book.
13. The user notes location.
14. The user quits catalog system.

Alternative courses:

4. If user password is not valid
   4.1 The system displays error message.
   4.2 The system returns to step 1.

5. If user knows the book details
   5.1 The user chooses to enter book details.
   5.2 The system displays book details screen.
   5.3 The user enters book details.
   5.4 The system goes to step 12.

7.6.3 Essential use cases

Essential use cases were developed by Constantine and Lockwood (1999) to combat what they see as the limitations of both scenarios and use cases as described arrangeMeeting

USER INTENTION	SYSTEM RESPONSIBILITY
arrange a meeting	
	request meeting attendees and constraints
identify meeting attendees and constraints	
	suggest potential dates
choose preferred date	
	book meeting

Figure 7.10 An essential use case for arranging a meeting in the shared calendar application.

above. Scenarios are concrete stories that concentrate on realistic and specific activities. They therefore can obscure broader issues concerned with the wider organizational view. On the other hand, traditional use cases contain certain assumptions, including the fact that there is a piece of technology to interact with, and also assumptions about the user interface and the kind of interaction to be designed.

Essential use cases represent abstractions from scenarios, i.e., they represent a more general case than a scenario embodies, and try to avoid the assumptions of a traditional use case. An essential use case is a structured narrative consisting of three parts: a name that expresses the overall user intention, a stepped description of user actions, and a stepped description of system responsibility. This division between user and system responsibilities can be very helpful during conceptual design when considering task allocation and system scope, i.e., what the user is responsible for and what the system is to do.

An example essential use case based on the library example given above is shown in Figure 7.10. Note that the steps are more generalized than those in the use case in Section 7.6.2, while they are more structured than the scenario in Section 7.6.1. For example, the first user intention does not say anything about typing in a list of names, it simply states that the user identifies meeting attendees. This could be done by identifying roles, rather than people's names, from an organizational or project chart, or by choosing names from a list of people whose calendars the system keeps, or by typing in the names. The point is that at the time of creating this essential use case, there is no commitment to a particular interaction design.

Instead of actors, essential use cases are associated with user roles. One of the differences is that an actor could be another system, whereas a user role is just that: not a particular person, and not another system, but a role that a number of different people may play when using the system. Just as with actors, though, producing an essential use case begins with identifying user roles.

ACTIVITY 7.5   Construct an essential use case "locateBook" for the user role "Library member" of the library catalog service discussed in Activity 7.4.

Comment   locateBook

USER INTENTION	SYSTEM RESPONSIBILITY
identify self	
	verify identity
	request appropriate details
offer known details	
	offer search results
note search results	
quit system	
	close

Note that here we don't talk about passwords, but merely state that the users need to identify themselves. This could be done using fingerprinting, or retinal scanning, or any other suitable technology. The essential use case does not commit us to technology at this point. Neither does it specify search options or details of how to initiate the search.

7.7 Task analysis

Task analysis is used mainly to investigate an existing situation, not to envision new systems or devices. It is used to analyze the underlying rationale and purpose of what people are doing: what are they trying to achieve, why are they trying to achieve it, and how are they going about it? The information gleaned from task analysis establishes a foundation of existing practices on which to build new requirements or to design new tasks.

Task analysis is an umbrella term that covers techniques for investigating cognitive processes and physical actions, at a high level of abstraction and in minute detail. In practice, task analysis techniques have had a mixed reception. The most widely used version is Hierarchical Task Analysis (HTA) and this is the technique we introduce in this chapter. Another well-known task analysis technique called GOMS (goals, operations, methods, and selection rules) that models procedural knowledge (Card et al., 1983) is described in Chapter 14.

7.7.1 Hierarchical task analysis

Hierarchical Task Analysis (HTA) was originally designed to identify training needs (Annett and Duncan, 1967). It involves breaking a task down into subtasks and then into sub-subtasks and so on. These are then grouped together as plans that specify how the tasks might be performed in an actual situation. HTA focuses on the physical and observable actions that are performed, and includes looking at actions that are not related to software or an interaction device at all. The starting point is a user goal. This is then examined and the main tasks associated with achieving that goal are identified. Where appropriate, these tasks are subdivided into subtasks.

Consider the library catalog service, and the task of borrowing a book. This task can be decomposed into other tasks such as accessing the library catalog, searching by name, title, subject, or whatever, making a note of the location of the book, going to the correct shelf taking it down off the shelf (provided it is there) and finally taking

0. In order to borrow a book from the library
   1. go to the library
   2. find the required book
      2.1 access library catalog
      2.2 access the search screen
      2.3 enter search criteria
      2.4 identify required book
      2.5 note location
   3. go to correct shelf and retrieve book
   4. take book to checkout counter
plan 0: do 1-3-4. If book isn't on the shelf expected, do 2-3-4.
plan 2: do 2.1-2.4-2.5. If book not identified do 2.2-2.3-2.4-2.5.

Figure 7.11 An HTA for borrowing a book from the library.

it to the check-out counter. This set of tasks and subtasks might be performed in a different order depending on how much is known about the book, and how familiar the user might be with the library and the book's likely location. Figure 7.11 shows these subtasks and some plans for different paths through those subtasks. Indentation shows the hierarchical relationship between tasks and subtasks.

Note how the numbering works for the task analysis: the number of the plan corresponds to the number of the step to which the plan relates. For example, plan 2 shows how the subtasks in step 2 can be ordered; there is no plan 1 because step 1 has no subtasks associated with it.

An alternative expression of an HTA is a graphical box-and-line notation. Figure 7.12 shows the graphical version of the HTA in Figure 7.11. Here the subtasks are represented by named boxes with identifying numbers. The hierarchical relationship between tasks is shown using a vertical line. If a task is not decomposed any further then a thick horizontal line is drawn underneath the corresponding box.

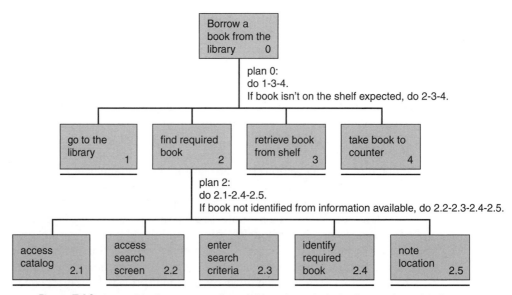

Figure 7.12 A graphical representation of the task analysis for borrowing a book.

Plans are also shown in this graphical form. They are written alongside the vertical line emitting from the task being decomposed. For example, in Figure 7.12 plan 2 is specified next to the vertical line from box 2 "find required book."

ACTIVITY 7.6 Look back at the scenario for arranging a meeting in the shared calendar application. Perform hierarchical task analysis for the goal of arranging a meeting. Include all plans in your answer. Express the task analysis textually and graphically.

Comment The main tasks involved in this are to find out who needs to be at the meeting, find out the constraints on the meeting such as length of meeting, range of dates, and location, find a suitable date, enter details into the calendar, and inform attendees. Finding a suitable date can be decomposed into other tasks such as looking in the departmental calendar, looking in individuals' calendars, and checking potential dates against constraints. The textual version of the HTA is shown below. Figure 7.13 shows the corresponding graphical representation.

> 0. In order to arrange a meeting
>   1. compile a list of meeting attendees
>   2. compile a list of meeting constraints
>   3. find a suitable date
>   3.1 identify potential dates from departmental calendar
>   3.2 identify potential dates from each individual's calendar
>   3.3 compare potential dates
>   3.4 choose one preferred date
>   4. enter meeting into calendars
>   5. inform meeting participants of calendar entry

plan 0: do 1-2-3. If potential dates are identified, do 4-5. If no potential dates can be identified, repeat 2-3.

plan 3: do 3.1-3.2-3.3-3.4 or do 3.2-3.1-3.3-3.4

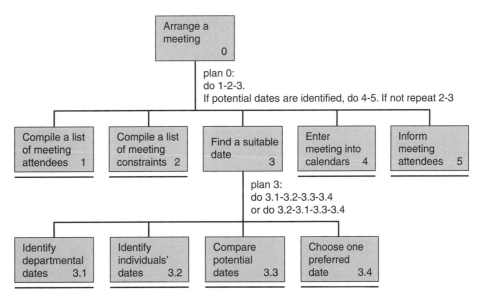

Figure 7.13 A graphical representation of the meeting HTA.

ACTIVITY 7.7	What do you think are the main problems with using task analysis on real problems? Think of more complex tasks such as scheduling delivery trucks, or organizing a large conference.
Comment	Real tasks are very complex. One of the main problems with task analysis is that it does not scale very well. The notation soon becomes unwieldy, making it difficult to follow. Imagine what it would be like to produce a task analysis in which there were hundreds or even thousands of subtasks.

A second problem is that task analysis is limited in the kind of tasks it can model. For example, it cannot model tasks that are overlapping or parallel, nor can it model interruptions. Most people work through interruptions of various kinds, and many significant tasks happen in parallel.

Assignment

This assignment is the first of four assignments that together take you through the complete development lifecycle for an interactive product. This assignment requires you to use techniques described in this chapter for identifying needs and establishing requirements. The further three assignments are at the end of Chapters 8, 13, and 14.

The overall assignment is for you to design and evaluate an interactive website for booking tickets online for events like concerts, the theatre and the cinema. This is currently an activity that in many instances, can be difficult or inconvenient to achieve using traditional means (e.g., waiting for ages on the phone to get hold of an agent, queuing for hours in the rain at a ticket office).

For this assignment, you should:

1. Identify users' needs for this website. You could do this in a number of ways. For example, you could observe people using ticket agents, think about your own experience of purchasing tickets, look at existing websites for booking tickets, talk to friends and family about their experiences, and so on. Record your data carefully.

2. Based on your user requirements, choose two different user profiles and produce one main scenario for each one, capturing how the user is expected to interact with the system.

3. Using the scenarios generated from your data gathering, perform a task analysis on the main task associated with the ticket booking system, i.e., booking a ticket.

4. Based on the data gathered in part 1 and your subsequent interpretation and analysis, identify different kinds of requirements for the website, according to the headings introduced in Section 7.3 above. Write up the requirements in the style of the Volere template.

Summary

In this chapter, we have looked in more detail at how to identify users' needs and establish requirements for interaction design. Various data-gathering techniques can be used to collect data for interpretation and analysis. The most common of these are questionnaires, interviews, focus groups, workshops, naturalistic observation, and studying documentation. Each of these has advantages and disadvantages that must be balanced against your constraints when choosing which techniques to use for a particular project. They can be combined in many different ways, and can be supported by props such as scenarios and prototypes. How to carry out these techniques is covered in Chapters 12 through 14. Scenarios, use cases, and essential use cases are helpful techniques for beginning to document the findings from the data-gathering sessions. Task analysis is a little more structured, but does not scale well.

Key points

- Getting the requirements right is crucial to the success of the interactive product.

- There are different kinds of requirements: functional, data, environmental, user, and usability. Every system will have requirements under each of these headings.

- The most commonly used data gathering techniques for this activity are: questionnaires, interviews, workshops or focus groups, naturalistic observation, and studying documentation.

- Descriptions of user tasks such as scenarios, use cases, and essential use cases help users to articulate existing work practices. They also help to express envisioned use for new devices.

- Task analysis techniques help to investigate existing systems and current practices.

Further reading

ROBERTSON, SUZANNE, AND ROBERTSON, JAMES (1999) *Mastering the Requirements Process*. Boston: Addison-Wesley. In this book, Robertson and Robertson explain a useful framework for software requirements work (see also the interview with Suzanne Robertson after this chapter).

CONSTANTINE, LARRY L., AND LOCKWOOD, LUCY A. D. (1999) *Software for Use*. Boston: Addison-Wesley. This very readable book provides a concrete approach for modeling and analyzing software systems. The approach has a user-centered focus and contains some useful detail. It also includes more information about essential use cases.

JACOBSON, I., BOOCH, G., AND RUMBAUGH, J. (1992) *The Unified Software Development Process*. Boston: Addison-Wesley. This is not an easy book to read, but it is the definitive guide for developing object-oriented systems using use cases and the modeling language Unified Modeling Language (UML).

BRUEGGE, BERND, AND DUTOIT, ALLEN H. (2000) *Object-oriented Software Engineering*. Upper Saddle River, NJ: Prentice-Hall. This book is a comprehensive treatment of the whole development process using object-oriented techniques such as use cases. The book is organized to help those involved in project work.

SOMMERVILLE, IAN (2001) *Software Engineering* (6th ed.). Boston: Addison-Wesley. If you are interested in pursuing notations for functional and data requirements, then this book introduces a variety of notations and techniques used in software engineering.

INTERVIEW   with Suzanne Robertson

Suzanne Roberston is a principal of The Atlantic Systems Guild, an international think tank producing numerous books and seminars whose aim is to make good ideas to do with systems engineering more accessible. Suzanne is particularly well known for her work in systems analysis and requirements gathering activities.

HS: What are requirements?

SR: Well the problem is that "requirements" has turned into an elastic term. Requirements is an enormously wide field and there are so many different types of requirements. One person may be talking about budget, somebody else may be talking about interfacing to an existing piece of software, somebody else may be talking about a performance requirement, somebody else may be talking about the calculation of an algorithm, somebody else may be talking about a data definition, and I could go on for hours as to what requirement means. What we advise people to do to start with is to look for something we call "linguistic integrity" within their own project. When all people who are connected with the project are talking about requirements, what do they mean? This gets very emotional, and that's why we came up with our framework. We gathered together all this experience of different types of requirements, tried to pick the most common organization, and then wrote them down in a framework.

HS: Please would you explain your framework? (The version discussed in this interview is shown in the figure on page 238. The most recent version may be downloaded from www.systemsguild.com.)

SR: Imagine a huge filing cabinet with 27 drawers, and in each drawer you've got a category of knowledge that is related to requirements. In the very first drawer for example you've got the goals, i.e., the reason for doing the project. In the second drawer you've got the stakeholders. These are roles because they could be played by more than one person, and one person may play more than one role. You've got the client who's going

to pay for the development, and the customer who's making the decision about buying it. Then you've got stakeholders like the project leader, the developers, the requirements engineers, the designers, the quality people, and the testers. Then you've got the less obvious stakeholders like surrounding organizations, professional bodies, and other people in the organization whose work might be affected by the project you're doing, even if they're never going to use the product.

HS: So do you find the stakeholders by just asking questions?

SR: Yes, partly that and partly by using the domain model of the subject matter, which is in drawer 9, as the driver to ask more questions about the stakeholders. For example, for each one of the subject matter areas, ask who have we got to represent this subject matter? For each one of the people that we come across, ask what subject matter are we expecting from them?

Drawer 3 contains the end users. I've put them in a separate drawer because an error that a lot of people make when they're looking for requirements is that the only stakeholder they talk about is the end user. They decide on the end user too quickly and they miss opportunities. So you end up building a product that is possibly less competitive. I keep them a bit fuzzy to start with, and as you start to fix on them then you can go into really deep analysis about them: What is their psychology? What are their characteristics? What's their subject-matter knowledge? How do they feel about their work? How do they feel about technology? All of these things help you to come up with the most competitive non-functional requirements for the product.

HS: How do you resolve conflict between stakeholders?

SR: Well, part of it is to get the conflicts out in the open up front, so people stop blaming each other, but that certainly doesn't resolve it. One of the ways is to make things very visible all the way through and to keep reminding people that conflict is respectable, that it's a sign of creativity, of people having ideas. The other thing that we do is that in our individual requirements (that is atomic requirements), which end up living in drawers 9 to 17 of this filing cabinet, we've got a place to say "Conflict: Which other requirement is this in conflict with?" and we encourage people to identify them. Sometimes these conflicts resolve themselves because they're on people's back burners, and some of the conflicts are resolved by people just talking to one another. We have a point at which we cross-check requirements and look for conflicts and if we find some that are just not sorting themselves out, then we stop and have a serious negotiation.

In essence, it's bubbling the conflicts up to the surface. Keep on talking about them and keep them visible. De-personalize it as much as you can. That helps.

HS: What other things are associated with these atomic requirements?

SR: Each one has a unique number and a description that is as close as you can get to what you think the thing means. It also has a rationale that helps you to figure out what it really is. Then the next component is the fit criterion, which is, "If somebody came up with a solution to this requirement, how would you know whether or not it satisfies the requirement?" So this means making the requirement quantifiable, measurable. And it's very powerful because it makes you think about the requirement. One requirement quite often turns into several when you really try and quantify it. It also provides a wonderful opportunity for involving testers, because at that point if you write the fit criterion you can get a tester and ask whether this can be used as input to writing a cost-effective test. Now this is different from the way we usually use the testers, which is to build tests that test our solutions. Here I want to get them in much earlier, I want them to test whether this requirement really is a requirement.

HS: So what's in drawers 18 through 27?

SR: Well here you can get into serious quarrels. The overall category is "project issues," and people often say they're not really requirements, and they aren't. But if the project is not being managed according to the real work that's being done, in other words the contents of the drawers, then the project goes off the rails. In project issues we create links so that a project manager can manage the project according to what's happening to the requirements.

In the last drawer we have design ideas. People say when you're gathering requirements you should not be concerned with how you're going to solve the problem. But mostly people tell you requirements in the form of a solution anyway. The key thing is to learn how to separate the real requirements from so-

lution ideas, and when you get a solution idea, pop it in this drawer. This helps requirements engineers, I think, because we are trained to think of solutions, not to dig behind and find the real problem.

HS: How do you go about identifying requirements?

SR: For too long we've been saying the stakeholders should give us their requirements: we'll ask them and they'll give them to us. We've realized that this is not practical—partly because there are many requirements people don't know they've got. Some requirements are conscious and they're usually because things have gone wrong or they'd like something extra. Some requirements are unconscious because maybe people are used to it, or maybe they haven't a clue because they don't see the overall picture. And then there are undreamed-of requirements that people just don't dream they could ever have, because we've all got boundaries based on what we think technology is capable of doing or what we know about technology or what our experience is. So it's not just asking people for things, it's also inventing requirements. I think that's where prototyping comes in and scenario modeling and storyboarding and all of those sorts of techniques to help people to imagine what they could have.

If you're building a product for the market and you want to be more competitive you should be inventing requirements. Instead of constricting yourself within the product boundary, say, "Can I push myself out a bit further? Is there something else I could do that isn't being done?"

HS: So what kinds of techniques can people use to push out further?

SR: One of the things is to learn how to imagine what it's like to be somebody else, and this is why going into other fields, for example family therapy, is helpful. They've learned an awful lot about how to imagine you might be somebody else. And that's not something that software engineers are taught in college normally and this is why it's very healthy for us to be bringing together the ideas of psychology and sociology and so on with software and systems engineering. Bringing in these human aspects—the performance, the usability features, the "look and feel" features—that's going to make our products more competitive. I always tell people to read a lot of novels. If you're having trouble relating to some stakeholders, for example, go and read some Jane Austen and then try to imagine what it would have been like to have been the heroine in *Pride and Prejudice*. What would it have been like to have to change your clothes three times a day? I find this helps me a lot, it frees your mind and then you can say, "OK, what's it really like to be that other person?" There's a lot to learn in that area.

HS: So what you're saying really is that it's not easy.

SR: It's not easy. I don't think there's any particular technique. But what we have done is we have come up with a lot of different "trawling" techniques, along with recommendations, that can help you.

HS: Do you have any other tips for gathering requirements?

SR: It's important for people to feel that they've been heard. The waiting room (drawer number 26)

was invented because of a very enthusiastic high-level stakeholder in a project we were doing. She was very enthusiastic and keen and very involved. Wonderful! She really gave us tremendous ideas and support. The problem was she kept having ideas, and we didn't know what to do. We didn't want to stop her having ideas, on the other hand we couldn't always include them because then we would never get anything built. So we invented the waiting room. All the good ideas we have we put in there and every so often we go into the waiting room and review the ideas. Some of them get added to the product, some are discarded, and some are left waiting. The psychology of it is very good because the idea's in the waiting room, everyone knows it's in there, but it's not being ignored. When people feel heard, they feel better and consequently they're more likely to cooperate and give you time.

The Template

PROJECT DRIVERS
1. The Purpose of the Product
2. Client, Customer and other Stakeholders
3. Users of the Product

PROJECT CONSTRAINTS
4. Mandated Constraints
5. Naming Conventions and Definitions
6. Relevant Facts and Assumptions

FUNCTIONAL REQUIREMENTS
7. The Scope of the Work
8. The Scope of the Product
9. Functional and Data Requirements

NON-FUNCTIONAL REQUIREMENTS
10. Look and Feel Requirements
11. Usability Requirements
12. Performance Requirements
13. Operational Requirements
14. Maintainability and Portability Requirements
15. Security Requirements
16. Cultural and Political Requirements
17. Legal Requirements

PROJECT ISSUES
18. Open Issues
19. Off-the-Shelf Solutions
20. New Problems
21. Tasks
22. Cutover
23. Risks
24. Costs
25. User Documentation and Training
26. Waiting Room
27. Ideas for Solutions

The Volere Requirements Specification Template (© 1995–2001 Atlantic Systems Guild).

Chapter 8

Design, prototyping and construction

8.1 Introduction

8.2 Prototyping and construction

    8.2.1 What is a prototype?

    8.2.2 Why prototype?

    8.2.3 Low-fidelity prototyping

    8.2.4 High-fidelity prototyping

    8.2.5 Compromises in prototyping

    8.2.6 Construction: from design to implementation

8.3 Conceptual design: moving from requirements to first design

    8.3.1 Three perspectives for developing a conceptual model

    8.3.2 Expanding the conceptual model

    8.3.3 Using scenarios in conceptual design

    8.3.4 Using prototypes in conceptual design

8.4 Physical design: getting concrete

    8.4.1 Guidelines for physical design

    8.4.2 Different kinds of widget

8.5 Tool support

8.1 Introduction

Design activities begin once a set of requirements has been established. Broadly speaking, there are two types of design: conceptual and physical. The former is concerned with developing a conceptual model that captures what the product will do and how it will behave, while the latter is concerned with details of the design such as screen and menu structures, icons, and graphics. The design emerges iteratively, through repeated design-evaluation-redesign cycles involving users.

For users to effectively evaluate the design of an interactive product, designers must produce an interactive version of their ideas. In the early stages of development, these interactive versions may be made of paper and cardboard, while as design progresses and ideas become more detailed, they may be polished pieces of software, metal, or plastic that resemble the final product. We have called the activity concerned with building this interactive version prototyping and construction.

There are two distinct circumstances for design: one where you're starting from scratch and one where you're modifying an existing product. A lot of design comes from the latter, and it may be tempting to think that additional features can be added, or existing ones tweaked, without extensive investigation, prototyping or evaluation. It is true that if changes are not significant then the prototyping and evaluation activities can be scaled down, but they are still invaluable activities that should not be skipped.

In Chapter 7, we discussed some ways to identify user needs and establish requirements. In this chapter, we look at the activities involved in progressing a set of requirements through the cycles of prototyping to construction. We begin by explaining the role and techniques of prototyping and then explain how prototypes may be used in the design process. Tool support plays an important part in development, but tool support changes so rapidly in this area that we do not attempt to provide a catalog of current support. Instead, we discuss the kinds of tools that may be of help and categories of tools that have been suggested.

The main aims of this chapter are to:

- Describe prototyping and different types of prototyping activities.
- Enable you to produce a simple prototype.
- Enable you to produce a conceptual model for a system and justify your choices.
- Enable you to attempt some aspects of physical design.
- Explain the use of scenarios and prototypes in conceptual design.
- Discuss standards, guidelines, and rules available to help interaction designers.
- Discuss the range of tool support available for interaction design.

8.2   Prototyping and construction

It is often said that users can't tell you what they want, but when they see something and get to use it, they soon know what they don't want. Having collected information about work practices and views about what a system should and shouldn't do, we then need to try out our ideas by building prototypes and iterating through several versions. And the more iterations, the better the final product will be.

8.2.1   What is a prototype?

When you hear the term *prototype*, you may imagine something like a scale model of a building or a bridge, or maybe a piece of software that crashes every few minutes. But a prototype can also be a paper-based outline of a screen or set of screens, an electronic "picture," a video simulation of a task, a three-dimensional paper and cardboard mockup of a whole workstation, or a simple stack of hyperlinked screen shots, among other things.

In fact, a prototype can be anything from a paper-based storyboard through to a complex piece of software, and from a cardboard mockup to a molded or pressed piece of metal. A prototype allows stakeholders to interact with an envisioned product, to gain some experience of using it in a realistic setting, and to explore imagined uses.

For example, when the idea for the PalmPilot was being developed, Jeff Hawkin (founder of the company) carved up a piece of wood about the size and shape of the device he had imagined. He used to carry this piece of wood around with him and pretend to enter information into it, just to see what it would be like to own such a device (Bergman and Haitani, 2000). This is an example of a very simple (some might even say bizarre) prototype, but it served its purpose of simulating scenarios of use.

Ehn and Kyng (1991) report on the use of a cardboard box with the label "Desktop Laser Printer" as a mockup. It did not matter that, in their setup, the printer was not real. The important point was that the intended users, journalists and typographers, could experience and envision what it would be like to have one of these machines on their desks. This may seem a little extreme, but in 1982 when this was done, desktop laser printers were expensive items of equipment and were not a common sight around the office.

So a prototype is a limited representation of a design that allows users to interact with it and to explore its suitability.

8.2.2 Why prototype?

Prototypes are a useful aid when discussing ideas with stakeholders; they are a communication device among team members, and are an effective way to test out ideas for yourself. The activity of building prototypes encourages reflection in design, as described by Schön (1983) and as recognized by designers from many disciplines as an important aspect of the design process. Liddle (1996), talking about software design, recommends that prototyping should always precede any writing of code.

Prototypes answer questions and support designers in choosing between alternatives. Hence, they serve a variety of purposes: for example, to test out the technical feasibility of an idea, to clarify some vague requirements, to do some user testing and evaluation, or to check that a certain design direction is compatible with the rest of the system development. Which of these is your purpose will influence the kind of prototype you build. So, for example, if you are trying to clarify how users might perform a set of tasks and whether your proposed device would support them in this, you might produce a paper-based mockup. Figure 8.1 shows a paper-based prototype of the design for a handheld device to help an autistic child communicate. This prototype shows the intended functions and buttons, their positioning and labeling, and the overall shape of the device, but none of the buttons actually work. This kind of prototype is sufficient to investigate scenarios of use and to decide, for example, whether the buttons are appropriate and the functions sufficient, but not to test whether the speech is loud enough or the response fast enough.

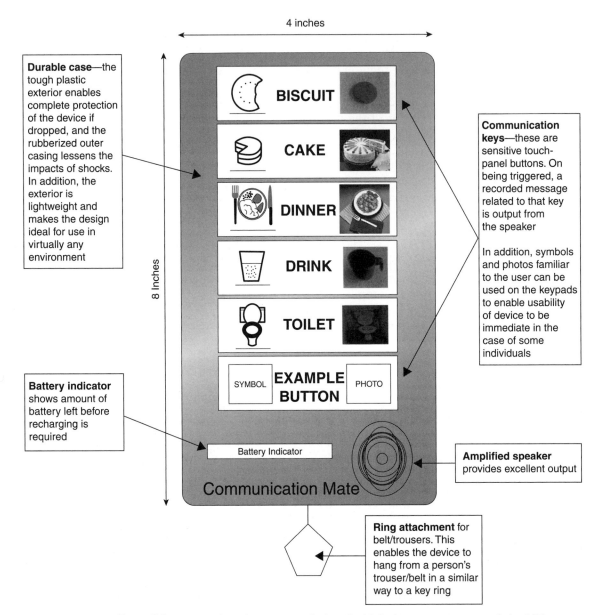

Figure 8.1 A paper-based prototype of a handheld device to support an autistic child.

Heather Martin and Bill Gaver (2000) describe a different kind of prototyping with a different purpose. When prototyping audiophotography products, they used a variety of different techniques including video scenarios similar to the scenarios we introduced in Chapter 7, but filmed rather than written. At each stage, the prototypes were minimally specified, deliberately leaving some aspects vague so as to stimulate further ideas and discussion.

8.2.3 Low-fidelity prototyping

A low-fidelity prototype is one that does not look very much like the final product. For example, it uses materials that are very different from the intended final version, such as paper and cardboard rather than electronic screens and metal. The lump of wood used to prototype the Palm Pilot described above is a low-fidelity prototype, as is the cardboard-box laser printer.

Low-fidelity prototypes are useful because they tend to be simple, cheap, and quick to produce. This also means that they are simple, cheap, and quick to modify so they support the exploration of alternative designs and ideas. This is particularly important in early stages of development, during conceptual design for example, because prototypes that are used for exploring ideas should be flexible and encourage rather than discourage exploration and modification. Low-fidelity prototypes are never intended to be kept and integrated into the final product. They are for exploration only.

Storyboarding Storyboarding is one example of low-fidelity prototyping that is often used in conjunction with scenarios, as described in Chapter 7. A storyboard consists of a series of sketches showing how a user might progress through a task using the device being developed. It can be a series of sketched screens for a GUI-based software system, or a series of scene sketches showing how a user can perform a task using the device. When used in conjunction with a scenario, the storyboard brings more detail to the written scenario and offers stakeholders a chance to role-play with the prototype, interacting with it by stepping through the scenario. The example storyboard shown in Figure 8.2 (Hartfield and Winograd,

Figure 8.2 An example storyboard.

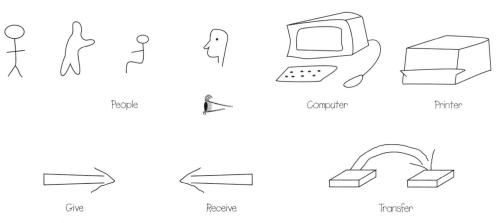

Figure 8.3 Some simple sketches for low-fidelity prototyping.

1996) depicts a person using a new system for digitizing images. This example doesn't show detailed drawings of the screens involved, but it describes the steps a user might go through in order to use the system.

Sketching   Low-fidelity prototyping often relies on sketching, and many people find it difficult to engage in this activity because they are inhibited about the quality of their drawing. Verplank (1989) suggests that you can teach yourself to get over this inhibition. He suggests that you should devise your own symbols and icons for elements you might want to sketch, and practice using them. They don't have to be anything more than simple boxes, stick figures, and stars. Elements you might require in a storyboard sketch, for example, include "things" such as people, parts of a computer, desks, books, etc., and actions such as give, find, transfer, and write. If you are sketching an interface design, then you might need to draw various icons, dialog boxes, and so on. Some simple examples are shown in Figure 8.3. Try copying these and using them. The next activity requires other sketching symbols, but they can still be drawn quite simply.

ACTIVITY 8.1   Produce a storyboard that depicts how to fill a car with gas (petrol).

Comment   Our attempt is shown in Figure 8.4.

Prototyping with Index Cards   Using index cards (small pieces of cardboard about 3×5 inches) is a successful and simple way to prototype an interaction, and is used quite commonly when developing websites. Each card represents one screen or one element of a task. In user evaluations, the user can step through the cards, pretending to perform the task while interacting with the cards. A more detailed example of this kind of prototyping is given in Section 8.3.4.

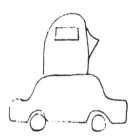

Drive car to gas pump

Take nozzle from pump...

...and put it into the
car's gas tank

Squeeze trigger on
the nozzle until
tank is full

Replace nozzle
when tank is full

Pay cashier

Figure 8.4 A storyboard depicting how to fill a car with gas.

Wizard of Oz   Another low-fidelity prototyping method called Wizard of Oz assumes that you have a software-based prototype. In this technique, the user sits at a computer screen and interacts with the software as though interacting with the product. In fact, however, the computer is connected to another machine where a human operator sits and simulates the software's response to the user. The method takes its name from the classic story of the little girl who is swept away in a storm and finds herself in the Land of Oz (Baum and Denslow, 1900).

8.2.4 High-fidelity prototyping

High-fidelity prototyping uses materials that you would expect to be in the final product and produces a prototype that looks much more like the final thing. For example, a prototype of a software system developed in Visual Basic is higher fidelity than a paper-based mockup; a molded piece of plastic with a dummy keyboard is a higher-fidelity prototype of the PalmPilot than the lump of wood.

If you are to build a prototype in software, then clearly you need a software tool to support this. Common prototyping tools include Macromedia Director, Visual Basic, and Smalltalk. These are also full-fledged development environments, so they are powerful tools, but building prototypes using them can also be very straightforward.

Table 8.1   Relative effectiveness of low- vs. high-fidelity prototypes (Rudd et al., 1996)

Type	Advantages	Disadvantages
Low-fidelity prototype	• Lower development cost. • Evaluate multiple design concepts. • Useful communication device. • Address screen layout issues. • Useful for identifying market requirements. • Proof-of-concept.	• Limited error checking. • Poor detailed specification to code to. • Facilitator-driven. • Limited utility after requirements established. • Limited usefulness for usability tests. • Navigational and flow limitations.
High-fidelity prototype	• Complete functionality. • Fully interactive. • User-driven. • Clearly defines navigational scheme. • Use for exploration and test. • Look and feel of final product. • Serves as a living specification. • Marketing and sales tool.	• More expensive to develop. • Time-consuming to create. • Inefficient for proof-of-concept designs. • Not effective for requirements gathering.

Marc Rettig (1994) argues that more projects should use low-fidelity prototyping because of the inherent problems with high-fidelity prototyping. He identifies these problems as:

• They take too long to build.

• Reviewers and testers tend to comment on superficial aspects rather than content.

• Developers are reluctant to change something they have crafted for hours.

• A software prototype can set expectations too high.

• Just one bug in a high-fidelity prototype can bring the testing to a halt.

High-fidelity prototyping is useful for selling ideas to people and for testing out technical issues. However, the use of paper prototyping and other ideas should be actively encouraged for exploring issues of content and structure. Further advantages and disadvantages of the two types of prototyping are listed in Table 8.1.

8.2.5   Compromises in prototyping

By their very nature, prototypes involve compromises: the intention is to produce something quickly to test an aspect of the product. The kind of questions or choices

BOX 8.1   Prototyping Cultures (Schrage, 1996)

"The culture of an organization has a strong influence on the quality of the innovations that the organization can produce." (Schrage, 1996, p. 193)

This observation is drawn mainly from product-related organizations, but also applies to software development. There are primarily two kinds of organizational culture for innovation: the specification culture and the prototyping culture. In the former, new products and development are driven by written specifications, i.e., by a collection of documented requirements. In the latter, understanding requirements and developing the new product are driven by prototyping. Large companies such as IBM or AT&T that have to gather and coordinate a large amount of information tend to be specification-driven, while smaller entrepreneurial companies tend to be prototype-driven. Both approaches have potential disadvantages. A carefully prepared specification may prove completely infeasible once prototyping begins. Similarly, a wonderful prototype may prove to be too expensive to produce on a large scale.

The medium used for developing the prototype affects the process itself. Schrage puts forward the example of General Motors, which used to produce clay prototypes of new cars and then try to capture these in CAD tools. On the other hand, Toyota designs its cars using CAD tools first and produces a clay prototype once the design has stabilized. The medium used also determines in part the questions that a prototype can answer. As a simple example, a horizontal software prototype will not be able to answer questions about the detailed operation of a function since it is not designed to model that level of detail.

The speed of prototype development and the time between prototyping iterations is often a product of organizational culture and tradition. Some companies have a set number of prototypes embodied in their development method and use this number irrespective of the technical needs of any particular product. Generally speaking, the more prototyping cycles there are, the more polished the final product will be.

The corporate prototyping culture is most starkly revealed by who is involved in prototyping and when. For example, who owns the prototype? Is there a special prototyping department? Who gets to see and evaluate the prototype? Sometimes designers are happy to show emerging prototypes to their peers but not to managers, for fear of being misunderstood and either having the project cancelled or finding an order to ship the prototype when it is not ready. Prototype demonstrations to senior managers often happen too late in the development cycle to have any real impact because of these fears.

David Kelley (Schrage, p.195) claims that organizations wanting to be innovative need to move to a prototype-driven culture. Schrage sees that there are two cultural aspects to this shift. First, scheduled prototyping cycles that force designers to build many prototypes are more likely to lead to a prototype-driven culture than allowing designers to produce *ad hoc* prototypes when they think it appropriate. Second, rather than innovative teams being needed for innovative prototypes, it is now recognized that innovative prototypes lead to innovating teams! This can be especially significant when the teams are cross-functional, i.e., multidisciplinary.

---

that any one prototype allows the designer to answer is therefore limited, and the prototype must be designed and built with the key issues in mind. In low-fidelity prototyping, it is fairly clear that compromises have been made. For example, with a paper-based prototype an obvious compromise is that the device doesn't actually work! For software-based prototyping, some of the compromises will still be fairly clear; for example, the response speed may be slow, or the exact icons may be sketchy, or only a limited amount of functionality may be available.

Two common compromises that often must be traded against each other are breadth of functionality provided versus depth. These two kinds of prototyping

"THEN IN HERE WE DO A CLAY MOCK-UP
OF THE COMPUTER MODEL"

are called *horizontal prototyping* (providing a wide range of functions but with little detail) and *vertical prototyping* (providing a lot of detail for only a few functions).

Other compromises won't be obvious to a user of the system. For example, the internal structure of the system may not have been carefully designed, and the prototype may contain "spaghetti code" or may be badly partitioned. One of the dangers of producing running prototypes, i.e., ones that users can interact with automatically, is that they may believe that the prototype is the system. The danger for developers is that it may lead them to consider fewer alternatives because they have found one that works and that the users like. However, the compromises made in order to produce the prototype must not be ignored, particularly the ones that are less obvious from the outside. We still must produce a good-quality system and good engineering principles must be adhered to.

8.2.6 Construction: from design to implementation

When the design has been around the iteration cycle enough times to feel confident that it fits requirements, everything that has been learned through the iterated steps of prototyping and evaluation must be integrated to produce the final product.

Although prototypes will have undergone extensive user evaluation, they will not necessarily have been subjected to rigorous quality testing for other characteristics such as robustness and error-free operation. Constructing a product to be used by thousands or millions of people running on various platforms and under a wide range of circumstances requires a different testing regime than producing a quick prototype to answer specific questions.

The dilemma box below discusses two different development philosophies. One approach, called *evolutionary prototyping*, involves evolving a prototype into the final product. An alternative approach, called *throwaway prototyping*, uses the prototypes as stepping stones towards the final design. In this case, the

---

DILEMMA Prototyping to Throw Away

Low-fidelity prototypes are never intended to be kept and integrated into the final product. But when building a software-based system, developers can choose to do one of two things: either build a prototype with the intention of throwing it away after it has fulfilled its immediate purpose, or build a prototype with the intention of evolving it into the final product.

Above, we talked about the compromises made when producing a prototype, and we commented that the "invisible" compromises, concerned with the structure of the underlying software must not be ignored. However, when a project team is under pressure to produce the final product and a complex prototype exists that fulfills many of the requirements, or maybe a set of vertical prototypes exists that together fulfill the requirements, it can become very tempting to pull them together and issue the result as the final product. After all, many hours of development have probably gone into developing the prototypes, and evaluation with the client has gone well,

so isn't it a waste to throw it all away? Basing the final product on prototypes in this way will simply store up testing and maintenance problems for later on: in short, this is likely to compromise the quality of the product.

Evolving the final prototype into the final product through a defined process of evolutionary prototyping can lead to a robust final product, but this must be clearly planned and designed for from the beginning. Building directly on prototypes that have been used to answer specific questions through the development process will not yield a robust product. As Constantine and Lockwood (1999) observe, "Software is the only engineering field that throws together prototypes and then attempts to sell them as delivered goods".

On the other hand, if your device is an innovation, then being first to market with a "good enough" product may be more important for securing your market position than having a very high-quality product that reaches the market two months after your competitors'.

---

prototypes are thrown away and the final product is built from scratch. If an evolutionary prototyping approach is to be taken, the prototypes should be subjected to rigorous testing along the way; for throw-away prototyping such testing is not necessary.

8.3 Conceptual design: moving from requirements to first design

Conceptual design is concerned with transforming the user requirements and needs into a conceptual model. Conceptual models were introduced in Chapter 2, and here we provide more detail and discuss how to go about developing one. We defined conceptual model as "a description of the proposed system in terms of a set of integrated ideas and concepts about what it should do, behave, and look like, that will be understandable by the users in the manner intended." The basis for designing this model is the set of user tasks the product will support. There is no easy transformation to apply to a set of requirements data that will produce "the best" or even a "good enough" conceptual model. Steeping yourself in the data and trying to empathize with the users while considering the issues raised in this section is one of the best ways to proceed. From the requirements and this experience, a picture of what you want the users' experience to be when using the new product will emerge.

Beyer and Holtzblatt (1998), in their method *Contextual Design* discussed in Chapter 9, recommend holding review meetings within the team to get different peoples' perspectives on the data and what they observed. This helps to deepen understanding and to expose the whole team to different aspects. Ideas will emerge as this extended understanding of the requirements is established, and these can be tested against other data and scenarios, discussed with other design team members and prototyped for testing with users. Other ways to understand the users' experience are described in Box 8.2.

Ideas for a conceptual model may emerge during data gathering, but remember what Suzanne Robertson said in her interview at the end of Chapter 7: you must separate the real requirements from solution ideas.

Key guiding principles of conceptual design are:

- Keep an open mind but never forget the users and their context.
- Discuss ideas with other stakeholders as much as possible.
- Use low-fidelity prototyping to get rapid feedback.
- Iterate, iterate, and iterate. Remember Fudd's first law of creativity: "To get a good idea, get lots of ideas" (Rettig, 1994).

Considering alternatives and repeatedly thinking about different perspectives helps to expand the solution space and can help prompt insights. Prototyping (introduced in Section 8.2) and scenarios (introduced in Chapter 7) are two techniques to help you explore ideas and make design decisions. But before explaining how these can help, we need to explore in more detail how to go about envisioning the product.

8.3.1   Three perspectives for developing a conceptual model

Chapter 2 introduced three ways of thinking about a conceptual model: Which interaction mode would best support the users' activities? Is there a suitable interface metaphor to help users understand the product? Which interaction paradigm will the product follow? In this section, we discuss each of these in more detail. In all the discussions that follow, we are not suggesting that one way of approaching a conceptual design is right for one situation and wrong for another; they all provide different ways of thinking about the product and hence aid in generating alternatives.

Which interaction mode?   Which interaction mode is most suitable for the product depends on the activities the user will engage in while using it. This information is identified through the requirements activity. The interaction mode refers to how the user invokes actions when interacting with the device. In Chapter 2 we introduced two different types of interaction mode: those based on activities and those based on objects. For those based on activities, we introduced four general styles: instructing, conversing, manipulating and navigating, and exploring and browsing. Which is best suited to your current design depends on the application domain and the kind of system being developed. For example, a computer game is most likely to suit a manipulating and navigating style, while a drawing package has aspects of instructing and conversing.

BOX 8.2   How to Really Understand the Users' Experience

Some design teams go to great lengths to ensure that they come to empathize with, not just understand, the users' experience. We know from learning things ourselves that "learning by doing" is more effective than being told something or just seeing something. Buchenau and Suri (2000) describe an approach they call experience prototyping that is intended to give designers some of the insight into a user's experience that comes only from first-hand knowledge. For example, they describe a team designing a chest-implanted automatic defibrillator. A defibrillator is used with victims of cardiac arrest when their heart muscle goes into a chaotic arrythmia and fails to pump blood, a state called fibrillation. A defibrillator delivers an electric shock to the heart, often through paddle electrodes applied externally through the chest wall; an implanted defibrillator does this through leads that connect directly to the heart muscle. In either case, it's a big electric shock intended to restore the heart muscle to its regular rhythm that can be powerful enough to knock people off their feet.

This kind of event is completely outside most people's experience, and so it is difficult really to understand what the user's experience is likely to be for this kind of device. You can't fit a proto-

type pacemaker to each member of the design team and simulate fibrillation in them! This makes it difficult for designers to gain the insight they need. However, you can simulate some critical aspects of the experience, one of which is the random occurrence of a defibrillating shock. To achieve this, each team member was given a pager to take home over the weekend (elements of the pack are shown in Figure 8.5). The pager message simulated the occurrence of a defibrillating shock. Messages were sent at random, and team members were asked to record where they were, who they were with, what they were doing, and what they thought and felt knowing that this represented a shock. Experiences were shared the following week, and example insights ranged from anxiety around everyday happenings such as holding a child and operating power tools, to being in social situations at a loss how to communicate to onlookers what was happening. This first-hand experience brought new insights to the design effort.

Another instance in which designers tried hard to come to terms with the user experience is the Third Age suit, developed at ICE, Loughborough University (see Figure 8.6). This suit was designed so that car designers could experience what it

Figure 8.5 The patient kit for experience prototyping.

might be like to be in an older body. The suit restricts movement in the neck, arms, legs, and ankles in a way that simulates the mobility problems typically experienced by someone over 55 years of age. For example, when operating the foot pedals in a car, many "third agers" (as they are called), lack the flexibility in their ankles to be able to rest their heel on the floor and operate the pedals by flexing their ankle. Thus they have to lift their whole foot up and push it down each time they operate the pedal, which puts much more stress on their leg muscles.

Figure 8.6 The Third Age suit: (a) riding a bike and (b) using a mobile phone.

Most conceptual models will be a combination of modes, and it is necessary to associate different parts of the interaction with different modes. For example, consider the shared calendar example introduced in Chapter 7. One of the user tasks is finding out what is happening on a particular day. In this instance, instructing is an appropriate mode of interaction. No dialog is necessary for the system to show the required information. On the other hand, the user task of trying to arrange a meeting among a set of people may be conducted more like a conversation. We can imagine that the user begins by selecting the people for the meeting and setting some constraints on the arrangements such as time limit, urgency, length of meeting, etc. Then the system might respond with a set of possible times and dates for the user to select. This is much more like a conversation. (You may like to refer back to the scenario of this task in Chapter 7 and consider how well it matches this interaction mode.) For the task of planning, the user is likely to want to scan through pages and browse the days.

ACTIVITY 8.2   Consider the library catalog system introduced in Chapter 7. Identify tasks associated with this product that would be best supported by each of the interaction modes instructing, conversing, manipulating and navigating, and exploring and browsing.

Comment   Here are some suggestions. You may have identified others:

(a) Instructing: the user wants to see details of a particular book, such as publisher and location.

(b) Conversing: the user wants to identify a book on a particular topic but doesn't know exactly what is required.

(c) Manipulating and navigating: the library books could be represented as icons that could be interrogated for information or manipulated to represent the book being reserved or borrowed.

(d) Exploring and browsing: the user is looking for interesting books, with no particular topic or author in mind.

Models based on objects provide a different perspective since they are structured around real-world objects. For example, the shared calendar system can be thought of as an electronic version of a paper calendar, which is a book kept by each person on their desk or in their bag. Alternatively, it could be thought of as a planner, a large flat piece of paper that is often pinned up on the wall in offices and is far more public. The choice of which objects to choose as a basis for the conceptual model is related to the choice of interface metaphor, which we consider below.

Mayhew (1999) identifies a similar distinction between conceptual models: process-oriented or product-oriented. The former kind of model best fits "an application in which there are no clearly identifiable primary work products. In these applications the main point is to support some work process." Examples of this might be software to control a chemical processing plant, a financial management package, or a customer care call-center. On the other hand, a product-oriented model "will best fit an application in which there are clear, identifiable work products that users individually create, modify and maintain." Examples of this are Microsoft products such as Excel, Powerpoint, Word, etc. More information about these kinds of conceptual model is given in Box 8.3.

Is there a suitable interface metaphor?   Interface metaphors are another way to think about conceptual models. They are intended to combine familiar knowledge with new knowledge in a way that will help the user understand the system. Choosing suitable metaphors and combining new and familiar concepts requires a careful balance and is based on a sound understanding of the users and their context. For example, consider an educational system to teach six-year-olds mathematics. You could use the metaphor of a classroom with a teacher standing at the blackboard. But if you consider the users of the system and what is likely to engage them, you will be more likely to choose a metaphor that reminds the children of something they enjoy, such as a ball game, the circus, a playroom, etc.

Erickson (1990) suggests a three-step process for choosing a good interface metaphor. The first step is to understand what the system will do. Identifying functional requirements was discussed in Chapter 7. Developing partial conceptual models and trying them out may be part of the process. The second step is to understand which bits of the system are likely to cause users problems. Another way of looking at this is to identify which tasks or subtasks cause problems, are complicated, or are critical. A metaphor is only a partial mapping between the software and the real thing upon which the metaphor is based. Understanding areas in which users are likely to have difficulties means that the metaphor can be chosen to support those aspects. The third step is to generate metaphors. Looking for metaphors in the users' description of the tasks is a good starting point. Also, any

BOX 8.3 Process-Oriented versus Product-Oriented Conceptual Models *(Mayhew, 1999)*

Mayhew (1999) characterizes conceptual models in terms of their focus on products or on process. This is similar to our characterization of models that focus on objects or ones that focus on activities.

The difference between these two kinds of conceptual model is the drivers for the design activity. For a product-oriented system, the main products and the tools needed to create them form the main structure of the application. For a process-oriented application, it is the list of process steps that forms the system's basis. Mayhew suggests the following issues must be addressed during conceptual design, whether the application is primarily product-oriented or process-oriented:

- Products or processes must be clearly identified. For example, what documents are to be generated and what other tools are required to produce them? In a process-oriented model, what processes are to be supported?

- A set of presentation rules must be designed. For example, urgent tasks must be placed on the desktop, while less urgent

tasks may be accessible through the menu bar. If designing for a GUI, design rules and guidelines come with the particular platform (see Box 8.5 below).

- Design a set of rules for how windows will be used.

- Identify how major information and functionality will be divided across displays.

- Define and design major navigational pathways. This will draw on the task analysis earlier, and leads to a structure for the tasks. Don't over-constrain users, make navigation easy, and provide facilities so that they always know where they are.

- Document alternative conceptual design models in sketches and explanatory notes.

An example conceptual model based on this approach is shown in Figure 8.7. This is a process-based model, and so it is structured around the processes and subprocesses the system is to support.

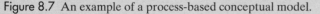

UTILITY CUSTOMER SERVICE

Service Requests	Billing Questions	Info Requests
Change Service	View Bill	Show Services
Cancel Service	Change Bill	Show Products
Install New Service	Sales—Offer Service Options	Show Installation Procedures
Add Customer Info		
Select Service	Maintenance Requests	
Check Credit	Report Problem	
Quote Rate	Schedule Maintenance	
Schedule Install	View Maintenance History	

Figure 8.7 An example of a process-based conceptual model.

metaphors used in the application domain with which the users may be familiar may be suitable.

When suitable metaphors have been generated, they need to be evaluated. Again, Erickson (1990) suggests five questions to ask.

1. How much structure does the metaphor provide? A good metaphor will require structure, and preferrably familiar structure.

Here is a possible design presenting these work processes through a process-oriented Conceptual Model:

Conceptual Model—Utility Customer Service

Application Window

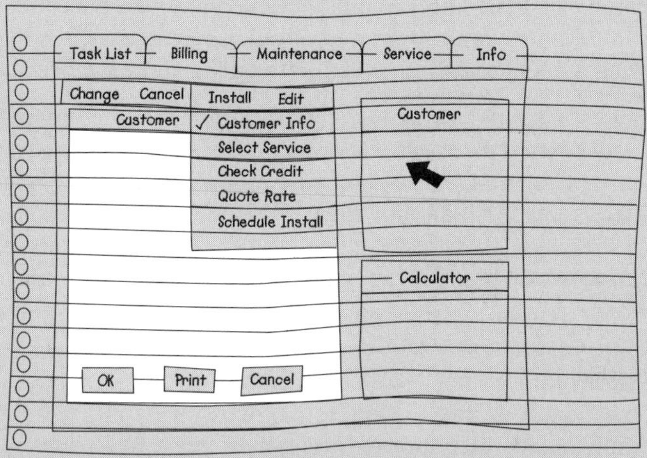

The presentation rules followed in this design are as follows (this design is not complete; it simply provides some examples of components of a process-oriented model).

The overall application is represented as a tab metaphor. *Highest-level processes* (e.g., billing questions, maintenance requests) are represented by tabs, and each tab represents a work space for that process. The tabbed work space includes a main window where that process is carried out, plus two "common windows," where tools common across all highest-level processes are maintained.

Second-level subprocesses are represented by selections in the menu bar within each tab, and *third-level subprocesses* by selections in pull-downs from the menu bar.

Structured subprocesses are controlled through dimming of subprocesses in pull-downs (until earlier subprocesses are completed, later subprocesses are dimmed out and unselectable).

Completed subprocesses are designated with a check mark.

Common activities available across highest-level processes (i.e., Customer, Calculator) are presented as separate, dedicated windows within the tabbed work spaces.

All windows are dialog boxes—that is, they cannot be minimized. They are all unresizable and unscrollable, but are movable and modeless. The main dialog box represents subprocesses, and it changes contents as the user moves through the subprocesses in sequence to complete a given subprocess and process.

Different windows have different background colors. Note that the active tabbed workspace itself is dark gray, the main subprocess dialog box is white, and all common dialog boxes are light gray.

Figure 8.7 (*continued*).

2. How much of the metaphor is relevant to the problem? One of the difficulties of using metaphors is that users may think they understand more than they do and start applying inappropriate elements of the metaphor to the system, leading to confusion or false expectations.

3. Is the interface metaphor easy to represent? A good metaphor will be associated with particular visual and audio elements, as well as words.

4. Will your audience understand the metaphor?

5. How extensible is the metaphor? Does it have extra aspects that may be useful later on?

In the calendar system, one obvious metaphor we could use is the individual's paper-based calendar. This is familiar to everyone, and we could combine that familiarity with facilities suitable for an electronic document such as hyperlinks and searching. Having thought of this metaphor, we need to apply the five questions listed above.

1. Does it supply structure? Yes, it supplies structure based on the familiar paper-based calendar. However, it does not supply structure for the notion of sharing information, i.e., other people looking in the calendar, because of two issues: first, an individual's calendar is very personal, and second, even if there is a paper-based calendar for a set of people, it can be closed and the information hidden from casual observers.

2. How much of the metaphor is relevant i.e., how many properties of the paper-based calendar are applicable to the electronic version? Well, in the electronic version it isn't appropriate to think of physically turning pages, but then a facility for looking at one "page" after another is required. The individual's calendar can be carried around from place to place. Whether or not we want to encourage that aspect of the metaphor depends on the kind of interaction paradigm we might consider. Finally, this is a shared calendar, and normally our personal calendars are not shared.

3. Is the metaphor easy to represent? Yes.

4. Will your audience understand the metaphor? Yes.

5. How extensible is the metaphor? The functionality of a paper-based calendar is fairly limited. However, it is also a book, and we could borrow facilities from electronic books (which are also familiar objects to most of our audience), so yes, it can be extended.

ACTIVITY 8.3   Another possible interface metaphor for the shared calendar system is the wall planner. Ask the five questions above of this metaphor.

Comment   (a) Does it supply structure? Yes, it supplies structure based on the wall-planner. This metaphor embodies the notion of public access more than the paper-based calendar. In particular, the wall planner is never "closed" to those who are near it.

(b) How much of the metaphor is relevant? Most of this metaphor is relevant. Individuals don't walk around with the wall planner, though, so the answer depends on how the calendar is to be used.

(c) Is the metaphor easy to represent? Yes, it could be represented as a spreadsheet.

(d) Will your audience understand the metaphor? Yes.

(e) How extensible is the metaphor? The functionality of a wall planner is also fairly limited. There are no obvious ways in which to extend the metaphor to help with this application.

Which interaction paradigm?   Interaction paradigms are design philosophies that help you think about the product being developed. Interaction paradigms include the now traditional desktop paradigm, with WIMP interface (windows, icons, menus and pointers), ubiquitous computing, pervasive computing, wearable computing, tangible bits, attentive environments, and the Workaday World. Thinking about the user tasks with these different paradigms in mind can help provide insight both to choose the interaction paradigm and to inspire a different perspective on the problem.

Thinking about environmental requirements is particularly relevant when considering interaction paradigms. For example, consider the shared calendar in the context of the following paradigms:

- *Ubiquitous computing.* Combining some of our earlier discussions, we could perhaps imagine the shared calendar as being like a planner on the wall, but in an electronic form with which people could interact.

- *Pervasive computing.* Carrying around our own copy of the shared calendar builds directly upon current expectations and experience of personal calendars. We can imagine a system that allows individuals to keep a copy of the system on their own palmtop computers or PDAs, while also being linked to a central server somewhere that allows access to other information that is shared.

- *Wearable computing.* Imagine having an earring or a tie pin telling you that you have an appointment in an hour's time at a client's office and that you need to book a taxi? Or maybe asking you whether it is all right to book a meeting with your colleague on a particular date. What other possibilities can this model conjure up?

ACTIVITY 8.4   Consider the library catalog system and think about each of the paradigms listed above. Choose two of them and suggest different kinds of interaction that these paradigms imply.

Comment   We had the following thoughts, but you may have others. The library catalog is likely to be used only in certain places, such as the library or perhaps in an office. The idea of wearable computers is not as attractive in this situation as pervasive computing would be, since people would have to put on the wearable when they arrived at the library. Alternatively, the library system might be designed to "cut in" on an existing wearable. Both of these solutions seem a little intrusive. Pervasive computing, on the other hand, would allow users to interact with the catalog wherever in the library they were, rather than having to go to a place where the PC or card catalog sits. You could possibly have digital books at the end of each library shelf that gave access to the catalog.

8.3.2   Expanding the conceptual model

Considering the issues in the previous section helps the designer to envision a product. These ideas must be thought through in more detail before being prototyped or tested with users. One aspect that will need to be decided is what technologies to use, e.g., mutimedia, virtual reality, or web-based materials, and what input and output devices best suit the situation, e.g., pen-based, touch screen, speech, keyboard, and so on. These decisions will depend on the constraints on the system, arising from the requirements you have established. For example, input and output devices will be influenced particularly by user and environmental requirements.

You also have to decide what concepts need to be communicated between the user and the product and how they are to be structured, related, and presented. This means deciding which functions the product will support, how those functions are related, and what information is required to support them. Although these decisions must be made, remember that they are made only tentatively to begin with and may change after prototyping and evaluation.

What functions will the product perform?   Understanding the tasks the product will support is a fundamental aspect of developing the conceptual model, but it is also important to consider more specifically what functions the product will perform, i.e., how the task will be divided up between the human and the machine. For example, in the shared calendar example, the system may suggest dates when a set of people are able to meet, but is that as far as it should go? Should it automatically book the dates, or should it email the people concerned informing them of the meeting or asking if this is acceptable? Or is the human user or the meeting attendee responsible for checking this out? Developing scenarios, essential use cases, and use cases for the system will help clarify the answers to these questions. Deciding what the system will do and what must be left for the user is sometimes called *task allocation*. The trade-off between what to hand over to the device and what to keep in the control of the user has cognitive implications (see Chapter 3), and is linked to social aspects of collaboration (see Chapter 4). An example relating to our shared calendar system was discussed in Box 4.2 of Chapter 4: should the system allow users to book meetings in others' calendars without asking their consent first? In addition, if the cognitive load is too high for the user, then the device may be too stressful to use. On the other hand, if the device takes on too much and is too inflexible, then it may not be used at all.

Another aspect concerns the functions the hardware will perform, i.e., what functions will be hard-wired into the device and what will be left under software control, and thereby possibly indirectly in the control of the human user? This leads to considerations of the architecture of the device, although you would not expect necessarily to have a clear architectural design at this stage of development.

How are the functions related to each other?   Functions may be related temporally, e.g., one must be performed before another, or two can be performed in parallel. They may also be related through any number of possible categorizations, e.g., all functions relating to telephone memory storage in a cell phone, or all options for accessing files in a word processor. The relationships between tasks may constrain use or may indicate suitable task structures within the device. For example, if a task is dependent on completion of another task, then you may want to restrict the user to performing the tasks in strict order. An instance in which this has been put into practice is in some CASE (Computer-Aided Software Engineering) tools designed to support a specific development approach. Often these tools will insist that certain diagrams must be drawn before others. For example, in object-oriented software development you normally draw class diagrams before sequence diagrams, and some tools do not allow you to draw a sequence diagram until the relevant class diagram is in place. If you're working on a small project that doesn't require this kind of discipline, this can be very frustrating, but from the perspective of a manager in charge of a large project, having these restrictions in place may be advantageous.

If task analysis has been performed on relevant tasks, the breakdown will support these kinds of decisions. For example, in the shared calendar example, the task analysis performed in Section 7.1 shows the subtasks involved and the order in which the subtasks can be performed. Thus, the system could allow meeting constraints to be found before or after the list of people, and the potential dates could be identified in the individuals' calendars before checking with the departmental calendar. It is, however, important to get both the list of attendees and meeting constraints before looking for potential dates.

What information needs to be available?   What data is required to perform a task? How is this data to be transformed by the system? Data is one of the categories of requirements we aim to identify and capture through the requirements activity. During conceptual design, we need to consider the information requirements and ensure that our model caters for the necessary data and that information is available as required to perform the task. Detailed issues of structure and display, such as whether to use an analog display or a digital display, will more likely be dealt with in the later, physical design activity, but implications arising from the type of data to be displayed may impact conceptual design issues.

For example, in the task of booking a meeting among a set of people using the shared calendar, the system needs to be told who is to be at the meeting, how long the meeting is to take, what its location should be, and what is the latest date on which the meeting should be booked, e.g., in the next week, next two weeks, etc. In order to perform the function, the system must have this information and also must have calendar information for each of the people in the meeting, the set of locations where the meeting may take place, and ideally some way of knowing how long a person would have to travel to the location.

8.3.3 Using scenarios in conceptual design

In Chapter 7, we introduced scenarios as informal stories about user tasks and activities. They are a powerful mechanism for communicating among team members and with users. We stated in Chapter 7 that scenarios could be used and refined through different data gathering sessions, and they can indeed be used to check out potential conceptual models.

Scenarios can be used to explicate existing work situations, but they are more commonly used for expressing proposed or imagined situations to help in conceptual design. Often, stakeholders are actively involved in producing and checking through scenarios for a product. Bødker identifies four roles that have been suggested for scenarios (Bødker, 2000, p. 63):

- as a basis for the overall design
- for technical implementation
- as a means of cooperation within design teams
- as a means of cooperation across professional boundaries, i.e., as a basis of communication in a multidisciplinary team

In any one project, scenarios may be used for any or all of these. Box 8.4 details how different scenarios were used throughout the development of a speech-

Scenario 3: Hyper-wonderland
This scenario addresses the positive aspects of how a hypermedia solution will work.
The setting is the Lindholm construction site sometime in the future.
Kurt has access to a portable PC. The portables are hooked up to the computer at the site office via a wireless modem connection, through which the supervisors run the hypermedia application.
Action: During inspection of one of the caissons Kurt takes his portable PC, switches it on and places the cursor on the required information. He clicks the mouse button and gets the master file index together with an overview of links. He chooses the links of relevance for the caisson he is inspecting.
Kurt is pleased that he no longer needs to plan his inspections in advance. This is a great help because due to the 'event-driven' nature of inspection, constructors never know where and when an inspection is taking place. Moreover, it has become much easier to keep track of personal notes, reports etc. because they can be entered directly on the spot.
The access via the construction site interface does not force him to deal with complicated keywords either. Instead, he can access the relevant information right away, literally from where he is standing.
A positive side effect concerns his reachability. As long as he has logged in on the computer, he is within reach of the secretaries and can be contacted when guests arrive or when he is needed somewhere else on the site. Moreover, he can see at a glance where his colleagues are working and get in touch with them when he needs their help or advice.
All in all, Kurt feels that the new computer application has put him more in control of things.
Scenario 4: Panopticon
This scenario addresses the negative aspects of how a hypermedia solution will work.
The setting is the Lindholm construction site sometime in the future.
Kurt has access to a portable PC. The portables are hooked up to the computer at the site office via a wireless modem connection, through which the supervisors run the hypermedia application.
Action: During inspecting one of the caissons Kurt starts talking to one of the builders about some reinforcement problem. They argue about the recent lab tests, and he takes out his portable PC in order to provide some data which justify his arguments. It takes quite a while before he finds a spot where he can place the PC: either there is too much light, or there is no level surface at a suitable height. Finally, he puts the laptop on a big box and switches it on. He positions the cursor on the caisson he is currently inspecting and clicks the mouse to get into the master file. The table of contents pops up and from the overview of links he chooses those of relevance - but no lab test appears on the screen. Obviously, the file has not been updated as planned.
Kurt is rather upset. This loss of prestige in front of a contractor engineer would not have happened if he had planned his inspection as he had in the old days.
Sometimes, he feels like a hunted fox especially in situations where he is drifting around thinking about what kind of action to take in a particular case. If he has forgotten to log out, he suddenly has a secretary on the phone: "I see you are right at caisson 39, so could you not just drop by and take a message?"
All in all Kurt feels that the new computer application has put him under control.

Figure 8.8 Example plus and minus scenarios.

recognition system. More specifically, scenarios have been used as scripts for user evaluation of prototypes, providing a concrete example of a task the user will perform with the product. Scenarios can also be used to build a shared understanding among team members of the kind of system being developed. Scenarios are good at selling ideas to users, managers, and potential customers. For example the scenario presented in Figure 7.7 was designed to sell ideas to potential customers on how a product might enhance their lifestyles.

An interesting idea also proposed by Bødker is the notion of *plus* and *minus* *scenarios*. These attempt to capture the most positive and the most negative consequences of a particular proposed design solution (see Figure 8.8) thereby helping designers to gain a more comprehensive view of the proposal.

ACTIVITY 8.5   Consider an in-car navigation device for planning routes, and suggest one plus and one minus scenario. For the plus scenario, try to think of all the possible benefits of the device. For the minus scenario, try to imagine everything that could go wrong.

Comment   *Scenario 1* This plus scenario shows some potential positive aspects of an in-car navigation system.

> *"Beth is in a hurry to get to her friend's house. She jumps into the car and switches on her in-car navigation system. The display appears quickly, showing her local area and indicating the current location of her car with a bright white dot. She calls up the memory function of the device and chooses her friend's address. A number of her frequent destinations are stored like this in the device, ready for her to pick the one she wants. She chooses the "shortest route" option and the device thinks for a few seconds before showing her a bird's-eye view of her route. This feature is very useful because she can get an overall view of where she is going.*
>
> *Once the engine is started, the display reverts to a close-up view to show the details of her journey. As she pulls away from the pavement, a calm voice tells her to "drive straight on for half a mile, then turn left." After half a mile, the voice says again "turn left at the next junction." As Beth has traveled this route many times before, she doesn't need to be told when to turn left or right, so she turns off the voice output and relies only on the display, which shows sufficient detail for her to see the location of her car, her destination and the roads she needs to use."*

Scenario 2 This minus scenario shows some potential negative aspects of an in-car navigation system.

> *"Beth is in a hurry to get to her friend's house. She gets in her car and turns on the in-car navigation system. The car's battery is faulty so all the information she had entered into the device has been lost. She has to tell the device her destination by choosing from a long list of towns and roads. Eventually, she finds the right address and asks for the quickest route. The device takes ages to respond, but after a couple of minutes displays an overall view of the route it has found. To Beth's dismay, the route chosen includes one of the main roads that is being dug up over this weekend, so she cannot use the route. She needs to find another route, so she presses the cancel button and tries again to search for her friend's address through the long list of towns and roads. By this time, she is very late."*

BOX 8.4   Using Scenarios throughout Design

Scenarios were used throughout the design of a speech-recognition system (Karat, 1995). The goal of the project was to produce a product that used speech-recognition technology, so there was no defined set of user requirements to start with. The system offered speech-to-text dictation capabilities and also speech command capabilities for an application running on the same platform.

Initially, scenarios were used to set the direction of the project: discussions revolved around whether the scenario was correct or not, i.e., whether people would want to use the device to achieve the suggested task. Then scenarios were used to sketch out screens and an early user guide. Discussions at this point included checking what information was needed on the screen at what time, and also deciding what components needed to be built. Use-oriented scenarios, i.e., scenarios suggesting how the device might be used, formed the basis of early design meetings that resulted in a shared understanding of what facilities the system might include. An example scenario from basic direction setting was, "Imagine taking away the keyboard and mouse from your current workstation and describe doing everything through voice commands."

Once the basic direction was agreed, further scenarios were generated to discuss the components of the system. These scenarios focused on typical use of speech commands so that vocabulary could be tracked. An example scenario for discussing vocabulary and system components was:

> *Overall task*: Open system editor, find file REPORT.TXT, change font to Times 16, save changes, and exit the editor.

This scenario was then broken down into a specific word list as follows:

> *Voice scenario steps*: "system_editor" "open" "open" "file" "find" "r" "e" "p" "open" "font" "times" "16" "ok" "save" "close"

A short user guide was developed early on, in parallel with the initial scenario development. User guide scenarios were generated by thinking about the kinds of questions a user might need to answer, for example, What is a speech manager? How do I know what I can say?

Once early prototypes were developed, scenarios together with additional tasks were used as a basis for user testing. One of the problems was that people were unsure of what they could say, and although the system included a "What can I say?" module, this itself proved difficult to use. An example scenario used in testing was "Change the background color of the icon for the communications folder to red."

Scenarios in the form of video prototypes were taken to potential customers later in the project for feedback. The feedback they received was mostly in scenario form too, and the scenarios extracted were fed back into the design process. For example, one of the scenarios collected was, "I would like to walk around while I dictate." This could be accommodated by making mobility a factor when selecting the microphone.

Collecting feedback in the form of scenarios continued later in the project, and these informed both the design of the product and the associated documentation.

8.3.4   Using prototypes in conceptual design

The whole point of producing a prototype is to allow some evaluation of the emerging ideas to take place. As pointed out above, prototypes are built in order to answer questions. Producing anything concrete requires some consideration of the details of the design. If the prototype is to be evaluated seriously by users, then they must be able to see how their tasks might be supported by the product, and this will require consideration of more detailed aspects.

Prototyping is used to get feedback on emerging designs. This feedback may be from users, or from colleagues, or it may be feedback telling you that the idea is not technically feasible. Different kinds of prototype are therefore used at different points in the development iterations and with different people. Generally speaking, low-fidelity prototypes (such as paper-based scenarios) are used earlier in design and higher-fidelity prototypes (such as limited software implementations) are used later in design. However, low-fidelity prototypes are not very impressive to look at, so if the feedback you're looking for is approval from people who will be basing their judgment on first impressions, then a horizontal, high-fidelity prototype might suit the job better than one based on post-its or cards.

Figure 8.9 shows a card-based prototype for the shared calendar system created for a user testing session to check that the task flow and the information requirements were correct for the task of arranging a meeting. The first card shows the screen that asks the user for relevant information to find a suitable meeting date. The second card shows the screen after the system has found some potentially suitable dates and displays the results. Finally, the third screen depicts the situation

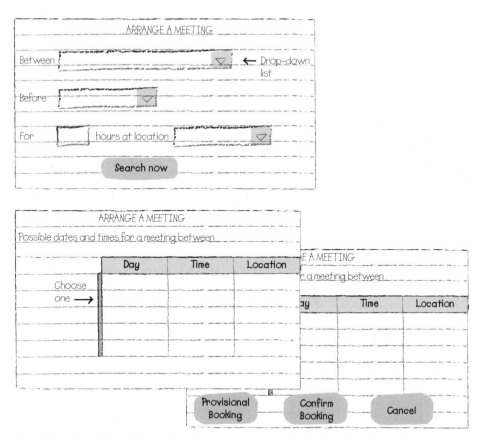

Figure 8.9 A card-based prototype for booking a meeting in the shared calendar system.

after a user has chosen one of the dates and is asked to provisionally book the chosen option, to confirm that this should be booked, or to cancel.

Note that at this point we have not decided how the navigation will work, i.e., whether there will be a tool bar, menus, etc. But we have included some detailed aspects of the design, in order to provide enough detail for users to interact with the prototype.

To illustrate how these cards can be used and the kind of information they can yield, we held a prototyping session with a potential user of the calendar. The session was informal (a kind of "quick and dirty" evaluation that you'll learn more about in Chapter 11) and lasted about 20 minutes. The user was walked through the task to see if the work flow was appropriate for the task of booking a meeting. Generally, the work flow agreed with the user's model of the task, but the session also highlighted some further considerations that did not arise in the original data gathering. Some of these had to do with work flow, but others were concerned with more detailed design. For example, the user suggested that it should be possible to state a range of dates rather than just a "before" date; he also thought that the people attending the meeting should have a chance to confirm the date through the system, and then when everyone had confirmed, the booking could be confirmed and placed in the calendar. On the detailed design, he thought that date entry through a matrix rather than a drop-down list would be more comfortable, and he asked how the possible meeting dates would be ordered. There were many more comments, all of which would be food for thought in the design. We considered only the one task, and yet it yielded a lot of very useful information.

ACTIVITY 8.6   Produce a card-based prototype for the library catalog system and the task of borrowing a book as described by the scenario, use case, and HTA in Chapter 7. You may also like to ask one of your peers to act as a user and step through the task using the prototype.

Comment   Our version of the prototype is shown in Figure 8.10.

8.4 Physical design: getting concrete

Physical design involves considering more concrete, detailed issues of designing the interface, such as screen or keypad design, which icons to use, how to structure menus, etc.

There is no rigid border between conceptual design and physical design. As you saw above, producing a prototype inevitably means making some detailed decisions, albeit tentatively. Interaction design is inherently iterative, and so some detailed issues will come up during conceptual design; similarly, during physical design it will be necessary to revisit decisions made during conceptual design. Exactly where the border lies is not relevant. What is relevant is that the conceptual design should be allowed to develop freely without being tied to physical constraints too early, as this might inhibit creativity.

Design is about making choices and decisions, and the designer must strive to balance environmental, user, data and usability requirements with functional

Figure 1.2 Novel forms of interactive products embedded with computational power (clockwise from top left):

(ii) an IBM prototype of a color electronic ink page, is intended for e-newspapers that can 'typeset' themselves and update while being light enough to carry around.

(i) Electrolux screen-fridge that provides a range of functionality, including food management where recipes are displayed, based on the food stored in the fridge.

(iv) Barney, an interactive cuddly toy that makes learning enjoyable.

(iii) 'geek chic', a Levi jacket equipped with a fully integrated computer network (body area network), enabling the wearer to be fully connected to the web.

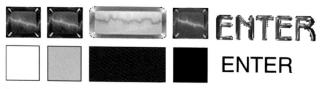

Figure 1.11 2D and 3D buttons. Which are easier to distinguish between?

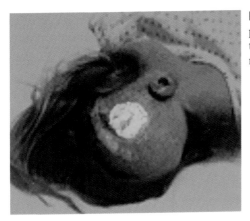

Figure 2.1 An example of augmented reality. Virtual and physical worlds have been combined so that a digital image of the brain is superimposed on the person's head, providing a new form of medical visualization.

Figure 2.14 The i-room project at Stanford: a graphical rendering of the Interactive Room Terry Winograd's group is researching, which is an innovative technology-rich prototype workspace, integrating a variety of displays and devices. An overarching aim is to explore new possibilities for people to work together (see http://graphics.stanford.EDU/projects/iwork/).

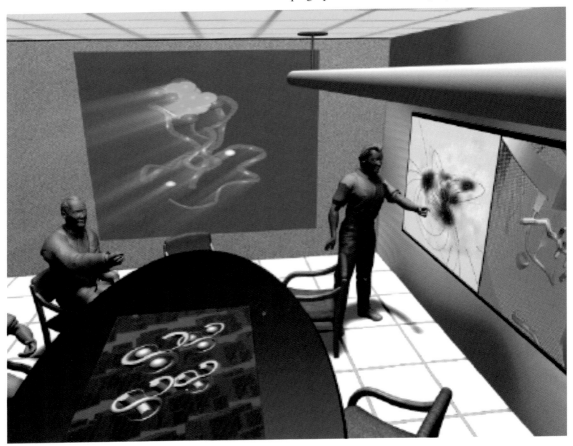

Figure 2.6 Recent direct-manipulation virtual environments

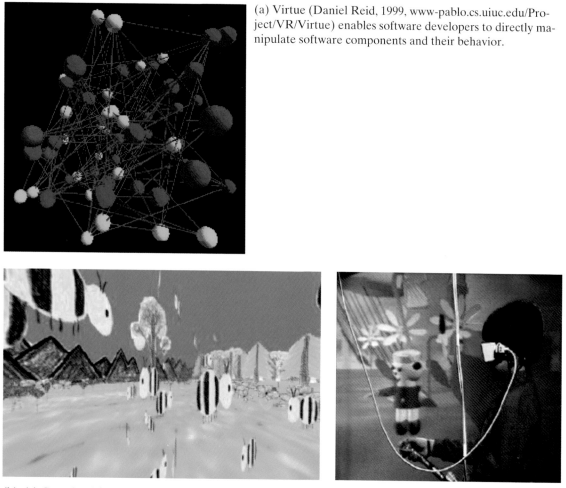

(a) Virtue (Daniel Reid, 1999, www-pablo.cs.uiuc.edu/Project/VR/Virtue) enables software developers to directly manipulate software components and their behavior.

(b), (c) Crayoland (Dave Pape, www.ncsa.uiuc.edu/Vis/) is an interactive virtual environment where the child in the image on the right uses a joystick to navigate through the space. The child is interacting with an avatar in the flower world.

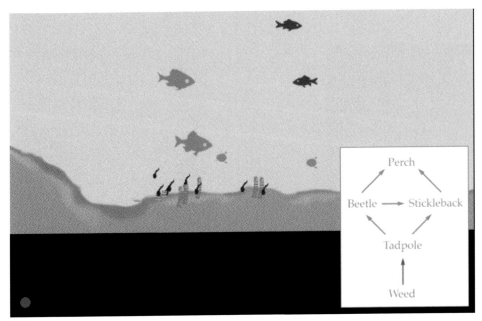

Figure 3.7 Dynalinking used in the PondWorld software. In the background is a simulation of a pond ecosystem, comprising perch, stickleback, beetles, tadpoles, and weeds. In the foreground is a food web diagram representing the same ecosystem but at a more abstract level. The two are dynalinked: changes made to one representation are reflected in the other. Here the user has clicked on the arrow between the tadpole and the weed represented in the diagram. This is shown in the PondWorld simulation as the tadpole eating the weed. The dynalinking is accompanied by a narrative explaining what is happening and sounds of dying organisms.

Figure 3.9 A see-through handset—transparency does not mean simply showing the insides of a machine but involves providing a good system image.

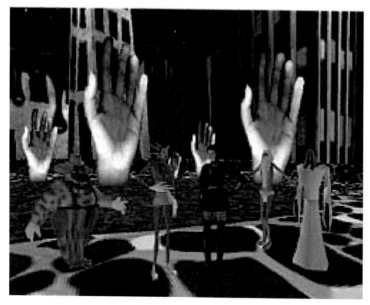

Figure 4.1 The rooftop garden in BowieWorld, a collaborative virtual environment (CVE) supported by Worlds.com. The User takes part by "dressing up" as an avatar. There are hundreds of avatars to choose from, including penguins and real people. Once avatars have entered a world, they can explore it and chat with other avatars.

Figure 5.3 Examples of aesthetically pleasing interactive products: iMac, Nokia cell phone and IDEO's digital radio for the BBC.

Figure 5.9 Virtual screen characters:

(a) Aibo, the interactive dog.

(b) Ananova, the virtual newscaster.

(c) Ecyas, the German virtual pop star.

Figure 5.11 Herman the bug watches as a student chooses roots for a plant in an Alpine meadow.

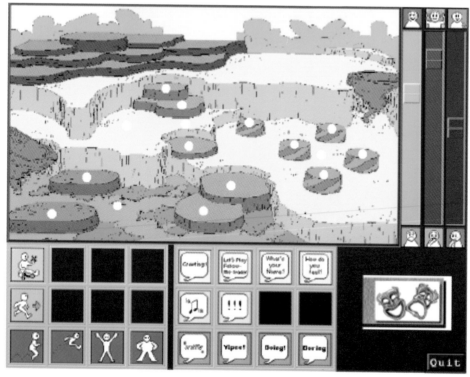

Figure 5.12 The Woggles interface, with icons and slider bars representing emotions, speech and actions.

Figure 5.13 Rea the real estate agent welcoming the user to look at a condo.

Figure 7.3(b) The KordGrip being used underwater.

Figure 15.8 The first foam models of a mobile communicator for children.

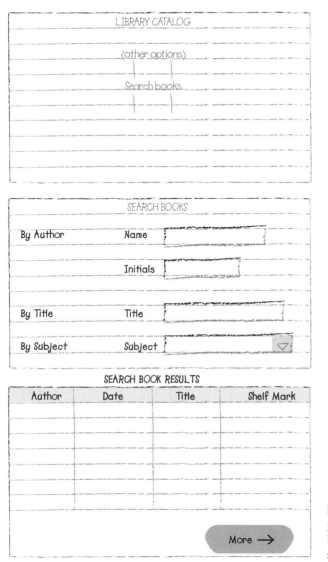

Figure 8.10 A card-based prototype for borrowing a book in the library catalog system.

requirements. These are often in conflict. For example, a cell phone must provide a lot of functionality but is constrained by having only a small screen and a small keyboard. This means that the display of information is limited and the number of unique function keys is also limited, resulting in restricted views of information and the need to associate multiple functions with function keys. Figure 8.11 shows the number of words it can display.

There are many aspects to the physical design of interactive products, and we can't cover them all in this book. Instead, we introduce some principles of

Figure 8.11 An average cell phone screen can display only a short message legibly.

good design in the context of some common interface elements. On our website (www.ID-book.com), you will find more activities and concrete examples of physical design.

8.4.1   Guidelines for physical design

The way we design the physical interface of the interactive product must not conflict with the user's cognitive processes involved in achieving the task. In Chapter 3, we introduced a number of these processes, such as attention, perception, memory, and so on, and we must design the physical form with these human characteristics very much in mind. For example, to help avoid memory overload, the interface should list options for us instead of making us remember a long list of possibilities. A wide range of guidelines, principles, and rules has been developed to help designers ensure that their products are usable, many of which are embodied in style guides and standards (see Box 8.5 for more information on this). Nielsen's set of guidelines were introduced in Chapter 1 in the form of heuristics. Another well-known set intended for informing design is Shneiderman's eight golden rules of interface design (Shneiderman, 1998):

1. *Strive for consistency.* For example, in every screen have a 'File' menu in the top left-hand corner. For every action that results in the loss of data, ask for confirmation of the action to give users a chance to change their minds.

2. *Enable frequent users to use shortcuts.* For example, in most word-processing packages, users may move around the functions using menus or shortcut "quick keys," or function buttons.

3. *Offer informative feedback.* Instead of simply saying "Error 404," make it clear what the error means: "The URL is unknown." This feedback is also influenced by the kinds of users, since what is meaningful to a scientist may not be meaningful to a manager or an architect.

4. *Design dialogs to yield closure.* For example, make it clear when an action has completed successfully: "printing completed."

5. *Offer error prevention and simple error handling.* It is better for the user not to make any errors, i.e., for the interface to prevent users from making mistakes. However, mistakes are inevitable and the system should be forgiving about the errors made and support the user in getting back on track.

6. *Permit easy reversal of actions.* For example, provide an "undo" key where possible.

7. *Support internal locus of control.* Users feel more comfortable if they feel in control of the interaction rather than the device being in control.

8. *Reduce short-term memory load.* For example, wherever possible, offer users options rather than ask them to remember information from one screen to another.

Other guidelines that have been suggested include keeping the interaction simple and clear, organizing interface elements to aid understanding and use through suitable groupings, and designing images to be immediate and generalizable. All of

---

BOX 8.5 Design Guidelines and Standards

Design guidelines and standards exist to help designers create better designs by learning from others' experience. Some guidelines are at a detailed level and are called design rules, while others are more abstract, require interpretation before being applied, and are called design principles. For example, one very general but very pertinent guideline for website design is, "Keep it simple." This is relevant throughout design but must be interpreted within the specific context before it can be applied. On the other hand, a design rule for web design might be, "Don't offer an option to search the whole web from your own website." This is a very specific rule that requires no interpretation to apply. These terms were introduced in Box 1.5 of Chapter 1, together with some others commonly used in this context.

Design Principles

Principles often embody design-related information derived from theory, and so this is one way in which the cognitive models and processes introduced in Chapter 3 can be put to practical use in your designs. For example, the guideline to use "recognition rather than recall" is based on theories of memory that say people find it is easier to recognize things than to remember them with no prompting. However carefully names are chosen to reflect function, it is easier to choose the right option from a list of options than to remember the name of a command. Shneiderman's (1998) eight golden rules of interface design are an example set of principles.

Rules

Rules are more specific versions of design guidelines and provide more detailed guidance. A classic example of design rules is the collection by Smith and Mosier (1986). These rules are quite detailed and prescriptive. For example, one dealing

with consistent format states, "Adopt a consistent format for the location of various display features from one display to another." Each rule is accompanied by explanatory notes such as examples, exceptions, comments that provide detailed guidance, and any useful references to their own or others' work.

Style guides

A style guide is a collection of specific design rules and principles from which the rules are derived. They are used to ensure a consistent look and feel across a set of applications. The most widely known style guides are those for Windows development (Microsoft Corporation, 1992) and for Macintosh development (Apple Computer Inc., 1993). An example from the Macintosh human interface guidelines concerned with designing color icons states that "When you design an icon, you should start by creating the black-and-white version first, then the color should be added."

Style guides tailored to a specific company can be used to deliver a particular corporate image. Such style guides are called corporate style guides.

Standards

Some international standards govern the development of interactive systems. These are also collections of principles and rules to provide designers with a framework based on others' experience. The most pertinent standards are:

- ISO9241 Ergonomic Requirements for Office Work with Visual Display Terminals (VDTs)
- ISO 13407 Human-centered Design Processes for Interactive Systems.
- ISO 14915 Design of the User Interface of Multimedia Applications.

these focus on making the communication between user and product as clear as possible.

Extensive experience in the art of communication (through posters, text, books, images, advertising, etc.) is relevant to interaction design. In her interview at the end of Chapter 6, Gillian Crampton Smith identifies the roles that traditional designers can play in interaction design; one of them she highlights is the fact that designers are trained to produce a coherent design that delivers the desired message to the intended audience. Including such designers on the team can bring this experience to bear. Mullet and Sano (1995) identify a number of useful design principles arising from the visual arts.

To see how these can be translated into the context of interaction design, we consider their application to different widgets, i.e., screen elements, in the next section.

8.4.2   Different kinds of widget

Interfaces are made up of widgets, elements such as dialog boxes, menus, icons, toolbars, etc. Each element must be designed or chosen from a predesigned set of widgets. Sometimes these decisions are made for you through the use of a style guide. Style guides may be commercially produced, such as the Windows style guide (called commercial style guides), or they may be internal to a company (called corporate style guides). A style guide dictates the look and feel of the interface, i.e., which widgets should be used for which purpose and what they look like. For example, study your favorite Windows applications. Which menu is always on the right-hand side of the toolbar? What icon is used to represent "close" or "print"? Which typeface is used in menus and dialog boxes? Each Windows product has the same look and feel, and this is specified in the Windows style guide. If you go to a commercial website, you may find that each screen also has the same look and feel to it. This kind of corporate identity can be captured in a corporate style guide. More information about standard and style guides is in Box 8.5.

We consider here briefly three main aspects of interface design: menu design, icon design, and screen layout. These are applicable to a wide range of interactive products, from standard desktop interfaces for PC software, to mobile communicator functions and microwave ovens.

Menu design   Menus provide users with a choice that can be a choice of commands or a choice of options related to a command. They provide the means by which the user can perform actions related to the task in hand and therefore are based on task structure and the information required to perform a task.

Menus may be designed as drop-down, pop-up or single-dialog menus. It may seem obvious how to design a menu, but if you want to make the application easy to use and provide user satisfaction, some important points must be taken into account. For example, for pull-down and pop-up menus, the most commonly used functions should be at the top, to avoid frequent long scans and scrolls. The principle of grouping can be used to good effect in menu design. For example, the menu can be divided into collections of items that are related, with each collection being

5.2 Grouping options in a menu

Menu options should be grouped within a menu to reflect user expectations and facilitate option search.

5.2.1 Logical groups

If the menu option contains a large number of options (eight or more) and these options can be logically grouped, options should be grouped by function or into other logical categories which are meaningful to users.

EXAMPLE: Grouping the commands in a word-processing system into such categories as customise, compose, edit, print.

5.2.2 Arbitrary groups

If 8 or more options are arranged arbitrarily in a menu panel, they should be arranged into equally distributed groups utilising the following equation:

$$g = \sqrt{n}$$

where

g is the number of groups
n is the number of options on the panel.

EXAMPLE: Given 19 options in a menu panel, arrange them into 4 groups of about 5 options each.

Figure 8.12 An excerpt from ISO 9241 concerning how to group items in a menu.

separated from others. Opposite operations such as "quit" and "save" should be clearly separated to avoid accidentally losing work instead of saving it (See Figure 1.6 in Chapter 1).

An excerpt from ISO 9241, a major international standard for interaction design, considers grouping in menu design, as shown in Figure 8.12.

To show how the design of menus may proceed, we return to the shared calendar. In our initial data gathering, we identified a number of possible tasks that the user might want to perform using the calendar. These included making an entry, arranging a meeting among a number of people, entering contact details, and finding out other people's engagements. Tied to these would also be a number of administrative and housekeeping actions such as deleting entries, moving entries, editing entries, and so on. Suppose we stick with just this list. The first question is what to call the menu entries. Menu names need to be short, clear, and unambiguous. The space for listing them will be restricted, so they must be short, and you want them to be distinguishable, i.e., not easily confused with one another so that the user won't choose the wrong one by mistake. Our current descriptions are really too long. For example, instead of "find out other people's engagements" we could have *Query entry* as a menu option, following through to a dialog box that asks for relevant details.

We need to consider logical groupings. In this case, we could group according to user goal, i.e., have *Query entry*, *Add entry*, *Edit entry*, *Move entry*, and *Delete entry* grouped together (see Figure 8.13). Similarly, we could group *Add contact*,

Calendar Entry	Contacts	Arrange Meeting
Add Entry	Add Contact	
Edit Entry	Edit Contact	
Move Entry	Delete Contact	
Delete Entry		

Figure 8.13 Possible menu groupings for the shared calendar system.

Edit contact and *Delete contact* together. Finding other people's engagements could be generalized to a simple *Search* option that led to a dialog box in which the search parameters are specified. Arranging a meeting is also an option that doesn't clearly group with other commands. This and the *Search* option may be better represented as options on a toolbar than as menu items on their own.

Icon design   Designing a good icon takes more than a few minutes. You may be able to think up good icons in a matter of seconds, but such examples are unlikely to be widely acceptable to your user group. When symbols for representing ladies' and gents' toilets first appeared in the UK, a number of confused tourists did not understand the culturally specific icons of a woman wearing a skirt and a man wearing trousers. For example, Americans protested that they thought the male icon was a woman wearing a trouser suit. We are now all used to these symbols, and indeed internationally recognized symbols for how to wash clothes, fire exits, road signs, etc. now exist. However, icons are cultural and context-specific. Designing a good icon takes time.

At a simple level, designers should always draw on existing traditions or standards, and certainly should not contradict them. Concrete objects or things are easier to represent as an icon; since they can be just a picture of the item. Actions are harder but can sometimes be captured. For example, using a picture of a pair of scissors to represent "cut" in a word-processing application provides sufficient clues as long as the user understands the convention of "cut" for deleting text.

In our shared calendar, if we are going to have the *Search* and *Arrange a Meeting* commands on a tool bar, we need to identify a suitable icon for each of them. A number of possible icons spring to mind for the *Search* option, mainly because searching is a fairly common action in many interactive products: a magnifying glass or a pair of binoculars are commonly used for such options. Arranging a meeting is a little difficult, though. It's probably easier to focus on the meeting itself than the act of arranging the meeting, but how do you capture a meeting? You want the icon to be immediately recognizable, yet it must be small and simple. What characteristic(s) of a meeting might you capture? One of the things that comes to mind is a group of people, so maybe we could consider a collection of stick people? Another element of a meeting is usually a table, but a table on its own isn't enough, so maybe having a table with a number of people around it would work?

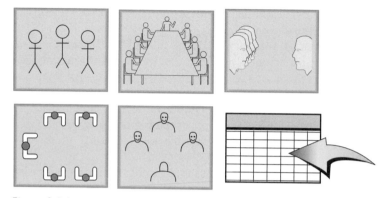

Figure 8.14 A variety of possible icons to represent the "arrange a meeting" function.

ACTIVITY 8.7   Sketch a simple, small icon to represent a set of people around the table, or suggest an icon of your own. Show it to your peers or friends, tell them that it's an icon for a shared calendar application, and see if they can understand what it represents.

Comment   A variety of attempts are shown in Figure 8.14. The last icon is the icon that palm.net uses for arranging meetings. This is a different possibility that tries to capture the fact that you're entering data into the planner.

We discussed some cognitive aspects relevant to icon design in Chapter 3. For example, icons must be designed so that users can readily perceive their meaning and so that they are distinguishable one from another. Since the size of icons on the screen is often very small, this can be difficult to achieve, but users must be able to tell them apart. Look back again at Figure 3.4 and the activity associated with it. How easy do you think it would be to tell some of these icons apart if they were just a little smaller, or the screen resolution was lower?

Screen design.   There are two aspects to screen design: how the task is split across a number of screens, and how the individual screens are designed.

The first aspect can be supported by reference to the task analysis, which broke down the user's task into subtasks and plans of action. One starting point for screen design is to translate the task analysis into screens, so that each task or subtask has its own screen. This will require redesign and adjustment, but it is a starting point. The interaction could be divided into simple steps, each involving a decision or simple data entry. However, this can become idiotic, and having too many simple screens can become just as frustrating as having information all crammed into one screen. This is one of the balances to be drawn in screen design. Tasks that are more complicated than this (and are usually unsuited to simple task analysis) may require a different model of interaction in which a number of screens are open at the same time and the user is allowed to switch among them.

Another issue affecting the division of a task across screens is that all pertinent information must be easily available at relevant times.

Guidelines for the second aspect, individual screen design, draw more clearly from some of the visual communication principles we mentioned above: for example, designing the screen so that users' attention is drawn immediately to the salient points, and using color, motion, boxing and grouping to aid understanding and clarity. Each screen should be designed so that when users first see it, their attention is focused on something that is appropriate and useful to the task at hand. Animations can be very distracting if they are not relevant to the task, but are effective if used judiciously.

Good organization helps users to make sense of an interaction and to interpret it within their own context (as discussed in Chapter 3). This is another example where principles of good grouping can be applied, for example, grouping similar things together or providing separation between dissimilar or unrelated items. Grouping can be achieved in different ways: by placing things close together, using colors, boxes, or frames to segregate items, or using shapes to indicate relationships among elements. There is a trade-off between sparsely populated screens with a lot of open space and overcrowded screens with too many and too complicated sets of icons. If the screen is overcrowded, then users

BOX 8.6   Design Patterns for HCI

Design patterns have become popular in software engineering since the early 1990s. Patterns capture experience, but they have a different structure and a different philosophy from other forms of guidance, such as the guidelines we introduced earlier, or specific methods. One of the intentions of the patterns community is to create a vocabulary, based on the names of the patterns, that designers can use to communicate with one another and with users. Another is to produce a literature in the field that documents experience in a compelling form.

The idea of patterns was first proposed by Christopher Alexander, a British architect who described patterns in architecture. His hope was to capture the "quality without a name" that is recognizable in something when you know it's good.

But what is a pattern? One simple definition is that it is a solution to a problem in a context. What this means is that a pattern describes a problem, a solution, and where this solution has been found to work. This means that users of the pattern can see not only the problem and solution, but also understand when and where it has worked before and access a rationale for why it worked. This helps designers in adopting it (or not) for themselves.

Patterns on their own are interesting, but are not as powerful as a pattern language. A pattern language is a network of patterns that reference one another and work together to create a complete structure.

The application of patterns in HCI is still in its infancy. But some work has been done in the area, and some pattern languages have been proposed. One of the most mature languages is that described by Jan Borchers (2001) for interactive music exhibits. Borchers presents three related languages: one for music, one for HCI and one for software engineering, all of which have arisen from his experience of designing music exhibits. The HCI language addresses issues such as accommodating groups as well as single users, handling complexity, content structure, and interaction devices.

BOX 8.7 Designing for the Web

Web pages need to exhibit the kinds of good interaction design we talk about in this chapter, but they also have some specific requirements. For example, Nielsen (2000) has suggested a set of evaluation criteria specifically for the web (see Chapter 13 for more detail).

The key design issues for websites that are different from other interaction designs are captured very well by three questions proposed by Keith Instone (quoted in Veen, 2001): Where am I? What's here? Where can I go? Each web page should be designed with these three questions in mind. The answers must be clear to users. Jeffrey Veen (2001) expands on these questions. He suggests that a simple way to view a web page is to deconstruct it into three areas (see Figure 8.15). Across the top would be the answer to "Where am I?" Because users can arrive at a site from any direction (and rarely through the front door, or home page), telling them where they are is critical. Having an area at the top of the page that "brands" the page instantly provides that information. Down the left-hand side is an area in which navigation or menus sit. This should allow users immediately to see what else is available on the site, and answers the question "Where can I go?'

The most important information, and the reason a user has come to the site in the first place, is provided in the third area, the content area, which answers the question "What's here?" Content for web pages must be designed differently from standard documents, since the way users read web pages is different. On web pages, content should be short and precise, with crisp sentences. Using headlines to capture the main points of a paragraph is one way to increase the chances of your message getting over to a user who will, perhaps, merely scan the page rather than look at it in detail.

Branding a page is important for the reasons given above. We have also talked about the need to keep screens uncluttered so that people can find their way around and see clearly what is available. However, there may be occasions when the need to maintain a brand overrides other design issues. For example, the website for the Swedish newspaper *Aftonbladet*, while quite busy and crowded (see Figure 8.16), was designed to continue the style of the paper-based version, which also has a busy and crowded appearance.

Download times are critical for the success of websites. Users who have to wait too long for a page will move on somewhere else. So although it's attractive to have graphics on your pages, use them sparingly. One suggestion from Nielsen (2000) is to have few graphics on the welcoming or more abstract pages, and offer users the chance to see pictures of products, or maps, or whatever, only when they explicitly ask for them. It is quite common to use thumbnails, miniaturized versions of the full picture, as links.

Traditionally, hyperlinks have been indicated by highlighting the text in blue and underlining it.

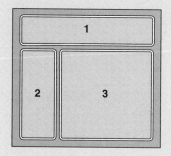

Figure 8.15 Any web page has three main areas.

Figure 8.16 The front web page of the Aftonbladet newspaper.

This is how links were indicated in hypertext, and this convention transferred to web pages. However, now that web pages can be designed more flexibly, some sites are moving away from the traditional blue underlining and are using "roll-over" tags to signify links, i.e., when the mouse rolls over a word then a label appears. This lack of consistency, together with the fact that underlined text is not always a link, can cause considerable confusion for the user.

will become confused and distracted. But too much open space and consequently many screens can lead to frequent screen changes, and a disjointed series of interactions.

Information display. Making sure that the relevant information is available for the task is one aspect of information display, but another concerns the format. Different types of information lend themselves to different kinds of display. For example, data that is discrete in nature, such as sales figures for the last month, could be displayed graphically using a digital technique, while data that is continuous in nature, such as the percentage increase in sales over the last month, is better displayed using an analog device.

If data is to be transferred to the device from a paper-based medium or *vice versa*, it makes sense to have the two consistent. This reduces user confusion and search time in reconciling data displayed with data on the paper.

In the shared calendar application, there is potentially a lot of information to display. If you have five members of the department, each with their own calendars, and the departmental calendar too, then you need to display six sets of engagement information. When we showed the prototype system to our user, he suggested that dates should be chosen through a matrix of some kind rather than a drop-down list. Displaying information appropriately can make communication a lot easier.

8.5 Tool support

The tools available to support the activities described here are wide-ranging and various. We mentioned development environments when talking about prototypes in Section 8.2, but other kinds of support are available.

Much research has been done into appropriate support for different kinds of design and software production, resulting in a huge variety of tools. Because technology moves so quickly, any discussion of specific tools would be quickly out of date. Up-to-date information about support tools can be found on our website (www.ID-book.com). Here we report on some general observations about software tools.

Brad Myers (1995) suggests nine facilities that user interface software tools might provide:

- help design the interface given a specification of the end users' tasks
- help implement the interface given a specification of the design
- create easy-to-use interfaces
- allow the designer to rapidly investigate different designs
- allow nonprogrammers to design and implement user interfaces
- automatically evaluate the interface and propose improvements
- allow the end user to customize the interface
- provide portability
- be easy to use

In a later paper Myers et al. (2000), look at the past, present, and future of user interface tools. Box 8.8 describes some types of tool that have been successful and some that have been unsuccessful.

BOX 8.8   Successes and Failures for User Interface Tools *(Myers et al., 2000)*

Looking at the history of user interface design tools, we can see some tools that have been successful and have withstood the test of time, and others that have fallen by the wayside. Understanding something of what works and what doesn't gives us lessons for the future of such tools.

Tools that have been successful are:

Window Managers and Toolkits. The idea of overlapping windows was first proposed by Alan Kay (Kay, 1969). These have been successful because they help to manage scarce resources: screen space and human perceptual and cognitive resources such as limited visual field and attention.

Event languages that are designed to program actions based on external events: for example, when the left mouse button is depressed, move the cursor here. These have worked because they map well to the direct manipulation graphical user interface style.

Interactive graphical tools or interface builders, such as Visual Basic. These allow the easy construction of user interfaces by placing interface elements on a screen using a mouse. They have been successful because they use a graphical means to design a graphical layout, i.e., you can build a graphical screen layout by grabbing and placing graphical elements without touching any program code.

Component systems are based on the idea of dynamically combining individual components that have been separately written and compiled. Sun's Java Beans uses this approach. One reason for its success is that it addresses the important software engineering goal of modularity.

Scripting languages have become popular because they support fast prototyping. Example scripting languages are Python and Perl.

Hypertext allows elements of a document to be linked in a multitude of ways, rather than the traditional linear layout. Most people are aware of hypertext links because of their use on the web.

Object-oriented programming. This programming approach is successful in interface development because the objects of an interface such as buttons and other widgets can so readily be cast as objects in the language.

Promising approaches that have not caught on are:

Technology has changed so fast that in some cases the tools to support the development of certain technologies have failed to keep up with the rapidly changing requirements. Good ideas that have fallen by the wayside include:

User interface management tools (UIMS). The idea behind UIMS was akin to the idea behind database management systems. Their purpose was to abstract away the details of interface implementation to allow developers to specify and manipulate interfaces at a higher level of abstraction. This separation turned out to be undesirable, as it is not always appropriate to be able to understand and manipulate interface elements only at a high level of abstraction.

Formal language based tools. Many systems in the 1980s were based on formal language concepts such as state transition diagrams and parsers for context-free grammars. These failed to catch on because: the dialog-based interfaces for which these tools were particularly suited were overtaken by direct manipulation interfaces; they were very good at producing sequential interfaces, but not at expressing unordered sequences of action; and they were difficult to learn even for programmers.

Constraints. Tools that were designed to maintain constraints, i.e., relationships among elements of an interface such as that the scroll bar should always be on the right of the window, or that the color of one item should be the same as the color of other items. These systems have not caught on because they can be unpredictable. Once constraints are set up, the tool must find a solution to maintain them, and since there is more than one solution, the tool may find a solution the user didn't expect.

Model-based and automatic techniques. The aim of these systems was to let developers specify interfaces at a high level of abstraction and then for an interface to be automatically generated according to a predefined set of interpretation rules. These too have suffered from problems of unpredictability, since the generation of the interfaces relies on heuristics and rules that are themselves unpredictable in concert.

Assignment

This Assignment continues work on the web-based ticket reservation system at the end of Chapter 7.

(a) Based on the information gleaned from the assignment in Chapter 7, suggest three different conceptual models for this system. You should consider each of the aspects of a conceptual model discussed in this chapter: interaction paradigm, interaction mode, metaphors, activities it will support, functions, relationships between functions, and information requirements. Of these, decide which one seems most appropriate and articulate the reasons why.

(b) Produce the following prototypes for your chosen conceptual model.

(i) Using the scenarios generated for the ticket reservation system, produce a storyboard for the task of buying a ticket for one of your conceptual models. Show it to two or three potential users and get some informal feedback.

(ii) Now develop a prototype based on cards and post-it notes to represent the structure of the ticket reservation task, incorporating the feedback from the first evaluation. Show this new prototype to a different set of potential users and get some more informal feedback.

(iii) Using a software-based prototyping tool (e.g., Visual Basic or Director) or web authoring tool (e.g., Dreamweaver), develop a software-based prototype that incorporates all the feedback you've had so far. If you do not have experience in using any of these, create a few web HTML pages to represent the basic structure of your website.

(c) Consider the web page's detailed design. Sketch out the application's main screen (home page or data entry). Consider the screen layout, use of colors, navigation audio, animation, etc. While doing this, use the three main questions introduced in Box 8.7 as guidance: Where am I? What's here? Where can I go? Write one or two sentences explaining your choices, and consider whether the choice is a usability consideration or a user experience consideration.

Summary

This chapter has explored the activities of design prototyping and construction. Prototyping and scenarios are used throughout the design process to test out ideas for feasibility and user acceptance. We have looked at the different forms of prototyping, and the activities have encouraged you to think about and apply prototyping techniques in the design process.

Key points

- Prototyping may be low fidelity (such as paper-based) or high fidelity (such as software-based).
- High-fidelity prototypes may be vertical or horizontal.
- Low-fidelity prototypes are quick and easy to produce and modify and are used in the early stages of design.
- There are two aspects to the design activity: conceptual design and physical design.
- Conceptual design develops a model of what the product will do and how it will behave, while physical design specifies the details of the design such as screen layout and menu structure.

- We have explored three perspectives to help you develop conceptual models: an interaction paradigm point of view, an interaction mode point of view, and a metaphor point of view.
- Scenarios and prototypes can be used effectively in conceptual design to explore ideas.
- We have discussed four areas of physical design: menu design, icon design, screen design, and information display.
- There is a wide variety of support tools available to interaction designers.

Further reading

WINOGRAD, TERRY (1996) *Bringing Design to Software*. Addison-Wesley and ACM Press. This book is a collection of articles all based on the theme of applying ideas from other design disciplines in software design. It has a good mixture of interviews, articles, and profiles of exemplary systems, projects or techniques. Anyone interested in software design will find it inspiring.

CARROLL, JOHN M. (ed.) (1995) *Scenario-based Design*. John Wiley & Sons, Inc. This volume is an edited collection of papers arising from a three-day workshop on use-oriented design. The book contains a variety of papers including case studies of scenario use within design, and techniques for using them with object-oriented development, task models and usability engineering. This is a good place to get a broad understanding of this form of development.

MULLET, KEVIN, AND SANO, DARELL (1995) *Designing Visual Interfaces*. SunSoft Press. This book is full of practical guidance for designing interactions that focus on communication. The ideas here come from communication-oriented visual designers. Mullet and Sano show how to apply these techniques to interaction design, and they also show some common errors made by interaction designers that contravene the principles.

VEEN, JEFFREY (2001) *The Art and Science of Web Design*. New Riders. A very bright book, providing a lot of practical information taken from the visual arts about how to design websites. It also includes sections on common mistakes to help you avoid these pitfalls.

MYERS, BRAD, HUDSON, S. E., AND PAUSCH, R. (2000) Past, present and future of user interface software tools. *ACM Transactions on Computer-Human Interaction*, 7(1), 3–28. This paper presents an interesting description of user interface tools, expanding on the information given in Box 8.8.

Chapter 9

User-centered approaches to interaction design

9.1 Introduction
9.2 Why is it important to involve users at all?
    9.2.1   Degrees of involvement
9.3   What is a user-centered approach?
9.4   Understanding users' work: applying ethnography in design
    9.4.1   Coherence
    9.4.2   Contextual Design
9.5   Involving users in design: participatory design
    9.5.1   PICTIVE
    9.5.2   CARD

9.1 Introduction

As you would expect, user-centered development involves finding out a lot about the users and their tasks, and using this information to inform design. In Chapter 7 we introduced some data-gathering techniques which can be used to collect this information, including naturalistic observation. Studying people in their "natural" surroundings as they go about their work can provide insights that other data-gathering techniques cannot, and so interaction designers are keen to use this approach where appropriate. One particular method that has been used successfully for naturalistic observation in the social sciences is ethnography. It has also been used with some success in product development but there have been some difficulties knowing how to interpret and present the data gathered this way so that it can be translated into practical design.

Another aspect of user-centered development is user involvement in the development process. There are different degrees of involvement, one of which is through evaluation studies, as discussed in Chapters 10 through 14. Another is for users to contribute actively to the design itself—to become co-designers. As Gillian Crampton Smith said in the interview at the end of Chapter 6, users are not designers, but the payoffs for allowing users to contribute to the design themselves are quite high in terms of user acceptance of the product. So techniques have been developed that engage users actively and productively in design.

In this chapter, we discuss some issues surrounding user involvement, and expand on the principles underlying a user-centered approach. Then we describe two approaches to using ethnographic data to inform design and two approaches to involving users actively in design.

The main aims of this chapter are to:

- Explain some advantages of involving users in development.
- Explain the main principles of a user-centered approach.
- Describe some ethnographic-based methods aimed at understanding users' work.
- Describe some participative design techniques that help users take an active part in design decisions.

9.2 Why is it important to involve users at all?

We talked in Chapter 6 about the importance of identifying stakeholders and of consulting the appropriate set of people. In the past, developers would often talk to managers or to "proxy-users," i.e., people who role-played as users, when eliciting requirements. But the best way to ensure that development continues to take users' activities into account is to involve real users throughout. In this way, developers can gain a better understanding of their needs and their goals, leading to a more appropriate, more useable product. However, two other aspects which have nothing to do with functionality are equally as important if the product is to be usable and used: expectation management and ownership.

Expectation management is the process of making sure that the users' views and expectations of the new product are realistic. The purpose of expectation management is to ensure that there are no surprises for users when the product arrives. If users feel they have been "cheated" by promises that have not been fulfilled, then this will cause resistance and maybe rejection. Expectation management is relevant whether you are dealing with an organization introducing a new software system or a company developing a new interactive toy. In both cases, the marketing of the new arrival must be careful not to misrepresent the product. How many times have you seen an advert for something you thought would be really good to have, but when you see one, discover that the marketing "hype" was a little exaggerated? I expect you felt quite disappointed and let down. Well, this is the kind of feeling that expectation management tries to avoid.

It is better to exceed users' expectations than to fall below them. This does not mean just adding more features, however, but that the product supports the users' work more effectively than they expect. Involving users throughout development helps with expectation management because they can see from an early stage what the product's capabilities are and what they are not. They will also understand better how it will affect their jobs and what they can expect to do with the product; they are less likely to be disappointed. Users can also see the capabilities develop and understand, at least to some extent, why the features are the way they are.

Adequate and timely training is another technique for managing expectations. If you give people the chance to work with the product before it is released, either by training them on the real system or by offering hands-on demonstrations of a prerelease version, then they will understand better what to expect when the final product is released.

A second reason for user involvement is ownership. Users who are involved and feel that they have contributed to a product's development, are more likely to feel a sense of "ownership" towards it and to be receptive to it when it finally emerges. Remember Suzanne Robertson's comment in her interview at the end of Chapter 7 about how important it is for people to feel heard? Well, this is true throughout development, not just at the requirements stage.

9.2.1 Degrees of involvement

Different degrees of user involvement may be implemented in order to manage expectations and to create a feeling of ownership. At one end of the spectrum, users may be co-opted to the design team so that they are major contributors. For any one user, this may be on a full-time basis or a part-time basis, and it may be for the duration of the project or for a limited time only. There are advantages and disadvantages to each situation. If a user is co-opted full-time for the whole project, their input will be consistent and they will become very familiar with the system and its rationale. However, if the project takes many years they may lose touch with the rest of the user group, making their input less valuable. If a user is co-opted part-time for the whole project, she will offer consistent input to development while remaining in touch with other users. Depending on the situation, this will need careful management as the user will be trying to learn new jargon and handle unfamiliar material as a member of the design team, yet concurrently trying to fulfill the demands of their original job. This can become very stressful for the individuals. If a number of users from each user group are co-opted part-time for a limited period, input is not necessarily consistent across the whole project, but careful coordination between users can alleviate this problem. In this case, one user may be part of the design team for six months, then another takes over for the next six months, and so on.

At the other end of the spectrum, users may be kept informed through regular newsletters or other channels of communication. Provided they are given a chance to feed into the development process through workshops or similar events, this can be an effective approach to expectation management and ownership. In a situation with hundreds or even thousands of users it would not be feasible to involve them all as members of the team, and so this might be the only viable option.

If you have a large number of users, then a compromise situation is probably the best. Representatives from each user group may be co-opted onto the team on a full-time basis, while other users are involved through design workshops, evaluation sessions, and other data-gathering activities.

The individual circumstances of the particular project affect what is realistic and appropriate. If your end user groups are identifiable, e.g., you are developing a product for a particular company, then it is easier to involve them. If, however, you are developing a product for the open market, it is unlikely that you will be able to co-opt a user to your design team. Box 9.1 explains how Microsoft involves users in its developments.

One of the reasons often cited for not involving users in development is the amount of time it takes to organize, manage, and control such involvement. This issue may appear particularly acute in developing systems to run on the Internet where ever-shorter timescales are being forced on teams—in this fast-moving area, projects lasting three months or less are common. You might think, therefore, that it would be particularly difficult to involve users in such projects. However, Braiterman et al. (2000) report two case studies showing how to involve users successfully in large-scale but very short multidisciplinary projects, belying the claim that involving users can waste valuable development time.

The first case study was a three-week project to develop the interaction for a new web shopping application. The team included a usability designer, an information architect, a project manager, content strategists, and two graphic designers. In such a short timeframe, long research and prototyping sessions were impossible, so the team produced a hand-drawn paper prototype of the application that was

---

BOX 9.1   *How Microsoft Involves Users (Cusumano and Selby, 1995)*

The synch-and-stabilize process of development used by Microsoft was described in Chapter 6. Here we look at some of the main ways in which users are involved in the development process.

Users are involved throughout development in a variety of ways, from product and feature identification to feature development and testing, and via the customer support call centers.

Microsoft bases feature selection and prioritization on a technique called "activity-based planning." This technique involves studying what users do to achieve a certain activity like writing a letter, and using the results of the study to choose product features. Each new release of a software product is limited to supporting about four new major activities. Each of these proposed new activities can be broken down into subactivities and these mapped against features already existing in the software. Any new features required are noted. If a feature can support more than one activity, then it is placed higher in the priority list. The techniques used to gather customer data for activity-based planning do not appear to be prescribed in any way, and can vary from visiting customers through to asking them to use an instrumented version of the software, i.e., a version that records the actions they take. Microsoft also employs contextual inquiry (see below) to learn about their customers' work, although they find that it can be time-consuming and the results ambiguous.

Because the world of applications software changes so rapidly, developers need to continually observe and test with users. Throughout the development phase, usability tests are carried out in Microsoft's usability lab. Each time a developer believes that a feature is finished, then it is scheduled for testing in the usability lab. A group of about 10 users "off the street" are invited into the lab to perform certain tasks while their behavior is observed and their performance recorded. The data is then analyzed and the findings fed back into development. This results in thorough testing of all features. As an example, Office 4.0 (incorporating Word, Excel, PowerPoint, and other common office software) went through over 8000 hours of usability testing.

Once a product is complete, it is used internally by Microsoft staff (who are selected users and atypical, but are using it in a realistic working environment); then it may be released in a beta version to selected customers.

Microsoft has millions of customers around the world, about 30% of whom call their customer support lines with problems and frustrations resulting from poor features or software errors. This data about customer behavior and their problems with the products is a further source of information that is fed back into product development and improvement.

revised daily in response to customer testing. The customers were asked to perform tasks with the prototype, which was manipulated by one of the team in order to simulate interaction, e.g., changing screens. After half the sessions were conducted, the team produced a more formal version of the prototype in Adobe Illustrator. They found that customers were enthusiastic about using the paper prototype and were keen to offer improvements.

The second case study involved the development of a website for a video game publisher over three months. In order to understand what attracts people to such gaming sites, the multidisciplinary team felt they needed to understand the essence of gaming. To do this, they met 32 teenage gamers over a ten-day period, during which they observed and interviewed them in groups and individually. This allowed the team to understand something of the social nature of gaming and gave insights into the gamers themselves. During design, the team also conducted research and testing sessions in their office lab. This led them to develop new strategies and web designs based on the gamers' habits, likes, and dislikes.

Box 9.2 describes a situation in which users were asked to manage a software development project. There were hundreds of potential users, and so in addition,

---

DILEMMA   Too Much of a Good Thing?

Involving users in development is a good thing. Or is it? And how much should they become involved? Box 9.2 describes a project in which users were appointed as project managers and were actively involved in development throughout. But are users qualified to lead a technical development project? And does this matter, provided there is sufficient technical expertise in the team?

Involving users at any level incurs costs, whether in terms of time for communication, or for workshops, or time spent explaining technical issues. Detailed user studies may also require the use of recording equipment and the subsequent cost of transcription and analysis. What evidence is there that user involvement is productive, or that it is worth putting the required level of resources into development? Research by Keil and Carmel (1995) indicates that the more successful projects do have direct links to users and customers. Kujala and Mäntylä (2000) performed some empirical work to investigate the costs and benefits of user studies early in product development. They concluded that user studies do in fact produce benefits that outweigh the costs of conducting them.

On the other hand, Heinbokel et al. (1996) suggest that a high user involvement has some nega-

tive effects. They found that projects with high user participation showed lower overall success, fewer innovations, a lower degree of flexibility, and low team effectiveness, although these effects were noticeable only later in the project (at least 6–12 months into the project). In short, projects with a high level of user participation tended to run less smoothly. They identified four issues related to communication among users and developers that they suggest caused problems. First, as the project progressed, users developed more sophisticated ideas, and they wanted them to be incorporated late in the project. Second, users were fearful of job losses or worsening job conditions and this led to a tendency for participation to be not constructive. Third, users were unpredictable and not always sympathetic to software development matters. For example, they asked for significant changes to be made just as testing was due to start. Fourth, user orientation in the designers may lead to higher aspirations and hence higher levels of stress.

Webb (1996) too has concerns about user involvement, but Scaife et al. (1997) suggest that it is not the fact of user involvement that is in question, but how and at what stage in development they should get involved.

BOX 9.2 Users as Project Team Leaders *(M880, 2000)*

The Open University (OU) in the UK is a large distance education university with many thousands of students enrolled each year in a variety of courses (undergraduate, graduate, vocational and professional) in a variety of subjects (Technology, Languages, Education, etc.). The courses are presented through paper-based and electronic media including video and audio tapes. It has a network of centers through which it supports and distributes courses to students throughout the UK and Europe (and more recently to the US). The OU employs about 3000 academic and other full-time staff and about 6000 part-time and counseling staff. In 1998/9 the university had over 200,000 students and customers for its education packs, and distributed materials throughout the UK and Europe to over 165,000 students. Such an operation requires considerable computerized support: in 1993 approximately 54 major systems of varying sizes were held on mainframe UN-LX host/workstations, VAX hosts, or PCs.

Traditionally, the systems had been built by an in-house software development team, who, due to resource constraints, sometimes needed to make business decisions although their expertise was in technical issues, not in the business side of the university. When it was time to re-develop these information systems, the OU decided that a new approach to development was required: users were to have a much more significant role.

Development was divided into a number of areas, each with its own project team and its own schedule. Consistency across the areas was maintained through the development of a GUI interface standard style guide that ensured that all systems had the same look and feel (style guides are discussed in Chapter 8). Users were involved in development on a number of different levels, typically 30–80% of their time. For example, in one area (Area E), one user was appointed full-time to manage the project team, two others joined the project team part-time for a limited period (about 18 months each), one user was consulted on a regular basis, and a wider set of users were involved through workshops and prototyping sessions. The project team also included technically trained analysts and developers.

When asked for the most successful and the least successful aspects of the project, both users and technical developers agreed that the most successful had been getting users involved in development. They said that this had made the system closer to what the users wanted. One user commented that, because users were part of the team for only a limited time, they did not see the development through from the beginning, but saw only some of the phases, and that this led to lack of continuity. Another user commented on the fact that the business had changed faster than the software could be developed, and hence the system had to be changed. The users' reactions were not all favorable, however. Another group of users who were consulted mainly through workshops and prototyping sessions did not feel that their needs had been adequately addressed.

One of the user project managers had this to say:

> The most successful thing has been getting people to go back to basics. We didn't look at existing systems and say, "We want the same thing but with go-faster stripes." We've examined what the University wants from the area. The most disappointing part has been that increased user involvement has not brought about ownership of the system by user areas. There was an expectation that we could move away from the traditional view of, "This is a computer system devised by computer people for you to use." In practice it's been far more difficult to get users to make decisions; they tend to say, "That's part of development. You decide."

This lack of ownership was commented upon by users and developers alike. One of the analysts commented:

> The user-led aspect has resulted in [the system's][1] greatest successes and greatest failures. User project managers do not have a systems background. Depending on their character they can be open to ideas or very blinkered. If they come from a user area with a system already it can be hard for them to see beyond their current system.

---

[1]When reporting raw data such as quotations anonymously, it is common practice to replace specific words or phrases that might compromise anonymity with similar words enclosed in square brackets to indicate that they are not the speaker's original words.

users became design team members on a full- and part-time basis; regular design workshops, debriefings, and training sessions were also held.

How actively users should be involved is a matter for debate. Some studies have shown that too much user involvement can lead to problems. This issue is discussed in the Dilemma box.

9.3 What is a user-centered approach?

Throughout this book, we have emphasized the need for a user-centered approach to development. By this we mean that the real users and their goals, not just technology, should be the driving force behind development of a product. As a consequence, a well-designed system should make the most of human skill and judgment, should be directly relevant to the work in hand, and should support rather than constrain the user. This is less a technique and more a philosophy.

In 1985, Gould et al. (1985) laid down three principles they believed would lead to a "useful and easy to use computer system." These are very similar to the three key characteristics of interaction design introduced in Chapter 6.

1. *Early focus on users and tasks*. This means first understanding *who* the users will be by directly studying their cognitive, behavioral, anthropomorphic, and attitudinal characteristics. This required observing users doing their normal tasks, studying the nature of those tasks, and then involving users in the design process.

2. *Empirical measurement*. Early in development, the reactions and performance of intended users to printed scenarios, manuals, etc. is observed and measured. Later on, users interact with simulations and prototypes and their performance and reactions are observed, recorded, and analyzed.

3. *Iterative design*. When problems are found in user testing, they are fixed and then more tests and observations are carried out to see the effects of the fixes. This means that design and development is iterative, with cycles of "design, test, measure, and redesign" being repeated as often as necessary.

Iteration is something we have emphasized throughout these chapters on design, and it is now widely accepted that iteration is required. When Gould et al. wrote their paper, however, the iterative nature of design was not accepted by most developers. In fact, they comment in their paper how "obvious" these principles are, and remark that when they started recommending these to designers, the designers' reactions implied that these principles were indeed obvious. However, when they asked designers at a human factors symposium for the major steps in software design, most of them did not cite most of the principles—in fact, only 2% mentioned all of them. So maybe they had "obvious" merit, but were not so easy to put into practice. The Olympic Messaging System (OMS) (Gould et al., 1987) was the first reported large computer-based system to be developed using these three principles. Here a combination of techniques was used to elicit users' reactions to designs, from the earliest prototypes through to the final product. In this case, users were mainly involved in evaluating designs. The OMS is discussed further in Chapter 10.

The iterative nature of design and the need to develop usability goals have been discussed in Chapter 6. Here, we focus on the first principle, early focus on users and tasks, and suggest five further principles that expand and clarify what this means:

1. *User's tasks and goals are the driving force behind the development.* In a user-centered approach to design, while technology will inform design options and choices, it should not be the driving force. Instead of saying, "Where can we deploy this new technology?," say, "What technologies are available to provide better support for users' goals?"

2. *Users' behavior and context of use are studied and the system is designed to support them.* This is about more than just capturing the tasks and the users' goals. How people perform their tasks is also significant. Understanding behavior highlights priorities, preferences, and implicit intentions. One argument against studying current behavior is that we are looking to improve work, not to capture bad habits in automation. The implication is that exposing designers to users is likely to stifle innovation and creativity, but experience tells us that the opposite is true (Beyer and Holtzblatt, 1998). In addition, if something is designed to support an activity with little understanding of the real work involved, it is likely to be incompatible with current practice, and users don't like to deviate from their learned habits if operating a new device with similar properties (Norman, 1988).

3. *Users' characteristics are captured and designed for.* When things go wrong with technology, we often say that it is our fault. But as humans, we are prone to making errors and we have certain limitations, both cognitive and physical. Products designed to support humans should take these limitations into account and should limit the mistakes we make. Cognitive aspects such as attention, memory, and perception issues were introduced in Chapter 3. Physical aspects include height, mobility, and strength. Some characteristics are general, such as that about one man in 12 has some form of color blindness, but some characteristics may be associated more with the job or particular task at hand. So as well as general characteristics, we need to capture those specific to the intended user group.

4. *Users are consulted throughout development from earliest phases to the latest and their input is seriously taken into account.* As discussed above, there are different levels of user involvement and there are different ways in which to consult users. However involvement is organized, it is important that users are respected by designers.

5. *All design decisions are taken within the context of the users, their work, and their environment.* This does not necessarily mean that users are actively involved in design decisions. As you read in Gillian Crampton Smith's interview at the end of Chapter 6, not everyone believes that it is a good idea for users to be designers. As long as designers remain aware of the users while

making their decisions, then this principle will be upheld. Keeping this context in mind can be difficult, but an easily accessible collection of gathered data is one way to achieve this. Some design teams set up a specific design room for the project where data and informal records of brainstorming sessions are pinned on the walls or left on the table. (This is discussed again in Section 9.4.2 on Contextual Design.)

ACTIVITY 9.1 Assume that you are involved in developing a new e-commerce site for selling garden plants. Suggest ways of applying the above principles in this task.

Comment To address the first three principles, we would need to find out about potential users of the site. As this is a new site, there is no immediate set of users to consult. However, the tasks and goals, behavior, and characteristics of potential users of this site can be identified by investigating how people shop in existing online and physical shopping situations—for example, shopping through interactive television, through other online sites, in a garden center, in the local corner shop, and so on. For each of these, you will find advantages and disadvantages to the shopping environment and you will observe different behaviors. By investigating behavior and patterns in a physical garden center, you can find out a lot about who might be interested in buying plants, how these people choose plants, what criteria are important, and what their buying habits are. From existing online shopping behavior, you could determine likely contexts of use for the new site.

For the fourth principle, because we don't have an easily tapped set of users available, we could follow a similar route to the Internet company described in Section 9.2, and try to recruit people we believe to be representative of the group. These people may be involved in workshops or in evaluation sessions, possibly in a physical shopping environment. Valuable input can be gained in targeted workshops, focus groups, and evaluation sessions. The last principle could be supported through the creation of a design room to house all the data collected.

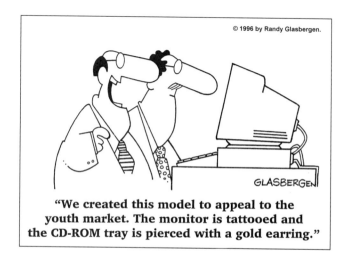

© 1996 by Randy Glasbergen.

GLASBERGEN

"We created this model to appeal to the youth market. The monitor is tattooed and the CD-ROM tray is pierced with a gold earring."

9.4 Understanding users' work: applying ethnography in design

Kuhn (1996) provides a good example illustrating the importance of understanding users' work. She describes a case where a computer system was introduced to cut down the amount of time spent on conversations between telephone-company repair personnel. Such conversations were regarded as inefficient and "off-task." What management had failed to realize was that in the conversations workers were often consulting one another about problems, and were pooling their knowledge to solve them. By removing the need for conversation, they removed a key mechanism for solving problems. If only the designers had understood the work properly, they would not have considered removing it.

Ethnography is a method that comes originally from anthropology and literally means "writing the culture" (Hammersley and Atkinson, 1983). It has been used in the social sciences to display the social organization of activities, and hence to understand work. It aims to find the order within an activity rather than impose any framework of interpretation on it. It is a broad-based approach in which users are observed as they go about their normal activities. The observers immerse themselves in the users' environment and participate in their day-to-day work, joining in conversations, attending meetings, reading documents, and so on. The aim of an ethnographic study is to make the implicit explicit. Those in the situation, the users in this case, are so familiar with their surroundings and their daily tasks that they often don't see the importance of familiar actions or happenings, and hence don't remark upon them in interviews or other data-gathering sessions.

There are different ways in which this method can be associated with design. Beynon-Davies (1997) has suggested that ethnography can be associated with development as "ethnography *of,*" "ethnography *for,*" and "ethnography *within.*" Ethnography *of* development refers to studies of developers themselves and their workplace, with the aim of understanding the practices of development (e.g. Button and Sharrock, 1994; Sharp et al., 1999). Ethnography *for* development yields ethnographic studies that can be used as a resource for development, e.g., studies of organizational work. Ethnography *within* software development is the most common form of study (e.g., Hughes et al., 1993a); here the techniques associated with ethnography are integrated into methods and approaches for development (e.g., Viller and Sommerville, 1999).

Because of the very nature of the ethnographic experience, it is very difficult to describe explicitly what data is collected through such an exercise. It is an experience rather than a data-collection exercise. However, the experience must be shared with other team members, and therefore needs to be documented and rationalized. Box 9.3 provides an example ethnographic account in the form of a description of an ethnographic study of a new media company. In this case, the intention was not explicitly concerned with designing an interactive product, but was a business-oriented ethnography. The style and content of the piece, however, are typical of ethnographies.

Studying the context of work and watching work being done reveals information that might be missed by other methods that concentrate on asking about work away from its natural setting. For example, it can shed light on how people do the "real" work as opposed to the formal procedures that you'd find in documentation;

BOX 9.3   An Example Ethnography
(printed with the permission of Fiona Hovenden, the ethnographer)

Background   I was asked to design a retreat for a new media company. They were about to shift from working in a very open, unstructured way to formalizing their working processes. The main reason for this was that they had signed a deal with a large organization that was going to act as a financial patron in return for first options on the new media company's ideas and designs.

This proposed shift was causing some tension and anxiety within the company, with people feeling that their current working practices worked very well and that imposing structure would stifle the creativity on which their work depended.

Method   Over a four-day period, I had a desk in the office and observed the rhythm of work and the working practices. I spent two days just observing, and then I conducted one-to-one face-to-face interviews with every member of the company, and one-to-one email interviews with the three people from the patron organization who were coming to join the company.

The account (excerpted here) is my notes, built up over the four-day period. The structure, and the content are built up iteratively. For example, on the surface the company appeared very collectivist—everyone in the company treated everyone else as a peer, people invited other people to work with them on projects, nobody seemed to tell anyone else what to do. But during the interviews it became clear that everyone waited for the opinion of the leader before doing anything. In fact, on the surface it looked as though there were three joint partners, but the entire company implicitly and explicitly deferred to one of them. So, although in the account the surface appearance is what I first noted, the interviews indicated that there was no attempt at consensus.

Brief Characterization of User Community   This is an apparently loose aggregation of artists, artist-technicians, information designers, producers, and a small, nontraditional operations team. There is a commitment to an open, collectivist way of working that seems to translate as anyone can say anything or ask for anything, and they will be given a hearing. However, the way things actually get done does not seem to be by consensus. There are obvious and accepted loci of power associated with particular individuals. It is these individuals who bestow a hearing.

The commitment to collectivism is currently undergoing a shift, as a new patron relationship requires a more formal business structure. A business focus will also be a more explicit part of the new community life. There is a shifting power structure, in which [current leader] will be joined by [X] and [Y] from the patron organization at a more formal managerial level. They will become the gateway through which all project ideas will have to pass, and will also control the financing of projects. They will therefore have a great deal of power, while the power available to the collective will be the power to persuade/seduce.

Community Practices and Productions   This community creates new media products. The nature of this work means that there is a strong visual bias and a highly developed visual sophistication.

One of the original motives for the formation of the company was to explore non-traditional narratives. The lead information designer has also described the work the company does as "story telling." Both in the ongoing championing of the exploration of narrative and in the actual client work done, the community is practiced at the visual presentation and production of stories. This (dominant) aspect of work could be described as translating speech to visuals.

Client projects seem primarily visual, sound is not the focus. In terms of workspace noise, sound bleeds in from the surrounding environment. Public music is played sporadically, and usually by [designer], who has brought in a CD player that sits on his desk. However, many community members wear headphones for significant periods of their time at work.

Many of the working practices seem informal and fluidly structured. This is according to the accounts given by community members in interviews, and the specific request from [leader] to make formalizing working practices a major part of the retreat. More information on working practices is not covered by the available data. However, there is one community practice I would like to briefly mention here. I attended one Tuesday morning production meeting. This seems to be the one set time in the week when everyone in the community comes together to create a community-wide status report. However, not everyone showed up. Meetings are possibly the most significant rituals of modern business practice, and seem to work best when the form and function of the ritual is known, understood, and felt to be relevant by everyone involved. The most startling aspect of the production meeting to me was that individuals seemed to decide when the meeting was over and left the room accordingly, without marking the fact in any other way, say by announcing their departure. The effect (to me) was of the meeting dribbling away, which seemed to lessen its importance.

the nature and purposes of collaboration, awareness of other's work, and implicit goals that may not even be recognized by the workers themselves. For example, Heath et al. (1993) have been exploring the implications of ethnographic studies of real-world settings for the design of cooperative systems. We described their underground control room study in Chapter 4, but they have also studied medical centers, architects' practices, and TV and radio studios.

In one of their studies Heath et al. (1993) looked at how dealers in a stock exchange work together. A main motivation was to see whether proposed technological support for market trading was indeed suitable for that particular setting. One of the tasks examined in detail was the process of writing tickets to record deals. It had been commented upon earlier by others that this process of deal capture, using "old-fashioned" paper and pencil technology, was currently time-consuming and prone to error. Based on this finding, it had been further suggested that the existing way of making deals could be improved by introducing new technologies, including touch screens to input the details of transactions, and headphones to eliminate distracting external noise.

However, when Heath et al. began observing the deal capture in practice, they quickly discovered that these proposals were misguided. In particular, they warned that these new technologies would destroy the very means by which the traders currently communicate and keep informed of what others are up to. The touch screens would reduce the availability of information to others on how deals were progressing, while headphones would impede the dealers' ability to inadvertently monitor one another's conversations. They pointed out how this kind of peripheral monitoring of other dealers' actions was central to the way deals are done. Moreover, if any dealers failed to keep up with what the other dealers were doing by continuously monitoring them, it was likely to affect their position in the market, which ultimately could prove very costly to the bank they were working for.

Hence, the ethnographic study proved to be very useful in warning against attempts to integrate new technologies into a workplace without thinking through the implications for the work practice. As an alternative, Heath et al. suggested pen-based mobile systems with gestural recognition that could allow deals to be made efficiently while also allowing the other dealers to continue to monitor one another unobtrusively.

Hughes et al (1993b) state that "doing" ethnography is about being reasonable, courteous and unthreatening, and interested in what's happening. This is particularly important when trying to perform studies in people's homes, such as those described in Box 9.4. There is, of course, more to it than this. Training and practice are required to produce good ethnographies.

BOX 9.4   Ethnographies of the Home

Home use of technology such as the personal computer, wireless telephones, cell phones, remote controls, and so on has grown over the last decade. Although consumer surveys and similar questionnaires may be able to gather some information about this market, ethnographic studies have been used to gain that extra insight that ensures that products do not just perform needed functions but are also pleasurable and easy to use.

Dray and Mrazek (1996) report on an international study of families' use of technology in which they visited 20 families in America, Germany and France. They spent at least four hours in each of the homes, talking with all members of the family, including children of all ages, about their use of computer technology. They give no details of the data collected, but assert that the study was extremely useful, that "there is no substitute for contextual studies," and that the results have influenced many design decisions and specifications for new products. One aspect of the study they emphasize is the need to develop a rapport with the family. They focused their attention on building a strong positive rapport in the first few minutes of the visit. In all cases, they used food as an icebreaker, by either bringing dinner with them for themselves and the family, or by ordering food to be delivered. This provided a mundane topic of conversation that allowed a natural conversation to be held.

After dinner, they moved to the location of the computer and began by asking the children about their use of the technology. Each family member was engaged in conversation about the technology, and printed samples of work were gathered by the researchers. A protocol designed by the marketing and engineering departments of the company were used to guide the conduct of this part of the study, but after all of the protocol had been covered, families were encouraged to discuss topics they were interested in. Immediately after a visit, the team held a formal debriefing session during which all photos, videotapes, products, and notes were reviewed and a summary debriefing questionnaire was completed. A thank-you letter was later sent to the families.

From this description you can see that a huge amount of preparation was required in order to ensure that the study resulted in getting the right data, i.e., in collecting data that was going to answer the relevant questions.

Mateas et al. (1996) report on a pilot ethnographic study that was also aimed at informing the design and development of domestic computing systems. They visited ten families and also emphasize the importance of making families feel comfortable with them. In their study, this was partly achieved by bringing a pizza dinner for everyone. After dinner, the adults and the children were separated. The researchers wanted to get an understanding of a typical day in the home. To do this, each family member was asked to walk through a typical day, using a felt board with a layout of their house, and felt rooms, products, activities, and people that could be moved around on the felt house.

From their work they derived a model of space, time, and social communication that differed from the model implied by the standard PC. For example, the standard PC is designed to be used in one location by one user for long periods of uninterrupted time. The studies revealed that on the other hand, family activity is distributed throughout multiple spaces, is rarely conducted alone, and is not partitioned into long periods of uninterrupted use. In addition, the PC does not support communication among co-located members of family, which is a key element of family life. They conclude that small, integrated computational appliances supporting multiple co-located users are more appropriate to domestic activity than the single PC.

Collecting ethnographic data is not hard although it may seem a little bewildering to those accustomed to using a frame of reference to focus the data collection rather than letting the frame of reference arise from the available data. You collect what is available, what is "ordinary," what it is that people do, say, how they work. The data collected therefore has many forms: documents, notes of your own, pictures, room layouts. Notebook notes may include snippets of conversation and descriptions of rooms, meetings, what someone did, or how people reacted to a situation. It is opportunistic in that you collect what you can collect and make the most of opportunities presented to you. You don't go in with a firm plan, and so the data you collect is not specifiable in advance. You have to do it rather than read about it. What you record can become more focused after being in the field for a while.

ACTIVITY 9.2   Look up from reading this book and observe your surroundings. Wherever you are, the chances are that you can see and hear lots of things, and probably other people too. Start to make a list of what you observe, and when things change or people move, write down what has happened and how it happened. For example, if someone spoke, what did his voice sound like? Angry, calm, whispering, happy? Spend just a few minutes observing what you can see.

Now think about the same observations but begin to interpret them: imagine that you have to place the main items or people that you can see into categories. For example, on a train you might consider who might be getting off at which station, in a bedroom you might think about how to tidy up the items lying around.

How easy is it to go from the detailed description to the more abstracted one?

Comment   As I am writing this, I am in a room on my own. I therefore don't have people to observe, but my desk is covered with things: a pen, a boarding pass from a recent trip abroad, a rosette from a parcel wrapping, and many books, papers, disks etc. If I look around then I can see the wallpaper and the curtains, clothes hanging and in piles on the bed. In the background I can hear cars moving along the road, and the television downstairs. To spend any length of time really describing any one of the things I observe would take up a lot of words, and that's a lot of data.

If I now consider how to file the things I can see, then I would start to think of categories such as which are books, which are research papers, what can be thrown away, and so on. It becomes easier to feel like I'm making progress. The other thing to notice is that some things I can observe are blocked out of my sphere of interest, such as the cars outside.

In some ways, the goals of design and the goals of ethnography are at opposite ends of a spectrum. Design is concerned with abstraction and rationalization. Ethnography, on the other hand, is about detail. An ethnographer's account will be concerned with the minutiae of observation, while a designer is looking for useful abstractions that can be used to inform design. One of the difficulties faced by those wishing to use this very powerful technique is how to harness the data gathered in a form that can be used in design.

Below, we introduce one framework that has been developed specifically to help structure the presentation of ethnographies in a way that enables designers to use them (other frameworks to help orient observers and how to organize this kind

DILEMMA What To Lose When You Abstract?

In Chapter 7, we discussed the need for data interpretation and analysis. This involves structuring and abstracting from the data, so that important aspects of a situation can be reasoned about at a higher level of generalisation without getting bogged down in details. It is inevitable that when moving from a more detailed description to a more abstract one, information will be lost. But what is important and what is irrelevant? This is a key question to answer if ethnographic data is to be used to inform design.

of study are described in Chapter 12). This framework has three main dimensions (Hughes et al, 1997):

1. The *distributed co-ordination* dimension focuses on the distributed nature of the tasks and activities, and the means and mechanisms by which they are co-ordinated. This has implications for the kind of automated support required.

2. The *plans and procedures* dimension focuses on the organizational support for the work, such as workflow models and organizational charts, and how these are used to support the work. Understanding this aspect impacts on how the system is designed to utilize this kind of support.

3. The *awareness of work* dimension focuses on how people keep themselves aware of others' work. No-one works in isolation, and it has been shown that being aware of others' actions and work activities can be a crucial element of doing a good job. In the stock market example described above, this was one aspect that ethnographers identified. Implications here relate to the sharing of information.

Rather than taking data from ethnographers and interpreting this in design, an alternative approach is to train developers to collect ethnographic data themselves. This has the advantage of giving the designers first-hand experience of the situation. Telling someone how to perform a task, or explaining what an experience is like, is very different from showing them or even gaining the experience themselves. Finding people with the skills of ethnographers and interaction designers may be difficult, but it is possible to provide notational and procedural mechanisms to allow designers to gain some of the insights first-hand. The two methods described below provide such support.

9.4.1 Coherence

The Coherence method (Viller and Sommerville, 1999) combines experiences of using ethnography to inform design with developments in requirements engineering. Specifically, it is intended to integrate social analysis with object-oriented analysis from software engineering (which includes producing use cases as described in Chapter 7). Coherence does not prescribe how to move from the social analysis to use cases, but claims that presenting the data from an ethnographic study based around a set of "viewpoints" and "concerns" facilitates the identification of the product's most important use cases.

Viewpoints and concerns

Coherence builds upon the framework introduced above and provides a set of focus questions for each of the three dimensions, here called "viewpoints". The focus questions (see Figure 9.1) are intended to guide the observer to particular aspects of the workplace. They can be used as a starting point to which other questions may be added as experience in the domain and the method increases.

In addition to viewpoints, Coherence has a set of concerns and associated questions. Concerns are a kind of goal, and they represent criteria that guide the requirements activity. These concerns are addressed within each appropriate viewpoint. One of the first tasks is to determine whether the concern is indeed relevant to the viewpoint. If it is relevant, then a set of elaboration questions is used to explore the concern further. The concerns, which have arisen from experience of using ethnography in systems design, are:

1. *Paperwork and computer work*. These are embodiments of plans and procedures, and at the same time are a mechanism for developing and sharing an awareness of work.

2. *Skill and the use of local knowledge*. This refers to the "workarounds" that are developed in organizations and are at the heart of how the real work gets done.

Distributed coordination

- How is the division of labor manifest through the work of individuals and its coordination with others?
- How clear are the boundaries between one person's responsibilities and another's?
- What appreciation do people have of the work/tasks/roles of others?
- How is the work of individuals oriented towards the others?

Plans and procedures

- How do plans and procedures function in the workplace?
- Do they always work?
- How do they fail?
- What happens when they fail?
- How, and in what situations, are they circumvented?

Awareness of work

- How does the spatial organization of the workplace facilitate interaction between workers and with the objects they use?
- How do workers organize the space around them? Which artifacts that are kept to hand are likely to be important to the achievement of everyday work?
- What are the notes and lists that the workers regularly refer to?
- What are the location(s) of objects, who uses them, how often?

Figure 9.1 Focus questions for the three viewpoints.

Paperwork and computer work

- How do forms and other artifacts on paper or screen act as embodiments of the process?
- To what extent do the paper and computer work make it clear to others what stage people are at in their work?
- How flexible is the technology at supporting the work process—is a particular process enforced, or are alternatives permitted?

Skill and the use of local knowledge

- What are the everyday skills employed by individuals and teams in order to get the work done?
- How is local knowledge used and made available, e.g., through the use of personalized checklists, asking experts, etc.?
- To what extent have standard procedures been adapted to take local factors into account?

Spatial and temporal organization

- How does the spatial organization of the workplace reflect how the work is performed?
- Which aspects of the work to be supported are time-dependent?
- Does any data have a "use-by-date"?
- How do workers make sure that they make use of the most up-to-date information?

Organizational memory

- How do people learn and remember how to perform their work?
- How well do formal records match the reality of how work is done?

Figure 9.2 Focus questions for the four concerns.

3. *Spatial and temporal organization.* This concern looks at the physical layout of the workplace and areas where time is important.

4. *Organizational memory.* Formal documents are not the only way in which things are remembered within an organization. Individuals may keep their own records, or there may be local gurus.

The focus questions associated with these concerns are listed in Figure 9.2 and a sample social concern from the air traffic control domain, together with resultant requirements, is shown in Figure 9.3.

9.4.2 Contextual Design

Contextual Design is another technique that was developed to handle the collection and interpretation of data from fieldwork with the intention of building a software-based product. It provides a structured approach to gathering and representing information from fieldwork such as ethnography, with the purpose

Paperwork and computer work

Flight strips embody the process of an aircraft's progress through the sector of airspace controlled by a suite. As an aircraft approaches the sector, its strip is moved progressively to the bottom of the rack until it becomes the current strip for the controller to deal with. The work of the controller can therefore be viewed in terms of dealing with the flow of strips as aircraft enter, traverse, and leave the controller's sector.

The collection of strips in various racks in a suite provide an 'at a glance' means of determining the current and future workload of a particular controller. The practice of 'cocking out' strips, i.e., raising them slightly in the racks, informs the controller that there is something non-standard about the flight concerned. This may be done by the assistant controller when inserting the strip, or by the controller as a reminder. Glancing at the strips provides a controller with an indication of their current and future workload, in the same way as it allows other controllers to see the relative loading on other sectors. This feature of the organization of the strips is used in particular at change over of shifts, where the incoming controller will spend up to 10 minutes looking over the shoulder of the out-going controller in order to 'get the picture' of the current state of the sector.

Flight strips provide incredibly flexible support for the work of controllers. Different practices exist regarding whether strips are placed into the racks in a top to bottom sequence or vice versa. All instructions given by controllers to pilots, and the pilots' acknowledgements, are recorded onto the relevant flight strip. These annotations are made using a standard set of symbols, and different coloured pens according to the annotator's role within the controlling team. In this way, flight strips constitute a record of a flight's progress through a sector.

Requirement 1. The system shall support controllers 'getting the picture' by providing the ability to determine current and future load for a sector 'at a glance'

Requirement 2. The system shall provide a facility to mark exceptional or non-standard flights requiring special attention

Requirement 3. Annotations to flight records shall be recoded and presented in such a way that they identify the person who made them.

Figure 9.3 Elaboration of paperwork and computer work screen.

of feeding it into design. It has been used on a number of projects, e.g., see Box 9.5.

Contextual Design has seven parts: Contextual Inquiry, Work Modeling, Consolidation, Work Redesign, User Environment Design, Mockup and Test with Customers, and Putting It into Practice. In this chapter we are focusing on understanding users' work, and so shall discuss only the first three steps. Step 4 involves changing work practices, which is outside our scope here. Step 5 produces a prototype that is used with customers, and the final step concerns the practicality of the working system. The activities involved in these last two steps have been discussed in general terms in Section 8.2.

Contextual inquiry

Contextual inquiry is an approach to ethnographic study used for design that follows an apprenticeship model: the designer works as an apprentice to the user. The

BOX 9.5   Using Contextual Design for Office Products

Page (1996) reports on the use of Contextual Design in customer research for a new version of the word processor WordPerfect. The company already had some experience of field research, since the initial version of WordPerfect had been based on informal user observation and user testing, although it wasn't seen as such at the time.

The scope of this study was quite wide, with the team wanting to learn about "the making of documents": how they were conceived, created, reviewed, approved, and distributed. To cover this scope, the team was multidisciplinary, involving expertise in word-processor development, human factors, documentation, marketing, and usability. Contextual Design was chosen because it leads the team systematically through the data-gathering and interpretation activities to product design.

The team undertook three weeks of training, organized as one week of training, four weeks of work, one week of training, four weeks of work, etc. Users were chosen carefully so as to reflect different types, including those who use the existing version of the product and those who don't use computers at all, and to be sure that they were representative of the company's main client base. The set of users was refined as data collection progressed and it became obvious where gaps were, and what kinds of user were needed to fill them.

Even though the intentions of the researchers had been communicated to the collaborators, they often arrived at sites to find that a focus group had been arranged rather than an opportunity for observation. Also, some people thought that the researchers were there to help solve their software problems and expected them to spend time on this rather than on data collection. Observing people at higher levels of management proved difficult at times, despite arranging visits well in advance. The result of this was that they collected more data about support roles, such as administrators and secretaries, than about others.

All team members took part in observation, and interviews were conducted in the worker's workplace, as laid down in the Contextual Design method. Generally, the interviews were taped, although if the data was interpreted within 24 to 48 hours there was no need to listen to the tape again.

Data was interpreted with the entire team. The observer would review her notes while other team members asked questions to draw out information. One team member was charged with writing down each important factor identified by the team, and others were responsible for drawing the workflow, sequence, physical, and context models. It was felt, however, that the contextual model did not represent the cultural influences in any useful way, and so it was not used. To structure the data, they used the affinity model, consolidated models, redesigned work models, user environment model, and user interface designs. A portion of the affinity diagram is shown in Figure 9.4.

When producing the redesigned work model, the goal was to streamline processes and eliminate breakdowns. Any emerging technologies that might help in this were identified and studied. For the user interface designs, paper prototypes were developed early on and tested with users, and as concepts became more certain, running prototypes were created in ToolBook and Delphi.

The researchers knew they might have problems in selling the idea to the implementors. To overcome this, they started having open days for the rest of the company as soon as they had their first affinity diagram and consolidated models. In some cases, members of the development teams were on the Contextual Design teams; where developers were not on the teams, they were invited to contribute to the design ideas before products were in their final form. This involvement helped to increase the developers' sense of ownership.

Page (1996) gives two examples of actual features that were a result of their fieldwork: "Make It Fit" and "QuickTasks." Make It Fit is a feature in WordPerfect 6.1 for Windows that takes the text and makes it fit into the available space, either by expanding it to fill blank space or by shrinking it. QuickTasks is a feature of PerfectOffice 3.0 that automates a series of steps across multiple applications, prompting users for information as it is needed.

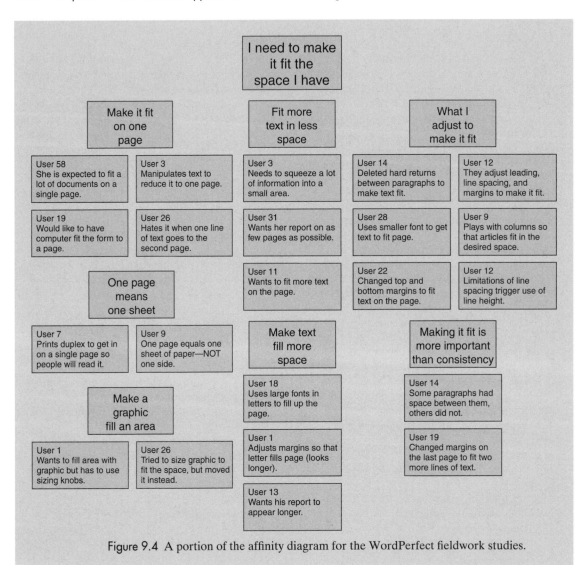

Figure 9.4 A portion of the affinity diagram for the WordPerfect fieldwork studies.

most typical format for contextual inquiry is a contextual interview, which is a combination of observation, discussion, and reconstruction of past events. Contextual inquiry rests on four main principles: context, partnership, interpretation and focus.

The context principle emphasizes the importance of going to the workplace and seeing what happens. The partnership principle states that the developer and the user should collaborate in understanding the work; in a traditional interviewing or workshop situation, the interviewer or workshop leader is in control, but in contextual inquiry the spirit of partnership means that the understanding is developed through cooperation.

The interpretation principle says that the observations must be interpreted in order to be used in design, and this interpretation should also be developed in cooperation between the user and the developer. For example, I have a set of paper cards stuck on my screen at work. They are covered in notes; some list telephone numbers and some list commands for the software I use. Someone coming into my office might interpret these facts in a number of ways: that I don't have access to a telephone directory; that I don't have a user manual for my software; that I use the software infrequently; that the commands are particularly difficult to remember. The best way to interpret these facts is to discuss them with me. In fact, I do have a telephone directory, but I keep the numbers on a note to save me the trouble of looking them up in the directory. I also have a telephone with a memory, but it isn't clear to me how to put the numbers in memory, so I use the notes instead. The commands are there because I often forget them and waste time searching through menu structures.

The fourth principle, the focus principle, was touched upon above in our discussion of ethnography and was also addressed in Coherence: how do you know what to look for? In contextual inquiry, it is important that the discussion remains pertinent for the design being developed. To this end, a project focus is established to guide the interviewer, which will then be augmented by the individual's own focus that arises from their perspective and background. The contextual inquiry interview differs from ethnographic studies in a number of ways:

1. It is much shorter than a typical ethnographic study. A contextual inquiry interview lasts about two or three hours, while an ethnographic study tends to be longer, probably weeks or months.

2. The interview is much more intense and focused than an ethnographic study, which takes in a wide view of the environment.

3. In the interview, the designer is not taking on a role of participant observer, but is inquiring about the work. The designer is observing, and is questioning behavior, but is not participating.

4. In the interview, the intention is to design a new system, but when conducting an ethnography, there is no particular agenda to be followed.

ACTIVITY 9.3	How does the contextual inquiry interview compare with the interviews introduced in Chapter 7?
Comment	We introduced structured, unstructured, and semi-structured interviews in Chapter 7. Contextual inquiry could be viewed as an unstructured interview, but is more wide-ranging than this. The interviewer does not have a set list of questions to ask, and can be guided by the interviewee. Contextual inquiry, however, is to be conducted at the interviewee's place of work, while normal work continues. It incorporates other data-gathering techniques such as observation although other interviews too may be used in conjunction with other techniques.

Normally, each team member conducts at least one contextual inquiry session. Data is collected in the form of notes and perhaps audio and video recording, but a lot of information is in the observer's head. It is important to review the experience and to start documenting the findings as soon as possible after the session. Contextual Design includes an interpretation session in which a number of models are generated (see below). Figures 9.5 to 9.8 show flow, sequence, cultural, and physical models focused around the system manager of an organization (Holtzblatt and Beyer, 1996).

Work Modeling

For customer-centered design, the first task of a design team is to shift focus from the system that the team is chartered to build and redirect it to the work of potential customers. Work, and understanding work becomes the primary consideration. But "work" is a slippery concept. What is work? (Beyer and Holtzblatt, 1998, p. 81)

Contextual design identifies five aspects to modeling "work," each of which guides the team to take a different perspective on what they have observed:

- The *work flow model* (Figure 9.5) represents the people involved in the work and the communication and coordination that takes place among them in order to achieve the work.

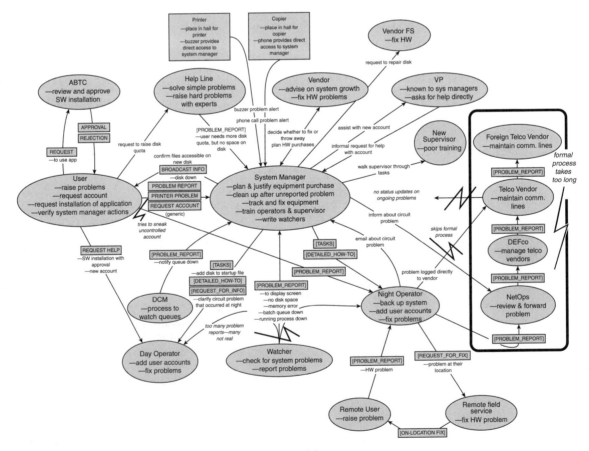

Figure 9.5 An example work flow model.

U1: Move user to larger disk

Intent: Give user more disk quota

Trigger: User requests higher disk quota

⇩

Requests more quota of customer support

⇩

Customer support discovers there's no more room on the user's disk

⇩

Customer support calls U1

⇩

**Intent: Relocate user to a disk with more free
space without losing any user data**

U1 looks for a scratch disk

⇩

Initializes and mounts scratch disk

⇩

Creates user directory

⇩

Moves user's files to the new disk

⇩

Uses DIR to check that files are there

⇩

Call user to confirm the user agrees all files are there

⇩

User checks and confirms

⇩

Delete user files from the old disk

⇩

Send mail to system manager to add new disk to regular startup

⇩

System manager adds new disk

⇩

Done

Figure 9.6 An example sequence model.

- The *sequence model* (Figure 9.6) shows the detailed work steps necessary to achieve a goal. Sequences are collected during the contextual interview, as the user works. However, understanding the steps alone is not sufficient, since although you may be able to streamline the steps themselves, if you do not understand the goals you may create a nonsensical work sequence. The sequence model also states the trigger for the set of steps.

- The *artifact model* represents the physical things created to do the work, such as the sticky notes at my desk, described above. The model consists of an annotated picture (or drawing) of each significant physical artifact used in achieving the work.

- The *cultural model* (Figure 9.7) represents constraints on the system caused by organizational culture. Organizations have cultures, teams build up their

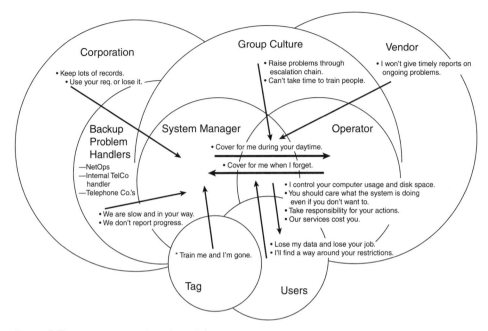

Figure 9.7 An example cultural model.

own culture, and work is performed in a cultural context. Culture influences the values and beliefs held by those taking part in the culture, and it determines rituals, expectations, and behavior. As a simple example, consider the dress codes for different situations in which you may find yourself. If you turn up at a baseball game in a three-piece suit, people will think you're a bit odd. On the other hand, if you turn up at a formal dinner in jeans and T-shirt, you will be refused entry. The cultural model aims to identify the main influencers on work, i.e., people or groups who constrain or affect work in some way.

- The *physical model* (Figure 9.8) shows the physical structure of the work. It may be a physical plan of the users' work environment, e.g., the office, or it may be a schematic of a communications network showing how components are linked together. The model captures the physical characteristics that constrain work and may make some work patterns infeasible.

The interpretation session

The work models are captured during an interpretation session. The team has to build an agreed view of the customers, their work, and the system to be built. Each developer therefore has to communicate to all the others on the team everything learned from her own interviewing experiences. So, after a contextual inquiry interview has been conducted, the team comes together to produce one consolidated view of the users' work.

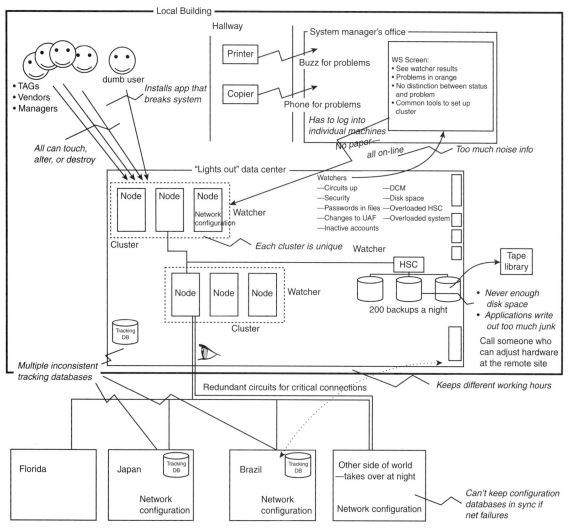

Figure 9.8 An example physical model.

Certain roles need to be adopted by the participants of this session. The *interviewer* is the person who has conducted the interviews and whose models are being examined. He must describe to the team what happened and in what order. During this recounting, the other members of the team can question the interviewer for clarification and extra information. *Work modelers* draw the work models as they emerge from the description given by the interviewer. The *recorder* keeps notes of the interpretation session that provide a sequential record of the meeting. The rest of the team (*participants*) listen to the description, ask questions, suggest design ideas (which are noted and not discussed at this time), observe, and contribute to the building of the models. The *moderator* stage-manages the meeting, keeps discussions focused on the main issue, keeps the pace of the meeting brisk, encourages everyone to take part, and notes where in the story the interviewer was in case of interruptions. The *rat-hole watcher* steers the conversation away from any distractions.

The output from this session is a set of models associated with the particular contextual inquiry interview. Each contextual inquiry interview generates its own set of models that is inevitably focused on the interviewee. These sets of models must be consolidated to gain a more general view of the work as described below.

ACTIVITY 9.4   The thick lightning marks in the flow models represent points at which breakdowns in communication or coordination occur. Alongside each lightning bolt is a description of the cause for this breakdown. Study the flow model in Figure 9.5 and identify all the breakdowns and their causes.

Comment   There are five breakdowns:

(a) too many problem reports—many not real

(b) the flow "problem logged directly to vendor" skips the formal process.

(c) no status updates on ongoing problems

(d) formal process takes too long

(e) tries to sneak uncontrolled account

Consolidating the models

The affinity diagram (see Figure 9.9) aims to organize the individual notes captured in the interpretation sessions into a hierarchy showing common structures and themes. Notes are grouped together because they are similar in some fashion. The groups are not predefined, but must emerge from the data. The process was originally introduced into the software quality community from Japan, where it is regarded as one of the seven quality processes. The affinity diagram is constructed after a cross-section of users has been interviewed and the corresponding interpretation sessions completed.

The affinity diagram is built by a process of induction. One note is put up first, and then the team searches for other notes that are related in some way.

The models produced during the interpretation session need to be consolidated so as to get a more general model of the work, one that is valid across individuals. The primary aim in consolidating flow models is to identify key roles. Any one individual may take on more than one role, and so it is necessary to identify and compare roles across and among individuals. For example, two different people may take on the role of quality assessor in different departments, and one of these may also be a production manager. To do this, the individuals' responsibilities are listed and a group of them that all lead towards one goal is identified. This goal and its set of responsibilities represents one role. Like the affinity diagram, this activity is concerned with grouping elements together along theme lines. Sometimes individuals use different names for the same role. The artifacts and communications among people need to be consolidated, too, in terms of flows between roles.

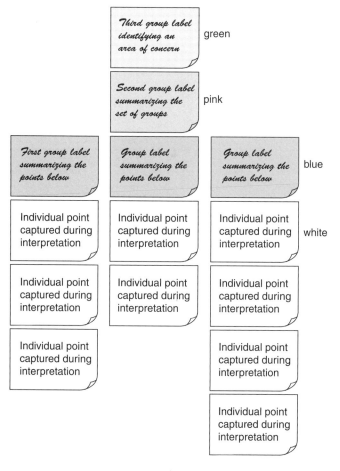

Figure 9.9 The structure of an affinity diagram.

Consolidated sequence models show the structure of a task and common strategies. The consolidated sequence model allows the team to identify what really needs to happen to accomplish the work, and hence what needs to be supported.

Artifact models show how people organize and structure their work, so a consolidated model shows common approaches to this across different people. The sequence models show the steps in the task, while the artifact model shows what is manipulated in order to achieve the task.

Physical space also has commonalities. For example, most companies have an entrance lobby with a receptionist or security guard, then beyond that personal offices and meeting rooms. Within one organization, even if it is distributed across different buildings, there is commonality of physical structure and hence constraints under which the work must be accomplished.

The cultural models help in identifying what matters to people who are doing the work. The cultural model identifies the influencers, so a consolidated model shows the set of common influencers within the organization.

All together, the consolidated models help designers to understand the users' intent, strategy to achieve that intent, structures to support the strategy, concepts to help manage and think about work, and the users' mind set.

The Design Room

An important element of Contextual Design is the design room, where all the work models are kept, pinned to the wall. The room is an environment that contains everything the team knows about the customer and their work. Design discussions held in the room can refer to data collected at the beginning of the project, and this can be used to support design ideas and decisions. This physical space in which the team is surrounded by the data is a key element of Contextual Design.

Contextual Design has been used successfully in a variety of situations from cell phone design (see Chapter 15) to office products (see Box 9.5). Its strength lies in the fact that it provides a clear route from observing users through to interpreting and structuring the data, prototyping and feeding the results into product development. This systematic approach means that, with suitable training, interaction designers can perform the observations and subsequent interpretation themselves, thus avoiding some of the misunderstandings that can happen if observations are conducted by others. Contextual Design is discussed further in the interview with Karen Holtzblatt at the end of this chapter.

9.5   Involving users in design: Participatory Design

Another approach to involving users is *Participatory Design*. In contrast to Contextual Design, users are actively involved in development. The intention is that they become an equal partner in the design team, and they design the product in cooperation with the designers.

The idea of participatory design emerged in Scandinavia in the late 1960s and early 1970s. There were two influences on this early work: the desire to be able to communicate information about complex systems, and the labor union movement pushing for workers to have democratic control over changes in their work. In the 1970s, new laws gave workers the right to have a say in how their working environment was changed, and such laws are still in force today. A fuller history of the movement is given in Ehn (1989) and Nygaard (1990).

Several projects at this time attempted to involve users in design and tried to focus on work rather than on simply producing a product. One of the most discussed is the UTOPIA project, a cooperative effort between the Nordic Graphics Workers Union and research institutions in Denmark and Sweden to design computer-based tools for text and image processing.

Involving users in design decisions is not simple, however. Cultural differences can become acute when users and designers are asked to work together to produce a specification for a system. Bødker et al. (1991) recount the following scene from the UTOPIA project:

> *Late one afternoon, when the designers were almost through with a long presentation of a proposal for the user interface of an integrated text and image processing system, one of the typographers commented on the lack of information about typographical code-*

Sort machine mock-up. The headline reads: "We did not understand the blueprints, so we made our own mock-ups."

Figure 9.10 A newspaper cutting showing a parcel-sorting machine mockup.

structure. He didn't think that it was a big error (he was a polite person), but he just wanted to point out that the computer scientists who had prepared the proposal had forgotten to specify how the codes were to be presented on the screen. Would it read "<bf/" or perhaps just "\b" when the text that followed was to be printed in boldface?

In fact, the system being described by the designers was a WYSIWIG system, and so text that needed to be in bold typeface would appear as bold (although most typographic systems at that time did require such codes). The typographer was unable to link his knowledge and experience with what he was being told. In response to this kind of problem, the project started using mockups (introduced in Chapter 8). Simulating the working situation helped workers to draw on their experience and tacit knowledge, and designers to get a better understanding of the actual work typographers needed to do. An example mockup for a computer-controlled parcel-sorting system, from another project, is shown in Figure 9.10 (Ehn and Kyng, 1991). The headline of this newspaper clipping reads, "We did not understand the blueprints, so we made our own mockups".

Mockups are one way to make effective use of the users' experience and knowledge. Other paper-based prototyping techniques that have been developed for participatory design are PICTIVE (Muller, 1991) and CARD (Tudor, 1993).

9.5.1   PICTIVE

PICTIVE (Plastic Interface for Collaborative Technology Initiatives through Video Exploration) uses low-fidelity office items, such as sticky notes and pens, and a collection of design objects to investigate specific screen and window layouts for a system. The motives for developing the techniques were to:

- empower users to act as full participants in the design process
- improve knowledge acquisition for design

A PICTIVE session may involve one-on-one collaboration or it may involve a small group. To perform a PICTIVE session you need video recording equipment, simple office supplies such as pens, pencils, paper, sticky notes, cards, etc., and some design components prepared by the design team such as dialog boxes, menu bars, and icons. These plastic design components may be generic or they may be specific to the system being developed, based on the development so far. The shared design surface is where the design will be created, jointly between the designers and the users, by manipulating and changing the design components and using the office supplies to create new elements. The video equipment records what happens on the shared design surface. Sample design objects and the layout for a PICTIVE session are shown in Figure 9.11 (Muller, 1991).

Before a session, each participant is asked to prepare a "homework assignment." Typically, users are asked to generate scenarios of use for the system, illustrating what they would like the system to do for them (along the lines of the scenarios we discussed in Chapter 7). Developers are asked to develop a set of system components that they think may be relevant to the system. These may be generic elements that will be used in many design exercises, they may be specifically for the system under discussion, or a combination of these.

The design session itself is divided roughly into four parts (Muller et al., 1995). First of all, the stakeholders all introduce themselves, specifically describing their personal and/or organizational stake in the project. Then there may be some brief tutorials about the different domains represented at the meeting. The third part of the meeting concentrates on brainstorming the designs, using the design objects and the homework assignments. The design objects are manipulated during the session to produce a synthesis of each participant's view. The scenarios developed by the users may help provide concrete detail about the work flow of the design. The final session is a walkthrough of the design and the decisions discussed. The role of the video recording is mainly that of record-keeper, so that there is a complete and informal record of the design decisions made and how they were made.

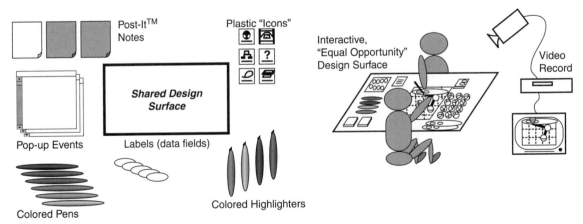

Figure 9.11 PICTIVE design objects and PICTIVE setting.

ACTIVITY 9.5   Describe a set of design components you would develop for a PICTIVE session for the shared calendar application discussed in Chapter 8.

Comment   From our earlier design activities, we know that having dialog boxes and icons for arranging a meeting would be appropriate. Also, different mechanisms for specifying the people to attend the meeting and for choosing dates, e.g., drop-down lists, free text entry, or planner-style date display. These components could be based on our preliminary designs. We will also need a menu bar and associated menu lists, calendar page display, and function button components. It would also be important to have some blank components that could be completed during the brainstorming session.

9.5.2 CARD

CARD (Collaborative Analysis of Requirements and Design) is similar to PICTIVE, but uses playing cards with pictures of computers and screen dumps on them to explore workflow options (see Figure 9.12 for an example set of cards

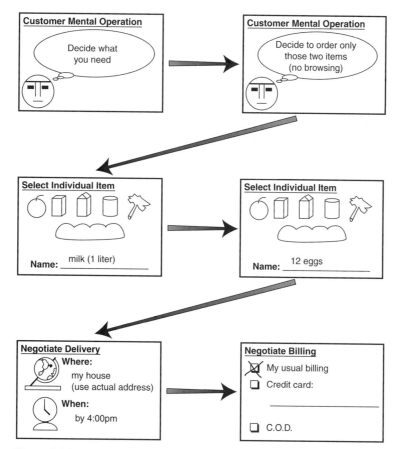

Figure 9.12 Example of CARD.

(Muller et al., 1995)). Whereas PICTIVE concentrates on detailed aspects of the system, CARD takes a more macroscopic view of the task flow. CARD is a form of storyboarding (see Chapter 8).

A CARD session could have the same format as that described for PICTIVE. During the design brainstorming part of the session, the playing cards are manipulated by the participants in order to show the work flow between computer screens or task decision points. The example in Figure 9.12 shows how the task of buying groceries through a computer screen such as the Internet can be represented by

Table 9.1   A comparison of techniques introduced in this chapter

	Ethnography	Coherence	Contextual Design	Participatory Design[2]
Active user involvement	Low level	Low level	Medium to low level	Equal partners, users can be very influential
Role of designer/ researcher	Uncover findings about work	Collect and present ethnographic data according to the viewpoints and concerns	Steer discussion Interpret findings	Equal partners
Length of study	Typically continuous and extensive.	N/A	A series of 2-hour interviews	A series of 2-hour design sessions
Benefits	Yields a good understanding of the work	Overcomes one of the disadvantages of ethnographic data	Systematic Is designed to feed into the design process	Users' sense of ownership is increased User contact is beneficial for designers
Drawbacks	Requires expertise Difficulties translating findings into design Requires a long lead-in time	Coverage limited to presenting ethnographic data Limited support currently for progressing to design	Involves many diagrams and notations May be complicated for users to understand the output	Users' thinking can be constrained by what they know If users are involved too much they get bored and it becomes counter-productive
When to use	Most settings where there is sufficient time and expertise	If an ethnographic study for interaction design is to be conducted (by ethnographer or designer)	When a user-centered focus is required Particularly useful for innovative product design	Whenever users are available and willing to become actively involved in design

[2]The main difference between CARD and PICTIVE lies in the level of detail at which design takes place. For the purpose of this comparison, they can be considered under the common title of Participatory Design.

playing cards. Note that the cards can be used to represent users' goals or intentions as well as specific computer screens or task elements. Participants can easily create new cards during the session as deemed appropriate.

CARD can be used to complement PICTIVE as it provides a different granularity of focus. Muller et al. (1995) characterized this as a bifocal view, CARD giving a macroscopic view, and PICTIVE the microscopic.

At the beginning of this chapter, we explained that there are different levels of user involvement, from newsletters and workshops through to full-time membership of the design team. Each project will need to decide on the level of user involvement required. To support this involvement, a project may also choose to use one or a combination of the techniques introduced in Sections 9.4 and 9.5. For example, Contextual Design could be used even if one of the users is a member of the design team; an ethnographic study might be running alongside a series of user workshops. These techniques expand the level of user involvement. However, each approach has advantages and disadvantages, and Table 9.1 provides a brief comparison between the main techniques introduced in this chapter.

Assignment

This assignment asks you to apply some elements of Coherence and Contextual Design to your own work or home circumstances.

(a) Using the questions for elaborating the viewpoints and concerns in Coherence, study the environment of your workplace, university library or somewhere similar that you know. Begin by deciding which concerns are relevant to each viewpoint, e.g., ask, "Are there paper artifacts used in the workplace?" or "Is local knowledge used?" Then answer the questions of elaboration for the three viewpoints and the four concerns.

Study your answers to the questions and see if you can identify priorities or constraints within the organization that you were not aware of before.

(b) Again using your workplace or similar location, attempt to draw the five Contextual Design work models introduced in Section 9.4.3.

First of all, identify a key player in the workplace. This may be one of the librarians, a clerk or secretary, or a manager. If possible, run a contextual inquiry interview by sitting with her while working and asking her to tell you about one major aspect of work. If this is not possible, then identify one of the main tasks that is visible to you, such as the librarian issuing books, and sit and watch how the task is performed.

Draw the models from the information you have collected. If you find that you need more data, go back and collect more. Once you feel that the models are complete, take them back to the person you interviewed (if possible) and ask for comments.

Summary

This chapter has elaborated on some issues surrounding the involvement of users in the design process. We have also introduced the method of ethnography as a useful source of information for a user-centered design process. One of the main disadvantages to using ethnography is finding a way to represent the output of the study so that it can be fed into the design process. We have described two approaches to design (Coherence and Contextual Design) that were derived from ethnography and other approaches, to address this problem.

Users may be involved passively or they may be more actively involved in making design decisions. Participatory design is an approach in which users are co-designers. We have described two techniques (PICTIVE and CARD) that have helped users' input to be more effective.

Key Points

- Involving users in the design process helps with expectation management and feelings of ownership, but how and when to involve users is a matter of dispute.
- Putting a user-centered approach into practice requires much information about the users to be gathered and interpreted.
- Ethnography is a good method for studying users in their natural surroundings.
- Representing the information gleaned from an ethnographic study so that it can be used in design has been problematic.
- The goals of ethnography are to study the details, while the goals of system design are to produce abstractions; hence they are not immediately compatible.
- Coherence is a method that provides focus questions to help guide the ethnographer towards issues that have proved to be important in systems development.
- Contextual Design is a method that provides models and techniques for gathering contextual data and representing it in a form suitable for practical design.
- PICTIVE and CARD are both participatory design techniques that empower users to take an active part in design decisions.

Further reading

GREENBAUM, JOAN, AND KYNG, MORTEN (eds.) (1991) *Design at Work: Co-operative Design of Computer Systems*. Hillsdale, NJ: Lawrence Erlbaum. This book is a good collection of papers about the co-design of software systems: both why it is worthwhile and experience of how to do it.

BEYER, HUGH AND HOLTZBLATT, KAREN (1998) *Contextual Design: Defining Customer-Centered Systems*. San Francisco: Morgan Kaufmann. This book will tell you more about contextual design and the rationale behind the steps and the models.

CUSUMANO, M.A., AND SELBY, R. W. (1995) *Microsoft Secrets*. London: Harper-Collins Business. This is a fascinating book based on a two-and-a-half-year study of Microsoft and how they build software. The book details findings about strategies to manage an innovative organization competing in a rapidly changing world, to develop and ship products that appeal to mass markets, and to continually build on and improve market position.

WIXON, DENNIS, AND RAMEY, JUDITH (eds.) (1996) *Field Methods Casebook for Software Design*. New York: John Wiley & Sons, Inc. This book is a collection of papers about practical use of field research methods in software design, some of which are directly mentioned in the present chapter. The three main approaches that these papers cover are ethnography, participative design, and contextual design. There are 14 chapters describing case studies and three chapters giving an overview of the main methods. For anyone interested in the practical use of these methods in software development, it's a fascinating read!

INTERVIEW   with Karen Holtzblatt

Karen Holtzblatt is the originator of Contextual Inquiry, a process for gathering field data on product use, which was the precursor to Contextual Design, a complete method for the design of systems. Together with Hugh Beyer, the codeveloper of Contextual Design, Karen Holtzblatt is co-founder of InContext Enterprises, which specializes in process and product design consulting.

HS: What is Contextual Design?

KH: If you're going to build something that people want, there are basically three large steps that you have to go through. The first question that you ask as a company is, "What in the world matters to the customer such that if we make something, they're likely to buy it?" So the question is "What matters?" Now once you identify what the issues are, every corporation will have the corporate response, or "vision." Then you have to work out the details and structure it into a product. In any design process, whether it's formalized or not, every company must do those things. They have to find out what matters, they have to vision their corporate response, and then they have to structure it into a system.

Contextual Design gives you team and individual activities that bring you through those processes in an orderly fashion so as to bring the cross-functions of an organization together. So you could say that Contextual Design is a set of techniques to be used in a customer-centered design process with design teams. It is also a set of practices that help people engage in creative and productive design thinking with customer data and it helps them co-operate and design together.

HS: What are the steps of Contextual Design?

KH: In the "what matters" piece, we go out into the field, we talk with people about their work as they do it: that's Contextual Inquiry and that's a one-on-one, two to two-and-a-half-hour field interview. Then we interpret that data with a cross-functional team, and we model the work with five work models: communication and coordination, the cultural environment,

the physical environment, task, and artifact. We also capture individual points on post-it notes. After the interpretation session, every person we interviewed has a set of models and a set of post-its. Our next step is to consolidate all that data because you don't want to be designing from one person, from yourself, or from any one interview; we need to look at the structure of the practice itself. The consolidation step means that we end up with an affinity diagram and five consolidated models showing the issues *across* the market.

At that point, we have modeled the work practice as it is and we have now six communication devices that the team can dialog with. Each one of them poses a point of view on which to have the conversation "what matters?"

Now the team moves into that second piece, which is "what should our corporate response be?" We have a visioning process that is a very large group story-telling about reinventing work practice given technological possibility and the core competency of the organization. And after that, we develop storyboards driven by the consolidated data and the vision. At this point we have not done a systems design; we want to design the work practice first, seeing the technology as it will appear within the work.

To structure the system we start by rolling the storyboards into a user environment design—the structure of the system itself, independent of the user interface and the object model. The user environment design operates like a software floor plan that structures the movement inside the product. This is used to drive the user interface design, which is mocked up in paper and tested and iterated with the user. When it has stabilized, the User Environment design, the storyboards, and the user interface drive development of the object model.

This is the whole process of Contextual Design, a full front-end design process. Because it is done with a cross-functional team, everyone in the organization knows what they're doing at each point: they know how to select the data, they know how to work in groups to get all these different steps done. So not only do you end up with a set of design thinking techniques that help you to design, you have an organizational process that helps the organization actually do it.

HS: How did the idea of contextual design emerge?

KH: Contextual Design started with the invention of Contextual Inquiry in a post-doctoral internship with John Whiteside at Digital. At the time, usability testing and usability issues had been around maybe eight years or so and he was asking the question, "Usability identifies about 10 to 20% of the fixes at the tail end of the process to make the frosting on the cake look a little better to the user. What would it take to really infuse usability?" Contextual Inquiry was my answer to that question. After that, I took a job with Lou Cohen's Quality group at DEC, where I picked up the affinity diagram idea. Also at that time, Pelle Ehn and Kim Madsen were talking about Morten Kyng's ideas on paper mock-ups and I added paper prototyping with post-its to check out the design. Hugh and I hooked up 13 years ago. He's a software and object-oriented developer. We started working with teams and we noticed that they didn't know how to go from the data to the design and they didn't know how to structure the system to think about it. So then we invented more of the work models and the user environment design.

So the Contextual Design method came from looking at the practice; we evolved every single step of this process based on what people needed. The whole process was worked out with real people doing real design in real companies. So, where did it come from? It came from dialog with the problem.

HS: What are the main problems that organizations face when putting Contextual Design into practice?

KH: The question is, "What does organizational change look like?" because that's what we're talking about. The problem is that people want to change and they don't want to change. What we communicate to people is that organizational change is piecemeal. In order to own a process you have to say what's wrong with it, you have to change it a little bit, you have to say how whoever invented the process is wrong and how the people in the organization want to fix it, you have to make it fit with your organizational culture and issues. Most people will adopt the field-data gathering first and that's all they'll do and they'll tell me that they don't have time for anything else and they don't need anything else, and that's fine. And then they'll wake up one day and they'll say, "We have all this qualitative stuff and nobody's using it . . . maybe we should have a debriefing session." So then they have debriefing sessions. Then they wake up later on and they say, "We don't have any way of structuring this information . . . models are a good idea." And basically they reconstruct the whole process as they hit the next problem.

Now it's not quite that clean, but my point is that organizational adoption is about people making it their own and taking on the parts, changing them, doing what they can. You have to get somebody to do something and then once they do something it snowballs.

What's nice about the Contextual Design way of doing everything on paper is that it creates a design room, the design room creates a talk event, and the talk event pulls everyone in because they want to know what you're doing. Then if they like the data, they feel left out, and because they feel left out they want to do a project and they want to have a room for themselves as well.

The biggest complaint about Contextual Design is that it takes too long. Some of that is about time, some of it is about thought. You have people who are used to coding and now have to think about field data. They're not used to that.

HS: What's the future direction of Contextual Design?

KH: Every process can always be tweaked. I think the primary parts of Contextual Design are there. There are interesting directions in which it can go, but there's only so much we can get our audience to buy.

I think that for us there are two key things that we're doing. One is we're starting to talk about design and what design is, so we can talk about the role of design in design thinking. And we are still helping train everyone who wants to learn. But the other thing we're finding is that sometimes the best way to support the client is to do the design work for them. So we have the design wing of the business where we put together the contextual design teams.

We're working with distributed teams, we're working with creativity and invention, we're working with how it impacts with business processes and marketing, we're working with the balance of all those things. But it's only going to be in the context of a team that's actually very advanced in the standard process that new process inventions will occur. Out of that will come lessons that can then be put back into the standard contextual design. For most organizations looking to adopt a customer-centered design process, the standard contextual design is enough for now, they have to get started. And because Contextual Design is a scaffolding, they can plug other processes into it. They take their usability testing and they can plug it here, if they have their special creativity thing they can plug it here; if they have a focus group they can plug it here. But most people haven't got a backbone for design, and Contextual Design is a good backbone to start with.

Chapter 10

Introducing evaluation

10.1   Introduction
10.2   What, why, and when to evaluate
    10.2.1   What to evaluate
    10.2.2   Why you need to evaluate
    10.2.3   When to evaluate
10.3   HutchWorld case study
    10.3.1   How the team got started: Early design ideas
    10.3.2   How was the testing done?
    10.3.3   Was it tested again?
    10.3.4   Looking to the future
10.4   Discussion

10.1   Introduction

Recently I met two web designers who, proud of their newest site, looked at me in astonishment when I asked if they had tested it with users. "No," they said "but we know it's OK." So, I probed further and discovered that they had asked the "web whiz-kids" in their company to look at it. These guys, I was told, knew all the tricks of web design.

The web's presence has heightened awareness about usability, but unfortunately this reaction is all too common. Designers assume that if they and their colleagues can use the software and find it attractive, others will too. Furthermore, they prefer to avoid doing evaluation because it adds development time and costs money. So why is evaluation important? Because without evaluation, designers cannot be sure that their software is usable and is what users want. But what do we mean by evaluation? There are many definitions and many different evaluation techniques, some of which involve users directly, while others call indirectly on an understanding of users' needs and psychology. In this book we define evaluation as the process of systematically collecting data that informs us about what it is like for a particular user or group of users to use a product for a particular task in a certain type of environment.

As you read in Chapter 9, the basic premise of user-centered design is that users' needs are taken into account throughout design and development. This is achieved by evaluating the design at various stages as it develops and by amending it to suit users' needs (Gould and Lewis, 1985). The design, therefore, progresses in iterative cycles of design-evaluate redesign. Being an effective interaction designer requires knowing how to evaluate different kinds of systems at different stages of development. Furthermore, developing systems in this way usually turns out to be less expensive than fixing problems that are discovered after the systems have been shipped to customers (Karat, 1993). Studies also suggest that the business case for using systems with good usability is compelling (Dumas and Redish, 1999; Mayhew, 1999): thousands of dollars can be saved.

Many techniques are available for supporting design and evaluation. Chapter 9 discussed techniques for involving users in design and part of this involvement comes through evaluation. In this and the next four chapters you will learn how different techniques are used at different stages of design to examine different aspects of the design. You will also meet some of the same techniques that are used for gathering user requirements, but this time used to collect data to evaluate the design. Another aim is to show you how to do evaluation.

This chapter begins by discussing *what* evaluation is, *why* evaluation is important, and *when* to use different evaluation techniques and approaches. Then a case study is presented about the evaluation techniques used by Microsoft researchers and the Fred Hutchinson Cancer Research Center in developing HutchWorld (Cheng et al., 2000), a virtual world to support cancer patients, their families, and friends. This case study is chosen because it illustrates how a range of techniques is used during the development of a new product. It introduces some of the practical problems that evaluators encounter and shows how iterative product development is informed by a series of evaluation studies. The HutchWorld study also lays the foundation for the evaluation framework that is discussed in Chapter 11.

The main aims of this chapter are to:

- Explain the key concepts and terms used to discuss evaluation.
- Discuss and critique the HutchWorld case study.
- Examine how different techniques are used at different stages in the development of HutchWorld.
- Show how developers cope with real-world constraints in the development of HutchWorld.

10.2   What, why, and when to evaluate

Users want systems that are easy to learn and to use as well as effective, efficient, safe, and satisfying. Being entertaining, attractive, and challenging, etc. is also essential for some products. So, knowing what to evaluate, why it is important, and when to evaluate are key skills for interaction designers.

10.2.1   What to evaluate

There is a huge variety of interactive products with a vast array of features that need to be evaluated. Some features, such as the sequence of links to be followed to find an item on a website, are often best evaluated in a laboratory, since such a setting allows the evaluators to control what they want to investigate. Other aspects, such as whether a collaborative toy is robust and whether children enjoy interacting with it, are better evaluated in natural settings, so that evaluators can see what children do when left to their own devices.

You may remember from Chapters 2, 6 and 9 that John Gould and his colleagues (Gould et al., 1990; Gould and Lewis, 1985) recommended three similar principles for developing the 1984 Olympic Message System:

- focus on users and their tasks
- observe, measure, and analyze their performance with the system
- design iteratively

Box 10.1 takes up the evaluation part of the 1984 Olympic Messaging System story and lists the many evaluation techniques used to examine different parts of the OMS during its development. Each technique supported Gould et al.'s three principles.

Since the OMS study, a number of new evaluation techniques have been developed. There has also been a growing trend towards observing how people interact with the system in their work, home, and other settings, the goal being to obtain a better understanding of how the product is (or will be) used in its intended setting. For example, at work people are frequently being interrupted by phone calls, others knocking at their door, email arriving, and so on—to the extent that many tasks are interrupt-driven. Only rarely does someone carry a task out from beginning to end without stopping to do something else. Hence the way people carry out an activity (e.g., preparing a report) in the real world is very different from how it may be observed in a laboratory. Furthermore, this observation has implications for the way products should be designed.

10.2.2 Why you need to evaluate

Just as designers shouldn't assume that everyone is like them, they also shouldn't presume that following design guidelines guarantees good usability. Evaluation is needed to check that users can use the product and like it. Furthermore, nowadays users look for much more than just a usable system, as the Nielsen Norman Group, a usability consultancy company, point out (www.nngroup.com):

> "User experience" encompasses all aspects of the end-user's interaction . . . the first requirement for an exemplary user experience is to meet the exact needs of the customer, without fuss or bother. Next comes simplicity and elegance that produce products that are a joy to own, a joy to use."

Bruce Tognazzini, another successful usability consultant, comments (www.asktog.com) that:

> "Iterative design, with its repeating cycle of design and testing, is the only validated methodology in existence that will consistently produce successful results. If you don't have user-testing as an integral part of your design process you are going to throw buckets of money down the drain."

BOX 10.1   The Story of the 1984 Olympic Messaging System

The 1984 Olympic Message System (OMS), a voice mail system, was developed by IBM so that Olympic Games contestants and their families and friends could send and receive messages (Gould et al., 1990). They could hear the message and the actual voice of the sender exactly as it was spoken. This system could be used from almost any push-button phone system around the world: this may not sound amazing when compared with today's technology, but in 1983 it was highly innovative.

Non-Olympians called their own country's National Olympic Committee using either push-button or dial telephones and spoke in their own language. They were helped to connect to OMS so that they could leave their messages. The voice message was immediately transferred by a central telephone operator to the message boxes of the Olympian for whom it was intended. The OMS worked in 12 languages. The kiosks looked like the one in Figure 10.1 and the dialog is shown in Figure 10.2

Figure 10.1 The Olympic message system kiosk.

During development, the evaluation activities included:

- Use of printed scenarios of the screens to get feedback from the Olympic committee and the Olympians themselves.
- Iteratively testing the user guides for the OMS with the Olympians, their families, and friends.
- Developing early simulations of a telephone keypad with a person speaking the commands back. These simulations really tested how much a user needed to know about the system, what feedback was needed, and any incorrect assumptions about user behavior made by the designers.
- Developing early demonstrations to test the reactions of people outside the US who did not know much about computers.
- An Olympian joining the design team to discuss ideas and provide feedback.
- Interviews with Olympians to make sure that the system being developed was what the users wanted.
- Overseas tests of the interface with friends and family.
- Free coffee and donut tests: 65 people were enticed to test the system in return for these treats.
- More traditional usability tests (discussed in Chapter 14) of the prototype involving about 100 participants.
- A 'try-to-destroy-it' test in which 24 computer science students were challenged to bring down the system. One of these tests involved all the students calling into the OMS at the same time. The students enjoyed the challenge and didn't need any other motivation!
- A pre-Olympic field test of the interface at an international event with competitors from 65 countries. The outcome of this test was surprising because, despite all the other testing, 57 different usability problems were recorded by the end of the five-day test period. The lesson for the design team was that the results of

field tests could be surprising. In this case they discovered that strong cultural differences affected how users from different countries used the OMS. Testers from Oman, Colombia, Pakistan, Japan, and Korea were unable to use the system. Gould and his colleagues comment that "watching this helplessness and hopelessness had a far greater impact than reading about it. It was embarrassing . . ." (Gould et al., 1990, p. 274).

- Two other tests examined the reliability of the system with heavy traffic generated by 2800 and 1000 people respectively.

This extensive evaluation was needed because the Olympics was such a high-profile event and IBM's reputation was at stake. Less intensive evaluation is more normal. However, the take-away message from this study is that the more evaluation with users, the better the final product.

Caller:	(Dials 213-888-8888.)
Operator:	Irish National Olympic Committee. Can I help you?
Caller:	I want to leave a message for my son, Michael.
Operator:	Is he from Ireland?
Caller:	Yes.
Operator:	How do you spell his name?
Caller:	*K-E-L-L-Y.*
Operator:	Thank you. Please hold for about 30 seconds while I connect you to the Olympic Message System.
Operator:	Are you ready?
Caller:	Yes.
OMS:	When you have completed your message, hang up and it will be automatically sent to Michael Kelly. Begin talking when you are ready.
Caller:	'Michael, your Mother and I will be hoping you win. Good luck.' (Caller hangs up.)

Figure 10.2 Parent leaving a voice message for an Olympian.

Tognazzini points out that there are five good reasons for investing in user testing:

1. Problems are fixed before the product is shipped, not after.
2. The team can concentrate on real problems, not imaginary ones.
3. Engineers code instead of debating.
4. Time to market is sharply reduced.
5. Finally, upon first release, your sales department has a rock-solid design it can sell without having to pepper their pitches with how it will all actually work in release 1.1 or 2.0.

Now that there is a diversity of interactive products, it is not surprising that the range of features to be evaluated is very broad. For example, developers of a new web browser may want to know if users find items faster with their product. Government authorities may ask if a computerized system for controlling traffic lights results in fewer accidents. Makers of a toy may ask if six-year-olds can manipulate the controls and whether they are engaged by its furry case and pixie face. A company that develops the casing for cell phones may ask if the shape, size, and color of the case is appealing to teenagers. A new dotcom company may want to assess market reaction to its new home page design.

This diversity of interactive products, coupled with new user expectations, poses interesting challenges for evaluators, who, armed with many well tried and tested techniques, must now adapt them and develop new ones. As well as usability, user experience goals can be extremely important for a product's success, as discussed in Chapter 1.

ACTIVITY 10.1   Think of examples of the following systems and write down the usability and user experience features that are important for the success of each:

(a) a word processor

(b) a cell phone

(c) a website that sells clothes

(d) an online patient support community

Comment   (a) It must be as easy as possible for the intended users to learn and to use and it must be satisfying. Note, that wrapped into this are characteristics such as consistency, reliability, predictability, etc., that are necessary for ease of use.

(b) A cell phone must also have all the above characteristics; in addition, the physical design (e.g., color, shape, size, position of keys, etc.) must be usable and attractive (e.g., pleasing feel, shape, and color).

(c) A website that sells clothes needs to have the basic usability features too. In particular, navigation through the system needs to be straightforward and well supported. You may have noticed, for example, that some sites always show a site map to indicate where you are. This is an important part of being easy to use. So at a deeper level you can see that the meaning of "easy to use and to learn" is different for different systems. In addition, the website must be attractive, with good graphics of the clothes—who would want to buy clothes they can't see or that look unattractive? Trust is also a big issue in online shopping, so a well-designed procedure for taking customer credit card details is essential: it must not only be clear but must take into account the need to provide feedback that engenders trust.

(d) An online patient support group must support the exchange of factual and emotional information. So as well as the standard usability features, it needs to enable patients to express emotions either publicly or privately, using emoticons. Some 3D environments enable users to show themselves on the screen as avatars that can jump, wave, look happy or sad, move close to another person, or move away. Designers have to identify the types of social interactions that users want to express (i.e., sociability) and then find ways to support them (Preece, 2000).

From this selection of examples, you can see that success of some interactive products depends on much more than just usability. Aesthetic, emotional, engaging, and motivating qualities are important too.

Usability testing involves measuring the performance of typical users on typical tasks. In addition, satisfaction can be evaluated through questionnaires and interviews. As mentioned in Chapter 1, there has been a growing trend towards developing ways of evaluating the more subjective user-experience goals, like emotionally satisfying, motivating, fun to use, etc.

10.2.3 When to evaluate

The product being developed may be a brand-new product or an upgrade of an existing product. If the product is new, then considerable time is usually invested in market research. Designers often support this process by developing mockups of the potential product that are used to elicit reactions from potential users. As well as helping to assess market need, this activity contributes to understanding users' needs and early requirements. As we said in Chapter 8, sketches, screen mockups, and other low-fidelity prototyping techniques are used to represent design ideas. Many of these same techniques are used to elicit users' opinions in evaluation (e.g., questionnaires and interviews), but the purpose and focus of evaluation is different. The goal of evaluation is to assess how well a design fulfills users' needs and whether users like it.

In the case of an upgrade, there is limited scope for change and attention is focused on improving the overall product. This type of design is well suited to usability engineering in which evaluations compare user performance and attitudes with those for previous versions. Some products, such as office systems, go through many versions, and successful products may reach double-digit version numbers. In contrast, new products do not have previous versions and there may be nothing comparable on the market, so more radical changes are possible if evaluation results indicate a problem.

Evaluations done during design to check that the product continues to meet users' needs are know as *formative evaluations*. Evaluations that are done to assess the success of a finished product, such as those to satisfy a sponsoring agency or to check that a standard is being upheld, are know as *summative evaluation*. Agencies such as National Institute of Standards and Technology (NIST) in the USA, the International Standards Organization (ISO) and the British Standards Institute (BSI) set standards by which products produced by others are evaluated.

ACTIVITY 10.2 Re-read the discussion of the 1984 Olympic Messaging System (OMS) in Box 10.1 and briefly describe some of the things that were evaluated, why it was necessary to do the evaluations, and when the evaluations were done.

Comment Because the Olympic Games is such a high-profile event and IBM's reputation was at stake, the OMS was intensively evaluated throughout its development. We're told that early evaluations included obtaining feedback from Olympic officials with scenarios that used printed screens and tests of the user guides with Olympians, their friends, and family. Early evaluations of simulations were done to test the usability of the human-computer dialog. These were done first in the US and then with people outside of the US. Later on, more formal tests investigated how well 100 participants could interact with the system. The system's robustness was also tested when used by many users simultaneously. Finally, tests were done with users from minority cultural groups to check that they could understand how to use the OMS.

So how do designers decide *which* evaluation techniques to use, *when* to use them, and *how* to use the findings? To address these concerns, we provide a case study showing how a range of evaluation techniques were used during the development of a new system. Based on this, we then discuss issues surrounding the "which, when, and how" questions relating to evaluation.

10.3   HutchWorld case study

HutchWorld is a distributed virtual community developed through collaboration between Microsoft's Virtual Worlds Research Group and librarians and clinicians at the Fred Hutchinson Cancer Research Center in Seattle, Washington. The system enables cancer patients, their caregivers, family, and friends to chat with one another, tell their stories, discuss their experiences and coping strategies, and gain emotional and practical support from one another (Cheng et. al., 2000). The design team decided to focus on this particular population because caregivers and cancer patients are socially isolated: cancer patients must often avoid physical contact with others because their treatments suppress their immune systems. Similarly, their caregivers have to be careful not to transmit infections to patients.

The big question for the team was how to make HutchWorld a useful, engaging, easy-to-use and emotionally satisfying environment for its users. It also had to provide privacy when needed and foster trust among participants. A common approach to evaluation in a large project like Hutchworld is to begin by carrying out a number of informal studies. Typically, this involves asking a small number of users to comment on early prototypes. These findings are then fed back into the iterative development of the prototypes. This process is then followed by more formal usability testing and field study techniques. Both aspects are illustrated in this case study. In addition, you will read about how the development team managed their work while dealing with the constraints of working with sick people in a hospital environment.

10.3.1   How the design team got started: early design ideas

Before developing this product, the team needed to learn about the patient experience at the Fred Hutchinson Center. For instance, what is the typical treatment process, what resources are available to the patient community, and what are the needs of the different user groups within this community? They had to be particularly careful about doing this because many patients were very sick. Cancer patients also typically go through bouts of low emotional and physical energy. Caregivers also may have difficult emotional times, including depression, exhaustion, and stress. Furthermore, users vary along other dimensions, such as education and experience with computers, age and gender and they come from different cultural backgrounds with different expectations.

It was clear from the onset that developing a virtual community for this population would be challenging, and there were many questions that needed to be answered. For example, what kind of world should it be and what should it provide? What exactly do users want to do there? How will people interact? What should it look like? To get answers, the team interviewed potential users from all the stakeholder groups—patients, caregivers, family, friends, clinicians, and social support staff—and observed their daily activity in the clinic and hospital. They also read the latest research literature, talked to experts and former patients, toured the Fred Hutchinson (Hutch) research facilities, read the Hutch web pages, and visited the Hutch school for pediatric patients and juvenile patient family members. No stone was left unturned.

The development team decided that HutchWorld should be available for patients any time of day or night, regardless of their geographical location. The team knew from reading the research literature that participants in virtual communities are often more open and uninhibited about themselves and will talk about problems and feelings in a way that would be difficult in face-to-face situations. On the downside, the team also knew that the potential for misunderstanding is higher in virtual communities when there is inadequate non-verbal feedback (e.g., facial expressions and other body language, tone of voice, etc.). On balance, however, research indicates that social support helps cancer patients both in the psychological adjustments needed to cope and in their physical wellbeing. For example, research showed that women with breast cancer who received group therapy lived on average twice as long as those who did not (Spiegel, et al., 1989). The team's motivation to create HutchWorld was therefore high. The combination of information from research literature and from observations and interviews with users convinced them that this was a worthwhile project. But what did they do then?

The team's informal visits to the Fred Hutchinson Center led to the development of an early prototype. They followed a user-centered development methodology. Having got a good feel for the users' needs, the team brainstormed different ideas for an organizing theme to shape the conceptual design—a conceptual model possibly based on a metaphor. After much discussion, they decided to make the design resemble the outpatient clinic lobby of the Fred Hutchinson Cancer Research Center. By using this real-world metaphor, they hoped that the users would easily infer what functionality was available in HutchWorld from their knowledge of the real clinic. The next step was to decide upon the kind of communication environment to use. Should it be synchronous or asynchronous? Which would support social and affective communications best? A synchronous chat environment was selected because the team thought that this would be more realistic and personal than an asynchronous environment. They also decided to include 3D photographic avatars so that users could enjoy having an identifiable online presence and could easily recognize each other.

Figure 10.3 shows the preliminary stages of this design with examples of the avatars. You can also see the outpatient clinic lobby, the auditorium, the virtual garden, and the school. Outside the world, at the top right-hand side of the screen, is a list of commands in a palette and a list of participants. On the right-hand side at the bottom is a picture of participants' avatars, and underneath the window is the textual chat window. Participants can move their avatars and make them gesture to tour the virtual environment. They can also click on objects such as pictures and interact with them.

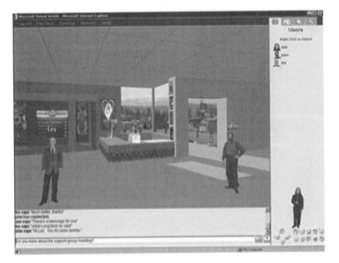

Figure 10.3 Preliminary design showing a view of the entrance into Hutch-World.

The prototype was reviewed with users throughout early development and was later tested more rigorously in the real environment of the Hutch Center using a variety of techniques. A Microsoft product called V-Chat was used to develop a second interactive prototype with the subset of the features in the preliminary design shown in Figure 10.3; however, only the lobby was fully developed, not the auditorium or school, as you can see in the new prototype in Figure 10.4.

Before testing could begin, the team had to solve some logistical issues. There were two key questions. Who would provide training for the testers and help for the patients? And how many systems were needed for testing and where should they be placed? As in many high-tech companies, the Microsoft team was used to short, market-driven production schedules, but this time they were in for a shock. Organizing the testing took *much* longer than they anticipated, but they soon

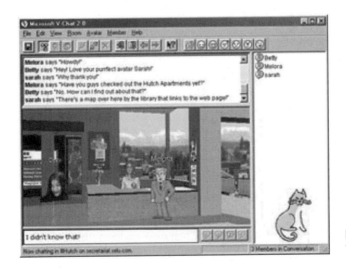

Figure 10.4 The Hutch V-Chat prototype.

learned to set realistic expectations that were in synch with hospital activity and the unexpected delays that occur when working with people who are unwell.

10.3.2   How was the testing done?

The team ran two main sets of user tests. The first set of tests was informally run onsite at the Fred Hutchinson Center in the hospital setting. After observing the system in use on computers located in the hospital setting, the team redesigned the software and then ran formal usability tests in the usability labs at Microsoft.

Test 1: Early observations onsite

In the informal test at the hospital, six computers were set up and maintained by Hutch staff members. A simple, scaled-back prototype of HutchWorld was built using the existing product, Microsoft V-Chat and was installed on the computers, which patients and their families from various hospital locations used. Over the course of several months, the team trained Hutch volunteers and hosted events in the V-Chat prototype. The team observed the usage of the space during unscheduled times, and they also observed the general usage of the prototype.

Test 1: What was learned?

This V-Chat test brought up major usability issues. First, the user community was relatively small, and there were never enough participants in the chat room for successful communication—a concept known as *critical mass*. In addition, many of the patients were not interested in or simultaneously available for chatting. Instead, they preferred asynchronous communication, which does not require an immediate response. Patients and their families used the computers for email, journals, discussion lists, and the bulletin boards largely because they could be used at any time and did not require others to be present at the same time. The team learned that a strong asynchronous base was essential for communication.

The team also observed that the users used the computers to play games and to search the web for cancer sites approved by Hutch clinicians. This information was not included in the virtual environment, and so users were forced to use many different applications. A more "unified" place to find all of the Hutch content was desired that let users rapidly swap among a variety of communication, information, and entertainment tasks.

Test 1: The redesign

Based on this trial, the team redesigned the software to support more asynchronous communication and to include a variety of communication, information, and entertainment areas. They did this by making HutchWorld function as a portal that provides access to information-retrieval tools, communication tools, games, and other types of entertainment. Other features were incorporated too, including email, a bulletin board, a text-chat, a web page creation tool, and a way of checking to see if anyone is around to chat with in the 3D world. The new portal version is show in Figure 10.5.

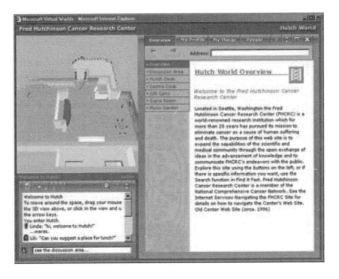

Figure 10.5 HutchWorld portal version.

Test 2: Usability tests

After redesigning the software, the team then ran usability tests in the Microsoft usability labs. Seven participants (four male and three female) were tested. Four of these participants had used chat rooms before and three were regular users. All had browsed the web and some used other communications software. The participants were told that they would use a program called HutchWorld that was designed to provide support for patients and their families. They were then given five minutes to explore HutchWorld. They worked independently and while they explored they provided a running commentary on what they were looking at, what they were thinking, and what they found confusing. This commentary was recorded on video and so were the screens that they visited, so that the Microsoft evaluator, who watched through a one-way mirror, had a record of what happened for later analysis. Participants and the evaluator interacted via a microphone and speakers. When the five-minute exploration period ended, the participants were asked to complete a series of *structured tasks* that were designed to test particular features of the HutchWorld interface.

These tasks focused on how participants:

- dealt with their virtual identity; that is, how they represented themselves and were perceived by others
- communicated with others
- got the information they wanted
- found entertainment

Figure 10.6 shows some of the structured tasks. Notice that the instructions are short, clearly written, and specific.

Welcome to the HutchWorld Usability Study

For this study we are interested in gaining a better understanding of the problems people have when using the program HutchWorld. HutchWorld is an all-purpose program created to offer information and social support to patients and their families at the Fred Hutchinson Cancer Research Center.

The following pages have tasks for you to complete that will help us achieve that better understanding.

While you are completing these tasks, it is important for us know what is going on inside your mind. Therefore, as you complete each task please tell us what you are looking at, what you are thinking about, what is confusing to you, and so forth.

Task #1: Explore HutchWorld

Your first task is to spend five minutes exploring HutchWorld.

A. First, open HutchWorld.

B. Now, explore!

Remember, tell us what you are looking at and what you are thinking about as you are exploring HutchWorld.

Task #2: All about Your Identity in HutchWorld

A. Point to the 3 dimensional (3D) view of HutchWorld.

B. Point at yourself in the 3D view of HutchWorld.

C. Get a map view in the 3D view of HutchWorld.

D. Walk around in the 3D view: go forward, turn left and turn right.

E. Change the color of your shirt.

F. Change some information about yourself, such as where you are from.

Task #3: All about Communicating with Others

A. Send someone an email.

B. Read a message on the HutchWorld Bulletin Board.

C. Post a message on the HutchWorld Bulletin Board.

D. Check to see who is currently in HutchWorld.

E. Find out where the other person in HutchWorld is from.

F. Make the other person in HutchWorld a friend.

G. Chat with the other person in HutchWorld

H. Wave to the other person in HutchWorld.

I. Whisper to the other person in HutchWorld.

Task #4: All about Getting Information

A. Imagine you have never been to Seattle before. Your task is to find something to do.

B. Find out how to get to the Fred Hutchinson Cancer Research Center.

C. Go to your favorite website. [Or go to Yahoo: www.yahoo.com]

D. Once you have found a website, resize the screen so you can see the whole web page.

Figure 10.6 A sample of the structured tasks used in the HutchWorld evaluation.

Task #5: All about Entertainment

  A. Find a game to play.

  B. Get a gift from a Gift Cart and send yourself a gift.

  C. Go and open your gift.

Figure 10.6 (*continued*).

During the study, a member of the development team role-played being a participant so that the real participants would be sure to have someone with whom to interact. The evaluator also asked the participants to fill out a short questionnaire after completing the tasks, with the aim of collecting their opinions about their experiences with HutchWorld. The questionnaire asked:

- What did you *like* about HutchWorld?
- What did you *not like* about HutchWorld?
- What did you find confusing or difficult to use in HutchWorld?
- How would you suggest improving HutchWorld?

Test 2: What was learned from the usability tests?

When running the usability tests, the team collected masses of data that they had to make sense of by systematical analysis. The following discussion offers a snapshot of their findings. Some participants' problems started right at the beginning of the five-minute exploration. The login page referred to "virtual worlds" rather than the expected HutchWorld and, even though this might seem trivial, it was enough to confuse some users. This isn't unusual; developers tend to overlook small things like this, which is why evaluation is so important. Even careful, highly skilled developers like this team tend to forget that users do not speak their language. Fortunately, finding the "go" button was fairly straightforward. Furthermore, most participants read the welcome message and used the navigation list, and over half used the chat buttons, managed to move around the 3D world, and read the overview. But only one-third chatted and used the navigation buttons. The five-minute free-exploration data was also analyzed to determine what people thought of HutchWorld and how they commented upon the 3D view, the chat area, and the browse area.

Users' performance on the structured tasks was analyzed in detail and participant ratings were tabulated. Participants rated the tasks on a scale of 1–3 where 1 = easy, 2 = OK, 3 = difficult, and bold = needed help. Any activity that received an average rating above 1.5 across participants was deemed to need detailed review by the team. Figure 10.7 shows a fragment of the summary of the analysis.

In addition, the team analyzed all the problems that they observed during the tests. They then looked at all their data and drew up a table of issues, noting whether they were a priority to fix and listing recommendations for changes.

Structured Tasks

Participant number:	1	2	3	4	5	6	7	Average
Background Information								
Sex	F	F	M	M	F	M	M	3F, 4M
Age	37	41	43	54	46	44	21	40.9
years of college	4	2	4	4	4	1	2	3.0
hours of chat use in past year	0	3	0	0	365	200	170	105.4
hours of web use in past year	9	11	36	208	391	571	771	285.3
Structured Tasks								
Identify 3D view	1	1	1	1	1	1	1	1.0
Identity self in 3D view	1	2	1	1	1	1	1	1.1
Get a map view of 3D view	1	2	2	1	2	3	1	1.7
Walk in 3D view	1	3	2	1	3	2	1	1.9
Change color of shirt	1	1	3	3	2	3	2	2.1
Change where self is from	1	1	3	1	1	3	1	1.6
Find place to send email	1	3	3	1	3	2	2	2.1
Read a bulletin board message	2	1	3	1	1	1	–	1.5
Post a bulletin board message	1	3	3	3	2	2	–	2.3
Check to see who is currently on	1	3	1	3	2	3	2	2.1
Find out where the other person is from	1	1	2	1	1	3	2	1.6
Make the other person a friend	1	1	3	1	1	2	1	1.4
Chat with the other person	3	1	3	1	1	3	1	1.9
Wave to the other person	1	1	1	1	1	1	1	1.0
Whisper to the other person	1	3	2	2	1	2	1	1.7
Find something to do in Seattle	2	1	2	1	1	1	2	1.4
Find out how to get to FHCRC	1	3	3	2	1	1	2	1.9
Go to a website	1	3	2	3	3	1	1	2.0
Resize web screen	1	3	2	2	2	3	1	2.0
Find a game to play	1	1	2	1	1	1	2	1.3
Send self a gift	1	3	3	3	3	3	3	2.7
Open gift	3	1	2	3	3	3	3	2.6
Participant Average:	1.3	1.9	2.2	1.7	1.7	2.0	1.6	

The following descriptions provide examples of some of the problems participants experience.

Get map view. People generally did not immediately know how to find the map view. However, they knew to look in the chat buttons, and by going through the buttons they found the map view.

Walk in 3D view. People found the use of the mouse to move the avatar awkward, especially when they were trying to turn around. However, once they were used to using the mouse they had no difficulty. For a couple of people, it was not clear to them that they should click on the avatar and drag it in the desired direction. A couple of people tried to move by clicking the place they wanted to move to.

Figure 10.7 Participant information and ratings of difficulty in completing the structured tasks. 1 = easy, 2 = okay, 3 = difficult and bold = needed help.

Issue#	Issue Priority	Issue	Recommendation
1	high	Back button sometimes not working.	Fix back button.
2	high	People are not paying attention to navigation buttons.	Make navigation buttons more prominent.
3	low	Fonts too small, hard to read for some people.	Make it possible to change fonts. Make the font colors more distinct from the background color.
4	low	When navigating, people were not aware overview button would take them back to the main page.	Change the overview button to a home button, change the wording of the overview page accordingly.
5	medium	"Virtual worlds" wording in login screen confusing.	Change wording to "HutchWorld".
6	high	People frequently clicking on objects in 3D view expecting something to happen.	Make the 3D view have links to web pages. For example, when people click on the help desk the browser area should show the help desk information.
7	low	People do not readily find map view button.	Make the icon on the map view button more map-like.
8	medium	Moving avatar with mouse took some getting used to.	Encourage the use of the keyboard. Mention clicking and dragging the avatar in the welcome.
9	low	People wanted to turn around in 3D view, but it was awkward to do so.	Make one of the chat buttons a button that lets you turn around.
10	medium	Confusion about the real world/virtual world distinction.	Change wording of overview description, to make clear Hutch-World is a "virtual" place made to "resemble" the FHCRC, and is a place where anybody can go.
11	high	People do not initially recognize that other real people could be in HutchWorld, that they can talk to them and see them.	Change wording of overview description, to make clear Hutch-World is a place to "chat" with others who are "currently in" the virtual HutchWorld.
12	high	People not seeing/finding the chat window. Trying to chat to people from the people list where other chat-like features are (whisper, etc.)	Make chat window more prominent. Somehow link chat-like features of navigation list to chat window. Change wording of chat window. Instead of type to speak here, type to chat here.

Figure 10.8 A fragment of the table showing problem rankings.

13	low	Who is here list and who has been here list confused.	Spread them apart more in the people list.
14	medium	Difficulty in finding who is here.	Change People button to "Who is On" button.
15	low	Went to own profile to make someone a friend.	Let people add friends at My profile
16	low	Not clear how to append/reply to a discussion in the bulletin board.	Make an append button pop up when double clicking on a topic. Change wording from "post a message" to "write a message" or "add a message".
17	low	Bulletin board language is inconsistent.	Change so it is either a bulletin board, or a discussion area.

Figure 10.8 (*continued*).

Figure 10.8 shows part of this table. Notice that issues were ranked in priority: low, medium, and high. There were just five high-ranking problems that absolutely had to be fixed:

- The back button did not always work.
- People were not paying attention to navigation buttons, so they needed to be more prominent.
- People frequently clicked on objects in the 3D view and expected something to happen. A suggestion for fixing this was to provide links to a web page.
- People did not realize that there could be other real people in the 3D world with whom they could chat, so the wording in the overview description had to be changed.
- People were not noticing the chat window and instead were trying to chat to people in the participant list. The team needed to clarify the instructions about where to chat.

In general, most users found the redesigned software easy to use with little instruction. By running a variety of tests, the informal onsite test, and the formal usability test, key problems were identified at an early stage and various usability issues could be fixed before the actual deployment of the software.

10.3.3 Was it tested again?

Following the usability testing, there were more rounds of observation and testing with six new participants, two males and four females. These tests followed the same general format as those just described but this time they tested multiple users at once, to ensure that the virtual world supported multiuser interactions. The tests were also more detailed and focused. This time the results were more positive, but

---

DILEMMA   When Is It Time to Stop Testing?

Was HutchWorld good enough after these evalua-tions? When has enough testing been done? This frequently asked question is difficult to answer. Few developers have the luxury of testing as thoroughly as John Gould and his colleagues when developing the 1984 Olympic Messaging System (Gould and Lewis, 1990), or even as much as Microsoft's Hutch-World team. Since every test you do will reveal some area where improvement can be made, you cannot assume that there will be a time when the system is perfect: no system is ever perfect. Nor-mally schedule and budget constraints determine when to stop. Joseph Dumas and Ginny Redish, es-tablished usability consultants, point out that for it-erative design and testing to be successful, each test should take as little time as possible while still yield-ing useful information and without burdening the team (Dumas and Redish, 1999).

---

of course there were still usability problems to be fixed. Then the question arose: what to do next? In particular, had they done enough testing (see Dilemma)?

After making a few more fixes, the team stopped usability testing with specific tasks. But the story didn't end here. The next step was to show HutchWorld to can-cer patients and caregivers in a focus-group setting at the Fred Hutchinson Cancer Research Center to get their feedback on the final version. Once the team made adjustments to HutchWorld in response to the focus-group feedback, the final step was to see how well HutchWorld worked in a real clinical environment. It was therefore taken to a residential building used for long-term patient and family stays that was fully wired for Internet access. Here, the team observed what happened when it was used in this natural setting. In particular, they wanted to find out how HutchWorld would integrate with other aspects of patients' lives, particularly with their medical care routines and their access to social support. This informal obser-vation allowed them to examine patterns of use and to see who used which parts of the system, when, and why.

10.3.4 Looking to the future

Future studies were planned to evaluate the effects of the computers and the soft-ware in the Fred Hutchinson Center. The focus of these studies will be the social support and wellbeing of patients and their caregivers in two different conditions. There will be a control condition in which users (i.e., patients) live in the residential building without computers and an experimental condition in which users live in similar conditions but with computers, Internet access, and HutchWorld. The team will evaluate the user data (performance and observation) and surveys collected in the study to investigate key questions, including:

- How does the computer and software impact the social wellbeing of patients and their caregivers?

- What type of computer-based communication best supports this patient community?

- What are the general usage patterns? i.e., which features were used and at what time of day were they used, etc.?

• How might any medical facility use computers and software like Hutch-World to provide social support for its patients and caregivers?

There is always more to learn about the efficacy of a design and how much users enjoy using a product, especially when designing innovative products like HutchWorld for new environments. This study will provide a longer-term view of how HutchWorld is used in its natural environment that is not provided by the other evaluations. It's an ambitious plan because it involves a comparison between two different environmental settings, one that has computers and HutchWorld and one that doesn't (see Chapter 13 for more on experimental design).

ACTIVITY 10.3

(a) The case study does not say much about early evaluation to test the conceptual design shown in Figure 10.5. What do you think happened?

(b) The evaluators recorded the gender of participants and noted their previous experience with similar systems. Why is this important?

(c) Why do you think it was important to give participants a five-minute exploration period?

(d) *Triangulation* is a term that describes how different perspectives are used to understand a problem or situation. Often different techniques are used in triangulation. Which techniques were triangulated in the evaluations of the HutchWorld prototype?

(e) The evaluators collected participants' opinions. What kinds of concerns do you think participants might have about using HutchWorld? Hints: personal information, medical information, communicating feelings, etc.

Comment

(a) There was probably much informal discussion with representative users: patients, medical staff, relatives, friends, and caregivers. The team also visited the clinic and hospital and observed what happened there. They may also have discussed this with the physicians and administrators.

(b) It is possible that our culture causes men and women to react differently in certain circumstances. Experience is an even more important influence than gender, so knowing how much previous experience users have had with various types of computer systems enables evaluators to make informed judgments about their performance. Experts and novices, for example, tend to behave very differently.

(c) The evaluators wanted to see how participants reacted to the system and whether or not they could log on and get started. The exploration period also gave the participants time to get used to the system before doing the set tasks.

(d) Data was collected from the five-minute exploration, from performance on the structured tasks, and from the user satisfaction questionnaire.

(e) Comments and medical details are personal and people want privacy. Patients might be concerned about whether the medical information they get via the computer and from one another is accurate. Participants might be concerned about how clearly and accurately they are communicating because non-verbal communication is reduced online.

10.4   Discussion

In both HutchWorld and the 1984 Olympic Messaging System, a variety of evaluation techniques were used at different stages of design to answer different questions. "Quick and dirty" observation, in which the evaluators informally examine how a prototype is used in the natural environment, was very useful in early design. Following this with rounds of usability testing and redesign revealed important usability problems. However, usability testing alone is not sufficient. Field studies were needed to see how users used the system in their natural environments, and sometimes the results were surprising. For example, in the OMS system users from different cultures behaved differently. A key issue in the HutchWorld study was how use of the system would fit with patients' medical routines and changes in their physical and emotional states. Users' opinions also offered valuable insights. After all, if users don't like a system, it doesn't matter how successful the usability testing is: they probably won't use it. Questionnaires and interviews were used to collect user's opinions.

An interesting point concerns not only how the different techniques can be used to address different issues at different stages of design, but also how these techniques complement each other. Together they provide a broad picture of the system's usability and reveal different perspectives. In addition, some techniques are better than others for getting around practical problems. This is a large part of being a successful evaluator. In the HutchWorld study, for example, there were not many users, so the evaluators needed to involve them sparingly. For example, a technique requiring 20 users to be available at the same time was not feasible in the HutchWorld study, whereas there was no problem with such an approach in the OMS study. Furthermore, the OMS study illustrated how many different techniques, some of which were highly opportunistic, can be brought into play depending on circumstances. Some practical issues that evaluators routinely have to address include:

- what to do when there are not many users
- how to observe users in their natural location (i.e., field studies) without disturbing them
- having appropriate equipment available
- dealing with short schedules and low budgets
- not disturbing users or causing them duress or doing anything unethical
- collecting "useful" data and being able to analyze it
- selecting techniques that match the evaluators' expertise

There are many evaluation techniques from which to choose and these practical issues play a large role in determining which are selected. Furthermore, selection depends strongly on the stage in the design and the particular questions to be answered. In addition, each of the disciplines that contributes to interaction design has preferred bodies of theory and techniques that can influence this choice. These issues are discussed further in the next chapter.

Assignment

1. Reconsider the HutchWorld design and evaluation case study and note *what* was evaluated, *why* and *when*, and *what* was learned at each stage?

2. How was the design advanced after each round of evaluation?

3. What were the main constraints that influenced the evaluation?

4. How did the stages and choice of techniques build on and complement each other (i.e., triangulate)?

5. Which parts of the evaluation were directed at usability goals and which at user experience goals? Which additional goals not mentioned in the study could the evaluations have focused upon?

Summary

The aim of this chapter was to introduce basic evaluation concepts that will be revisited and built on in the next four chapters. We selected the HutchWorld case study because it illustrates how a team of designers evaluated a novel system and coped with a variety of practical constraints. It also shows how different techniques are needed for different purposes and how techniques are used together to gain different perspectives on a product's usability. This study highlights how the development team paid careful attention to usability and user experience goals as they designed and evaluated their system.

Key points

- Evaluation and design are very closely integrated in user-centered design.

- Some of the same techniques are used in evaluation as in the activity of establishing requirements and identifying users' needs, but they are used differently (e.g., interviews and questionnaires, etc.).

- Triangulation involves using combinations of techniques in concert to get different perspectives or to examine data in different ways.

- Dealing with constraints, such as gaining access to users or accommodating users' routines, is an important skill for evaluators to develop.

Further reading

CHENG, L., STONE, L., FARNHAM, S., CLARK, A. M., AND ZANER-GODSEY, M. (2000) *Hutchworld: Lessons Learned. A Collaborative Project: Fred Hutchinson Cancer Research Center & Microsoft Research*. In the Proceedings of the Virtual Worlds Conference 2000, Paris, France. This paper describes the HutchWorld study and, as the title suggests, it discusses the design lessons that were learned. It also describes the evaluation studies in more detail.

GOULD, J. D., BOIES, S. J., LEVY, S., RICHARDS, J. T., AND SCHOONARD, J. (1990). The 1984 Olympic Message System: A test of behavioral principles of system design. In J. Preece and L. Keller (eds.), *Human-Computer Interaction (Readings)*. Prentice Hall International Ltd., Hemel Hempstead, UK: 260–283. This edited paper tells the story of the design and evaluation of the OMS.

GOULD, J. D., BOIES, S. J., LEVY, S., RICHARDS, J. T., AND SCHOONARD, J. (1987). The 1984 Olympic Message System: a test of behavioral principles of systems design. *Communications of the ACM*, 30(9), 758–769. This is the original, full version of the OMS paper.

Chapter 11

An evaluation framework

11.1 Introduction
11.2 Evaluation paradigms and techniques
    11.2.1 Evaluation paradigms
    11.2.2 Techniques
11.3 DECIDE: A framework to guide evaluation
    11.3.1 Determine the goals
    11.3.2 Explore the questions
    11.3.3 Choose the evaluation paradigm and techniques
    11.3.4 Identify the practical issues
    11.3.5 Decide how to deal with the ethical issues
    11.3.6 Evaluate, interpret and present the data
11.4 Pilot studies

11.1 Introduction

Designing useful and attractive products requires skill and creativity. As products evolve from initial ideas through conceptual design and prototypes, iterative cycles of design and evaluation help to ensure that they meet users' needs. But how do evaluators decide *what* and *when* to evaluate? The HutchWorld case study in the previous chapter described how one team did this, but the circumstances surrounding every product's development are different. Certain techniques work better for some than for others.

Identifying usability and user experience goals is essential for making every product successful, and this requires understanding users' needs. The role of evaluation is to make sure that this understanding occurs during all the stages of the product's development. The skillful and sometimes tricky part of doing this is knowing what to focus on at different stages. Initial requirements get the design process started, but, as you have seen, understanding requirements tends to happen by a process of negotiation between designers and users. As designers understand users' needs better, their designs reflect this understanding. Similarly, as users see and experience design ideas, they are able to give better feedback that enables the designers to improve their designs further. The process is cyclical, with evaluation playing a key role in facilitating understanding between designers and users.

Evaluation is driven by questions about how well the design or particular aspects of it satisfy users' needs. Some of these questions provide high-level goals to guide the evaluation. Others are much more specific. For example, can users find a particular menu item? Is a graphic useful and attractive? Is the product engaging? Practical constraints also play a big role in shaping evaluation plans: tight schedules, low budgets, or little access to users constrain what evaluators can do. You read in chapter 10 how the HutchWorld team had to plan its evaluation around hospital routines and patients' health.

Experienced designers get to know what works and what doesn't, but those with little experience can find doing their first evaluation daunting. However, with careful advance planning, problems can be spotted and ways of dealing with them can be found. Planning evaluation studies involves thinking about key issues and asking questions about the process. In this chapter we propose the DECIDE framework to help you do this.

The main aims of this chapter are to:

- Continue to explain the key concepts and terms used to discuss evaluation.
- Describe the evaluation paradigms and techniques used in interaction design.
- Discuss the conceptual, practical, and ethical issues to be considered when planning evaluation.
- Introduce the DECIDE framework to help you plan your own evaluation studies.

11.2   Evaluation paradigms and techniques

Before we describe the techniques used in evaluation studies, we shall start by proposing some key terms. Terminology in this field tends to be loose and often confusing so it is a good idea to be clear from the start what you mean. We start with the much-used term *user studies*, defined by Abigail Sellen in her interview at the end of Chapter 4 as follows: "user studies essentially involve looking at how people behave either in their natural [environments], or in the laboratory, both with old technologies and with new ones." Any kind of evaluation, whether it is a user study or not, is guided either explicitly or implicitly by a set of beliefs that may also be underpinned by theory. These beliefs and the practices (i.e., the methods or techniques) associated with them are known as an evaluation paradigm, which you should not confuse with the "interaction paradigms" discussed in Chapter 2. Often evaluation paradigms are related to a particular discipline in that they strongly influence how people from the discipline think about evaluation. Each paradigm has particular methods and techniques associated with it. So that you are not confused, we want to state explicitly that we will not be distinguishing between methods and techniques. We tend to talk about techniques, but you may find that other books call them methods. An example of the relationship between a paradigm and the techniques used by evaluators following that paradigm can be seen for usability testing, which is an applied science and engineering paradigm. The techniques associated with usability testing are: user testing in a controlled environment; observation of user activity in the controlled environment and the field; and questionnaires and interviews.

11.2.1   Evaluation paradigms

In this book we identify four core evaluation paradigms: (1) "quick and dirty" evaluations; (2) usability testing; (3) field studies; and (4) predictive evaluation. Other texts may use slightly different terms to refer to similar paradigms.

"Quick and dirty" evaluation

A "quick and dirty" evaluation is a common practice in which designers informally get feedback from users or consultants to confirm that their ideas are in line with users' needs and are liked. "Quick and dirty" evaluations can be done at any stage and the emphasis is on fast input rather than carefully documented findings. For example, early in design developers may meet informally with users to get feedback on ideas for a new product (Hughes et al., 1994). At later stages similar meetings may occur to try out an idea for an icon, check whether a graphic is liked, or confirm that information has been appropriately categorized on a webpage. This approach is often called "quick and dirty" because it is meant to be done in a short space of time. Getting this kind of feedback is an essential ingredient of successful design.

As discussed in Chapter 9, any involvement with users will be highly informative and you can learn a lot early in design by observing what people do and talking to them informally. The data collected is usually descriptive and informal and it is fed back into the design process as verbal or written notes, sketches and anecdotes, etc. Another source comes from consultants, who use their knowledge of user behavior, the market place and technical know-how, to review software quickly and provide suggestions for improvement. It is an approach that has become particularly popular in web design where the emphasis is usually on short timescales.

Usability testing

Usability testing was the dominant approach in the 1980s (Whiteside et al., 1998), and remains important, although, as you will see, field studies and heuristic evaluations have grown in prominence. Usability testing involves measuring typical users' performance on carefully prepared tasks that are typical of those for which the system was designed. Users' performance is generally measured in terms of number of errors and time to complete the task. As the users perform these tasks, they are watched and recorded on video and by logging their interactions with software. This observational data is used to calculate performance times, identify errors, and help explain why the users did what they did. User satisfaction questionnaires and interviews are also used to elicit users' opinions.

The defining characteristic of usability testing is that it is *strongly controlled* by the evaluator (Mayhew, 1999). There is no mistaking that the evaluator is in charge! Typically tests take place in laboratory-like conditions that are controlled. Casual visitors are not allowed and telephone calls are stopped, and there is no possibility of talking to colleagues, checking email, or doing any of the other tasks that most of us rapidly switch among in our normal lives. Everything that the participant does is recorded—every keypress, comment, pause, expression, etc., so that it can be used as data.

Quantifying users' performance is a dominant theme in usability testing. However, unlike research experiments, variables are not manipulated and the typical number of participants is too small for much statistical analysis. User satisfaction data from questionnaires tends to be categorized and average ratings are presented. Sometimes video or anecdotal evidence is also included to illustrate problems that users encounter. Some evaluators then summarize this data in a usability specification so that developers can use it to test future prototypes or versions of the product against it. Optimal performance levels and minimal levels of acceptance are often specified and current levels noted. Changes in the design can then be agreed and engineered—hence the term "usability engineering." User testing is explained further in Chapter 14, how to observe users is described in Chapter 12, and issues concerned with interviews and questionnaires are explored in Chapter 13.

Field studies

The distinguishing feature of field studies is that they are done in natural settings with the aim of increasing understanding about what users do naturally and how technology impacts them. In product design, field studies can be used to (1) help identify opportunities for new technology; (2) determine requirements for design; (3) facilitate the introduction of technology; and (4) evaluate technology (Bly, 1997).

Chapter 9 introduced qualitative techniques such as interviews, observation, participant observation, and ethnography that are used in field studies. The exact choice of techniques is often influenced by the theory used to analyze the data. The data takes the form of events and conversations that are recorded as notes, or by audio or video recording, and later analyzed using a variety of analysis techniques such as content, discourse, and conversational analysis. These techniques vary considerably. In content analysis, for example, the data is analyzed into content categories, whereas in discourse analysis the use of words and phrases is examined. Artifacts are also collected. In fact, anything that helps to show what people do in their natural contexts can be regarded as data.

In this text we distinguish between two overall approaches to field studies. The first involves observing explicitly and recording what is happening, as an *outsider* looking on. Qualitative techniques are used to collect the data, which may then be analyzed qualitatively or quantitatively. For example, the number of times a particular event is observed may be presented in a bar graph with means and standard deviations.

In some field studies the evaluator may be an *insider* or even a participant. Ethnography is a particular type of insider evaluation in which the aim is to explore the details of what happens in a particular social setting. "In the context of human-computer interaction, ethnography is a means of studying work (or other activities) in order to inform the design of information systems and understand aspects of their use" (Shapiro, 1995, p. 8).

Predictive evaluation

In predictive evaluations experts apply their knowledge of typical users, often guided by heuristics, to predict usability problems. Another approach involves theoretically-based models. The key feature of predictive evaluation is that users need *not* be present, which makes the process quick, relatively inexpensive, and thus attractive to companies; but it has limitations.

In recent years heuristic evaluation in which experts review the software product guided by tried and tested heuristics has become popular (Nielsen and Mack, 1994). As mentioned in Chapter 1, usability guidelines (e.g., always provide clearly marked exits) were designed primarily for evaluating screen-based products (e.g. form fill-ins, library catalogs, etc.). With the advent of a range of new interactive products (e.g., the web, mobiles, collaborative technologies), this original set of heuristics has been found insufficient. While some are still applicable (e.g., speak the users' language), others are inappropriate. New sets of heuristics are also needed that are aimed at evaluating different classes of interactive products. In particular, specific heuristics are needed that are tailored to evaluating web-based products, mobile devices, collaborative technologies, computerized toys, etc. These should be based on a combination of usability and user experience goals, new research findings and market research. Care is needed in using sets of heuristics. As you will see in Chapter 13, designers are sometimes led astray by findings from heuristic evaluations that turn out not to be as accurate as they at first seemed.

Table 11.1 summarizes the key aspects of each evaluation paradigm for the following issues:

- the role of users
- who controls the process and the relationship between evaluators and users during the evaluation
- the location of the evaluation
- when the evaluation is most useful
- the type of data collected and how it is analyzed
- how the evaluation findings are fed back into the design process
- the philosophy and theory that underlies the evaluation paradigms

Some other terms that you may encounter in your reading are shown in Box 11.1.

ACTIVITY 11.1 Think back to the HutchWorld case study.

  (a) Which evaluation paradigms were used in the study and which were not?

  (b) How could the missing evaluation paradigms have been used to inform the design and why might they not have been used?

Comment (a) The team did some "quick and dirty" evaluation during early development but this is not stressed in their report. Usability testing played a strong role, with some tests being carried out at the Fred Hutchinson Center and later tests in Microsoft's usability laboratories. Field studies are not strongly featured, but the team does mention

Table 11.1   Characteristics of different evaluation paradigms

Evaluation paradigms	"Quick and dirty"	Usability testing	Field studies	Predictive
Role of users	Natural behavior.	To carry out set tasks.	Natural behavior.	Users generally not involved.
Who controls	Evaluators take minimum control.	Evaluators strongly in control.	Evaluators try to develop relationships with users.	Expert evaluators.
Location	Natural environment or laboratory.	Laboratory.	Natural environment.	Laboratory-oriented but often happens on customer's premises.
When used	Any time you want to get feedback about a design quickly. Techniques from other evaluation paradigms can be used–e.g., experts review software.	With a prototype or product.	Most often used early in design to check that users' needs are being met or to assess problems or design opportunities.	Expert reviews (often done by consultants) with a prototype, but can occur at any time. Models are used to assess specific aspects of a potential design.
Type of data	Usually qualitative, informal descriptions.	Quantitative. Sometimes statistically validated. Users' opinions collected by questionnaire or interview.	Qualitative descriptions often accompanied with sketches, scenarios, quotes, other artifacts.	List of problems from expert reviews. Quantitative figures from model, e.g., how long it takes to perform a task using two designs.
Fed back into design by . . .	Sketches, quotes, descriptive report.	Report of performance measures, errors etc. Findings provide a benchmark for future versions.	Descriptions that include quotes, sketches, anecdotes, and sometimes time logs.	Reviewers provide a list of problems, often with suggested solutions. Times calculated from models are given to designers.
Philosophy	User-centered, highly practical approach.	Applied approach based on experimentation, i.e., usability engineering.	May be objective observation or ethnographic.	Practical heuristics and practitioner expertise underpin expert reviews. Theory underpins models.

BOX 11.1 Some Definitions to Help You

Objective and subjective Objective evaluations are based on techniques that use quantitative measurement rather than users' or experts' opinions. Subjective evaluations are based on opinions and anecdotes.

Quantitative and qualitative Quantitative evaluations involve measurements, whereas qualitative evaluations involve descriptions and anecdotes. Quantitative evaluations tend to be seen as objec-

tive and impartial, whereas many qualitative evaluations tend to be seen as subjective but this isn't necessarily true.

Laboratory and field or naturalistic studies Laboratory studies occur in controlled environments. They may be done in a specially built laboratory or in a space that is specially adapted for the purpose. Field or naturalistic studies are situated in the real-world context in which the system is or will be used.

---

observing how patients used HutchWorld in the Center. Field studies were planned in which patients, who have access to HutchWorld and the web, could be systematically compared with another group who does not have these facilities. However, distinguishing between evaluation paradigms isn't always clear-cut. In practice elements typically found in one may be transferred to another (e.g., the controlled approach the HutchWorld team planned to use in the field). The only evaluation paradigm that is not mentioned in the study is predictive evaluation.

(b) Expert reviews could have been done any time during its development but the team may have thought they were not needed, or there wasn't time, or perhaps they were performed but not reported.

11.2.2 Techniques

There are many evaluation techniques and they can be categorized in various ways, but in this text we will examine techniques for:

- observing users
- asking users their opinions
- asking experts their opinions
- testing users' performance
- modeling users' task performance to predict the efficacy of a user interface

The brief descriptions below offer an *overview* of each category, which we discuss in detail in the next three chapters. Be aware that some techniques are used in different ways in different evaluation paradigms.

Observing users

Observation techniques help to identify needs leading to new types of products and help to evaluate prototypes. Notes, audio, video, and interaction logs are well-known ways of recording observations and each has benefits and drawbacks. Obvious challenges for evaluators are how to observe without disturbing the people being observed and how to analyze the data, particularly when large quantities of video data are collected or when several different types must be integrated to tell the story (e.g., notes, pictures, sketches from observers). You met several observation techniques in Chapter 7 in the context of the requirements activity; in Chapter 12 we will focus on how they are used in evaluation.

Asking users

Asking users what they think of a product—whether it does what they want; whether they like it; whether the aesthetic design appeals; whether they had problems using it; whether they want to use it again—is an obvious way of getting feedback. Interviews and questionnaires are the main techniques for doing this. The questions asked can be unstructured or tightly structured. They can be asked of a few people or of hundreds. Interview and questionnaire techniques are also being developed for use with email and the web. We discuss these techniques in Chapter 13.

Asking experts

Software inspections and reviews are long established techniques for evaluating software code and structure. During the 1980s versions of similar techniques were developed for evaluating usability. Guided by heuristics, experts step through tasks role-playing typical users and identify problems. Developers like this approach because it is usually relatively inexpensive and quick to perform compared with laboratory and field evaluations that involve users. In addition, experts frequently suggest solutions to problems. In Chapter 13 you will learn a few inspection techniques for evaluating usability.

User testing

Measuring user performance to compare two or more designs has been the bedrock of usability testing. As we said earlier when discussing usability testing, these tests are usually conducted in controlled settings and involve typical users performing typical, well-defined tasks. Data is collected so that performance can be analyzed. Generally the time taken to complete a task, the number of errors made, and the navigation path through the product are recorded. Descriptive statistical measures such as means and standard deviations are commonly used to report the results. In Chapter 14 you will learn the basics of user testing and how it differs from scientific experiments.

Modeling users' task performance

There have been various attempts to model human-computer interaction so as to predict the efficiency and problems associated with different designs at an early stage without building elaborate prototypes. These techniques are successful for systems with limited functionality such as telephone systems. GOMS and the keystroke model are the best known techniques. They have already been mentioned in Chapter 3 and in Chapter 14 we examine their role in evaluation.

Table 11.2 summarizes the categories of techniques and indicates how they are commonly used in the four evaluation paradigms.

Table 11.2 The relationship between evaluation paradigms and techniques.

Techniques	Evaluation paradigms			
	"Quick and dirty"	Usability testing	Field studies	Predictive
Observing users	Important for seeing how users behave in their natural environments.	Video and interaction logging, which can be analyzed to identify errors, investigate routes through the software, or calculate performance time.	Observation is the central part of any field study. In ethnographic studies evaluators immerse themselves in the environment. In other types of studies the evaluator looks on objectively.	N/A
Asking users	Discussions with users and potential users individually, in groups or focus groups.	User satisfaction questionnaires are administered to collect users' opinions. Interviews may also be used to get more details.	The evaluator may interview or discuss what she sees with participants. Ethnographic interviews are used in ethnographic studies.	N/A
Asking experts	To provide critiques (called "crit reports") of the usability of a prototype.	N/A	N/A	Experts use heuristics early in design to predict the efficacy of an interface.
User testing	N/A	Testing typical users on typical tasks in a controlled laboratory-like setting is the cornerstone of usability testing.	N/A	N/A
Modeling users' task performance	N/A	N/A	N/A	Models are used to predict the efficacy of an interface or compare performance times between versions.

© 1999 Randy Glasbergen.

GLASBERGEN

"It's the latest innovation in office safety.
When your computer crashes, an air bag is activated
so you won't bang your head in frustration."

11.3   DECIDE: A framework to guide evaluation

Well-planned evaluations are driven by clear goals and appropriate questions (Basili et al., 1994). To guide our evaluations we use the DECIDE framework, which provides the following checklist to help novice evaluators:

1. Determine the overall *goals* that the evaluation addresses.

2. Explore the specific *questions* to be answered.

3. Choose the *evaluation paradigm* and *techniques* to answer the questions.

4. Identify the *practical issues* that must be addressed, such as selecting participants.

5. Decide how to deal with the *ethical issues*.

6. Evaluate, interpret, and present the *data*.

11.3.1   Determine the goals

What are the high-level goals of the evaluation? Who wants it and why? An evaluation to help clarify user needs has different goals from an evaluation to determine the best metaphor for a conceptual design, or to fine-tune an interface, or to examine how technology changes working practices, or to inform how the next version of a product should be changed.

Goals should guide an evaluation, so determining what these goals are is the first step in planning an evaluation. For example, we can restate the general goal statements just mentioned more clearly as:

- Check that the evaluators have understood the users' needs.

- Identify the metaphor on which to base the design.

- Check to ensure that the final interface is consistent.
- Investigate the degree to which technology influences working practices.
- Identify how the interface of an existing product could be engineered to improve its usability.

These goals influence the evaluation approach, that is, which evaluation paradigm guides the study. For example, engineering a user interface involves a quantitative engineering style of working in which measurements are used to judge the quality of the interface. Hence usability testing would be appropriate. Exploring how children talk together in order to see if an innovative new groupware product would help them to be more engaged would probably be better informed by a field study.

11.3.2 Explore the questions

In order to make goals operational, questions that must be answered to satisfy them have to be identified. For example, the goal of finding out why many customers prefer to purchase paper airline tickets over the counter rather than e-tickets can be broken down into a number of relevant questions for investigation. What are customers' attitudes to these new tickets? Perhaps they don't trust the system and are not sure that they will actually get on the flight without a ticket in their hand. Do customers have adequate access to computers to make bookings? Are they concerned about security? Does this electronic system have a bad reputation? Is the user interface to the ticketing system so poor that they can't use it? Maybe very few people managed to complete the transaction.

Questions can be broken down into very specific sub-questions to make the evaluation even more specific. For example, what does it mean to ask, "Is the user interface poor?": Is the system difficult to navigate? Is the terminology confusing because it is inconsistent? Is response time too slow? Is the feedback confusing or maybe insufficient? Sub-questions can, in turn, be further decomposed into even finer-grained questions, and so on.

11.3.3 Choose the evaluation paradigm and techniques

Having identified the goals and main questions, the next step is to choose the evaluation paradigm and techniques. As discussed in the previous section, the evaluation paradigm determines the kinds of techniques that are used. Practical and ethical issues (discussed next) must also be considered and trade-offs made. For example, what seems to be the most appropriate set of techniques may be too expensive, or may take too long, or may require equipment or expertise that is not available, so compromises are needed.

As you saw in the HutchWorld case study, combinations of techniques can be used to obtain different perspectives. Each type of data tells the story from a different point of view. Using this triangulation reveals a broad picture.

11.3.4   Identify the practical issues

There are many practical issues to consider when doing any kind of evaluation and it is important to identify them *before* starting. Some issues that should be considered include users, facilities and equipment, schedules and budgets, and evaluators' expertise. Depending on the availability of resources, compromises may involve adapting or substituting techniques.

Users

It goes without saying that a key aspect of an evaluation is involving *appropriate* users. For laboratory studies, users must be found and screened to ensure that they represent the user population to which the product is targeted. For example, usability tests often need to involve users with a particular level of experience e.g., novices or experts, or users with a range of expertise. The number of men and women within a particular age range, cultural diversity, educational experience, and personality differences may also need to be taken into account, depending on the kind of product being evaluated. In usability tests participants are typically screened to ensure that they meet some predetermined characteristic. For example, they might be tested to ensure that they have attained a certain skill level or fall within a particular demographic range. Questionnaire surveys require large numbers of participants so ways of identifying and reaching a representative sample of participants are needed. For field studies to be successful, an appropriate and accessible site must be found where the evaluator can work with the users in their natural setting.

Another issue to consider is how the users will be involved. The tasks used in a laboratory study should be representative of those for which the product is designed. However, there are no written rules about the length of time that a user should be expected to spend on an evaluation task. Ten minutes is too short for most tasks and two hours is a long time, but what is reasonable? Task times will vary according to the type of evaluation, but when tasks go on for more than 20 minutes, consider offering breaks. It is accepted that people using computers should stop, move around and change their position regularly after every 20 minutes spent at the keyboard to avoid repetitive strain injury. Evaluators also need to put users at ease so they are not anxious and will perform normally. Even when users are paid to participate, it is important to treat them courteously. At no time should users be treated condescendingly or made to feel uncomfortable when they make mistakes. Greeting users, explaining that it is the system that is being tested and not them, and planning an activity to familiarize them with the system before starting the task all help to put users at ease.

Facilities and equipment

There are many practical issues concerned with using equipment in an evaluation. For example, when using video you need to think about how you will do the recording: how many cameras and where do you put them? Some people are disturbed by having a camera pointed at them and will not perform normally, so how can you avoid making them feel uncomfortable? Spare film and batteries may also be needed.

Schedule and budget constraints

Time and budget constraints are important considerations to keep in mind. It might seem ideal to have 20 users test your interface, but if you need to pay them, then it could get costly. Planning evaluations that can be completed on schedule is also important, particularly in commercial settings. However, as you will see in the interview with Sara Bly in the next chapter, there is never enough time to do evaluations as you would ideally like, so you have to compromise and plan to do a good job with the resources and time available.

Expertise

Does the evaluation team have the expertise needed to do the evaluation? For example, if no one has used models to evaluate systems before, then basing an evaluation on this approach is not sensible. It is no use planning to use experts to review an interface if none are available. Similarly, running usability tests requires expertise. Analyzing video can take many hours, so someone with appropriate expertise and equipment must be available to do it. If statistics are to be used, then a statistician should be consulted before starting the evaluation and then again later for analysis, if appropriate.

ACTIVITY 11.2   Informal observation, user performance testing, and questionnaires were used in the Hutch-World case study. What practical issues are mentioned in the case study? What other issues do you think the developers had to take into account?

Comment   No particular practical issues are mentioned for the informal observation, but there probably were restrictions on where and what the team could observe. For example, it is likely that access would be denied to very sick patients and during treatment times. Not surprisingly, user testing posed more problems, such as finding participants, putting equipment in place, managing the tests, and underestimation of the time needed to work in a hospital setting compared with the fast production times at Microsoft.

11.3.5   Decide how to deal with the ethical issues

The Association for Computing Machinery (ACM) and many other professional organizations provide ethical codes (Box 11.2) that they expect their members to uphold, particularly if their activities involve other human beings. For example, people's privacy should be protected, which means that their name should not be associated with data collected about them or disclosed in written reports (unless they give permission). Personal records containing details about health, employment, education, financial status, and where participants live should be confidential. Similarly,

---

BOX 11.2   ACM Code of Ethics

The ACM code outlines many ethical issues that professionals are likely to face. Section 1 outlines fundamental ethical considerations, while section 2 addresses additional, more specific considerations of professional conduct. Statements in section 3 pertain more specifically to individuals who have a leadership role. Principles involving compliance with the code are given in section 4. Two principles of particular relevance to this discussion are:

- Ensure that users and those who will be affected by a system have their needs clearly articulated during the assessment of requirements; later the system must be validated to meet requirements.

- Articulate and support policies that protect the dignity of users and others affected by a computing system.

---

it should not be possible to identify individuals from comments written in reports. For example, if a focus group involves nine men and one woman, the pronoun "she" should not be used in the report because it will be obvious to whom it refers.

Most professional societies, universities, government and other research offices require researchers to provide information about activities in which human participants will be involved. This documentation is reviewed by a panel and the researchers are notified whether their plan of work, particularly the details about how human participants will be treated, is acceptable.

People give their time and their trust when they agree to participate in an evaluation study and both should be respected. But what does it mean to be respectful to users? What should participants be told about the evaluation? What are participants' rights? Many institutions and project managers require participants to read and sign an informed consent form similar to the one in Box 11.3. This form explains the aim of the tests or research and promises participants that their personal details and performance will not be made public and will be used only for the purpose stated. It is an

---

BOX 11.3   Informed Consent Form

I state that I am over 18 years of age and wish to participate in a program of research being conducted by Dr. Hoo Hah and his colleagues at the College of Extraordinary Research, University of Highland, College Estate.

The purpose of the research is to assess the usability of HighFly, a website developed at the National Library to provide information to the general public.

The procedures involve the monitored use of HighFly. I will be asked to perform specific tasks using HighFly. I will also be asked open-ended questions about HighFly and my experience using it.

All information collected in the study is confidential, and my name will not be identified at any time.

I understand that I am free to ask questions or to withdraw from participation at any time without penalty.

_____   _____
Signature of Participant   Date

(Adapted from Cogdill, 1999.)

agreement between the evaluator and the evaluation participants that helps to confirm the professional relationship that exists between them. If your university or organization does not provide such a form it is advisable to develop one, partly to protect yourself in the unhappy event of litigation and partly because the act of constructing it will remind you what you should consider.

The following guidelines will help ensure that evaluations are done ethically and that adequate steps to protect users' rights have been taken.

- Tell participants the goals of the study and exactly what they should expect if they participate. The information given to them should include outlining the process, the approximate amount of time the study will take, the kind of data that will be collected, and how that data will be analyzed. The form of the final report should be described and, if possible, a copy offered to them. Any payment offered should also be clearly stated.

- Be sure to explain that demographic, financial, health, or other sensitive information that users disclose or is discovered from the tests is confidential. A coding system should be used to record each user and, if a user must be identified for a follow-up interview, the code and the person's demographic details should be stored separately from the data. Anonymity should also be promised if audio and video are used.

- Make sure users know that they are free to stop the evaluation at any time if they feel uncomfortable with the procedure.

- Pay users when possible because this creates a formal relationship in which mutual commitment and responsibility are expected.

- Avoid including quotes or descriptions that inadvertently reveal a person's identity, as in the example mentioned above, of avoiding use of the pronoun "she" in the focus group. If quotes need to be reported, e.g., to justify conclusions, then it is convention to replace words that would reveal the source with representative words, in square brackets. We used this convention in Boxes 9.2 and 9.3.

- Ask users' permission in advance to quote them, promise them anonymity, and offer to show them a copy of the report before it is distributed.

The general rule to remember when doing evaluations is *do unto others only what you would not mind being done to you*.

ACTIVITY 11.3 Think back to the HutchWorld case study. What ethical issues did the developers have to consider?

Comment The developers of HutchWorld considered all the issues listed above. In addition, because the study involved patients, they had to be particularly careful that medical and other personal information was kept confidential. They were also sensitive to the fact that cancer patients may become too tired or sick to participate so they reassured them that they could stop at any time if the task became onerous.

ACTIVITY 11.4   Usability laboratories often have a one-way mirror that allows evaluators to watch users doing their tasks in the laboratory without the users seeing the evaluators. Should users be told that they are being watched?

Comment   Yes, users should be told that they will be observed through a one-way mirror. It is unethical not to. This honest approach will not compromise the study because users forget about the mirror as they get more absorbed in their tasks. Telling users what is happening helps to build trust.

The recent explosion in Internet and web usage has resulted in more research on how people use these technologies and their effects on everyday life. Consequently, there are many projects in which developers and researchers are logging users' interactions, analyzing web traffic, or examining conversations in chatrooms, bulletin boards, or on email. Unlike most previous evaluations in human-computer interaction, these studies can be done without users knowing that they are being studied. This raises ethical concerns, chief among which are issues of privacy, confidentiality, informed consent, and appropriation of others' personal stories (Sharf, 1999). People often say things online that they would not say face to face. Furthermore, many people are unaware that personal information they share online can be read by someone with technical know-how years later, even after they have deleted it from their personal mailbox (Erickson et al., 1999).

ACTIVITY 11.5   Studies of user behavior on the Internet may involve logging users' interactions and keeping a copy of their conversations with others. Should users be told that this is happening?

DILEMMA   What Would You Do?

There is a famous and controversial story about a 1961–62 experiment by Yale social psychologist Stanley Milgram to investigate how people respond to orders given by people in authority. Much has been written about this experiment and details have been changed and embellished over the years, but the basic ethical issues it raises are still worth considering, even if the details of the actual study have been distorted.

The subjects were ordinary residents of New Haven who were asked to administer increasingly high levels of electric shocks to victims when they made errors in the tasks they were given. As the electric shocks got more and more severe, so did the apparent pain of the victims receiving them, to the extent that some appeared to be on the verge of dying. Not surprisingly, those administering the shocks became more and more disturbed by what they were being asked to do, but several continued, believing that they should do as their superiors told them. What they did not realize was that the so-called victims were, in fact, very convincing actors who were not being injured at all. Instead, the shock administrators were themselves the real subjects. It was their responses to authority that were being studied in this deceptive experiment.

This story raises several important ethical issues. First, this experiment reveals how power relationships can be used to control others. Second and equally important, this experiment relied on deception. The experimenters were, in fact, the subjects and the fake subjects colluded with the real scientists to deceive them. Without this deception the experiment would not have worked.

Is it acceptable to deceive subjects to this extent? What do you think?

Comment   Yes, it is better to tell users in advance that they are being logged. As in the previous exam-
              ple, the users' knowledge that they are being logged often ceases to be an issue as they be-
              come involved in what they are doing.

11.3.6   Evaluate, interpret, and present the data

Choosing the evaluation paradigm and techniques to answer the questions that satisfy the evaluation goal is an important step. So is identifying the practical and ethical issues to be resolved. However, decisions are also needed about what data to
collect, how to analyze it, and how to present the findings to the development team.
To a great extent the technique used determines the type of data collected, but
there are still some choices. For example, should the data be treated statistically? If
qualitative data is collected, how should it be analyzed and represented? Some general questions also need to be asked (Preece et al., 1994): Is the technique reliable?
Will the approach measure what is intended, i.e., what is its validity? Are biases
creeping in that will distort the results? Are the results generalizable, i.e., what is
their scope? Is the evaluation ecologically valid or is the fundamental nature of the
process being changed by studying it?

Reliability

The reliability or consistency of a technique is how well it produces the *same* results
on separate occasions under the *same* circumstances. Different evaluation
processes have different degrees of reliability. For example, a carefully controlled
experiment will have high reliability. Another evaluator or researcher who follows
exactly the same procedure should get similar results. In contrast, an informal, unstructured interview will have low reliability: it would be difficult if not impossible
to repeat exactly the same discussion.

Validity

Validity is concerned with whether the evaluation technique measures what it is
supposed to measure. This encompasses both the technique itself and the way it is
performed. If for example, the goal of an evaluation is to find out how users use a
new product in their homes, then it is not appropriate to plan a laboratory experiment. An ethnographic study in users' homes would be more appropriate. If the
goal is to find average performance times for completing a task, then counting only
the number of user errors would be invalid.

Biases

Bias occurs when the results are distorted. For example, expert evaluators performing a heuristic evaluation may be much more sensitive to certain kinds of design flaws than others. Evaluators collecting observational data may consistently
fail to notice certain types of behavior because they do not deem them important.

Put another way, they may selectively gather data that they think is important. Interviewers may unconsciously influence responses from interviewees by their tone of voice, their facial expressions, or the way questions are phrased, so it is important to be sensitive to the possibility of biases.

Scope

The scope of an evaluation study refers to how much its findings can be generalized. For example, some modeling techniques, like the keystroke model, have a narrow, precise scope. The model predicts expert, error-free behavior so, for example, the results cannot be used to describe novices learning to use the system.

Ecological validity

Ecological validity concerns how the environment in which an evaluation is conducted influences or even distorts the results. For example, laboratory experiments are strongly controlled and are quite different from workplace, home, or leisure environments. Laboratory experiments therefore have low ecological validity because the results are unlikely to represent what happens in the real world. In contrast, ethnographic studies do not impact the environment, so they have high ecological validity.

Ecological validity is also affected when participants are aware of being studied. This is sometimes called the *Hawthorne effect* after a series of experiments at the Western Electric Company's Hawthorne factory in the US in the 1920s and 1930s. The studies investigated changes in length of working day, heating, lighting, etc., but eventually it was discovered that the workers were reacting positively to being given special treatment rather than just to the experimental conditions.

11.4   Pilot studies

It is always worth testing plans for an evaluation by doing a pilot study before launching into the main study. A pilot study is a small trial run of the main study. The aim is to make sure that the plan is viable before embarking on the real study. For example, the equipment and instructions for its use can be checked. It is also an opportunity to practice interviewing skills, or to check that the questions in a questionnaire are clear or that an experimental procedure works properly. A pilot study will identify potential problems in advance so that they can be corrected. Sending out 500 questionnaires and then being told that two of the questions were very confusing wastes time, annoys participants, and is expensive.

Many evaluators run several pilot studies. As in iterative design, they get feedback, amend the procedure, and test it again until they know they have a good study. If it is difficult to find people to participate or if access to participants is limited, colleagues or peers can be asked to comment. Getting comments from peers is quick and inexpensive and can save a lot of trouble later. In theory, at least, there is no limit to the number of pilot studies that can be run, although there will be practical constraints.

Assignment

Find a journal or conference publication that describes an interesting evaluation study or select one using www.hcibib.org. Then use the DECIDE framework to determine which paradigms and techniques were used. Also consider how well it fared on ethical and practical issues.

(a) Which evaluation paradigms and techniques are used?

(b) Is triangulation used? How?

(c) Comment on the reliability, validity, ecological validity, biases and scope of the techniques described.

(d) Is there evidence of one or more pilot studies?

(e) What are the strengths and weakness of the study report? Write a 50–100 word critique that would help the author(s) improve their report.

Summary

This chapter has introduced four core evaluation paradigms and five categories of techniques and has shown how they relate to each other. The DECIDE framework identifies the main issues that need to be considered when planning an evaluation. It also introduces many of the basic concepts that will be revisited and built upon in the next three chapters: Chapter 12, which discusses observation techniques; Chapter 13, which examines techniques for gathering users' and experts' opinions; and Chapter 14, which discusses user testing and techniques for modeling users' task performance.

Key points

- An evaluation paradigm is an approach in which the methods used are influenced by particular theories and philosophies. Four evaluation paradigms were identified:
  1. "quick and dirty"
  2. usability testing
  3. field studies
  4. predictive evaluation
- Methods are combinations of techniques used to answer a question but in this book we often use the terms "methods" and "techniques" interchangeably. Five categories were identified:
  1. observing users
  2. asking users
  3. asking experts
  4. user testing
  5. modeling users' task performance
- The DECIDE framework has six parts:
  1. Determine the overall goals of the evaluation.
  2. Explore the questions that need to be answered to satisfy the goals.
  3. Choose the evaluation paradigm and techniques to answer the questions.
  4. Identify the practical issues that need to be considered.
  5. Decide on the ethical issues and how to ensure high ethical standards.
  6. Evaluate, interpret, and present the data.
- Drawing up a schedule for your evaluation study and doing one or several pilot studies will help to ensure that the study is well designed and likely to be successful.

Further reading

DENZIN, N. K. AND LINCOLN, Y. S. (1994) *Handbook of Qualitative Research*. London: Sage. This book is a collection of chapters by experts in qualitative research. It is an excellent reference source.

DIX, A., FINLAY, J., ABOWD, G. AND BEALE, R. (1998) *Human-Computer Interaction* (2d ed.). London: Prentice Hall Europe. This book provides a useful introduction to evaluation.

SHNEIDERMAN, B. (1998) *Designing the User Interface: Strategies for Effective Human-Computer Interaction* (3rd ed.). Reading, MA: Addison-Wesley. This text provides an alternative way of categorizing evaluation techniques and offers a good overview.

ROBSON, C. (1993) *Real World Research*. Oxford, UK: Blackwell. This book offers a practical introduction to applied research and evaluation. It is very readable.

WHITESIDE, J., BENNETT, J., AND HOLTZBLATT, K. (1998) Usability engineering: our experience and evolution. In M. Helander (ed.), *Handbook of Human-Computer Interaction*. Amsterdam: North Holland. This chapter reviews the strengths and weakness of usability engineering and explains why ethnographic techniques can provide a useful alternative in some circumstances, 791–817.

Chapter 12

Observing users

12.1   Introduction
12.2   Goals, questions, and paradigms
    12.2.1   What and when to observe
    12.2.2   Approaches to observation
12.3   How to observe
    12.3.1   In controlled environments
    12.3.2   In the field
    12.3.3   Participant observation and ethnography
12.4   Data collection
    12.4.1   Notes plus still camera
    12.4.2   Audio recording plus still camera
    12.4.3   Video
12.5   Indirect Observation: tracking user's activities
    12.5.1   Diaries
    12.5.2   Interaction logging
12.6   Analyzing, interpreting, and presenting data
    12.6.1   Qualitative analysis to tell a story
    12.6.2   Qualitative analysis for categorization
    12.6.3   Quantitative data analysis
    12.6.4   Feeding the findings back into design

12.1   Introduction

Observation involves watching and listening to users. Observing users interacting with software, even casual observing, can tell you an enormous amount about what they do, the context in which they do it, how well technology supports them, and what other support is needed. In Chapter 9 we discussed the role of observation and ethnography in informing design, particularly early in the process. In this chapter we describe how to observe and do ethnography and discuss their role in evaluation.

Users can be observed in controlled laboratory-like conditions, as in usability testing, or in the natural environments in which the products are used—i.e., the field. How the observation is done depends on why it is being done and the approach adopted. There is a variety of structured, less structured, and descriptive observation techniques for evaluators to choose from. Which they select and how their findings are interpreted will depend upon the evaluation goals, the specific questions being addressed, and practical constraints. This chapter focuses on how to select appropriate observation techniques, how to do observation, and how to analyze the data and present findings from it. We also discuss the benefits and practicalities associated with each technique. An interview with interaction design consultant Sara Bly at the end of the chapter discusses how she uses observation in her work.

The main aims of this chapter are to:

- Discuss the benefits and challenges of different types of observation.
- Describe how to observe as an on-looker, a participant, and an ethnographer.
- Discuss how to collect, analyze and present data from observational evaluation.
- Examine key issues for doing think-aloud evaluation, diary studies and interaction logging.
- Give you experience in selecting and doing observational evaluation.

In general, observing and talking to users usually go together, but we leave the details of interview techniques until Chapter 13.

12.2 Goals, questions, and paradigms

Goals and questions provide a focus for observation, as the DECIDE framework points out. Even studies that use "quick and dirty" observations have a goal; for example, to identify or confirm usability and user experience goals in a prototype. *Goals and questions should guide all evaluation studies.* Just because some evaluators do not make their goals obvious does not mean that they don't have goals. Expert evaluators sometimes don't articulate their goals, but as you will read in Sara Bly's interview they do have them. Even in field studies and ethnography there is a careful balance between being guided by goals and being open to modifying, shaping, or refocusing the study as you learn about the situation. Being able to keep this balance is a skill that develops with experience.

ACTIVITY 12.1   (a) Find a small group of people who are using any kind of technology (e.g., computers, household or entertainment appliances, etc.) and try to answer the question, "What are these people doing?" Watch for three to five minutes and write down what you observe. When you have finished, note how you felt doing this.

(b) If you were to repeat the exercise what would you look for when you next observe the group? How would you refine your goals?

Comment   (a) What was the group doing? Were they talking, working, playing or something else? How were you able to decide? Did you feel awkward or embarrassed watching? Did you wonder whether you should tell them that you were observing them? What problems did you encounter doing this exercise? Was it hard to watch everything and remember what happened? What were the most important things? Did you wonder if you should be trying to identify and remember just those things? Was remembering the order of events tricky? Perhaps you naturally picked up a pen and paper and took notes. If so, was it difficult to record fast enough? How do you think the people being watched felt? Did they know they were being watched? Did knowing affect the way they behaved? Perhaps some of them objected and walked away. If you didn't tell them, do you think you should have?

(b) Your questions should be more focused. For example, you might ask, what are the people specifically trying to do and how is the technology being used? Is everyone in the group using the technology? Is it supporting or hindering the users' goals?

Having a goal, even a very general goal, helps to guide the observation because there is always so much going on.

12.2.1 What and when to observe

Observing is useful at any time during product development. Early in design, observation helps designers understand users' needs. Other types of observation are done later to examine whether the developing prototype meets users' needs.

Depending on the type of study, evaluators may be onlookers, participant observers, or ethnographers. Remember Christian Heath's and Paul Luff's ethnographic study of the London Underground discussed in Chapter 4 (Heath and Luff, 1992)? This study demonstrates the power of insightful observation to improve the redesign of a system. However, in order to understand how London Underground workers do their jobs the authors needed "insider" knowledge. The degree of immersion that evaluators adopt varies across a broad outsider-insider spectrum. Where a particular study falls along this spectrum depends on its goal and on the practical and ethical issues that constrain and shape it.

ACTIVITY 12.2   To understand this notion of an outsider-insider spectrum better, read the scenarios below and answer the questions that follow.

Scenario 1. A usability consultant joins a group who have been given WAP phones to test on a visit to Washington, DC. Not knowing the restaurants in the area, they use the WAP phone to find a list of restaurants within a five-mile radius of their hotel. Several are listed and while the group waits for a taxi, they find the telephone numbers of a couple, call them to ask about their menus, select one, make a booking, and head off to the restaurant. The usability consultant observes some problems keying instructions because the buttons seem small. She also notices that the screen seems rather small, but the person using it is able to get the information needed and call the restaurant, etc. Discussion with the group supports the evaluator's impression that there are problems with the interface, but on balance the device is useful and the group is pleased to get a table at a good restaurant nearby.

Scenario 2. A usability consultant observes how participants perform a pre-planned task using the WAP phone in a usability laboratory. The task requires the participants to find the telephone number of a restaurant called Matisse. It takes them several minutes to do this and they appear to have problems. The video recording and interaction log suggest that the screen is too small for the amount of information they need to access and this is supported by participants' answers on a user satisfaction questionnaire.

(a) In which situation does the observer take the most control?

(b) What are the advantages and disadvantages of these two types of observation?

(c) When might each type of observation be useful?

Comment

(a) The observer takes most control in the second study. The task is predetermined, the participant is instructed what to do, and she is located in a controlled laboratory environment.

(b) The advantages of the field study are that the observer got to see how the device could be used in a real situation to solve a real problem. She experienced the delight expressed with the overall concept and the frustration with the interface. By watching how the group used the device "on the move," she gained an understanding of what they liked and needed. The disadvantage is that the observer was an "insider" in the group, so how objective could she be? The data is qualitative and while anecdotes can be very persuasive, how useful are they in evaluation? Maybe she was having such a good time that her judgment was clouded and she missed hearing negative comments and didn't notice some people's annoyance. Another study could be done to find out more, but it is not possible to replicate the exact situation, whereas the laboratory study is easier to replicate.

The advantages of the laboratory are that several users performed the same task, so different users' performance could be compared and averages calculated. The observer could also be more objective because she was more of an outsider. The disadvantage is that the study is artificial and says nothing about how the device would be used in the real environment.

(c) Both types of studies have merits. Which is better depends on the goals of the study. The laboratory study is useful for examining details of the interaction style to make sure that usability problems with the interface and button design are diagnosed and corrected. The field study reveals how the phone is used in a real world context and how it integrates with or changes users' behavior. Without this study, it is possible that developers might not have discovered the enthusiasm for the phone because the reward for doing laboratory tasks is not as compelling as a good meal!

Table 12.1   Type of observation

Observation	Controlled environment (i.e., lab-like)	Field environment (i.e., natural)
Outsider looking on	"Quick and dirty" In usability testing	"Quick and dirty" In field studies
Insider	(Not applicable)	Participant observation (e.g., in ethnography)

Table 12.1 summarizes this insider-outsider discussion, how it relates to different types of environments, and how much control evaluators take over the evaluation process.

12.2.2 Approaches to observation

Observers can be outsiders in the field and in the controlled environments, but they can't be insiders in a controlled environment. In the field it is possible to have varying degrees of "insider-outsiderness." In practice these distinctions are more difficult to describe than to experience!

"Quick and dirty" observation

"Quick and dirty" observations can occur anywhere, anytime. For example, evaluators often go into a school, home, or office to watch and talk to users in a casual way to get immediate feedback about a prototype or product. Evaluators can also join a group for a short time, which gives them a slightly more insider role. Quick and dirty observations are just that, ways of finding out what is happening quickly and with little formality.

Observation in usability testing

Video and interaction logs capture everything that the user does during a usability test including keystrokes, mouse clicks, and their conversations. In addition, observers can watch through a one-way mirror or via a remote TV screen. The observational data is used to see and analyze what users do and how long they spend on different aspects of the task. It also provides insights into users' affective reactions. For example, sighs, tense shoulders, frowns, and scowls speak of users' dissatisfaction and frustrations. The environment is controlled but users often forget that they are being observed. In addition, many evaluators also supplement findings from the laboratory with observations in the field.

Observation in field studies

In field studies, as we have said, observers may be anywhere along the outsider-insider spectrum. Looking on as an outsider, being a participant observer, or being an ethnographer brings a philosophy and practices that influence what data is collected, how data collection is done, and how the data is analyzed and reported. Colin Robson (1993) summarizes the possible levels of participation as: complete participants, more marginal participants, observers who also participate, and people who observe from the outside and do not participate.

Whether and in what ways observers influence those being observed depends on the type of observation and the observer's skills. The goal is to cause as little disruption as possible. An example of outsider observation is when an observer is interested only in the presence of certain types of behavior. For instance, in a study of the time spent by boys and girls using technology in the classroom, an observer may go into the classroom to note when technology is used by boys and when by girls. She could do this by standing at the back of the room with a data sheet on which she notes the gender of the children who use the computer and how long they spend using it. In contrast, if the goal is to understand how the computer integrates with other artifacts and social interactions in the classroom, a more holistic approach would be better. In this situation the evaluator might take more of an insider perspective in which she talks to participants as well as observes. The observer mixes and integrates with participants more, but there is no illusion that she is anything other than an observer.

Inside observers may be participant observers or ethnographers. In participant observation evaluators participate with users in order to learn what they do and how and why they do it. A fully participant observer observes from the inside as a member of the group, which means she must not only be present to share experiences, but also learn the social conventions of the group, including beliefs and protocols, dress codes, communication conventions, use of language, and non-verbal communication. "Participant observation combines participation in the lives of the people under study with maintenance of a professional distance that allows adequate observation and recording of data" (Fetterman, 1998, p. 34–35).

Ethnographers can be thought of as participant observers or not, depending on your point of view. Ethnographers themselves debate this issue. Some see participant observation as virtually synonymous with ethnography (Atkinson and Hammersley, 1994). Others view participant observation as a technique that is used in ethnography along with informants from the community, interviews with community members, and the study of community artifacts (Fetterman, 1998). Ethnographic evaluation is derived from ethnography. Ethnographic studies typically take weeks, months, or even longer to gain an "inside" understanding of what is going on in a community. Much shorter studies are usual in interaction design because of the time constraints imposed by development schedules.

As in any evaluation study, goals and questions determine whether the observation will be "quick and dirty," in a controlled environment or in the field, and the extent to which the observers are outsiders or insiders. Determining goals, exploring questions, and choosing techniques are necessary steps in the DECIDE framework. Practical and ethical issues also have to be identified and decisions made about how to handle them.

12.3   How to observe

The same basic data-collection tools are used for laboratory and field studies (i.e., direct observation, taking notes, collecting video, etc.) but the way in which they are used is different. In the laboratory the emphasis is on the details of what individuals do, while in the field the context is important and the focus is on how people interact with each other, the technology, and their environment. Furthermore, the equipment in the laboratory is usually set up in advance and is relatively static, whereas in the field it usually must be moved around. In this section we discuss how to observe, and then examine the practicalities and compare data-collection tools.

12.3.1 In controlled environments

The role of the observer is to first collect and then make sense of the stream of data on video, audiotapes, or notes made while watching users in a controlled environment. Many practical issues have to be thought about in advance, including the following.

- It is necessary to decide where users will be located so that the equipment can be set up. Many usability laboratories, for example, have two or three wall-mounted, adjustable cameras to record users' activities while they work on test tasks. One camera might record facial expressions, another might focus on mouse and keyboard activity, and another might record a broad view of the participant and capture body language. The stream of data from the cameras is fed into a video editing and analysis suite where it is annotated and partially edited. Another form of data that can be collected is an interaction log. This records all the user's key presses. Mobile usability laboratories, as the name suggests, are intended to be moved around, but the equipment can be bulky. Usually it is taken to a customer's site where a temporary laboratory environment is created.

- The equipment needs testing to make sure that it is set up and works as expected, e.g., it is advisable that the audio is set at the right level to record the user's voice.

- An informed consent form should be available for users to read and sign at the beginning of the study. A script is also needed to guide how users are greeted, and to tell them the goals of the study, how long it will last, and to explain their rights. It is also important to make users feel comfortable and at ease.

Whether in a real or make-do laboratory one of the problems with this type of observation is that the observer doesn't know what users are thinking, and can only guess from what she sees.

Think-aloud technique   Imagine observing someone who has been asked to evaluate the interface of the web search engine Northernlight. The user, who has used the web only once before, is told to find a list of the books written by the well-known biologist Stephen Jay Gould. He is told to type http://www.northernlight.com and then proceed however he thinks best. He types the URL and gets a screen similar to the one in Figure 12.1.

Next he goes to the search box but types Stephen Jay Gouild without realizing that he has made a typing error and added an 'i'. He presses return and gets a screen similar to the one in Figure 12.2.

He is silent. What is going on, you wonder? What is he thinking? One way around this problem is to collect a think-aloud protocol, using a technique developed by Erikson and Simon for examining people's problem-solving strategies (Erikson and Simon, 1985). The technique requires people to say out loud everything that they are thinking and trying to do, so that their thought processes are externalized.

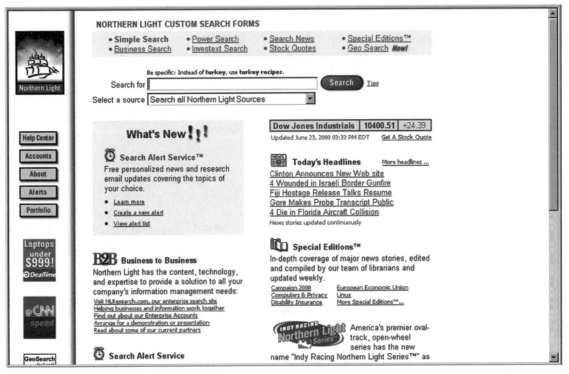

Figure 12.1 Home page of Northernlight search engine (www.northernlight.com).

So, let's imagine an action replay of the situation just described, but this time the user has been instructed to think aloud:

> *I'm typing in http://www.northernlight.com as you told me.* (types)
> *Now I press the enter key, right?* (presses enter key)
> (pause and silence)
> *It's taking a few moments to respond.*
> *Oh! Here it is.* (Figure 12.1 appears)
> *Gosh, there's a lot of stuff on this screen, hmmm, I wonder what I do next.* (pauses and looks at the screen) *Probably a simple search. What's a power search and there's all these others too?*
> *I just want to find Stephen Jay Gould, right, and then it's bound to have a list of his books?* (pause) *Well, it looks like I should type his name in this box here.* (moves cursor towards the search box. Positions cursor. Types 'Stephen Jay Gouild'. Pauses, but does not notice that he has incorrectly included an "i" in Gould, then clicks the search button.) *Well, something seems to be happening . . .* (Watches) *something is happening. Ah! What's this . . .* (Looks at screen and Figure 12.2 appears)
> *Silence . . .*

Now you know more about what the user is trying to achieve but he is silent again. You can see that he has spelled Gould incorrectly and that he doesn't realize that he has typed Gouild. What you don't know is what he is thinking now or what is he

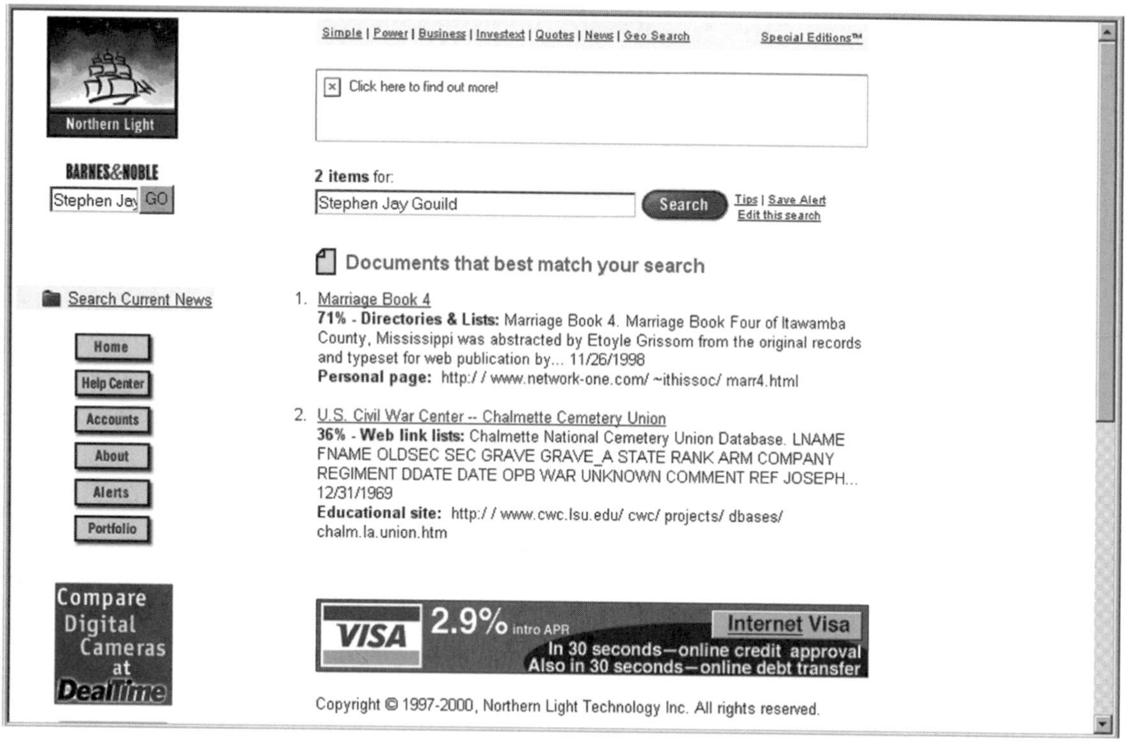

Figure 12.2 The screen that appears in response to searching for Stephen Jay Gouild.

looking at. Has he noticed his typing error or the Barnes and Noble box at the top left that says "Stephen Jay"?

ACTIVITY 12.3   Try a think-aloud exercise yourself. Go to an e-commerce website, such as Amazon.com or BarnesandNoble.com, and look for something that you want to buy. Think aloud as you search and notice how you feel and behave. Did you find it difficult to keep speaking all the way through the task? Did you feel awkward? Did you stop when you got stuck?

Comment   You probably felt self-conscious and awkward doing this. Some people say they feel really embarrassed. At times you may also have started to forget to speak out loud because it feels like talking to yourself, which most of us don't do. You may also have found it difficult to think aloud when the task got difficult. In fact, you probably stopped speaking when the task became demanding, and that is exactly the time when an evaluator is most eager to hear your comments.

The occurrence of these silences is one of the biggest problems with the think-aloud technique.

If a user is silent during a think-aloud protocol, the evaluator could interrupt and remind him to think out loud, but that would be intrusive. Another solution is to have two people work together so that they talk to each other. Working with another person is often more natural and revealing because they talk in order to help each other along. This technique has been found particularly successful with children. It is also very effective when evaluating systems intended to be used synchronously by groups of users, e.g., shared whiteboards.

12.3.2   In the field

Whether the observer sets out to be an outsider or an insider, events in the field can be complex and rapidly changing. There is a lot for evaluators to think about, so many experts have a framework to structure and focus their observation. The framework can be quite simple. For example, this is a practitioner's framework that focuses on just three easy-to-remember items to look for:

- *The person.* Who is using the technology at any particular time?
- *The place.* Where are they using it?
- *The thing.* What are they doing with it?

Frameworks like the one above help observers to keep their goals and questions in sight. Experienced observers may, however, prefer more detailed frameworks, such as the one suggested by Goetz and LeCompte (1984) below, which encourages observers to pay greater attention to the context of events, the people and the technology:

- *Who* is present? How would you characterize them? What is their role?
- *What* is happening? What are people doing and saying and how are they behaving? Does any of this behavior appear routine? What is their tone and body language?
- *When* does the activity occur? How is it related to other activities?
- *Where* is it happening? Do physical conditions play a role?
- *Why* is it happening? What precipitated the event or interaction? Do people have different perspectives?
- *How* is the activity organized? What rules or norms influence behavior?

Colin Robson (1993) suggests a slightly longer but similar set of items:

- *Space.* What is the physical space like and how is it laid out?
- *Actors.* What are the names and relevant details of the people involved?
- *Activities.* What are the actors doing and why?
- *Objects.* What physical objects are present, such as furniture?
- *Acts.* What are specific individuals doing?
- *Events.* Is what you observe part of a special event?
- *Goals.* What are the actors trying to accomplish?
- *Feelings.* What is the mood of the group and of individuals?

ACTIVITY 12.4

(a) Look at Goetz's and LeCompte's framework. Apart from there being more items than in the first framework, what is the other main difference?

(b) Now compare this framework with Robson's. What does Robson's attend to that is not obvious in Goetz's and LeCompte's framework?

(c) Which of the three frameworks do you think would be easiest to remember and why?

Comment

(a) The Goetz and LeCompte framework pays much more attention to the context of the observation.

(b) There is considerable overlap between the two frameworks despite differences in wording. The main difference is that Robson's framework pays attention to the mood of the group.

(c) The three-item framework is likely to be easy, but so is the Goetz and LeCompte framework because it adopts the much used organizing principle "who, what, when, where, why, how." Robson's framework has two extra items and no obvious way of remembering them. However, having said that, to me it is more explicit. Which is used for a particular study depends on the study goals and how much detail is needed, and to a degree, it is also a matter of personal preference.

These frameworks are useful not only for providing focus but also for organizing the observation and data-collection activity. Below is a checklist of things to plan before going into the field:

- State the initial study goal and questions clearly.
- Select a framework to guide your activity in the field.
- Decide how to record events—i.e., as notes, on audio, or on video, or using a combination of all three. Make sure you have the appropriate equipment and that it works. You need a suitable notebook and pens. A laptop computer might be useful but could be cumbersome. Although this is called observation, photographs, video, interview transcripts and the like will help to explain what you see and are useful for reporting the story to others.
- Be prepared to go through your notes and other records as soon as possible after each evaluation session to flesh out detail and check ambiguities with other observers or with the people being observed. This should be done routinely because human memory is unreliable. A basic rule is to do it within 24 hours, but sooner is better!
- As you make and review your notes, try to highlight and separate personal opinion from what happens. Also clearly note anything you want to go back to. Data collection and analysis go hand in hand to a large extent in fieldwork.
- Be prepared to refocus your study as you analyze and reflect upon what you see. Having observed for a while, you will start to identify interesting

phenomena that seem relevant. Gradually you will sharpen your ideas into questions that guide further observation, either with the same group or with a new but similar group.

- Think about how you will gain the acceptance and trust of those you observe. Adopting a similar style of dress and finding out what interests the group and showing enthusiasm for what they do will help. Allow time to develop relationships. Fixing regular times and venues to meet is also helpful, so everyone knows what to expect. Also, be aware that it will be easier to relate to some people than others, and it will be tempting to pay attention to those who receive you well, so make sure you attend to everyone in the group.

- Think about how to handle sensitive issues, such as negotiating where you can go. For example, imagine you are observing the usability of a portable home communication device. Observing in the living room, study, and kitchen is likely to be acceptable, but bedrooms and bathrooms are probably out of bounds. Take time to check what participants are comfortable with and be accommodating and flexible. Your choice of equipment for data collection will also influence how intrusive you are in people's lives.

- Consider working as a team. This can have several benefits; for instance, you can compare your observations. Alternatively, you can agree to focus on different people or different parts of the context. Working as a team is also likely to generate more reliable data because you can compare notes among different evaluators.

- Consider checking your notes with an informant or members of the group to ensure that you are understanding what is happening and that you are making good interpretations.

- Plan to look at the situation from different perspectives. For example, you may focus on particular activities or people. If the situation has a hierarchical structure, as in many companies, you will get different perspectives from different layers of management—e.g., end-users, marketing, product developers, product managers, etc.

12.3.3 Participant observation and ethnography

Being a participant observer or an ethnographer involves all the practical steps just mentioned, but especially that the evaluator must be accepted into the group. An interesting example of participant observation is provided by Nancy Baym's work (1997) in which she joined an online community interested in soap operas for over a year in order to understand how the community functioned. She told the community what she was doing and offered to share her findings with them. This honest approach gained her their trust, and they offered support and helpful comments. As Baym participated she learned about the community, who the key characters were, how people interacted, their values, and the types of discussions that were generated. She kept all the messages as data to be referred to later. She also adapted interviewing and questionnaire techniques to collect additional information. She summarizes her data collection as follows (Baym, 1997, p. 104):

> *The data for this study were obtained from three sources. In October 1991, I saved all the messages that appeared. . . . I collected more messages in 1993. Eighteen participants responded to a questionnaire I posted. . . . Personal email correspondence with 10 other . . . participants provided further information. I posted two notices to the group explaining the project and offering to exclude posts by those who preferred not to be involved. No one declined to participate.*

Using this data, Baym examined the group's technical and participatory structure, its emergent traditions, and its usage with the technology. As the work evolved, she shared its progress with the group members, who were supportive and helpful.

ACTIVITY 12.5   Drawing on your experience of using email, bulletin boards, UseNet News, or chat rooms, how might participant observation online differ from face-to-face participant observation?

Comment   In online participant observation you don't have to look people in the eye, deal with their skepticism, or wonder what they think of you, as you do in face-to-face situations. What you wear, how you look, or the tone of your voice don't matter. However, what you say or don't say and how you say it are central to the way others will respond to you. Online you only see part of people's context. You usually can't see how they behave off line, how they present themselves, their body language, how they spend their day, their personalities, who is present but not participating, etc.

As we said the distinction between ethnography and participant observation is blurred. Some ethnographers believe that ethnography is an open interpretivist approach in which evaluators keep an open mind about what they will see. Others, such as David Fetterman from Stanford University, see a stronger role for a theoretical underpinning: "before asking the first question in the field the ethnographer begins with a problem, a theory or model, a research design, specific data collection techniques, tools for analysis, and a specific writing style" (Fetterman, 1998, p. 1). This may sound as if ethnographers have biases, but by making assumptions explicit and moving between different perspectives, biases are at least reduced. Ethnographic study allows *multiple* interpretations of reality; it is *interpretivist*. Data collection and analysis often occur simultaneously in ethnography, with analysis happening at many different levels throughout the study. The question being investigated is refined as more understanding about the situation is gained.

The checklist below (Fetterman, 1998) for doing ethnography is similar to the general list just mentioned:

- Identify a problem or goal and then ask good questions to be answered by the study, which may or may not invoke theory depending on your philosophy of ethnography. The observation framework such as those mentioned above can help to focus the study and stimulate questions.

- The most important part of fieldwork is just being there to observe, ask questions, and record what is seen and heard. You need to be aware of people's feelings and sensitive to where you should not go.

- Collect a variety of data, if possible, such as notes, still pictures, audio and video, and artifacts as appropriate. Interviews are one of the most important data-gathering techniques and can be structured, semi-structured, or open. So-called *retrospective interviews* are used after the fact to check that interpretations are correct.

- As you work in the field, be prepared to move backwards and forwards between the broad picture and specific questions. Look at the situation holistically and then from the perspectives of different stakeholder groups and participants. Early questions are likely to be broad, but as you get to know the situation ask more specific questions.

- Analyze the data using a *holistic* approach in which observations are understood within the broad context—i.e., they are *contextualized*. To do this, first synthesize your notes, which is best done at the end of each day, and then check with someone from the community that you have described the situation accurately. Analysis is usually iterative, building on ideas with each pass.

ACTIVITY 12.6   Look at the steps listed for doing ethnography and compare them with the earlier generic set for field observation (see Section 12.3.2). What is the main difference?

Comment   Both sets of steps involve structuring observations and refining goals and questions through knowledge gained during the study. Both use similar data collection techniques and rely on the trust and cooperation of those being observed. Ethnographers tend to be deeply immersed in the group, whereas not everyone doing field studies takes this approach. Some ethnographers, such as David Fetterman, are guided by theory; others are strongly against this and believe that ethnography should be approached open-mindedly.

During the last ten years ethnography has gained credibility in interaction design because if products are to be used in a wide variety of environments designers must know the context and ecology of those environments (Nardi and O'Day, 1999). However, for those unfamiliar with ethnography and general field observation there are two dilemmas. The first dilemma is, "When have I observed

DILEMMA   When Should I Stop Observing?

Knowing when to stop doing *any* type of evaluation can be difficult for novice evaluators, but it is particularly tricky in observational studies and ethnography because there is no obvious ending. Schedules often dictate when your study ends. Otherwise, stop when you stop learning new things. Two indications of having done enough are when you start to see similar patterns of behavior being repeated, or when you have listened to all the main stakeholder groups and understand their perspectives.

DILEMMA   How Can I Adapt Ethnography to Fit the Development Process?

Many developers are unsure how to integrate ethnographic evaluation into development cycles. In addition, most developers have a technical training that does not encourage them to value qualitative data. We discussed the use of ethnography to inform design in Chapter 9. Here is an example where it has been adapted for evaluation.

In a project for the Department of Juvenile Justice, Ann Rose and her colleagues developed a procedure to be used by technical design teams with limited ethnographic training (Rose et al., 1995). This applied form of ethnography acknowledges the comparatively small amounts of time available for any kind of user study. By making the process more structured the amount of time needed for the study can be reduced. It also emphasizes that taking time to become familiar with the intricacies of a system enhances the evaluator's credibility during the field study and promotes productive fieldwork. The procedures this group advocates are highly structured, and while they may seem contrary to ethnographic practice, this structure helps to make it possible for some development teams to benefit from an applied ethnographic approach. There are four stages, as follows:

1 Preparation

Understand organization policies and work culture.

Familiarize yourself with the system and its history.

Set initial goals and prepare questions.

Gain access and permission to observe and interview.

2 Field study

Establish a rapport with managers and users.

Observe and interview users in their workplace and collect data.

Follow any leads that emerge from the visits.

Record your visits.

3 Analysis

Compile the collected data in numerical, textual, and multimedia databases.

Quantify data and compile statistics.

Reduce and interpret the data.

Refine the goals and processes used.

4 Reporting

Consider multiple audiences and goals.

Prepare a report and present the findings.

---

enough?" The second dilemma is, "How can I adapt ethnography so that it better fits the short development cycles and the mindset of the developers?"

ACTIVITY 12.7   What are the main differences between the stages that Rose et al. (1995) describe and the steps suggested by Fetterman (1998)?

Comment   The list in the "How Can I Adapt Ethnography" dilemma suggests that the evaluators are not as immersed in the study as Fetterman's process suggests. One aim of the Rose procedure is radically to reduce the time needed to do a study so that it is compatible with system development. Another aim is to reduce the data to a quantifiable form so that it is familiar and acceptable to the developers.

12.4 Data collection

Data collection techniques (i.e., taking notes, audio recording, and video recording) are used individually or in combination and are often supplemented with photos from a still camera. When different kinds of data are collected, evaluators have to coordinate them; this requires additional effort but has the advantage of providing more information and different perspectives. Interaction logging and participant diary studies are also used, as we discuss later in Section 12.5. Which techniques are used will depend on the context, time available, and the sensitivity of what is being observed. In most settings, audio, photos, and notes will be sufficient. In others it is essential to collect video data so as to observe in detail the intricacies of what is going on.

12.4.1   Notes plus still camera

Taking notes is the least technical way of collecting data, but it can be difficult and tiring to write and observe at the same time. Observers also get bored and the speed at which they write is limited. Working with another person solves some of these problems and provides another perspective. Handwritten notes are flexible in the field but must be transcribed. However, this transcription can be the first step in data analysis, as the evaluator must go through the data and organize it. A laptop computer can be a useful alternative but it is more obtrusive and cumbersome, and its batteries need recharging every few hours. If a record of images is needed, photographs, digital images, or sketches are easily collected.

12.4.2   Audio recording plus still camera

Audio can be a useful alternative to note taking and is less intrusive than video. It allows evaluators to be more mobile than with even the lightest, battery-driven video cameras, and so is very flexible. Tapes, batteries, and the recorder are now relatively inexpensive but there are two main problems with audio recording. One is the lack of a visual record, although this can be dealt with by carrying a small camera. The second drawback is transcribing the data, which can be onerous if the contents of many hours of recording have to be transcribed; often, however, only sections are needed. Using a headset with foot control makes transcribing less onerous. Many studies do not need this level of detail; instead, evaluators use the recording to remind them about important details and as a source of anecdotes for reports.

12.4.3   Video

Video has the advantage of capturing both visual and audio data but can be intrusive. However, the small, handheld, battery-driven digicams are fairly mobile, inexpensive and are commonly used.

A problem with using video is that attention becomes focused on what is seen through the lens. It is easy to miss other things going on outside of the camera view. When recording in noisy conditions, e.g., in rooms with many computers running or outside when it is windy, the sound may get muffled.

Analysis of video data can be very time-consuming as there is so much to take note of. Over 100 hours of analysis time for one hour of video recording is common for detailed analyses in which every gesture and utterance is analyzed. However, this

6-10   © 1989 Jim Unger

**"This is a video of you two watching
the video of our vacation."**

level of detail is usually not needed because evaluators often focus on particular episodes and use the whole recording only for contextual information and reference.

In Table 12.2 we summarize the key features, advantages and drawbacks of these three combinations of data collection techniques.

ACTIVITY 12.8   Imagine you are a consultant who is employed to help develop a new computerized garden-planning tool to be used by amateur and professional garden designers. Your goal is to find out how garden designers use an early prototype as they walk around their clients' gardens sketching design ideas, taking notes, and asking the clients about what they like and how they and their families use the garden. What are the advantages and disadvantages of the three types of data-collection techniques in this environment?

Comment   Handwritten notes do not require specialist equipment. They are unobtrusive and very flexible but difficult to do while walking around a garden. If it starts to rain there is no equipment to get wet, but taking notes is tiring, people lose concentration, biases creep in, and handwriting can be difficult to decipher. Video captures more information (e.g., the landscape, where the designers are looking, sketches, comments, etc.) but it is more intrusive, you must also carry equipment and film and what happens if it starts to rain? You also need access to

Table 12.2   Comparison of the three main data-collection techniques used in observation

Criterion	Notes plus camera	Audio plus camera	Video
Equipment	Paper, pencil and camera are easily available.	Inexpensive, handheld recorder with a good microphone. Headset useful for easy transcription.	More expensive. Editing, mixing and analysis equipment needed.
Flexibility of use	Very flexible. Unobtrusive.	Flexible. Relatively unobtrusive.	Needs positioning and focusing camera lens. Even portable versions can be bulky.
Completeness of data	Only get what note-taker thinks is important and can record in the time available. Problem with inexperienced evaluators.	Can obtain complete audio recording but visual data is missing. Notes, photographs, sketches can augment recording but need coordinating with the recording.	Most complete method of data collecting, especially if more than one camera used, but coordination of video material is needed.
Disturbance to users	Very low.	Low but cassette must be changed and microphone positioned.	Can be very obtrusive. Care needed to avoid Hawthorne effect.
Reliability of data	May be low. Relies on humans making a good record and knowing what to record.	High but external noise, e.g. fans in computers can muffle what is said.	Can be high but depends on what camera is focused on.
Analysis	Relatively easy to transcribe. Rich descriptions can be produced. Transcribing data can be onerous or a useful first step in data analysis.	Critical discussions can be identified. Transcription needed for detailed analysis. Permanent original record that can be revisited.	Critical incidents can be identified and tagged. Automated support needed for detailed analysis. Permanent original record that can be revisited.
Feedback to design team	Relies strongly on the authority of the evaluator.	Material captured on tape is more convincing than notes but feedback relies on authority of evaluator.	Hard to dispute material captured on video. Video clips are very powerful for communicating ideas.

playback and editing facilities. Audio could be a good compromise, but integrating sketches and other artifacts later can be a burden and garden planning is a highly visual, aesthetic activity. You could also supplement notes and audio with a still camera.

12.5 Indirect observation: tracking users' activities

Sometimes direct observation is not possible because it is obtrusive or evaluators cannot be present over the duration of the study, and so users' activities are tracked indirectly. Diaries and interaction logs are two techniques for doing this. From the records collected evaluators reconstruct what happened and look for usability and user experience problems.

12.5.1 Diaries

Diaries provide a record of what users did, when they did it, and what they thought about their interactions with the technology. They are useful when users are scattered and unreachable in person, as in many Internet and web evaluations. Diaries are inexpensive, require no special equipment or expertise, and are suitable for long-term studies. Templates can also be created online to standardize entry format and enable the data to go straight into a database for analysis. These templates are like those used in open-ended online questionnaires. However, diary studies rely on participants being reliable and remembering to complete them, so incentives are needed and the process has to be straightforward and quick. Another problem is that participants often remember events as being better or worse than they really were, or taking more or less time than they actually did.

Robinson and Godbey (1997) asked participants in their study to record how much time Americans spent on various activities. These diaries were completed at the end of each day and the data was later analyzed to investigate the impact of television on people's lives. In another diary study, Barry Brown and his colleagues from Hewlett Packard collected diaries form 22 people to examine when, how, and why they capture different types of information, such as notes, marks on paper, scenes, sounds, moving images, etc. (Brown, et al., 2000). The participants were each given a small handheld camera and told to take a picture every time they captured information in any form. The study lasted for seven days and the pictures were used as memory joggers in a subsequent semi-structured interview used to get participants to elaborate on their activities. Three hundred and eighty-one activities were recorded. The pictures provided useful contextual information. From this data the evaluators constructed a framework to inform the design of new digital cameras and handheld scanners.

12.5.2 Interaction logging

Interaction logging in which key presses, mouse or other device movements are recorded has been used in usability testing for many years. Collecting this data is usually synchronized with video and audio logs to help evaluators analyze users' behavior and understand how users worked on the tasks they set. Specialist software tools are used to collect and analyze the data. The log is also time-stamped so it can be used to calculate how long a user spends on a particular task or lingered in a certain part of a website or software application.

Explicit counters that record visits to a website were once a familiar sight. Recording the number of visitors to a site can be used to justify maintenance and upgrades to it. For example, if you want to find out whether adding a bulletin board to an e-commerce website increases the number of visits, being able to compare traffic before and after the addition of the bulletin board is useful. You can also track how long people stayed at the site, which areas they visited, where they came from, and where they went next by tracking their Internet Service Provider (I.S.P.) address. For example, in a study of an interactive art museum by researchers at the University of Southern California, server logs were analyzed by tracking visitors in this way (McLaughlin et al., 1999). Records of when people came to the site, what they requested, how long they looked at each page, what browser they were using, and what country they were from, etc., were collected over a seven-month period. The data was analyzed using Webtrends, a commercial analysis tool, and the evaluators discovered that the site was busiest on weekday evenings. In another study that investigated lurking behavior in listserver discussion groups, the number of messages posted was compared with list membership over a three-month period to see how lurking behavior differed among groups (Nonnecke and Preece, 2000).

An advantage of logging user activity is that it is unobtrusive, but this also raises ethical concerns that need careful consideration (see the dilemma about observing without being seen). Another advantage is that large volumes of data can be logged automatically. However, powerful tools are needed to explore and analyze this data quantitatively and qualitatively. An increasing number of visualization tools are being developed for this purpose; one example is WebLog, which dynamically shows visits to websites, as illustrated in Figure 12.3 (Hochheiser and Shneiderman, 2000).

DILEMMA   They Don't Know We Are Watching. Shall We Tell Them?

If you have appropriate algorithms and sufficient computer storage, large quantities of data about Internet usage can be collected and users need never know. Furthermore, if we tell users that we are logging their behavior they may react or change their behavior. So, what should we do? It depends on the context, how much personal information is collected, and how the information will be used. Many companies now tell you that your computer activity and phone calls may be logged for quality assurance and other purposes. Most people do not object to this practice. However, should we be concerned about logging personal information (e.g., discussions about health or financial information)? Should users be worried? How can we exploit the ability to log user behavior when visiting websites without overstepping a person's civil rights? Where should we draw the line?

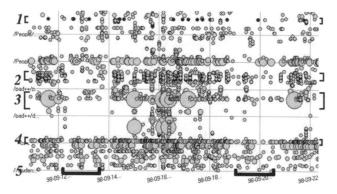

Figure 12.3 A display from WebLog, time vs. URL (Hochheiser and Shneiderman, 2001). The requested URL is on the *y*-axis, with the date and time on the *x*-axis. The dark lines on the *x*-axis correspond to weekends. Each circle represents a request for a single page, and the size of the circle indicates the number of bytes delivered for a given request. (Color, which is not shown here, indicates the Http status response.)

12.6 Analyzing, interpreting, and presenting the data

By now you should know that many, indeed most observational evaluations generate a lot of data in the form of notes, sketches, photographs, audio and video records of interviews and events, various artifacts, diaries, and logs. Most observational data is qualitative and analysis often involves interpreting what users were doing or saying by looking for patterns in the data. Sometimes qualitative data is categorized so that it can be quantified and in some studies events are counted.

Dealing with large volumes of data, such as several hours of video, is daunting, which is why it is particularly important to plan observation studies very carefully before starting them. The DECIDE framework suggests identifying goals and questions first before selecting techniques for the study, because the goals and questions help determine which data is collected and how it will be analyzed.

When analyzing any kind of data, the first thing to do is to "eyeball" the data to see what stands out. Are there patterns or significant events? Is there obvious evidence that appears to answer a question or support a theory? Then proceed to analyze it according to the goals and questions. The discussion that follows focuses on three types of data:

- *Qualitative data* that is *interpreted* and used to tell "the story" about what was observed.

- *Qualitative data* that is *categorized* using techniques such as content analysis.

- *Quantitative data* that is collected from interaction and video logs and presented as values, tables, charts and graphs and is treated statistically.

12.6.1   Qualitative analysis to tell a story

Much of the power of analyzing descriptive data lies in being able to tell a convincing story, illustrated with powerful examples that help to confirm the main points and will be credible to the development team. It is hard to argue with well-chosen video excerpts of users interacting with technology or anecdotes from transcripts.

In the interview with Sara Bly you will read about how she and her colleagues use data from several sources. At the end of each observation period they review their data, discuss what they observed, and construct a story from the data. This story evolves as more data is collected and more insights are generated. Teamwork plays an important role in this process because it provides different perspectives that can be compared. A large part of the analysis involves making "collections" of incidents or anecdotes that illustrate similar issues. For example, if several people comment at different times that it is hard to track down a manager in a particular work setting, these examples are powerful evidence of the need for better communication.

To summarize, the main activities involved in working with qualitative data to tell a story are:

- Review the data after each observation session to synthesize and identify key themes and make collections.
- Record the themes in a coherent yet flexible form, with examples. While post-its enable you to move ideas around and group similar ones, they can fall off and get lost and are not easily transported, so capture the main points in another form, either on paper or on a laptop, or make an audio recording.
- Record the date and time of each data analysis session. (The raw data should already be systematically logged with dates.)
- As themes emerge, you may want to check your understanding with the people you observe or your informants.
- Iterate this process until you are sure that your story faithfully represents what you observed and that you have illustrated it with appropriate examples from the data.
- Report your findings to the development team, preferably in an oral presentation as well as in a written report. Reports vary in form, but it is always helpful to have a clear, concise overview of the main findings presented at the beginning.

Analyzing and reporting ethnographic data   Ethnographers work in a similar way but emphasize understanding events within the context in which they happen. Data is collected from participant observation, interviews, and artifacts, and analysis is continuous with great attention to detail. Ethnographers reconstruct knowledge to produce detailed descriptions known as *rich* or *thick descriptions*. In these descriptions, quotes, pictures, and anecdotes play a convincing role in communicating the findings to others. The main activities in analyzing ethnographic data are similar to those just mentioned but notice the emphasis on detail (Fetterman, 1998):

- Look for key events within a group that speak about what drives the group's activity.

- Look for patterns of behavior in various situations and among different players. With experience, ethnographers build up sets of knowledge from various sources, asking questions, listening, probing, comparing and contrasting, synthesizing, and evaluating information.

- Compare sources of data against each other to provide consistent explanations.

- Finally, report your findings in a convincing and honest way. Writing is part of the analysis since it helps to crystallize ideas.

Software tools, such as NUDIST and Ethnograph, allow ethnographers to code their notes and artifact descriptions so that they can be sorted, searched, and retrieved. For example, using NUDIST, field notes can be searched for key words or phrases and a report printed listing every occasion the word or phrase is used. The information can also be printed out as a tree showing the relationship of occurrences. Similarly, NUDIST can be used to search a body of text to identify specific predetermined categories or words for content analysis. The more copious the notes, the more useful tools like NUDIST are. Furthermore, many exploratory searches can be done to test hypotheses among different categories of data.

Other computerized tools support basic statistical analysis. For example, some data can be analyzed using statistical tests (such as chi-square contingency table analysis or rank correlation) to determine whether particular trends are significant.

12.6.2 Qualitative analysis for categorization

Data from think-aloud protocols, video, or audio transcripts can be analyzed in different ways. These can be coarse-grained or detailed analyses of excerpts from a protocol in which each word, phrase, utterance, or gesture is analyzed. Sometimes examining the comment or action in the context of other behavior is sufficient. In this section we discuss a selection of techniques. Some are used more often in research while others are used more for product development.

Looking for incidents or patterns

Analyzing even a short half-hour videotape would be very time-consuming if evaluators studied every comment or action in detail. Furthermore, such fine-grained analyses are often not necessary. A common strategy is to look for critical incidents, such as times when users were obviously stuck. Such incidents are usually marked by a comment, silence, looks of puzzlement, etc. Evaluators focus on these incidents and review them in detail, using the rest of the video as context to inform their analysis. For example, Jurgen Koenemann-Belliveau et al. (1994) used this approach to compare the efficacy of two versions of a Smalltalk programming manual for supporting novice programmers. They used a form of critical incident analysis to examine breakdowns or problems in achieving a programming task and also to identify possible threats of incidents. This enabled them to identify specific problems that might otherwise have been overlooked. Taking this approach, they were able to trace through a sequence of incidents and achieve a more holistic understanding of the problem. For example they found that they needed to emphasize how objects interact in teaching object-oriented programming.

Theory may also be used to guide the study. Wendy Mackay et al. (2000) took this approach in analyzing a four-minute excerpt from a video of users working with a new software tool. Using Activity Theory to guide their analysis, they identified 19 shifts in attention between different parts of the tool interface and the task at hand. (In fact, some users spent so much time engaged in these shifts that they lost track of their original task.) Using the theory helped the evaluators to focus on relevant incidents.

Whether your analysis is coarse-grained or finer, whether you are guided by theory or are just looking for incidents and patterns of behavior, you need a way of handling your data and recording your analysis. For example, in another part of their study, Wendy Mackay et al. (2000) collected and analyzed video excerpts of users interacting with their tool and constructed a form of paper storyboards. The series of images taken from the video illustrated the changes made through the task, while the accompanying text descriptions provided details about the precise operations performed and the difficulties encountered.

A variety of tools are available to record, manipulate and search the data. NUDIST was mentioned above and Box 12.1 briefly describes the Observer Video-Pro tool. Typically reports from these analyses are fed back to the development team, often accompanied by video clips.

BOX 12.1   The Observer Video-Pro: An Automated Data Analysis Tool

The Observer Video-Pro provides the following features (Noldus, 2000):

- During preparation of a video tape recording, a *time code generator* adds an invisible time code to each video frame.

- During a data-collection session, a *time code reader* retrieves the time code from the tape, allowing frame-accurate event timing independent of the playback speed of the video cassette recorder (VCR).

- Each keyboard entry is firmly anchored to the video frame displayed at the instant the evaluator presses the first key of a behavior code or free-format note. The evaluator can also use a mouse to score events.

- Observational data can be reviewed and edited, with synchronized display of the corresponding video images.

- For optimal visual feedback during coding, the evaluator can display the video image in a window on the computer screen.

- The VCR can be controlled by the computer, allowing software-controlled "jog", "shuttle", and "search" functions.

- Video images can be captured and saved as disk files for use as illustrations in documents, slides for presentations, etc.

- Marked video episodes can be copied to an Edit Decision List for easy creation of highlight tapes.

ACTIVITY 12.9   What does the Observer Video-Pro tool allow you to search for in the data collected?

Comment   Depending on how the logs have been annotated, using the Observer Video-Pro product, you can search the data for various things including the following:

Video time—A specific time, e.g., 02:24:36.04 (hh:mm: ss.dd).

Marker—A previously entered free-format annotation.

Event—A combination of actor, behavior, and modifiers, with optional wildcards (e.g., the first occurrence of "glazed look" or "Sarah approaches Janice").

Text—Any word or alphanumeric text string occurring in the coded event records or free-format notes.

Analyzing data into categories

Content analysis provides another fine grain way of analyzing video data. It is a systematic, reliable way of coding content into a meaningful set of mutually exclusive categories (Williams et al., 1988). The content categories are determined by the evaluation questions and one of its most challenging aspects is determining meaningful categories that are orthogonal—i.e., do not overlap each other in any way.

Deciding on the appropriate granularity is another issue to be addressed. The content categories must also be reliable so that the analysis can be replicated. This can be demonstrated by training a second person to use the categories. When training is complete, both researchers analyze the same data sample. If there is a large discrepancy between the two analyses, either training was inadequate or the categorization is not working and needs to be refined. By talking to the researchers you can determine the source of the problem, which is usually with the categorization. If so, then a better categorization scheme needs to be devised and re-tested by doing more inter-researcher reliability tests. However, if the researchers do not seem to know how to carry out the process then they probably need more training.

When a high level of reliability is reached, it can be quantified by calculating an *inter-research reliability rating*. This is the percentage of agreement between the two researchers, defined as the number of items that both categorized in the same way expressed as a percentage of the total number of items examined. It provides a measure of the efficacy of the technique and the categories.

Content analysis *per se* is not used very often in evaluations because it is very labor-intensive and time-consuming but a study by Maria Ebling and Bonnie John (2000) showed how useful it can be. They developed a hierarchical content classification for analyzing data when evaluating a graphical interface for a distributed file system.

Analyzing discourse

Another approach to video, and audio analysis is to focus on the dialog, i.e., the meaning of what is said, rather than the content. Discourse analysis is strongly interpretive, pays great attention to context, and views language not only as reflecting psychological and social aspects but also as constructing it (Coyle, 1995). An underlying assumption of discourse analysis is that there is no objective scientific truth. Language is a form of social reality that is open to interpretation from different perspectives. In this sense, the underlying philosophy of discourse analysis is similar to that of ethnography. Language is viewed as a constructive tool and discourse analysis provides a way of focusing upon how people use language to construct versions of their worlds (Fiske, 1994).

Small changes in wording can change meaning, as the following excerpts indicate (Coyle, 1995):

> *Discourse analysis is what you do when you are saying that you are doing discourse analysis. . . .*
> *According to Coyle, discourse analysis is what you do when you are saying that you are doing discourse analysis. . . .*

By adding just three words "According to Coyle," the sense of authority changes, depending on what the reader knows about Coyle's work and reputation. Some analysts also suggest that a useful approach is to look for variability either within or between individuals.

Analyzing discourse on the Internet (e.g., in chatrooms, bulletin boards, and virtual worlds) has started to influence designers' understanding about users' needs in these environments. Conversation analysis is a very fine-grained form of discourse analysis that can be used for this purpose. In conversational analysis the semantics of the discourse are examined in fine detail. The focus is on how conversations are conducted. This technique is used in sociological studies and examines how conversations start, how turntaking is structured, and other rules of conversation. It can also be very useful when comparing conversations that take place during video-mediated sessions or in computer-mediated communication such as chatrooms as discussed in Chapter 4.

12.6.3　Quantitative data analysis

Video data collected in usability laboratories is usually annotated as it is observed. Small teams of evaluators watch monitors showing what is being recorded in a control room out of the users' sight. As they see errors or unusual behavior, one of the evaluators marks the video and records a brief remark. When the test is finished evaluators can use the annotated recording to calculate performance times so they can compared users' performance on different prototypes. The data stream from the interaction log is used in a similar way to calculate performance times. Typically this data is further analyzed using simple statistics such as means, standard deviations, T-tests, etc. Categorized data may also be quantified and analyzed statistically, as we have said.

12.6.4　Feeding the findings back into design

The results from an evaluation can be reported to the design team in several ways, as we have indicated. Clearly written reports with an overview at the beginning and detailed content list make for easy reading and a good reference document. Including anecdotes, quotations, pictures, and video clips helps to bring the study to life, stimulate interest, and make the written description more meaningful. Some teams like quantitative data, but its value depends on the type of study and its goals. Verbal presentations that include video clips can also be very powerful. Often both qualitative and quantitative data analysis are useful becuase they provide alternative perspectives.

Assignment

The aim of this assignment is for you to learn to do field observation. To do the assignment you will need to find a group of people or a single individual engaged in using one of the following: a mobile phone, a VCR, a photocopying machine, computer software, or some other type of technology that interests you. Assume that you have been employed to improve the product, either by doing a redesign or by creating a completely new product. You can observe people in your family, your friends, or people in your class or local community group.

For this assignment you should:

(a) Consider what the basic goal of "improving the product" means. What initial questions might you ask?

(b) Watch the group (or person) casually to get an understanding of issues that might create challenges for you doing this assignment and information that might enable you to refine your questions.

(c) Then plan your study:

   (i) Think again about what questions will help direct your observation. What are you evaluating?

   (ii) Decide where on the outsider-insider spectrum of observers you wish to be.

   (iii) Prepare an informed consent form and any scripts that you need to introduce yourself and your study.

   (iv) Decide how you will collect data and prepare any data-collection sheets needed; acquire and test any equipment needed.

   (v) Decide how you will analyze the data that you collect.

   (vi) Think through the DECIDE framework. Is everything covered?

   (vii) If so, do a pilot study to check your preparation.

(d) Carry out your study but limit its scope. For example, plan two half-hour observation periods.

(e) Now analyze your data using the method chosen above.

(f) Write a report about what you did and why; describe your data, how you analyzed it, and your findings.

(g) Suggest some ways in which the product might be improved.

Summary

Observing users in the field enables designers to see how technology is used in context. It is valuable for confirming designers' understanding of users' needs and for exploring new design ideas. Various amounts of control, intervention, and involvement with users are possible.

At one end of the spectrum, laboratory studies offer a strongly controlled environment with little evaluator involvement; at the other, participant observation and ethnography require deeper involvement with users and understanding of context. Diaries and data-logging techniques provide a way of tracking user activity without intruding.

Key points

- Observation in usability testing tends to be objective, from the outside. The observer watches and analyzes what happens.

- In contrast, in participant observation the evaluator works with users to understand their activities, beliefs and feelings within the context in which the technology is used.

- Ethnography uses a set of techniques that include participant observation and interviews. Ethnographers immerse themselves in the culture that they study.

- The way that observational data is collected and analyzed depends on the paradigm in which it is used: quick and dirty, user testing, or field studies.

- Combinations of video, audio and paper records, data logging, and diaries can be used to collect observation data.

- In participant observation, collections of comments, incidents, and artifacts are made during the observation period. Evaluators are advised to discuss and summarize their findings as soon after the observation session as possible.

- Analyzing video and data logs can be difficult because of the sheer volume of data. It is important to have clearly specified questions to guide the process and also access to appropriate tools.

- Evaluators often flag events in real time and return to examine them in more detail later. Identifying key events is an effective approach. Fine-grained analyses can be very time-consuming.

Further reading

BLY, S. (1997) Field work: Is it product work? *Interactions*, January and February, 25–30. This article provides additional information to supplement the interview with Sara Bly. It gives a broad perspective on the role of participant observation in product development.

BOGDEWIC, S. P. (1992) Participant observation. In B. F. Crabtree and W. L. Miller (eds.), *Doing Qualitative Research*. Newbury Park, CA: Sage, 45–69. This chapter provides an introduction to participant observation.

BROWN, B. A., SELLEN, A. J., AND O'HARA, K. P. (2000). *A diary study of information capture in working life*. In the Proceedings of CHI2000, The Hague, Holland, 438–445. This paper discusses how cameras were used in a diary study, fol-

lowed by semi-structured interviews, to inform the design of handheld storage devices.

FETTERMAN, D. M. (1998). *Ethnography: Step by Step* (2nd ed.). (Vol. 17). Thousand Oaks, CA: SAGE. This book provides an introduction to the theory and practice of ethnography and is an excellent guide for beginners. In addition, it has a useful section on computerized tools for ethnography.

ROBSON, C. (1993). *Real World Research*. Oxford, UK: Blackwell. Chapter 8 discusses a range of observation methods. There is a section on doing participant observation and also on observing from the outside using coding schemes.

INTERVIEW   with Sara Bly

Sara Bly is a user-centered design consultant who specializes in the design and evaluation of distributed group technologies and practices. As well as having a Ph.D. in computer science, Sara pioneers the development of rich, qualitative observational techniques for analyzing group interactions and activities that inform technology design. Prior to becoming a consultant, Sara managed the Collaborative Systems Group at Xerox Palo Alto Research Center (PARC). While at PARC, Sara also contributed to ground-breaking work on shared drawing, awareness systems, and systems that used non-speech audio to represent information, and to the PARC Media Space project, in which video, audio, and computing technologies are uniquely combined to create a trans-geographical laboratory.

JP: Sara, tell us about your work and what especially interests you.

SB: I'm interested in the ways that qualitative studies, particularly based on ethnographic methods, can inform design and development of technologies. My work spans the full gamut of user-centered design, from early conceptual design through iterative prototypes to final product deployment. I've worked on a wide range of projects from complex collaborative systems to straightforward desktop applications, and a variety of new technologies. My recent projects include a cell phone enhancement, a web-based video application, and the integration of text-based virtual environments with documents.

JP: Why do you think qualitative methods are so important for evaluating usability?

SB: I strongly believe that technical systems are closely bound with the social setting in which they are used. An important part of evaluation is to look "beyond the task." Too often we think of computer systems in isolation from the rest of the activities in which the people are involved. It's important to be able to see the interface in the context of ongoing practice. Usually the complexities and "messiness" of

everyday life do not lend themselves to constraining the evaluation to only a few variables for testing. Qualitative methods are particularly helpful for evaluating complex systems that involve several tasks, embedded in other activities that include multiple users.

JP: Can you give me an example?

SB: Recently I was asked to design and evaluate an application for setting up personal preferences and purchasing services on the web. I was told it would be hard to test the interface "in the field" because it was difficult to get a 45–60 minute test period when the user wasn't being interrupted. When I pointed out that interruptions were normal in the environment in which the product would be used and therefore should occur in the evaluation too, the client looked aghast. There was a moment of silence as he realized, for the first time, that this hadn't been taken into account in the design and that the interface timed out after 60 seconds. It was unusable because the user would have to start all over again after each timeout. This should have been noticed at the requirements stage. So why wasn't it? It sounds like such an obvious thing, but the team was so busy with the intricacies of the design that they failed to realize what the real world would be like in which the system would be used. This might sound extreme, but you'd be surprised how often such things happen.

JP: Collaborative applications seem particularly difficult to evaluate out of context.

SB: Yes, you have to evaluate collaborative systems integrated within an organizational culture in which working relationships are taken into account. We know that work practice impacts system design and that the introduction of a new system impacts work practice. Consequently, the system and the practice have to evolve together. Understanding the task or the interface is impossible without understanding the environment in which the system will be used.

JP: Much of what you've described involves various forms of observation. How do you collect and analyze this data?

SB: It's important that qualitative methods are not seen as just *watching*. Any method we use has at least three critical phases. First, there is the initial assessment of the domain and/or technology and the determination of the questions to address in the evaluation. Second is the data collection, analysis, and representation, and third, the communication of the findings with the development team. I try to start with a clear understanding of what I need to focus on in the field. However, I also try hard not to start with assumptions about what will be true. So, I start with a *well-defined* focus but *not* a hypothesis. In the field (or even in the lab), I primarily use interviews and observations with some self-reporting that often takes the form of diaries, etc. The data typically consists of my notes, the audio and/or videotapes from interviews and observation time, still pictures, and as many artifacts as I can appropriately gather (e.g., a work document covered with post-its, a page from an old calendar). I also prefer to work with at least one other colleague so that there is a minimum of two perspectives on the events and data.

JP: It sounds like keeping track of all this data could be a problem. How do you organize and analyze it?

SB: Obviously it's critical not to end with the data collection. Whenever possible, I do immediate debriefs after each session in the field with my colleague, noting individually and collectively whatever jumped out at us. Subsequently, I use the interview notes (from everyone involved) and the tapes and artifacts to construct as much of a picture of what happened as possible, without putting any judgment on it. For example, in a recent study six of us were involved in interviews and observations. We worked in pairs and tried to vary the pairings as often as possible. Thus, we had lots of conversations about the data and the situations before we ever came together. First, we wrote up the notes from each session (something I try to do as soon as possible). Next we got together and began looking across the data. That is, we created representations of important events (tables, maps, charts) together. Because we collectively had observed all the events and because we could draw upon our notes, we could feed the data from each observation into each finding. Oftentimes, we create collections, looking for common behaviors or events across multiple sessions. A collection will highlight activities that are crucial to the design of the system being evaluated. Whatever techniques we use, we always come back to the data as a reality and validity check.

JP: Is it difficult to get development teams and managers to listen to you? How do you feed your findings back?

SB: As often as possible, development teams are involved in the process along the way. They participate in setting the initial goals of the evaluation, occasionally in observation sessions, and as recipients of a final report. My goal with any project is to ensure that the final report is not a handoff but rather an interaction that offers a chance to work together on what we've found.

JP: What are the main challenges you face?

SB: It's always difficult to conduct a field study with as much time and participation as would be ideal. Most product cycles are short and the evaluation is just one of many necessary steps. So it's always a challenge to do an evaluation that is timely, useful, and yet based on solid methodology.

A gnawing question for me is how to evaluate a system in the context of the customer's own environment and experience when the system is not fully developed and ready to deploy? If we can't bring a product to the field, can we bring the field to the product? For example, a client recently had a prototype interface for a system that was intended to provide a new approach to person-to-person calls. But using the interface made sense only in the context of actual real-world interactions. So, while we certainly could do a standard usability study of the interface, this approach wouldn't get at the questions of how well the product would fit into an actual work situation.

JP: Finally, what about the future? Any comments?

SB: I think the explosion of computing technology is both exciting and overwhelming. We now have so much new information constantly available and so many new devices to master that it's hard to keep up. Evaluation is going to become ever more critical and complex and we should use all the techniques at our disposal as appropriate. I think an increasingly important aspect of new interfaces will be not only how well they support performance, satisfaction, and experience, but the way in which a user is able to grasp a conceptual model that is compatible with, but does not overwhelm their ongoing practice.

Chapter 13

Asking users and experts

13.1 Introduction
13.2 Asking users: interviews
    13.2.1 Developing questions and planning an interview
    13.2.2 Unstructured interviews
    13.2.3 Structured interviews
    13.2.4 Semi-structured interviews
    13.2.5 Group interviews
    13.2.6 Other sources of interview-like feedback
    13.2.7 Data analysis and interpretation
13.3 Asking users: questionnaires
    13.3.1 Designing questionnaires
    13.3.2 Question and response format
    13.3.3 Administering questionnaires
    13.3.4 Online questionnaires
    13.3.5 Analyzing questionnaire data
13.4 Asking experts: inspections
    13.4.1 Heuristic evaluation
    13.4.2 Doing heuristic evaluation
    13.4.3 Heuristic evaluation of websites
    13.4.4 Heuristics for other devices
13.5 Asking experts: walkthroughs
    13.5.1 Cognitive walkthroughs
    13.5.2 Pluralistic walkthroughs

13.1 Introduction

In the last chapter we looked at observing users. Another way of finding out what users do, what they want to do, like, or don't like is to ask them. Interviews and questionnaires are well-established techniques in social science research, market research, and human-computer interaction. They are used in "quick and dirty" evaluation, in usability testing, and in field studies to ask about *facts*, *behavior*, *beliefs*, and *attitudes*. Interviews and questionnaires can be structured (as in the HutchWorld case study in Chapter 10), or flexible and more like a discussion, as in field studies. Often interviews and observation go together in field studies, but in this chapter we focus specifically on interviewing techniques.

The first part of this chapter discusses interviews and questionnaires. As with observation, these techniques can be used in the requirements activity (as we described in Chapter 7), but in this chapter we focus on their use in evaluation. Another way of finding out how well a system is designed is by asking experts for their opinions. In the second part of the chapter, we look at the techniques of heuristic evaluation and cognitive walkthrough. These methods involve predicting how usable interfaces are (or are not). As in the previous chapter, we draw on the DECIDE framework from Chapter 11 to help structure studies that use these techniques.

The main aims of this chapter are to:

- Discuss when it is appropriate to use different types of interviews and questionnaires.

- Teach you the basics of questionnaire design.

- Describe how to do interviews, heuristic evaluation, and walkthroughs.

- Describe how to collect, analyze, and present data collected by the techniques mentioned above.

- Enable you to discuss the strengths and limitations of the techniques and select appropriate ones for your own use.

13.2   Asking users: interviews

Interviews can be thought of as a "conversation with a purpose" (Kahn and Cannell, 1957). How like an ordinary conversation the interview is depends on the questions to be answered and the type of interview method used. There are four main types of interviews: *open-ended or unstructured*, *structured*, *semi-structured*, and *group* interviews (Fontana and Frey, 1994). The first three types are named according to how much control the interviewer imposes on the conversation by following a *predetermined set of questions*. The fourth involves a small group guided by an interviewer who facilitates discussion of a specified set of topics.

The most appropriate approach to interviewing depends on the evaluation goals, the questions to be addressed, and the paradigm adopted. For example, if the goal is to gain first impressions about how users react to a new design idea, such as an interactive sign, then an informal, open-ended interview is often the best approach. But if the goal is to get feedback about a particular design feature, such as the layout of a new web browser, then a structured interview or questionnaire is often better. This is because the goals and questions are more specific in the latter case.

13.2.1   Developing questions and planning an interview

When developing interview questions, plan to keep them short, straightforward and avoid asking too many. Here are some guidelines (Robson, 1993):

- Avoid long questions because they are difficult to remember.

- Avoid compound sentences by splitting them into two separate questions. For example, instead of, "How do you like this cell phone compared with previous ones that you have owned?" Say, "How do you like this cell phone? Have you owned other cell phones? If so, How did you like it?" This is easier for the interviewee and easier for the interviewer to record.

- Avoid using jargon and language that the interviewee may not understand but would be too embarrassed to admit.

- Avoid leading questions such as, "Why do you like this style of interaction?" If used on its own, this question assumes that the person did like it.

- Be alert to unconscious biases. Be sensitive to your own biases and strive for neutrality in your questions.

Asking colleagues to review the questions and running a pilot study will help to identify problems in advance and gain practice in interviewing.

When planning an interview, think about interviewees who may be reticent to answer questions or who are in a hurry. They are doing *you* a favor, so try to make it as pleasant for them as possible and try to make the interviewee feel comfortable. Including the following steps will help you to achieve this (Robson, 1993):

1. An *Introduction* in which the interviewer introduces himself and explains why he is doing the interview, reassures interviewees about the ethical issues, and asks if they mind being recorded, if appropriate. This should be exactly the same for each interviewee.

2. A *warmup* session where easy, non-threatening questions come first. These may include questions about demographic information, such as "Where do you live?"

3. A *main* session in which the questions are presented in a logical sequence, with the more difficult ones at the end.

4. A *cool-off period* consisting of a few easy questions (to defuse tension if it has arisen).

5. A *closing* session in which the interviewer thanks the interviewee and switches off the recorder or puts her notebook away, signaling that the interview has ended.

The golden rule is to be professional. Here is some further advice about conducting interviews (Robson, 1993):

- Dress in a similar way to the interviewees if possible. If in doubt, dress neatly and avoid standing out.

- Prepare an informed consent form and ask the interviewee to sign it.

- If you are recording the interview, which is advisable, make sure your equipment works in advance and you know how to use it.

- Record answers exactly; do not make cosmetic adjustments, correct, or change answers in any way.

13.2.2 Unstructured interviews

Open-ended or unstructured interviews are at one end of a spectrum of how much control the interviewer has on the process. They are more like conversations that focus on a particular topic and may often go into considerable depth. Questions posed by the interviewer are *open*, meaning that the format and content of answers is not predetermined. The interviewee is free to answer as fully or as briefly as she wishes. Both interviewer and interviewee can steer the interview. Thus one of the skills necessary for this type of interviewing is to make sure that answers to relevant questions are obtained. It is therefore advisable to be organized and have a plan of the main things to be covered. Going in without an agenda to accomplish a goal is *not* advisable, and should not to be confused with being open to new information and ideas.

A benefit of unstructured interviews is that they generate rich data. Interviewees often mention things that the interviewer may not have considered and can be further explored. But this benefit often comes at a cost. A lot of unstructured data is generated, which can be very time-consuming and difficult to analyze. It is also impossible to replicate the process, since each interview takes on its own format. Typically in evaluation, there is no attempt to analyze these interviews in detail. Instead, the evaluator makes notes or records the session and then goes back later to note the main issues of interest.

The main points to remember when conducting an unstructured interview are:

- Make sure you have an interview agenda that supports the study goals and questions (identified through the DECIDE framework).
- Be prepared to follow new lines of enquiry that contribute to your agenda.
- Pay attention to ethical issues, particularly the need to get informed consent.
- Work on gaining acceptance and putting the interviewees at ease. For example, dress as they do and take the time to learn about their world.
- Respond with sympathy if appropriate, but be careful not to put ideas into the heads of respondents.
- Always indicate to the interviewee the beginning and end of the interview session.
- Start to order and analyze your data as soon as possible after the interview.

ACTIVITY 13.1   Ananova is a virtual news reporter created by the British Press Association on the website www.ananova.com, which is similar to the picture in Figure 13.1. Viewers who wish to hear Ananova report the news must select from the menu beneath her picture and must have downloaded software that enables them to receive streaming video. Those who wish to read text may do so.

The idea is that Ananova is a life-like, i.e., an 'anthropomorphic' news presenter. She is designed to speak, move her lips, and blink, and she has some human facial expressions. She reads news edited from news reports. Ananova's face, her voice tone, her hair, in fact everything about her was tested with users before the site was launched so that she would appeal to as many users as possible. She is fashionable and looks as though she is in her twenties or

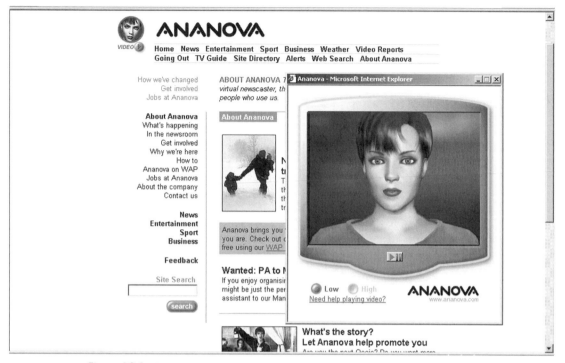

Figure 13.1 Ananova.com showing Ananova, a virtual news presenter.

early thirties—presumably the age that market researchers determined fits the profile of the majority of users—and she is also designed to appeal to older people too.

To see Ananova in action, go to the website (www.annanova.com) and follow the directions for downloading the software. Alternatively you can do the activity by just looking at the figure and thinking about the questions.

(a) Suggest unstructured interview questions that seek opinions about whether Ananova improves the quality of the news service.

(b) Suggest ways of collecting the interview data.

(c) Identify practical and ethical issues that need to be considered.

Comment

(a) Possible questions include: Do you think Ananova reading the news is good? Is it better than having to read it yourself from a news bulletin? In what ways does having Ananova read the news influence your satisfaction with the service?

(b) Taking notes might be cumbersome and distracting to the interviewee, and it would be easy to miss important points. An alternative is to audio record the session. Video recording is not needed as it isn't necessary to see the interviewee. However, it would be useful to have a camera at hand to take shots of the interface in case the interviewee wanted to refer to aspects of Ananova.

(c) The obvious practical issues are obtaining a cassette recorder, finding participants, scheduling times for the interviews and finding a quiet place to conduct them. Having a computer available for the interviewee to refer to is important. The ethical issues include telling the interviewees why you are doing the interviews and what you will do with the information, and guaranteeing them anonymity. An informed consent form may be needed.

13.2.3   Structured interviews

Structured interviews pose predetermined questions similar to those in a questionnaire (see Section 13.3). Structured interviews are useful when the study's goals are clearly understood and specific questions can be identified. To work best, the questions need to be short and clearly worded. Responses may involve selecting from a set of options that are read aloud or presented on paper. The questions should be refined by asking another evaluator to review them and by running a small pilot study. Typically the questions are *closed*, which means that they require a precise answer. The same questions are used with each participant so the study is standardized.

13.2.4   Semi-structured interviews

Semi-structured interviews combine features of structured and unstructured interviews and use both closed and open questions. For consistency the interviewer has a basic script for guidance, so that the same topics are covered with each interviewee. The interviewer starts with preplanned questions and then probes the interviewee to say more until no new relevant information is forthcoming. For example:

> Which websites do you visit most frequently? *<Answer>* Why? *<Answer mentions several but stresses that she prefers hottestmusic.com>* And why do you like it? *<Answer>* Tell me more about x? *<silence, followed by an answer>* Anything else? *<Answer>* Thanks. Are there any other reasons that you haven't mentioned?

It is important not to preempt an answer by phrasing a question to suggest that a particular answer is expected. For example, "You seemed to like this use of color . . ." assumes that this is the case and will probably encourage the interviewee to answer that this is true so as not to offend the interviewer. Children are particularly prone to behave in this way. The body language of the interviewer, for example, whether she is smiling, scowling, looking disapproving, etc., can have a strong influence.

Also the interviewer needs to accommodate *silences* and not to move on too quickly. Give the person time to speak. *Probes* are a device for getting more information, especially neutral probes such as, "Do you want to tell me anything else?" You may also *prompt* the person to help her along. For example, if the interviewee is talking about a computer interface but has forgotten the name of a key menu item, you might want to remind her so that the interview can proceed productively. However, semi-structured interviews are intended to be broadly replicable, so probing and prompting should aim to help the interview along without introducing bias.

ACTIVITY 13.2   Write a semi-structured interview script to evaluate whether receiving news from Ananova is appealing and whether Ananova's presentation is realistic. Show two of your peers the

Ananova.com website or Figure 13.1. Then ask them to comment on your interview script. Refine the questions based on their comments.

Comment   You can use questions that have a predetermined set of answer choices. These work well for fast interviews when the range of answers is known, as in the airport studies where people tend to be in a rush. Alternatively, open-ended questions can also be used if you want to explore the range of opinions.

Some questions that you might ask include:

- Have you seen Ananova before?
- Would you like to receive news from Ananova?
- Why?
- In your opinion, does Ananova look like a real person?

Some of the questions in Exercise 13.2 have a predetermined range of answers, such as "yes," "no," "maybe." Others, such as the one about interviewees' attitudes, do not have an easily predicted range of responses. But it would help us in collecting answers if we list possible responses together with boxes that can just be checked (i.e., ticked). Here's how we could convert the questions from Activity 13.2.

- Have you seen Ananova before? (Explore previous knowledge)
  Interviewer checks box ☐ *Yes* ☐ *No* ☐ *Don't remember/know*
- Would you like to receive news from Ananova? (Explore initial reaction, then explore the response)
  Interviewer checks box ☐ *Yes* ☐ *No* ☐ *Don't know*
- Why?
  If response is "Yes" or "No," interviewer says, "Which of the following statements represents your feelings best?"
  For "Yes," Interviewer checks the box
     ☐ *I don't like typing*
     ☐ *This is fun/cool*
     ☐ *I've never seen a system like this before*
     ☐ *It's going to be the way of the future*
     ☐ *Another reason (Interviewer notes the reason)*
  For "No," Interviewer checks the box
     ☐ *I don't like speech systems*
     ☐ *I don't like systems that pretend to be people*
     ☐ *It's faster to read*
     ☐ *I can't control the pace of presentation*
     ☐ *I can't be bothered to download the software*
     ☐ *Another reason (Interviewer notes the reason)*
- In your opinion, does Ananova look like a real person?
  Interviewer checks box
     ☐ *Yes, she looks like a real person*
     ☐ *No, she doesn't look like a real person*

As you can probably guess, there are problems deciding on the range of possible answers. Maybe you thought of other ones. In order to get a good range of answers for the second question, a large number of people would have to be interviewed before the questionnaire is constructed to identify all the possible answers and then those could be used to determine what should be offered.

ACTIVITY 13.3   Write three or four semi-structured interview questions to find out if Ananova is popular with your friends. Make the questions general.

Comment   Here are some suggestions:

(a) Would you listen to the news using Ananova?
   If yes, then ask, why?
   If no, then ask, why not?

(b) Is Ananova's appearance attractive to you?
   If yes, then say, Tell me more, what did you like?
   If no, then say, What don't you find attractive?

(c) Is there anything else you want to say about Ananova?

ACTIVITY 13.4   Prepare the full interview script to evaluate Ananova, including a description of why you are doing the interview, and an informed consent form, and the exact questions. Use the DE-CIDE framework for guidance. Practice the interview on your own, audiotape yourself, and then listen to it and review your performance. Then interview two peers and be reflective. What did you learn from the experience?

Comment   You probably found it harder than you thought to interview smoothly and consistently. Did you notice an improvement when you did the second interview? Were some of the questions poorly worded. Piloting your interview often reveals poor or ambiguous questions that you then have a chance to refine before holding the first proper interview.

13.2.5   Group interviews

One form of group interview is the focus group that is frequently used in marketing, political campaigning, and social sciences research. Normally three to 10 people are involved. Participants are selected to provide a representative sample of typical users; they normally share certain characteristics. For example, in an evaluation of a university website, a group of administrators, faculty, and students may be called to form three separate focus groups because they use the web for different purposes.

The benefit of a focus group is that it allows diverse or sensitive issues to be raised that would otherwise be missed. The method assumes that individuals develop opinions within a social context by talking with others. Often questions posed to focus groups seem deceptively simple but the idea is to enable people to put forward their own opinions in a supportive environment. A preset agenda is developed to guide the discussion but there is sufficient flexibility for a facilitator to follow unanticipated issues as they are raised. The facilitator guides and prompts discussion and skillfully encourages quiet people to participate and stops verbose ones from dominating the discussion. The discussion is usually recorded for later analysis in which participants my be invited to explain their comments more fully.

Focus groups appear to have high validity because the method is readily understood and findings appear believable (Marshall and Rossman, 1999). Focus groups are also attractive because they are low-cost, provide quick results, and can easily be scaled to gather more data. Disadvantages are that the facilitator needs to be skillful so that time is not wasted on irrelevant issues. It can also be difficult to get people together in a suitable location. Getting time with any interviewees can be difficult, but the problem is compounded with focus groups because of the number of people involved. For example, in a study to evaluate a university website the evaluators did not expect that getting participants would be a problem. However, the study was scheduled near the end of a semester when students had to hand in their work, so strong incentives were needed to entice the students to participate in the study. It took an increase in the participation fee and a good lunch to convince students to participate.

13.2.6 Other sources of interview-like feedback

Telephone interviews are a good way of interviewing people with whom you cannot meet. You cannot see body language, but apart from this telephone interviews have much in common with face-to-face interviews.

Online interviews, using either asynchronous communication as in email or synchronous communication as in chats, can also be used. For interviews that involve sensitive issues, answering questions anonymously may be preferable to meeting face to face. If, however, face to face meetings are desirable but impossible because of geographical distance, video-conferencing systems can be used (but remember the drawbacks discussed in Chapter 4). Feedback about a product can also be obtained from customer help lines, consumer groups, and online customer communities that provide help and support.

At various stages of design, it is useful to get quick feedback from a few users. These short interviews are often more like conversations in which users are asked their opinions. Retrospective interviews can be done when doing field studies to check with participants that the interviewer has correctly understood what was happening.

DILEMMA **What They Say and What They Do!**

What users say isn't always what they do. People sometimes give the answers that they think show them in the best light, or they may just forget what happened or how long they spent on a particular activity.

So, can evaluators believe all the responses they get? Are the respondents giving "the truth" or are they simply giving the answers that they think the evaluator wants to hear?

It isn't possible to avoid this behavior, but it is important to be aware of it and to reduce such biases by getting a large number of participants or by using a combination of techniques. Questions that suggest particular responses should also be avoided.

13.2.7   Data analysis and interpretation

Analysis of unstructured interviews can be time-consuming, though their contents can be rich. Typically each interview question is examined in depth in a similar way to observation data discussed in Chapter 12. A coding form may be developed, which may be predetermined or may be developed during data collection as evaluators are exposed to the range of issues and learn about their relative importance. Alternatively, comments may be clustered along themes and anonymous quotes used to illustrate points of interest. Tools such a NUDIST and Ethnograph can be useful for qualitative analyses as mentioned in Chapter 12. Which type of analysis is done depends on the goals of the study, as does whether the whole interview is transcribed, only part of it, or none of it. Data from structured interviews is usually analyzed quantitatively as in questionnaires which we discuss next.

13.3   Asking users: questionnaires

Questionnaires are a well-established technique for collecting demographic data and users' opinions. They are similar to interviews and can have *closed* or *open* questions. Effort and skill are needed to ensure that questions are clearly worded and the data collected can be analyzed efficiently. Questionnaires can be used on their own or in conjunction with other methods to clarify or deepen understanding. In the HutchWorld study discussed in Chapter 10, for example, you read how questionnaires were used along with observation and usability testing. The methods and questions used depends on the context, interviewees and so on.

The questions asked in a questionnaire, and those used in a structured interview are similar, so how do you know when to use which technique? One advantage of questionnaires is that they can be distributed to a large number of people. Used in this way, they provide evidence of wide general opinion. On the other hand, structured interviews are easy and quick to conduct in situations in which people will not stop to complete a questionnaire.

13.3.1   Designing questionnaires

Many questionnaires start by asking for basic demographic information (e.g., gender, age) and details of user experience (e.g., the time or number of years spent using computers, level of expertise, etc.). This background information is useful in finding out the range within the sample group. For instance, a group of people who are using the web for the first time are likely to express different opinions to another group with five years of web experience. From knowing the sample range, a designer might develop two different versions or veer towards the needs of one of the groups more because it represents the target audience.

Following the general questions, specific questions that contribute to the evaluation goal are asked. If the questionnaire is long, the questions may be subdivided into related topics to make it easier and more logical to complete.

Box 13.1 contains an excerpt from a paper questionnaire designed to evaluate users' satisfaction with some specific features of a prototype website for career changers aged 34–59 years.

Notice that in the following excerpt users are asked to circle appropriate responses, and check the box that most closely describes their opinion: these are commonly used techniques. Fewer than fifty participants were involved in this study, so inviting them to write on open-ended comment suggesting recommendations for change was manageable. It would have been difficult to collect this information with closed questions, since good suggestions would undoubtedly have been missed because the evaluator would probably not have thought to ask about them.

Participant #: _____

Please circle the most appropriate selection:

Age Range: 34–39   40–49   50–59

Gender: Male   Female

Career-Change Status: Exploring   In-Progress   Completed

Internet/Web Experience

Research, Information Gathering	Daily	Weekly	Monthly	Never
Bulletin Board Posting	Daily	Weekly	Monthly	Never
Chat Room Usage	Daily	Weekly	Monthly	Never

Please rate (i.e., check the box to show) agreement or disagreement with the following statements:

Question	Strongly Agree	Agree	Neutral	Disagree	Strongly Disagree
The navigation language on the links is clear and easy to understand					
The website site contains information that would be useful to me					
Information on the website is easy to find					
The "Center Design" presents information in an aesthetically pleasing manner					
The website pages are confusing and difficult to read					
I prefer darker colors to lighter colors for display					
It is apparent from the first website page (homepage) what the purpose of the website is.					

Please add any recommendations for changes to the overall design, language or navigation of the website on the back of this paper.

Thanks for your participation in the testing of this prototype.

The following is a checklist of general advice for designing a questionnaire:

- Make questions clear and specific.
- When possible, ask closed questions and offer a range of answers.
- Consider including a "no-opinion" option for questions that seek opinions.
- Think about the ordering of questions. The impact of a question can be influenced by question order. General questions should precede specific ones.
- Avoid complex multiple questions.
- When scales are used, make sure the range is appropriate and does not overlap.
- Make sure that the ordering of scales (discussed below) is intuitive and consistent, and be careful with using negatives. For example, it is more intuitive in a scale of 1 to 5 for 1 to indicate low agreement and 5 to indicate high agreement. Also be consistent. For example, avoid using 1 as low on some scales and then as high on others. A subtler problem occurs when most questions are phrased as positive statements and a few are phrased as negatives. However, advice on this issue is more controversial as some evaluators argue that changing the direction of questions helps to check the users' intentions. Scales such as those used in Box 13.1 are also preferred by some evaluators.
- Avoid jargon and consider whether you need different versions of the questionnaire for different populations.
- Provide clear instructions on how to complete the questionnaire. For example, if you want a check put in one of the boxes, then say so. Questionnaires can make their message clear with careful wording and good typography.
- A balance must be struck between using white space and the need to keep the questionnaire as compact as possible. Long questionnaires cost more and deter participation.

13.3.2 Question and response format

Different types of questions require different types of responses. Sometimes discrete responses are required, such as "Yes" or "No." For other questions it is better to ask users to locate themselves within a range. Still others require a single preferred opinion. Selecting the most appropriate makes it easier for respondents to be able to answer. Furthermore, questions that accept a specific answer can be categorized more easily. Some commonly used formats are described below.

Check boxes and ranges

The range of answers to demographic questionnaires is predictable. Gender, for example, has two options, male or female, so providing two boxes and asking respondents to check the appropriate one, or circle a response, makes sense for collecting this information (as in Box 13.1). A similar approach can be adopted if details of age are needed. But since some people do not like to give their exact age, many questionnaires ask respondents to specify their age as a range (Box 13.1). A common design error arises when the ranges overlap. For example, specifying two ranges as 15–20, 20–25 will cause confusion: which box do people who are 20 years old check? Making the ranges 14–19, 20–24 avoids this problem.

A frequently asked question about ranges is whether the interval must be equal in all cases. The answer is that it depends on what you want to know. For example, if you want to collect information for the design of an e-commerce site to sell life insurance, the target population is going to be mostly people with jobs in the age range of, say, 21–65 years. You could, therefore, have just three ranges: under 21, 21–65 and over 65. In contrast, if you are interested in looking at ten-year *cohort* groups for people over 21 the following ranges would be best: under 21, 22–31, 32–41, etc.

There are a number of different types of rating scales that can be used, each with its own purpose (see Oppenheim, 1992). Here we describe two commonly used scales, Likert and semantic differential scales.

The purpose of these is to elicit a range of responses to a question that can be compared across respondents. They are good for getting people to make judgments about things, e.g. how easy, how usable etc., and therefore are important for usability studies.

Likert scales rely on identifying a set of statements representing a range of possible opinions, while semantic differential scales rely on choosing pairs of words that represent the range of possible opinions. Likert scales are the most commonly used scales because identifying suitable statements that respondents will understand is easier than identifying semantic pairs that respondents interpret as intended.

Likert Scales

Likert scales are used for measuring opinions, attitudes, and beliefs, and consequently they are widely used for evaluating user satisfaction with products as in the HutchWorld evaluation described in Chapter 10. For example, users' opinions about the use of color in a website could be evaluated with a Likert scale using a range of numbers (1) or with words (2):

(1) The use of color is excellent: (where 1 represents strongly agree and 5 represents strongly disagree)

1	2	3	4	5
☐	☐	☐	☐	☐

(2) The use of color is excellent:

strongly agree	agree	OK	disagree	strongly disagree
☐	☐	☐	☐	☐

Below are some steps for designing Likert scales:

- Gather a pool of short statements about the features of the product that are to be evaluated e.g., "This control panel is easy to use." A brainstorming session with peers in which you examine the product to be evaluated is a good way of doing this.

- Divide the items into groups with about the same number of positive and negative statements in each group. Some evaluators prefer to have all negative or all positive questions, while others use a mix of positive and negative questions, as we have suggested here. Deciding whether to phrase the questionnaire positively or negatively depends partly on the complexity of the questionnaire and partly on the evaluator's preferences. The designers of QUIS (Box 13.2) (Chin et al., 1988), for example, decided not to mix negative and positive statements because the questionnaire was already complex enough without forcing participants to pay attention to the direction of the argument.

- Decide on the scale. QUIS (Box 13.2) uses a 9-point scale, and because it is a general questionnaire that will be used with a wide variety of products it also includes N/A (not applicable,) as a category. Many questionnaires use 7- or 5-point scales and there are also 3-point scales. Arguments for the number of points go both ways. Advocates of long scales argue that they help to show discrimination, as advocated by the QUIS team (Chin et al., 1988). Rating features on an interface is more difficult for most people than, say, selecting among different flavors of ice cream, and when the task is difficult there is evidence to show that people "hedge their bets." Rather than selecting the poles of the scales if there is no right or wrong, respondents tend to select values nearer the center. The counter-argument is that people cannot be expected to discern accurately among points on a large scale, so any scale of more than five points is unnecessarily difficult to use.

    Another aspect to consider is whether the scale should have an even or odd number of points. An odd number provides a clear central point. On the

---

BOX 13.2 QUIS, Questionnaire for User Interaction Satisfaction

The Questionnaire for User Interaction Satisfaction (QUIS) developed by the University of Maryland Human-Computer Interaction Laboratory is one of the most widely used questionnaires for evaluating interfaces (Chin et al., 1988; Shneiderman, 1998a). Although developed for evaluating user satisfaction, it is frequently applied to other aspects of interaction design. An advantage of this questionnaire is that it has gone through many cycles of refinement and has been used for hundreds of evaluation studies, so it is well tried and tested. The questionnaire consists of the following 12 parts that can be used in total or in parts:

- system experience (i.e., time spent on this system)
- past experience (i.e., experience with other systems)

- overall user reactions
- screen design
- terminology and system information
- learning (i.e., to operate the system)
- system capabilities (i.e., the time it takes to perform operations)
- technical manuals and online help
- online tutorials
- multimedia
- teleconferencing
- software installation

Notice that the third part of QUIS assesses users' overall reactions. Evaluators often use this part on its own because it is short so people are likely to respond.

other hand, an even number forces participants to make a decision and prevents them from sitting on the fence.

• Select items for the final questionnaire and reword as necessary to make them clear.

Semantic differential scales

Semantic differential scales are used less frequently than Likert scales. They explore a range of bipolar attitudes about a particular item. Each pair of attitudes is represented as a pair of adjectives. The participant is asked to place a cross in one of a number of positions between the two extremes to indicate agreement with the poles, as shown in Figure 13.2. The score for the evaluation is found by summing the scores for each bipolar pair. Scores can then be computed across groups of participants. Notice that in this example the poles are mixed so that good and bad features are distributed on the right and the left. In this example there are seven positions on the scale.

Instructions: for each pair of adjectives, place a cross at the point between them that reflects the extent to which you believe the adjectives describe the home page. You should place *only one cross* between the marks on each line.

Attractive	└─┴─┴─┴─┴─┴─┴─┘	Ugly
Clear	└─┴─┴─┴─┴─┴─┴─┘	Confusing
Dull	└─┴─┴─┴─┴─┴─┴─┘	Colorful
Exciting	└─┴─┴─┴─┴─┴─┴─┘	Boring
Annoying	└─┴─┴─┴─┴─┴─┴─┘	Pleasing
Helpful	└─┴─┴─┴─┴─┴─┴─┘	Unhelpful
Poor	└─┴─┴─┴─┴─┴─┴─┘	Well designed

Figure 13.2 An example of a semantic differential scale.

ACTIVITY 13.5 Spot the four poorly designed features in Figure 13.3.

Comment Some of the features that could be improved include:

• Request for exact age. Many people prefer not to give this information and would rather position themselves in a range.

• Years of experience is indicated with overlapping scales, i.e., <1, 1–3, 3–5, etc. How do you answer if you have 1, 3, or 5 years of experience?

• The questionnaire doesn't tell you whether you should check one, two, or as many boxes as you wish.

• The space left for people to write their own information is too small, and this will annoy them and deter them from giving their opinions.

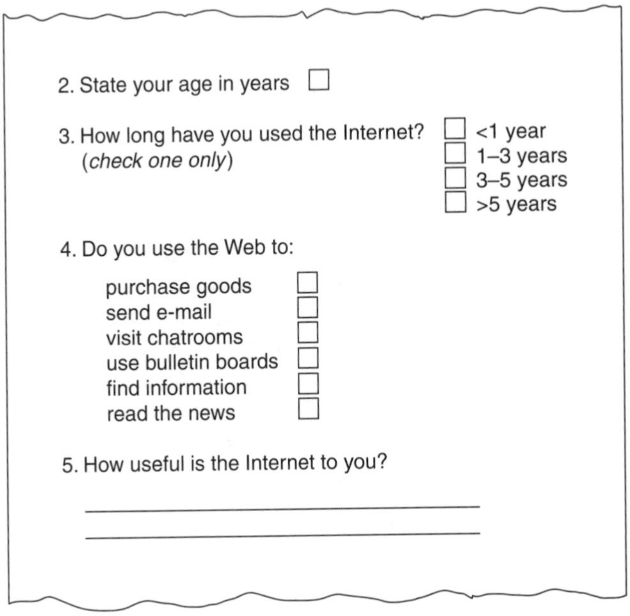

2. State your age in years ☐

3. How long have you used the Internet? ☐ <1 year
   (*check one only*)   ☐ 1–3 years
                                          ☐ 3–5 years
                                          ☐ >5 years

4. Do you use the Web to:

   purchase goods   ☐
   send e-mail   ☐
   visit chatrooms   ☐
   use bulletin boards ☐
   find information ☐
   read the news   ☐

5. How useful is the Internet to you?

   _____

   _____

Figure 13.3 A questionnaire with poorly designed features.

13.3.3 Administering questionnaires

Two important issues when using questionnaires are reaching a representative sample of participants and ensuring a reasonable response rate. For large surveys, potential respondents need to be selected using a sampling technique. However, interaction designers tend to use small numbers of participants, often fewer than twenty users. One hundred percent completion rates often are achieved with these small samples, but with larger, more remote populations, ensuring that surveys are returned is a well-known problem. Forty percent return is generally acceptable for many surveys but much lower rates are common.

Some ways of encouraging a good response include:

- Ensuring the questionnaire is well designed so that participants do not get annoyed and give up.
- Providing a short overview section, as in QUIS (Box 13.2), and telling respondents to complete just the short version if they do not have time to complete the whole thing. This ensures that you get something useful returned.
- Including a stamped, self-addressed envelope for its return.
- Explaining why you need the questionnaire to be completed and assuring anonymity.
- Contacting respondents through a follow-up letter, phone call or email.
- Offering incentives such as payments.

13.3.4 Online questionnaires

Online questionnaires are becoming increasingly common because they are effective for reaching large numbers of people quickly and easily. There are two types: email and web-based. The main advantage of email is that you can target specific users. However, email questionnaires are usually limited to text, whereas web-based questionnaires are more flexible and can include check boxes, pull-down and pop-up menus, help screens, and graphics (Figure 13.4). web-based questionnaires can also provide immediate data validation and can enforce rules such as select only one response, or certain types of answers such as numerical, which cannot be done in email or with paper. Other advantages of online questionnaires include (Lazar and Preece, 1999):

- Responses are usually received quickly.
- Copying and postage costs are lower than for paper surveys or often non-existent.
- Data can be transferred immediately into a database for analysis.
- The time required for data analysis is reduced.
- Errors in questionnaire design can be corrected easily (though it is better to avoid them in the first place).

A big problem with web-based questionnaires is obtaining a random sample of respondents. Few other disadvantages have been reported with online questionnaires, but there is some evidence suggesting that response rates may be lower online than with paper questionnaires (Witmer et al., 1999).

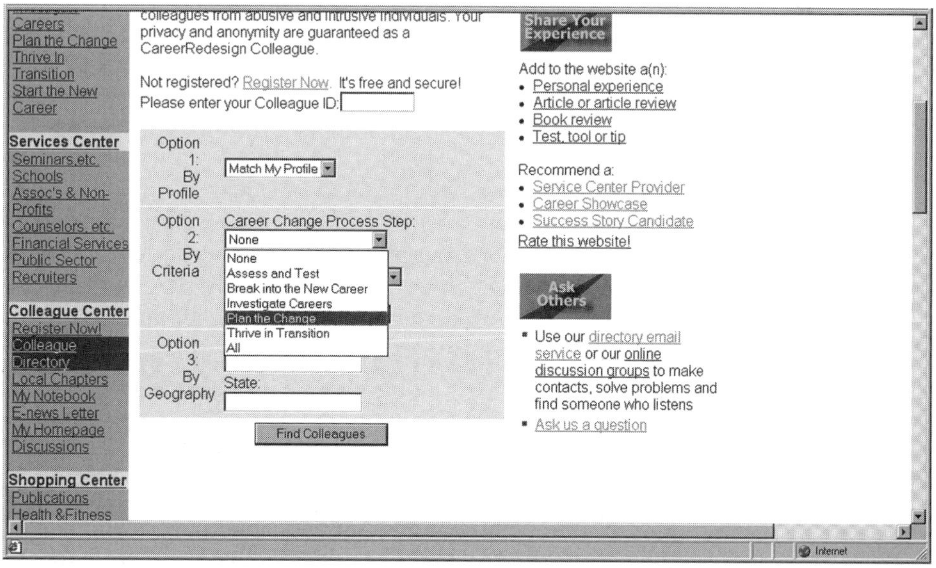

Figure 13.4 An excerpt from a web-based questionnaire showing pull-down menus.

Developing a web-based questionnaire

Developing a successful web-based questionnaire involves designing it on paper, developing strategies for reaching the target population, and then turning the paper version into a web-based version (Lazar and Preece, 1999).

It is important to devise the questionnaire on paper first, following the general guidelines introduced above, such as paying attention to the clarity and consistency of the questions, questionnaire layout, and so on. Only once the questionnaire has been reviewed and the questions refined adequately should it be translated into a web-based version. If reaching your target population is an issue, e.g., if some of them may not have access to the web, the paper version may be administered to them, but be careful to maintain consistency between the web-based version and the original paper version.

Identifying a random sample of a population so that the results are indicative of the whole population may be difficult, if not impossible, to achieve especially if the size and demography of the population is not known, as is often the case in Internet research. This has been a criticism of several online surveys including Georgia Tech's GVU survey, one of the first online surveys. This survey collects demographic and activity information from Internet users and has been distributed twice yearly since 1994. The policy that GVU employs to deal with this difficult sampling issue is to make as many people aware of the GVU survey as possible so that a wide variety of participants are encouraged to participate. However, even these efforts do not avoid biased sampling, since participants are self-selecting. Indeed, some survey experts are vehemently opposed to such methods and instead propose using national census records to sample offline (Nie & Ebring, 2000). In some countries, web-based questionnaires are used in conjunction with television to elicit viewers' opinions of programs and political events, and many such questionnaires now say that their results are "not scientific" when they cite them, meaning that unbiased sampling was not done. A term that is gaining popularity is *convenience sampling*, which is another way of saying that the sample includes those who were available rather than those selected using scientific sampling.

Turning the paper questionnaire into a web-based version requires four steps.

1. Produce an error-free interactive electronic version from the original paper-based one. This version should provide clear instructions and be free of input errors. For example, if just one box should be checked, the other attempts should be rejected automatically. It may also be useful to embed feedback and pop-up help within the questionnaire.

2. Make the questionnaire accessible from all common browsers and readable from different-size monitors and different network locations. Specialized software or hardware should be avoided. The need to download software also deters novice users and should be avoided.

3. Make sure information identifying each respondent will be captured and stored confidentially because the same person may submit several completed surveys. This can be done by recording the Internet domain name or the IP address of the respondent, which can then be transferred directly to a

database. However, this action could infringe people's privacy and the legal situation should be checked. Another way is to access the transfer and referrer logs from the web server, which provide information about the domains from which the web-based questionnaire was accessed. Unfortunately, people can still send from different accounts with different IP addresses, so additional identifying information may also be needed.

4. User-test the survey with pilot studies before distributing.

Commercial questionnaires are becoming available via the Internet. Two examples are SUMI and MUMMS, which are briefly discussed in Box 13.3.

BOX 13.3 Questionnaire Tools

SUMI (Software Usability Measurement Inventory) was developed in the early 1990s as part of a European project. The aim was to develop a standardized tool to evaluate users' reactions to a piece of software. More recently a new version has been developed, known as MUMMS (Measuring the Usability of Multi-Media Systems), that, as the name implies, is geared more towards current software in which multimedia is assumed to be a component. This questionnaire focuses on five concepts:

• how much the product captures the user's emotional responses

• how much the user feels in control of the software

• the degree to which the users can achieve their goals using the software

• the extent to which the product seems to assist the user

• the ease with which the user can learn to use the product

The developers are also planning to include a new concept that they are calling "excitement." This would address two kinds of user experience goals (emotional responses and excitement) and four usability goals. More information about SUMI and MUMMS can be found at www.ucc.ie/hfrg/questionnaires/

13.3.5 Analyzing questionnaire data

Having collected a set of questionnaire responses, you need to know what to do with the data. The first step is to identify any trends or patterns. Using a spreadsheet like Excel to hold the data can help in this initial analysis. Often only simple statistics are needed such as the number or percentage of responses in a particular category. If the number of participants is small, under ten for example, giving actual numbers is more honest, but for larger numbers of responses percentages are useful for standardizing the data, particularly if you want to compare two or more sets of responses. Bar charts can also be used to display data graphically. More advanced statistical techniques such as cluster analysis can also be used to show whether there is a relationship between question responses.

13.4 Asking experts: inspections

Sometimes users are not easily accessible or involving them is too expensive or takes too long. In such circumstances, experts or combinations of experts and users can provide feedback. Various inspection techniques began to be developed as alternatives to usability testing in the early 1990s. These included various kinds of expert evaluations or *reviews*, such as heuristic evaluations and walkthroughs, in which experts inspect the human-computer interface and predict problems users would have when interacting with it. Typically these techniques are relatively inexpensive and easy to learn as well as being effective, which makes them appealing. They are similar to some software engineering practices where code and other types of inspections have been conducted for years. In addition, they can be used at any stage of a design project, including early design before well-developed prototypes are available.

13.4.1   Heuristic evaluation

Heuristic evaluation is an informal usability inspection technique developed by Jakob Nielsen and his colleagues (Nielsen, 1994a) in which experts, guided by a set of usability principles known as *heuristics*, evaluate whether user-interface elements, such as dialog boxes, menus, navigation structure, online help, etc., conform to the principles. These heuristics closely resemble the high-level design principles and guidelines discussed in Chapters 1 and 8, e.g., making designs consistent, reducing memory load, and using terms that users understand. When used in evaluation, they are called heuristics. The original set of heuristics was derived empirically from an analysis of 249 usability problems (Nielsen, 1994b). We list the latest here (also in Chapter 1), this time expanding them to include some of the questions addressed when doing evaluation:

- *Visibility of system status*
  Are users kept informed about what is going on?
  Is appropriate feedback provided within reasonable time about a user's action?
- *Match between system and the real world*
  Is the language used at the interface simple?
  Are the words, phrases and concepts used familiar to the user?
- *User control and freedom*
  Are there ways of allowing users to easily escape from places they unexpectedly find themselves in?
- *Consistency and standards*
  Are the ways of performing similar actions consistent?
- *Help users recognize, diagnose, and recover from errors*
  Are error messages helpful?
  Do they use plain language to describe the nature of the problem and suggest a way of solving it?
- *Error prevention*
  Is it easy to make errors?
  If so where and why?
- *Recognition rather than recall*
  Are objects, actions and options always visible?

- *Flexibility and efficiency of use*
  Have accelerators (i.e., shortcuts) been provided that allow more experienced users to carry out tasks more quickly?
- *Aesthetic and minimalist design*
  Is any unnecessary and irrelevant information provided?
- *Help and documentation*
  Is help information provided that can be easily searched and easily followed?

However, some of these core heuristics are too general for evaluating new products coming onto the market and there is a strong need for heuristics that are more closely tailored to specific products. For example, Nielsen (1999) suggests that the following heuristics are more useful for evaluating commercial websites, and makes them memorable by introducing the acronym H O M E R U N:

- <u>H</u>igh-quality content
- <u>O</u>ften updated
- <u>M</u>inimal download time
- <u>E</u>ase of use
- <u>R</u>elevant to users' needs
- <u>U</u>nique to the online medium
- <u>N</u>etcentric corporate culture

Different sets of heuristics for evaluating toys, WAP devices, online communities, wearable computers, and other devices are needed, so evaluators must develop their own by tailoring Nielsen's heuristics and by referring to design guidelines, market research, and requirements documents. Exactly which heuristics are the best and how many are needed are debatable and depend on the product.

Using a set of heuristics, expert evaluators work with the product role-playing typical users and noting the problems they encounter. Although other numbers of experts can be used, empirical evidence suggests that five evaluators usually identify around 75% of the total usability problems, as shown in Figure 13.5 (Nielsen,

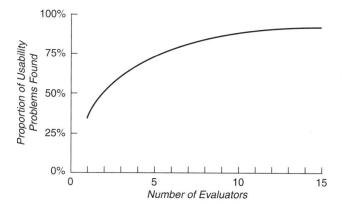

Figure 13.5 Curve showing the proportion of usability problems in an interface found by heuristic evaluation using various numbers of evaluators. The curve represents the average of six case studies of heuristic evaluation.

The 5th Wave　By Rich Tennant

"Just how accurately should my Web site reflect my place of business?"

1994a). However, skillful experts can capture many of the usability problems by themselves, and many consultants now use this technique as the basis for critiquing interactive devices—a process that has become know as an *expert crit* in some countries. Because users and special facilities are not needed for heuristic evaluation and it is comparatively inexpensive and quick, it is also known as *discount evaluation*.

13.4.2　Doing heuristic evaluation

Heuristic evaluation is one of the most straightforward evaluation methods. The evaluation has three stages:

1. The *briefing session* in which the experts are told what to do. A prepared script is useful as a guide and to ensure each person receives the same briefing.

2. The *evaluation period* in which each expert typically spends 1–2 hours *independently* inspecting the product, using the heuristics for guidance. The experts need to take *at least two* passes through the interface. The *first pass* gives a feel for the flow of the interaction and the product's scope. The *second pass* allows the evaluator to focus on specific interface ele-

ments in the context of the whole product, and to identify potential usability problems.

If the evaluation is for a functioning product, the evaluators need to have some specific user tasks in mind so that exploration is focused. Suggesting tasks may be helpful but many experts do this automatically. However, this approach is less easy if the evaluation is done early in design when there are only screen mockups or a specification; the approach needs to be adapted to the evaluation circumstances. While working through the interface, specification or mockups, a second person may record the problems identified, or the evaluator may think aloud. Alternatively, she may take notes herself. Experts should be encouraged to be as specific as possible and to record each problem clearly.

3. The *debriefing session* in which the experts come together to discuss their findings and to prioritize the problems they found and suggest solutions.

The heuristics focus the experts' attention on particular issues, so selecting appropriate heuristics is therefore critically important. Even so, there is sometimes less agreement among experts than is desirable, as discussed in the dilemma below.

There are fewer practical and ethical issues in heuristic evaluation than for other techniques because users are not involved. A week is often cited as the time needed to train experts to be evaluators (Nielsen and Mack, 1994), but this of course depends on the person's expertise. The best experts will have expertise in both interaction design and the product domain. Typical users can be taught to do

---

DILEMMA Problems or False Alarms?

You might think that heuristic evaluation is a panacea for designers, and that it can reveal all that is wrong with a design. However, it has problems. Several independent studies compare heuristic evaluation with other techniques, particularly user testing, indicating that the different approaches often identify *different* problems and that sometimes heuristic evaluation misses severe problems (Karat, 1994). This argues for using complementary techniques. Furthermore, heuristic evaluation should not be thought of as a replacement for user testing.

Another problem that Bill Bailey (2001) warns about is of experts reporting problems that don't exist. In other words, some of the experts' predictions are wrong. Bailey cites analyses from three published sources showing that about 33% of the problems reported were real usability problems, some of which were serious, others trivial. However, the heuristic evaluators missed about 21% of users' problems. Furthermore, about 43% of the

problems identified by the experts were *not* problems at all; they were false alarms! Bailey points out that if we do the arithmetic and round up the numbers, what this comes down to is that only about *half* the problems identified are true problems. "More specifically, for every true usability problem identified, there will be a little over one false alarm (1.2) and about one half of one missed problem (0.6). If this analysis is true, heuristic evaluators tend to identify more false alarms and miss more problems than they have true hits."

How can the number of false alarms or missed serious problems be reduced? Checking that experts really have the expertise that they claim would help but how can you do this? One way to over come biases is to have several evaluators. This helps to reduce the impact of one person's bias of poor performance. Using heuristic evaluation along with user testing and other techniques is also a good idea.

heuristic evaluation, although there have been claims that it is not very successful (Nielsen, 1994a). However, some closely related methods take a team approach that involves users (Bias, 1994).

13.4.3 Heuristic evaluation of websites

In this section we examine heuristics for evaluating websites. We begin by discussing MEDLINEplus, a medical information website created by the National Library of Medicine (NLM) to provide health information for patients, doctors, and researchers (Cogdill, 1999). The home page and two other screens are shown in Figures 13.6–13.8.

In 1999 usability consultant Keith Cogdill was commissioned by NLM to evaluate MEDLINEplus. Using a combination of his own knowledge of the users' tasks, problems that had already been reported by users, and advice from documented sources (Shneiderman, 1998a; Nielsen, 1993; Dumas and Redish, 1999), Cogdill identified the seven heuristics listed below. Some of the heuristics resemble Nielsen's original set, but have been tailored for evaluating MEDLINEplus.

- *Internal consistency.*
  The user should not have to speculate about whether different phrases or actions carry the same meaning.

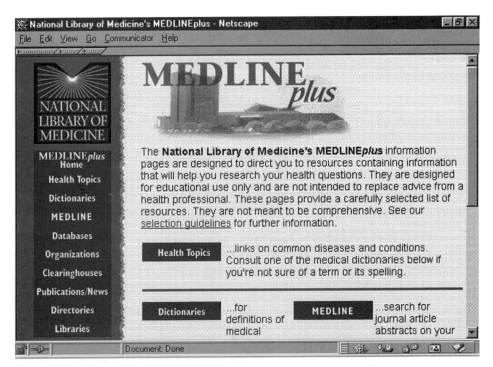

Figure 13.6 Home page of MEDLINEplus.

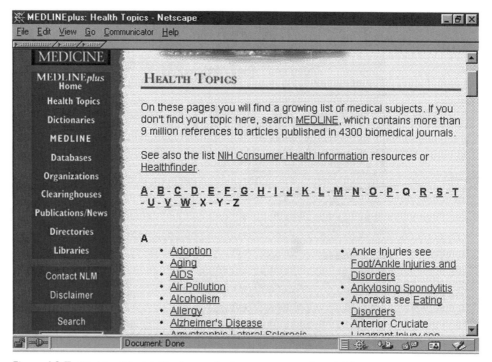

Figure 13.7 Clicking Health Topics on the home page produced this page.

- *Simple dialog.*
  The dialog with the user should not include information that is irrelevant, unnecessary, or rarely needed. The dialog should be presented in terms familiar to the user and not be system-oriented.

- *Shortcuts.*
  The interface should accommodate both novice and experienced users.

- *Minimizing the user's memory load.*
  The interface should not require the user to remember information from one part of the dialog to another.

- *Preventing errors.*
  The interface should prevent errors from occurring.

- *Feedback.*
  The system should keep the user informed about what is taking place.

- *Internal locus of control.*
  Users who choose system functions by mistake should have an "emergency exit" that lets them leave the unwanted state without having to engage in an extended dialog with the system.

These heuristics were given to three expert evaluators who independently evaluated MEDLINEplus. Their comments were then compiled and a meeting was

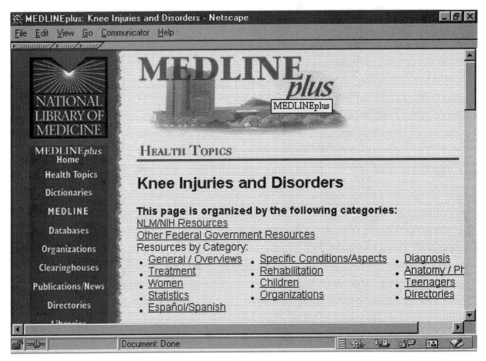

Figure 13.8 Categories of links within Health Topics for knee injuries.

called to discuss their findings and suggest strategies for addressing problems. The following points were among their findings:

- *Layout.*
  All pages within MEDLINEplus have a relatively uncomplicated vertical design. The home page is particularly compact, and all pages are well suited for printing. The use of graphics is conservative, minimizing the time needed to download pages.

- *Internal consistency.*
  The formatting of pages and presentation of the logo are consistent across the website. Justification of text, fonts, font sizes, font colors, use of terms, and links labels are also consistent.

The experts also suggested improvements, including:

- *Arrangement of health topics.*
  Topics should be arranged alphabetically as well as in categories. For example, health topics related to cardiovascular conditions could appear together.

- *Depth of navigation menu.*
  Having a higher "fan-out" in the navigation menu in the left margin would enhance usability. By this they mean that more topics should be listed on the

surface, giving many short menus rather than a few deep ones (see the experiment on breadth versus depth in Chapter 14 which provides evidence to justify this.)

Turning design guidelines into heuristics for the web

The following list of guidelines for evaluating websites was compiled from several sources and grouped into three categories: *navigation*, *access*, and *information design* (Preece, 2000). These guidelines provide a basis for developing heuristics by converting them into questions.

Navigation   One of the biggest problems for users of large websites is navigating around the site. The phrase "lost in cyberspace" is understood by every web user. The following six guidelines (from Nielsen (1998) and others) are intended to encourage good navigation design:

- *Avoid orphan pages i.e. pages that are not connected to the home page, because they lead users into dead ends.*
  Are there any orphan pages? Where do they go to?

- *Avoid long pages with excessive white space that force scrolling.*
  Are there any long pages? Do they have lots of white space or are they full of texts or lists?

- *Provide navigation support, such as a strong site map that is always present* (Shneiderman, 1998b).
  Is there any guidance, e.g. maps, navigation bar, menus, to help users find their way around the site?

- *Avoid narrow, deep, hierarchical menus that force users to burrow deep into the menu structure.*
  Empirical evidence indicates that broad shallow menus have better usability than a few deep menus (Larson and Czerwinski, 1998; Shneiderman, 1998b).

- *Avoid non-standard link colors.*
  What color is used for links? Is it blue or another color? If it is another color, then is it obvious to the user that it is a hyperlink?

- *Provide consistent look and feel for navigation and information design.*
  Are menus used, named, and positioned consistently? Are links used consistently?

Access   Accessing many websites can be a problem for people with slow Internet connections and limited processing power. In addition, browsers are often not sensitive to errors in URLs. Nielsen (1998) suggests the following guidelines:

- *Avoid complex URLs.*
  Are the URLs complex? Is it easy to make typing mistakes when entering them?

- *Avoid long download times that annoy users.*
Are there pages with lots of graphics? How long does it take to download each page?

Information design   Information design (i.e., content comprehension and aesthetics) contributes to users' understanding and impressions of the site as you can see in Activity 13.6.

ACTIVITY 13.6   Consider the following design guidelines for information design and for each one suggest a question that could be used in heuristic evaluation:

- *Outdated or incomplete information is to be avoided* (Nielsen, 1998). It creates a poor impression with users.
- *Good graphical design is important.* Reading long sentences, paragraphs, and documents is difficult on screen, so break material into discrete, meaningful chunks to give the website structure (Lynch and Horton, 1999).
- *Avoid excessive use of color.* Color is useful for indicating different kinds of information, i.e., cueing (Preece et al., 1994).
- *Avoid gratuitous use of graphics and animation.* In addition to increasing download time, graphics and animation soon become boring and annoying (Lynch and Horton, 1999).
- *Be consistent.* Consistency both within pages (e.g., use of fonts, numbering, terminology, etc.) and within the site (e.g., navigation, menu names, etc.) is important for usability and for aesthetically pleasing designs.

Comment   We suggest the following questions; you may have identified others:

- *Outdated or incomplete information.*
Do the pages have dates on them? How many pages are old and provide outdated information?
- *Good graphical design is important.*
Is the page layout structured meaningfully? Is there too much text on each page?
- *Avoid excessive use of color.*
How is color used? Is it used as a form of coding? Is it used to make the site bright and cheerful? Is it excessive and garish?
- *Avoid gratuitous use of graphics and animation.*
Are there any flashing banners? Are there complex introduction sequences? Can they be short-circuited? Do the graphics add to the site?
- *Be Consistent.*
Are the same buttons, fonts, numbers, menu styles, etc. used across the site? Are they used in the same way?

ACTIVITY 13.7   Look at the heuristics above and consider how you would use them to evaluate a website for purchasing clothes (e.g., REI.com, which has a home page similar to that in Figure 13.9).

Figure 13.9 The home page is similar to that of REI.com.

While you are doing this activity think about whether the grouping into three categories is useful.

(a) Does it help you focus on what is being evaluated?

(b) Might fewer heuristics be better? Which might be combined and what are the trade-offs?

Comment

(a) Informal evaluation in which the heuristics were categorized suggests that the three categories help evaluators to focus. However, 13 heuristics is still a lot.

(b) Some heuristics can be combined and given a more general description. For example, *providing navigation support* and *avoiding narrow, deep, hierarchical menus* could be replaced with "help users develop a good mental model," but this is a more abstract statement and some evaluators might not know what is packed into it. Producing questions suitable for heuristic evaluation often results in more of them, so there is a trade-off. An argument for keeping the detail is that it reminds evaluators of the issues to consider. At present, since the web is relatively new, we can argue that such reminders are needed. Perhaps in five years they will not be.

Heuristics for online communities

As we have already mentioned, different combinations and types of heuristics are needed to evaluate different types of applications and interactive products. Another kind of web application to which heuristics must be tailored is online communities. Here, a key concern is how to evaluate not merely usability but also how well social interaction (i.e., sociability) is supported. This topic has received less attention than the web but the following nine sets of example questions can be used as a starting point for developing heuristics to evaluate online communities (Preece, 2000):

- Sociability: Why should I join this community? (What are the benefits for me? Does the description of the group, its name, its location in the website, the graphics, etc., tell me about the purpose of the group?)
- Usability: How do I join (or leave) the community? (What do I do? Do I have to register or can I just post, and is this a good thing?)
- Sociability: What are the rules? (Is there anything I shouldn't do? Are the expectations for communal behavior made clear? Is there someone who checks that people are behaving reasonably?)
- Usability: How do I get, read and send messages? (Is there support for new-comers? Is it clear what I should do? Are templates provided? Can I send private messages?)
- Usability: Can I do what I want to do easily? (Can I navigate the site? Do I feel comfortable interacting with the software? Can I find the information and people I want?)
- Sociability: Is the community safe? (Are my comments treated with respect? Is my personal information secure? Do people make aggressive or unaccept-able remarks to each other?)
- Sociability: Can I express myself as I wish? (Is there a way of expressing emotions, such as using emoticons? Can I show people what I look like or re-veal aspects of my character? Can I see others? Can I determine who else is present—perhaps people are looking on but not sending messages?)
- Sociability: Do people reciprocate? (If I contribute will others contribute comments, support and answer my questions?)
- Sociability: Why should I come back? (What makes the experience worth-while? What's in it for me? Do I feel part of a thriving community? Are there interesting people with whom to communicate? Are there interesting events?)

ACTIVITY 13.8   Go to the communities in REI.com or to another site that has bulletin boards to which cus-tomers can send comments. Social interaction was discussed in Chapter 4, and this exercise involves picking up some of the concepts discussed there and developing heuristics to evalu-ate online communities. Before starting you will find it useful to familiarize yourself by car-rying out the following:

- read some of the messages
- send a message
- reply to a message
- search for information
- notice how many messages have been sent and how recently

- notice whether you can see the physical relationship between messages easily
- notice whether you can post to people privately using email
- notice whether you can gain a sense of what the other people are like and the emotional content of their messages
- notice whether there is a sense of community and of individuals being present, etc.

Then use the nine questions above as heuristics to evaluate the site:

(a) How well do the questions work as heuristics for evaluating the online community for both usability and sociability issues?

(b) Could these questions form the basis for heuristics for other online communities such as HutchWorld discussed in Chapter 10?

Comment

(a) You probably found that these questions helped focus your attention on the main issues of concern. You may also have noticed that some communities are more like ghost towns than communities; they get very few visitors. Unlike the website evaluation it is therefore important to pay attention to social interaction. A community without people is not a community no matter how good the software is that supports it.

(b) HutchWorld is designed to support social interaction and offers many additional features such as support for social presence by allowing participants to represent themselves as avatars, show pictures of themselves, tell stories, etc. The nine questions above are useful but may need adapting.

13.4.4   Heuristics for other devices

The examples in the previous activities start to show how heuristics can be tailored for specific applications. However, some products are even more different than those from the desktop world of the early 1990s that gave rise to Nielsen's original heuristics. For example, computerized toys are being developed that motivate, entice and challenge, in innovative ways. Handheld devices sell partly on size, color and other aesthetic qualities—features that can have a big impact on the user experience but are not covered by traditional heuristics. Little research has been done on developing heuristics for these products, but Activity 13.9 will start you thinking about them.

ACTIVITY 13.9   Allison Druin works with children to develop web applications and computerized toys (Druin, 1999). From doing this work Allison and her team know that children like to:

- be in control and not to be controlled
- create things
- express themselves
- be social
- collaborate with other children

(a) What kind of tasks should be considered in evaluating a fluffy robot toy dog that can be programmed to move and to tell personalized stories about itself and children? The target age group for the toy is 7–9 years.

(b) Suggest heuristics to evaluate the toy.

Comment   (a) Tasks that you could consider: making the toy tell a story about the owner and two friends, making the toy move across the room, turn, and speak. You probably thought of others.

(b) The heuristics could be written to cover: being in control, being flexible, supporting expression, being motivating, supporting collaboration and being engaging. These are based on the issues raised by Druin, but the last one is aesthetic and tactile. Several of the heuristics needed would be more concerned with user experience (e.g., motivating, engaging, etc.) than with usability.

13.5 Asking experts: walkthroughs

Walkthroughs are an alternative approach to heuristic evaluation for predicting users' problems without doing user testing. As the name suggests, they involve walking through a task with the system and noting problematic usability features. Most walkthrough techniques do not involve users. Others, such as pluralistic walkthroughs, involve a team that includes users, developers, and usability specialists.

In this section we consider cognitive and pluralistic walkthroughs. Both were originally developed for desktop systems but can be applied to web-based systems, handheld devices, and products such as VCRs.

13.5.1 Cognitive walkthroughs

"Cognitive walkthroughs involve simulating a user's problem-solving process at each step in the human-computer dialog, checking to see if the user's goals and memory for actions can be assumed to lead to the next correct action." (Nielsen and Mack, 1994, p. 6). The defining feature is that they focus on evaluating designs for ease of learning—a focus that is motivated by observations that users learn by exploration (Wharton et al., 1994). The steps involved in cognitive walkthroughs are:

1. The characteristics of typical users are identified and documented and sample tasks are developed that focus on the aspects of the design to be evaluated. A description or prototype of the interface to be developed is also produced, along with a clear sequence of the actions needed for the users to complete the task.

2. A designer and one or more expert evaluators then come together to do the analysis.

3. The evaluators walk through the action sequences for each task, placing it within the context of a typical scenario, and as they do this they try to answer the following questions:
   - Will the correct action be sufficiently evident to the user? (Will the user know what to do to achieve the task?)
   - Will the user notice that the correct action is available? (Can users see the button or menu item that they should use for the next action? Is it apparent when it is needed?)

- Will the user associate and interpret the response from the action correctly? (Will users know from the feedback that they have made a correct or incorrect choice of action?)

In other words: will users know what to do, see how to do it, and understand from feedback whether the action was correct or not?

4. As the walkthrough is being done, a record of critical information is compiled in which:
   - The assumptions about what would cause problems and why are recorded. This involves explaining why users would face difficulties.
   - Notes about side issues and design changes are made.
   - A summary of the results is compiled.

5. The design is then revised to fix the problems presented.

It is important to document the cognitive walkthrough, keeping account of what works and what doesn't. A standardized feedback form can be used in which answers are recorded to the three bulleted questions in step (3) above. The form can also record the details outlined in points 1–4 as well as the date of the evaluation. Negative answers to any of the questions are carefully documented on a separate form, along with details of the system, its version number, the date of the evaluation, and the evaluators' names. It is also useful to document the severity of the problems, for example, how likely a problem is to occur and how serious it will be for users.

The strengths of this technique are that it focuses on users' problems in detail, yet users do not need to be present, nor is a working prototype necessary. However, it is very time-consuming and laborious to do. Furthermore the technique has a narrow focus that can be useful for certain types of system but not others.

Example: Find a book at Amazon.com

This example shows a cognitive walkthrough of buying this book at Amazon.com.

> *Task: to buy a copy of this book from Amazon.com*
> *Typical users: students who use the web regularly*

The steps to complete the task are given below. Note that the interface for Amazon.com may have changed since we did our evaluation.

Step 1. Selecting the correct category of goods on the home page

Q. Will users know what to do?

Answer: Yes—they know that they must find "books."

Q. Will users see how to do it?

Answer: Yes—they have seen menus before and will know to select the appropriate item and click go.

Q. Will users understand from feedback whether the action was correct or not?

Answer: Yes—their action takes them to a form that they need to complete to search for the book.

Step 2. Completing the form

Q. Will users know what to do?

Answer: Yes—the online form is like a paper form so they know they have to complete it.

Answer: No—they may not realize that the form has defaults to prevent inappropriate answers because this is different from a paper form.

Q. Will users see how to do it?

Answer: Yes—it is clear where the information goes and there is a button to tell the system to search for the book.

Q. Will users understand from feedback whether the action was correct or not?

Answer: Yes—they are taken to a picture of the book, a description, and purchase details.

ACTIVITY 13.10   Activity 13.7 was about doing a heuristic evaluation of REI.com or a similar e-commerce retail site. Now go back to that site and do a cognitive walkthrough to buy something, say a pair of skis. When you have completed the evaluation, compare your findings from the cognitive walkthrough technique with those from heuristic evaluation.

Comment   You probably found that the cognitive walkthrough took longer than the heuristic evaluation for evaluating the same part of the site because it examines each step of a task. Consequently, you probably did not see as much of the website. It's likely that you also got much more detailed findings from the cognitive walkthrough. Cognitive walkthrough is a useful technique for examining a small part of a system in detail, whereas heuristic evaluation is useful for examining whole or parts of systems.

Variation of the cognitive walkthrough

A useful variation on this theme is provided by Rick Spencer of Microsoft, who adapted the cognitive walkthrough technique to make it more effective with a team who were developing an interactive development environment (IDE) (Spencer, 2000). When used in its original state, there were two major problems. First, answering the three questions in step (3) and discussing the answers took too long. Second, designers tended to be defensive, often invoking long explanations of cognitive theory to justify their designs. This second problem was particularly difficult because it undermined the efficacy of the technique and the social relationships of team members. In order to cope with these problems Rick Spencer adapted the technique by reducing the number of questions and curtailing discussion. This meant that the analysis was more coarse-grained but could be completed in much less time (about 2.5 hours). He also identified a leader, the usability specialist, and set strong ground rules for the session, including a ban on defending a design, debating cognitive theory, or doing designs on the fly.

These adaptations made the technique more usable, despite losing some of the detail from the analysis. Perhaps most important of all, he directed the social interactions of the design team so that they achieved their goal.

13.5.2 Pluralistic walkthroughs

"Pluralistic walkthroughs are another type of walkthrough in which users, developers and usability experts work together to step through a [task] scenario, discussing usability issues associated with dialog elements involved in the scenario steps" (Nielsen and Mack, 1994, p. 5). Each group of experts is asked to assume the role of typical users. The walkthroughs are then done by following a sequence of steps (Bias, 1994):

1. Scenarios are developed in the form of a series of hard-copy screens representing a single path through the interface. Often just two or a few screens are developed.

2. The scenarios are presented to the panel of evaluators and the panelists are asked to write down the sequence of actions they would take to move from one screen to another. They do this individually without conferring with one another.

3. When everyone has written down their actions, the panelists discuss the actions that they suggested for that round of the review. Usually, the representative users go first so that they are not influenced by the other panel members and are not deterred from speaking. Then the usability experts present their findings, and finally the developers offer their comments.

4. Then the panel moves on to the next round of screens. This process continues until all the scenarios have been evaluated.

The benefits of pluralistic walkthroughs include a strong focus on users' tasks. Performance data is produced and many designers like the apparent clarity of working with quantitative data. The approach also lends itself well to participatory design practices by involving a multidisciplinary team in which users play a key role. Limitations include having to get all the experts together at once and then proceed at the rate of the slowest. Furthermore, only a limited number of scenarios, and hence paths through the interface, can usually be explored because of time constraints.

Assignment

This assignment continues the work you did on the web-based ticketing system at the end of Chapters 7 and 8. The aim of this assignment is to evaluate the prototypes produced in the assignment of Chapter 8. The assignment takes an iterative form in which we ask you to evaluate and redesign your prototypes, following the iterative path in the interaction design process described in Chapter 6.

(a) For each prototype, return to the feedback you collected in Chapter 8 but this time perform open-ended interviews with a couple of potential users.

(b) Based on the feedback from this first evaluation, redesign the software/HTML prototype to take comments on all three prototypes into account.

(c) Decide on an appropriate set of heuristics and perform a heuristic evaluation of the redesigned prototype.

(d) Based on this evaluation, redesign the prototype to overcome the problems you encountered.

(e) Design a questionnaire to evaluate the system. The questionnaire may be paper-based or electronic. If it is electronic, make your software prototype and the questionnaire available to others and ask a selection of people to evaluate the system.

Summary

Techniques for asking users for their opinions vary from being unstructured and open-ended to tightly structured. The former enable exploration of concepts, while the latter provide structured information and can be replicated with large numbers of users, as in surveys. Predictive evaluation is done by experts who inspect the designs and offer their opinions. The value of these techniques is that they structure the evaluation process, which can in turn help to prevent problems from being overlooked. In practice, interviews and observations often go hand in hand, as part of a design process.

Key points

- There are three styles of interviews: structured, semi-structured and unstructured.
- Interview questions can be open or closed. Closed questions require the interviewee to select from a limited range of options. Open questions accept a free-range response.
- Many interviews are semi-structured. The evaluator has a predetermined agenda but will probe and follow interesting, relevant directions suggested by the interviewee. A few structured questions may also be included, for example to collect demographic information.
- Structured and semi-structured interviews are designed to be replicated.
- Focus groups are a form of group interview.
- Questionnaires are a comparatively low-cost, quick way of reaching large numbers of people.
- Various rating scales exist including selection boxes, Likert, and semantic scales.
- Inspections can be used for evaluating requirements, mockups, functional prototypes, or systems.
- Five experts typically find around 75% of the usability problems.
- Compared to user testing, heuristic evaluation is less expensive and more flexible.
- User testing and heuristic evaluation often reveal different usability problems.
- Other types of inspections include pluralistic and cognitive walkthroughs.
- Walkthroughs are very focused and so are suitable for evaluating small parts of systems.

Further reading

NIELSEN, J., AND MACK, R. L. (eds.) (1994) *Usability Inspection Methods*. New York: John Wiley & Sons. This book contains an edited collection of chapters on a variety of usability inspection methods. There is a detailed description of heuristic evaluation and walkthroughs and comparisons of these techniques with other evaluation techniques, particularly user testing. Jakob Nielsen's website *useit.com* provides additional information and advice on website design.

OPPENHEIM, A. N. (1992) *Questionnaire Design, Interviewing and Attitude Measurement*. London: Pinter Publishers. This text is useful for reference. It provides a detailed account of all aspects of questionnaire design, illustrated with many examples.

PREECE, J. (2000) *Online Communities: Designing Usability, Supporting Sociability*. Chichester, UK: John Wiley & Sons. This book is about the design of web-based online communities. It suggests guidelines for evaluating for sociability and usability that can be used as a basis for heuristics.

ROBSON, C. (1993) *Real World Research*. Blackwell. Oxford, UK. Chapter 9 provides basic practical guidance on how to interview and design questionnaires. It also contains many examples.

SHNEIDERMAN, B. (1998) *Designing the User Interface: Strategies for Effective Human-Computer Interaction (3rd Edition)* Reading, MA.: Addison-Wesley. Chapter 4 contains a discussion of the QUIS questionnaire.

INTERVIEW   with Jakob Nielsen

Jakob Nielsen is a pioneer of heuristic evaluation. He is currently principal of the Nielsen Norman Consultancy Group and the author of numerous articles and books, including his recent book, *Designing Web Usability* (New Riders Publishing). He is well-known for his regular sound bites on usability which for many years have appeared at useit.com. In this interview Jakob talks about heuristic evaluation, why he developed the technique, and how it can be applied to the web.

JP: Jakob, why did you create heuristic evaluation?

JN: It is part of a larger mission I was on in the mid-'80s, which was to simplify usability engineering, to get more people using what I call "discount usability engineering." The idea was to come up with several simplified methods that would be very easy and fast to use. Heuristic evaluation can be used for any design project or any stage in the design process, without budgetary constraints. To succeed it had to be fast, cheap, and useful.

JP: How can it be adapted for the web?

JN: I think it applies just as much to the web, actually if anything more, because a typical website will have tens of thousands of pages. A big one may have hundreds of thousands of pages, much too much to be assessed using traditional usability evaluation methods such as user testing. User testing is good for testing the home page or the main navigation system. But if you look at the individual pages, there is no way that you can really test them. Even with the discount approach, which would involve five users, it would still be hard to test all the pages. So all you are left with is the notion of doing a heuristic evaluation, where you just have a few people look at the majority of pages and judge them according to the heuristics. Now the heuristics are somewhat different, because people behave differently on the web. They are more ruthless

about getting a very quick glance at what is on a page and if they don't understand it then leaving it. Typically application users work a little harder at learning an application. The basic heuristics that I developed a long time ago are universal, so they apply to the web as well. But as well as these global heuristics that are always true, for example "consistency," there can be specialized heuristics that apply to particular systems. But most evaluators use the general heuristics because the web is still evolving and we are still in the process of determining what the web-specific heuristics should be.

JP: So how do you advise designers to go about evaluating a really large website?

JN: Well, you cannot actually test every page. Also, there is another problem: developing a large website is incredibly collaborative and involves a lot of different people. There may be a central team in charge of things like the home page, the overall appearance, and the overall navigation system. But when it comes to making a product page, it is the product-marketing manager of, say, Kentucky who is in charge of that. The division in Kentucky knows about the product line and the people back at headquarters have no clue about the details. That's why they have to do their own evaluations in that department. The big thing right now is that this is not being done, developers are not evaluating enough. That's one of the reasons I want to push the heuristic evaluation method even further to get it out to all the website contributors. The uptake of usability methods has dramatically improved from five years ago, when many companies didn't have a clue, but the need today is still great because of the phenomenal development of the web.

JP: When should you start doing heuristic evaluation?

JN: You should start quite early, maybe not quite as early as testing a very rough mockup, but as soon as there is a slightly more substantial prototype. For example, if you are building a website that might eventually have ten thousand pages, it would be appropriate to do a heuristic evaluation of, say, the first ten to twenty pages. By doing this you would catch quite a lot of usability problems.

JP: How do you combine user testing and heuristic evaluation?

JN: I suggest a sandwich model where you layer them on top of each other. Do some early user testing of two or three drawings. Develop the ideas somewhat, then do a heuristic evaluation. Then evolve the design further, do some user tests, evolve it and do heuristic evaluation, and so on. When the design is nearing completion, heuristic evaluation is very useful particularly for a very large design.

JP: So, do you have a story to tell us about your consulting experiences, something that opened your eyes or amused you?

JN: Well, my most interesting project started when I received an email from a co-founder of a large company who wanted my opinion on a new idea. We met and he explained his idea and because I know a lot about usability, including research studies, I could warn him that it wouldn't work—it was doomed. This was very satisfying and seems like the true role for a usability consultant. I think usability consultants should have this level of insight. It is not enough to just clean up after somebody makes the mistake of starting the wrong project or produces a poor design. We really should help define which projects should be done in the first place. Our role is to help identify options for really improving people's lives, for developing products that are considerably more efficient, easier or faster to learn, or whatever the criteria are. That is the ultimate goal of our entire field.

JP: One last question—how do you think the web will develop? What will we see next, what do you expect the future to bring?

JN: I hope we will abandon the page metaphor and reach back to the earlier days of hypertext. There are other ideas that would help people navigate the web better. The web is really an "article-reading" interface. My website useit.com, for example, is mainly articles, but for many other things people need a different interface, the current interface just does not work. I hope we will evolve a more interesting, useful interface that I'll call the "Internet desktop," which would have a control panel for your own environment, or another metaphor would be "your personal secretary." Instead of the old goal where the computer spits out more information, the goal would be for the computer to protect you from too much information. You shouldn't have to actually go and read all those webpages. You should have something that would help you prioritize your time so you would get the most out of the web. But, pragmatically speaking, these are not going to come any time soon. My prediction has been that Explorer Version 8 will be the first good web browser and that is still my prediction, but there are still a few versions to come before we reach that level. The more short-term prediction is really that designers will take much more responsibility for content and usability of the web. We need to write webpages so that people can read them. For instance, we need headlines that make sense. Even something as simple as a headline is a user interface, because it's now being used interactively, not as in a magazine where you just look at it. So writing the headline, writing the content, designing the navigation are jobs for the individual website designers. In combination, such decisions are really defining the user experience of the network economy. That's why we really have an obligation, every one of us, because we are building the new world and if the new world turns out to be miserable, we have only ourselves to blame, not Bill Gates. We've got to design the web for the way users behave.

Chapter 14

Testing and modeling users

14.1 Introduction
14.2 User testing
    14.2.1 Testing MEDLINEplus
14.3 Doing user testing
    14.3.1 Determine the goals and explore the questions
    14.3.2 Choose the paradigm and techniques
    14.3.3 Identify the practical issues: Design typical tasks
    14.3.4 Identify the practical issues: Select typical users
    14.3.5 Identify the practical issues: Prepare the testing conditions
    14.3.6 Identify the practical issues: Plan how to run the tests
    14.3.7 Deal with ethical issues
    14.3.8 Evaluate, analyze and present the data
14.4 Experiments
    14.4.1 Variables and conditions
    14.4.2 Allocation of participants to conditions
    14.4.3 Other practical issues
    14.4.4 Data collection and analysis
14.5 Predictive models
    14.5.1 The GOMS model
    14.5.2 The Keystroke level model
    14.5.3 Benefits and limitations of GOMS
    14.5.4 Fitts' Law

14.1 Introduction

A central aspect of interaction design is user testing. User testing involves measuring the performance of typical users doing typical tasks in controlled laboratory-like conditions. Its goal is to obtain objective performance data to show how usable a system or product is in terms of usability goals, such as ease of use or learnability. More generally, usability testing relies on a combination of techniques including observation, questionnaires and interviews as well as user testing, but user testing is of central concern, and in this chapter we focus upon it. We also examine key issues in experimental design because user testing has developed from experimental practice, and although there are important differences between them there is also commonality.

The last part of the chapter considers how user behavior can be modeled to predict usability. Here we examine two modeling approaches (based on psychological theory) that have been used to predict user performance. Both come from the well-known GOMS family of approaches: the GOMS model and the Keystroke level model. We also discuss Fitts' Law.

The main aims of this chapter are to:

- Explain how to do user testing.
- Discuss how and why a user test differs from an experiment.
- Discuss the contribution of user testing to usability testing.
- Discuss how to design simple experiments.
- Describe the GOMS model, the Keystroke level model and Fitts' law and discuss when these techniques are useful.
- Explain how to do a simple keystroke level analysis.

14.2   User testing

User testing is an applied form of experimentation used by developers to test whether the product they develop is usable by the intended user population to achieve their tasks (Dumas and Redish, 1999). In user testing the time it takes typical users to complete clearly defined, typical tasks is measured and the number and type of errors they make are recorded. Often the routes that users take through tasks are also noted, particularly in web-searching tasks. Making sense of this data is helped by observational data, answers to user-satisfaction questionnaires and interviews, and key stroke logs, which is why these techniques are used along with user testing in usability studies.

The aim of an experiment is to answer a question or hypothesis to discover new knowledge. The simplest way that scientists do this is by investigating the relationship between two things, known as *variables*. This is done by changing one of them and observing what happens to the other. To eliminate any other influences that could distort the results of this manipulation, the scientist attempts to control the experimental environment as much as possible.

In the early days, experiments were the cornerstone of research and development in user-centered design. For example, the Xerox Star team did experiments to determine how many buttons to put on a mouse, as described in Box 14.1. Other early experimental research in HCI examined such things as how many items to put in a menu and how to design icons.

Because user testing has features in common with scientific experiments, it is sometimes confused with experiments done for research purposes. Both measure performance. However, user testing is a systematic approach to evaluating user performance in order to inform and improve usability design, whereas research aims to discover new knowledge.

Research requires that the experimental procedure be rigorous and carefully documented so that it can be replicated by other researchers. User testing should

BOX 14.1 The Origins of User Testing

Xerox's Star office workstation was a landmark in interaction design. It was based on several user-centered design principles that are now well accepted, but at the time were revolutionary. The following principles guided the Star's development (Bewley et al., 1990):

- There should be an explicit, consistent conceptual model that draws on objects and activities already familiar to the user—the origins of the now familiar desktop metaphor.

- Seeing and pointing are easier than recalling and typing—the origins of the mouse and GUI.

- Commands should be uniform across similar domains—the important principle of consistency.

- The screen should show the state of the object the user is working on—what you see is what you get (WYSIWYG, pronounced "whizee-wig").

Even with these principles, the design space was still enormous and many proposed designs turned out to be unsatisfactory. Various tools and techniques were tried to support its development, including the keystroke level model discussed later (Card et al., 1983), but one of the most important decisions was to experiment and test design ideas intensively—i.e., to design and evaluate iteratively.

These tests included controlled experiments in which the evaluators describe their methodology in the language of science. For example, they tested six mouse selection schemes "using a between-subject paradigm, in which each of six groups was assigned one of the six schemes" (Bewley et al., 1990, p. 371). In addition, they also did more informal tests as the questions to be settled became less well defined, "... experiments took on a flavor of 'fishing expeditions' to see what we came up with" (Bewley et al., 1990, p. 380).

The design effort required for the Star, without doubt a mammoth undertaking, took more than six years. The implementation involved from 20–45 programmers over 3.5 years producing over 250,000 lines of high-level code. Over 15 human factor tests were performed using over 200 users and lasting over 400 hours. Each provided invaluable information about design decisions that were being made.

Two other early pioneers in usability testing were John Bennett, from IBM in the US, who helped to define usability and Brian Shackel from HUSAT in the UK, who worked to operationalize Bennet's definition so that it could be tested and measured. This involved taking vague notions such as "easy to use" and specifying what was meant. All this work paved the way for the development of current user testing practices.

be carefully planned and executed, but real-world constraints must be taken into account and compromises made. It is rarely exactly replicable, though it should be possible to repeat the tests and obtain similar findings. Experiments are usually validated using statistical tests, whereas user testing rarely employs statistics other than means and standard deviations.

Typically 5–12 users are involved in user testing (Dumas and Redish, 1999), but often there are fewer and compromises are made to work within budget and schedule constraints. "Quick and dirty" tests involving just one or two users are frequently done to get quick feedback about a design idea. Research experiments generally involve more participants, more tightly controlled conditions, and more extensive data analysis in which statistical analysis is essential.

14.2.1   Testing MEDLINEplus

In Chapter 13 we described how heuristic evaluation was used to identify usability problems in the National Library of Medicine (NLM) MEDLINEplus website (Figure 14.1 Cogdill, 1999). We now return to that study and focus on how the user testing was done to evaluate changes made after heuristic evaluation. This case study exemplifies the kinds of issues to be considered in user testing, including developing tasks and test procedures, and approaches to data collection and analysis.

Goals and questions

The goal of the study was to identify usability problems in the revised interface. More specifically, the evaluators wanted to know if the revised way of categorizing information, suggested by the expert evaluators, worked. They also wanted to check that users could navigate the system to find the information they needed. Navigating around large websites can be a major usability problem, so it was important to check that the design of MEDLINEplus supported users' navigation strategies.

Selection of participants

MEDLINEplus was tested with nine participants selected from primary health care practices in the Washington, DC metropolitan area. This was accomplished by

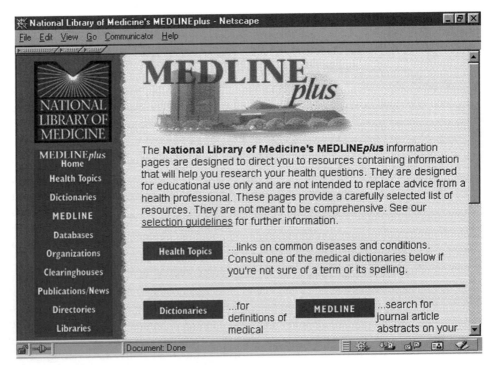

Figure 14.1 Home page of MEDLINEplus.

placing recruitment posters in the reception areas of two medical practices. People who wanted to participate were asked to complete a brief questionnaire, which asked about age, experience in using the web, and frequency of seeking health-related information. Dr. Cogdill, a usability specialist, then called all those who used the web more than twice a month. He explained that they would be involved in testing a product from the NLM, but did not mention MEDLINEplus so that potential testers would not review the site before doing the tests. Seven of the nine participants were women because balancing for gender was considered less important than web experience. It was important to find people in the Washington, DC region so that they could come to the test center and for the number of participants to fall within the range of 6–12 recommended by usability experts (Dumas and Redish, 1999).

Development of the tasks

The following five tasks were developed in collaboration with NLM staff to check the categorizing schemes suggested by the expert evaluators and navigation support. The topics chosen for the tasks were identified from questions most frequently asked by website users:

- Task 1: Find information about whether a dark bump on your shoulder might be skin cancer.
- Task 2: Find information about whether it's safe to use Prozac during pregnancy.
- Task 3: Find information about whether there is a vaccine for hepatitis C.
- Task 4: Find recommendations about the treatment of breast cancer, specifically the use of mastectomies.
- Task 5: Find information about the dangers associated with drinking alcohol during pregnancy.

The efficacy of each task was reviewed by colleagues and pilot tested.

The test procedure

The procedure involved five scripts that were prepared in advance and were used for each participant to ensure that all participants were given the same information and were treated in the same way. We present these scripts in figures to distinguish them from our own text. They are included here in their original form.

Testing was done in laboratory-like conditions. When the participants arrived they were greeted individually by the evaluator. He followed the script in Figure 14.2.

The participant was then asked to sit down at a monitor, and the goals of the study and test procedure were explained. Figure 14.3 shows the script used by the evaluator to explain the procedure to each participant (Cogdill, 1999), so that any performance differences that occurred among participants could not be attributed to different procedures.

Thank you very much for participating in this study.

The goal of this project is to evaluate the interface of MEDLINEplus. The results of our evaluation will be summarized and reported to the National Library of Medicine, the federal agency that has developed MEDLINEplus. Have you ever used MEDLINEplus before?

You will be asked to use MEDLINEplus to resolve a series of specific, health-related information needs. You will be asked to "think aloud" as you search for information with MEDLINEplus.

We will be videotaping only what appears on the computer screen. What you say as you search for information will also be recorded. Your face will not be videotaped, and your identity will remain confidential.

I'll need you to review and sign this statement of informed consent. Please let me know if you have any questions about it. (*He hands an informed consent form similar to the one in Box 11.3 to the participant.*)

Figure 14.2 The script used to greet participants in the MEDLINEplus study.

We'll start with a general overview of MEDLINEplus. It's a web-based product developed by the National Library of Medicine. Its purpose is to link users with sources of authoritative health information on the web.

The purpose of our work today is to explore the MEDLINEplus interface to identify features that could be improved. We're also interested in finding out about features that are particularly helpful.

In a few minutes I'll give you five tasks. For each task you'll use MEDLINEplus to find health-related information.

As you use MEDLINEplus to find the information for each task, please keep in mind that it is MEDLINEplus that is the subject of this evaluation—not you.

You should feel free to work on each task at a pace that is normal and comfortable for you. We *will* be keeping track of how long it takes you to complete each task, but you should not feel rushed. Please work on each task at a pace that is normal and comfortable for you. If any task takes you longer than *twenty* minutes, we will ask you to move on to the next task. The Home button on the browser menu has been set to the MEDLINEplus homepage. We'll ask you to return to this page before starting a new task.

As you work on each task, I'd like you to imagine that it's something you or someone close to you needs to know.

All answers can be found on MEDLINEplus or on one of the sites it points to. But if you feel you are unable to complete a task and would like to stop, please say so and we'll move on to the next task.

Before we proceed, do you have any questions at this point?

Figure 14.3 The script used to explain the procedure.

Before starting the main tasks the participants were invited to explore the website for up to 10 minutes and to think aloud as they moved through the site. Figure 14.4 contains the script used to describe how to do this exploration task.

Each participant was then asked to work through the five tasks and was allowed up to 20 minutes for each task. If they did not finish a task they were asked to stop and if they forgot to think out loud or appeared to be stuck they were prompted. The evaluator used the script in Figure 14.5 to direct participants' behavior (Cogdill, 1999).

Before we begin the tasks, I'd like you to explore MEDLINEplus independently for as long as ten minutes.

As you explore, please "think aloud." That is, please tell us your thoughts as you encounter the different features of MEDLINEplus.

Feel free to explore any topics that are of interest to you.

If you complete your independent exploration before the ten minutes are up, please let me know and we'll proceed with the tasks. Again, please remember to tell us what you're thinking as you explore MEDLINEplus.

Figure 14.4 The script used to introduce and describe the initial exploration task.

Please read aloud this task before beginning your use of MEDLINEplus to find the information.

After completing each task, please return to the MEDLINEplus home page by clicking on the "home" button.

Prompts: "What are you thinking?"
"Are you stuck?"
"Please tell me what you're thinking."
[*If time exceeds 20 minutes:* "I need to ask you to stop working on this task and proceed to the next one."]

Figure 14.5 The script used to direct participants' behavior.

When all the tasks were completed, the participant was given a post-test questionnaire consisting of items derived from the QUIS user satisfaction questionnaire (Chin et al., 1988) described in Chapter 13. Finally, when the questionnaire was completed, there was a debriefing (Figure 14.6) in which participants were asked for their opinions.

How did you feel about your performance on the tasks overall?
Tell me about what happened when [cite problem/error/excessive time].
What would you say was the best thing about the MEDLINEplus interface?
What would you say was the worst thing about the MEDLINEplus interface?

Figure 14.6 The debriefing script used in the MEDLINEplus study.

Data collection

Criteria for successfully completing each task were developed in advance. For example, participants had to find and access between 3–9 web page URLs. Each user's search moves were then recorded for each task. For example, the log revealed that Participant A visited the online resources shown in Table 14.1 while trying to complete the first task.

Completion times were automatically recorded and calculated from the video and interaction log data. The data from the questionnaire and the debriefing session

Table 14.1   The resources visited by participant A for the first task.

Databases
Home
MEDLINE/PubMed: "dark bump"
MEDLINE/PubMed: "bump"
Home
Dictionaries
External: Online Medical Dictionary
Home
Health Topics
Melanoma (HT)
External: American Cancer Society

were also used to help understand each participant's performance. The data collected contained the following:

- start time and completion time
- page count (i.e., pages accessed during the search task)
- external site count (i.e., number of external sites accessed during the search task)
- medical publications accessed during the search task
- the user's search path
- any negative comments or mannerisms observed during the search
- user satisfaction questionnaire data

ACTIVITY 14.1   What do you notice about how the user testing fits into the overall usability testing?

Comment   The user testing is closely integrated with the other techniques used in usability testing—questionnaires, interviews, thinkaloud, etc. In concert they provide a much broader picture of the user's interaction than any single technique would show.

Data analysis

Analysis of the data focused on such things as:

- website organization such as arrangement of topics, menu depth, organization of links, etc.
- browsing efficiency such as navigation menu location, text density, etc.
- the search features such as search interface consistency, feedback, terms, etc.

For example, Table 14.2 contains the performance data for the nine subjects for task 1. It shows the time to complete the task and the different kinds of searches undertaken. Similar tables were produced for each task. The exploration and questionnaire data was also analyzed to help explain the results.

Table 14.2 Performance data for task 1: Find information about whether a dark bump on your shoulder might be skin cancer. Mean (M) and standard deviation (SD) for all subjects are also shown.

Participant	Time to nearest minute	Reason for task termination	MEDLINEplus Pages	External sites accessed	MEDLINEplus searches	MEDLINE publication searches
A	12	Successful completion	5	2	0	2
B	12	Participant requested termination	3	2	3	0
C	14	Successful completion	2	1	0	0
D	13	Participant requested termination	5	2	1	0
E	10	Successful completion	5	3	1	0
F	9	Participant requested termination	3	1	0	0
G	5	Successful completion	2	1	0	0
H	12	Successful completion	3	1	0	6
I	6	Successful completion	3	1	0	0
M	10		3	2	1	1
SD	3		1	1	1	2

ACTIVITY 14.2   Examine Table 14.2.

(a) Why are letters used to indicate participants?

(b) What do you notice about the completion times when compared with the reasons for terminating tasks (i.e., completion records)?

(c) What does the rest of the data tell you?

Comment

(a) Participants' names should be kept confidential in reports, so a coding scheme is used.

(b) Completion times are not closely associated with successful completion of this task. For example, completion times range from 5–14 minutes for successful completion and from 9–13 minutes for those who asked to terminate the task.

(c) From the data it appears that there may have been several ways to complete the task successfully. For example, participants A and C both completed the task successfully but their records of visiting the different resources differ considerably.

Conclusions and reporting the findings

The main finding was that reaching external sites was often difficult. Furthermore, analysis of the search moves revealed that several participants experienced difficulty finding the health topics pages devoted to different types of cancer. The post-test questionnaire showed that participants' opinions of MEDLINEplus were fairly neutral. They rated it well for ease of learning but poorly for ease of use because there were problems in going back to previous screens. These results were fed back to the developers in an oral presentation and in a written report.

ACTIVITY 14.3

(a) Was the way in which participants were selected appropriate and were there enough participants? Justify your comments.

(b) Why do you think participants were asked to read each new task aloud before starting it and to return to the home page?

(c) Was the briefing material adequate? Justify your comment.

Comments

(a) This way of selecting participants was appropriate for user testing. The evaluator was careful to get a number of representative users across the user age range from both genders. Participants were screened to ensure that they were experienced web users. The evaluator decided to select from a local volunteer pool of participants, to ensure that he got people who wanted to be involved and who lived locally. Since using the web is voluntary, this is a reasonable approach. The number of participants was adequate for user testing.

(b) This was to make it easy for the evaluator to detect the beginning of a new task on the video log. Sending the participants back to the home page before starting each new task ensured that logging always started from the same place. It also helped to orient the participants.

(c) The briefing material was full and carefully prepared but not excessive. Participants were told what was expected of them and the prompts were preplanned to ensure that each participant was treated in the same way. An informed consent form was also included.

14.3 Doing user testing

There are many things to consider before doing user testing. Controlling the test conditions is central, so careful planning is necessary. This involves ensuring that the conditions are the same for each participant, that what is being measured is indicative of what is being tested and that assumptions are made explicit in the test design. Working through the DECIDE framework will help you identify the necessary steps for a successful study.

THE WALL STREET JOURNAL

"Oh, the commute in to work was a breeze, but I've been stuck in Internet traffic for four hours!"

14.3.1 Determine the goals and Explore the questions

User testing is most suitable for testing prototypes and working systems. Although the goal of a test can be broad, such as determining how usable a product is, more specific questions are needed to focus the study, such as, "can users complete a certain task within a certain time, or find a particular item, or find the answer to a question" as in the MEDLINEplus study?

14.3.2 Choose the paradigm and techniques

User testing falls in the usability testing paradigm and sometimes the term "user testing" is used synonymously with usability testing. It involves recording data using a combination of video and interaction logging, user satisfaction questionnaires, and interviews.

14.3.3 Identify the practical issues: Design typical tasks

Deciding on which tasks to test users' performance is critical. Typically, a number of "completion" tasks are set, such as finding a website, writing a document or creating a spreadsheet. Quantitative performance measures are obtained during the tests that produce the following types of data (Wixon and Wilson, 1997):

- time to complete a task
- time to complete a task after a specified time away from the product

- number and type of errors per task
- number of errors per unit of time
- number of navigations to online help or manuals
- number of users making a particular error
- number of users completing a task successfully

As Deborah Mayhew (1999) reports, these measures slot neatly into usability engineering specifications which specify:

- current level of performance
- minimum acceptable level of performance
- target level of performance

The type of test prepared will depend on the type of prototype available for testing as well as study goals and questions. For example, whether testing a paper prototype, a simulation, or a limited part of a system's functionality will influence the breadth and complexity of the tasks set.

Generally, each task lasts between 5 and 20 minutes and is designed to probe a problem. Tasks are often straightforward and require the user to find this or do that, but occasionally they are more complex, such as create a design, join an online community or solve a problem, like those described in the MEDLINEplus and HutchWorld studies. Easy tasks at the beginning of each testing session will help build users' confidence.

14.3.4 Identify practical issues: Select typical users

Knowing users' characteristics will help to identify typical users for the user testing. But what is a typical user? Some products are targeted at specific types of users, for example, seniors, children, novices, or experienced people. HutchWorld, for example, has a specific user audience, cancer patients, but their experience with the web differs so a range of users with different experience was important. It is usually advisable to have equal numbers of males and females unless the product is specifically being developed for the male or female market. One of the most important characteristics is previous experience with similar systems. If the user population is large you can use a short questionnaire to help identify testers, as in the MEDLINEplus study.

ACTIVITY 14.4   Why is it important to select a representative sample of users whenever possible?

Comment   It is important to have a representative sample to ensure that the findings of the user test can be generalized to the rest of the user population. Selecting participants according to clear objectives helps evaluators to avoid unwanted bias. For example, if 90% of the participants testing a product for 9–12 year-olds were 12, it would not be representative of the full age range. The results of the test would be distorted by the large group of users at the top-end of the age range.

DILEMMA   How Many Users are Enough?

Deciding how many users to test is partly a logistical issue that depends on schedules, budgets, participants and facilities available. Many professionals recommend that 5–12 testers is enough (Dumas and Redish, 1999). Others say that as soon as the same kinds of problems start being revealed and there is nothing new, it is time to stop. However, the more testers there are, the more representative the findings will be across the user population.

14.3.5   Identify practical issues: Prepare the testing conditions

User testing requires the testing environment to be controlled to prevent unwanted influences and noise that will distort the results. Many companies, such as Microsoft and IBM, test their products in specially designed usability laboratories to try to prevent this (Lund, 1994). These facilities often include a main testing laboratory, with recording equipment and the product being tested, and an observation room where the evaluators sit and subsequently analyze the data. There may also be a reception area for testers, a storage area, and a viewing room for observers. Such labs are very expensive and labor-intensive to run.

The space may be arranged to superficially mimic features of the real world. For example, if the product is an office product or for use in a hotel reception area, the laboratory can be set up to match. But in other respects it is artificial. Sound-proofing and lack of windows, telephones, fax machines, co-workers, etc. eliminate most of the normal sources of distraction. Typically there are two to three wall-mounted video cameras that record the user's behavior, such as hand movements, facial expression, and general body language. Utterances are also recorded and often a keystroke log.

The observation room is usually separated from the main laboratory by a one-way mirror so that evaluators can watch testers but testers cannot see them. Figure 14.7 shows a typical arrangement. Video and other data is fed through to monitors

Figure 14.7 A usability laboratory in which evaluators watch participants on a monitor and through a one-way mirror.

in the recording room. While the test is going on, the evaluators observe and annotate the video stream, indicating events for later more detailed analysis.

The viewing room is like a small auditorium with rows of seats at different levels. It is designed so that managers and others can watch the tests. Video monitors display video and the managers overlook the observation room and into the laboratory through one-way mirrors. Generally only large companies can afford this extra room and it is becoming less common.

The reception area also has bathroom facilities so that testers do not have to go into the outside world during a session. Similarly, telephones in the laboratory do not connect with the outside world, so there are no distractions. The only communication occurs between the tester and the evaluators. The laboratory can be modified to include other features of the environment in which the product will be used if necessary, but it is always tightly controlled.

Many companies and researchers cannot afford to have a usability laboratory, or even to rent one. Instead, they buy mobile usability equipment (e.g., video, interaction logging system) and convert a nearby room into a makeshift laboratory. The mobile laboratory can also be taken into companies and packed away when not needed. This kind of makeshift laboratory is more amenable to the needs of user testing. Modifications may have to be made to test different types of applications. For example, Chris Nodder and his colleagues at Microsoft had to partition the space when they were testing early versions of NetMeeting, a videoconferencing product, in the mid-1990s, as Figure 14.8 shows (Nodder et al., 1999).

Evaluation: Participants communicating
with each other using NetMeeting

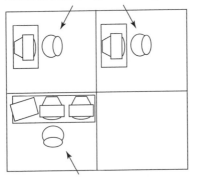

Usability engineer uses another PC to
become the third participant

Figure 14.8 The testing arrangement used for Net-Meeting videoconferencing system.

14.3.6   Identify practical issues: Plan how to run the tests

A schedule and scripts for running the tests, such as those used in MEDLINEplus, should be prepared beforehand. The equipment should be set up and a pilot test performed to make sure that everything is working, the instructions are clear, and there are no unforeseen glitches.

It's a good idea to start the session with a familiarization task, such as browsing a website in a web usability study, so that participants can get used to the equipment before testing starts. An easy first task encourages confidence; ending with a fairly easy one makes participants go away feeling good. A contingency plan is needed for dealing with people who spend too long on a task, as in MEDLINEplus.

A query from the evaluator asking if the participant is all right can help. If the participant gets really stuck then the evaluator should tell him to move on to the next part of the task.

Long tasks and a long testing procedure should be avoided. It is a good idea to keep the session under one hour. Remember, all the data that is collected has to be analyzed and if you have nine participants who together generate nine hours of video, there is a lot to review and analyze.

14.3.7 Deal with ethical issues

As in all types of evaluation, you need to prepare and plan to administer an informed consent form. If the study is situated in a usability laboratory, it is also necessary to point out the presence of one-way mirrors, video cameras, and use of interaction logging.

14.3.8 Evaluate, analyze, and present the data

Typically performance measures (time to complete specified actions, number of errors, etc.) are recorded from video and interaction logs. Since most user tests involve a small number of participants, only simple descriptive statistics can be used to present findings: maximum, minimum, average for the group and sometimes standard deviation, which is a measure of the spread around the mean value. These basic measures enable evaluators to compare performance on different prototypes or systems or across different tasks. An increasing number of analysis tools are also available to support web usability analysis, particularly video analysis as mentioned in Chapter 12.

14.4 Experiments

Although classically performed scientific experimentation is usually too expensive or just not practical for most usability evaluations, there are a few occasions when it is used. For example, in a case study about the testing of a voice response system discussed later in Chapter 15 plenty of participants were available. The development schedule was flexible, and the evaluators knew that quantitative results would be well received by their clients, so they adopted a more experimental approach than usual. For this reason, and because the roots of user testing are in scientific experimentation and many undergraduate projects involve experiments, we will discuss experimental design.

The aim of an experiment is to answer a question or to test a *hypothesis* that predicts a relationship between two or more events known as *variables*. For example,

"Will the time to read a screen of text be different if 12-point Helvetica font is used instead of 12-point Times New Roman?" Such hypotheses are tested by manipulating one or some of the variables involved. The variable that the researcher manipulates is known as the *independent variable*, because the conditions to test this variable are set up independently before the experiment starts. In the example above, type font is the independent variable. The other variable, time to read the text, is called the *dependent variable* because the time to read the text *depends* on the way the experimenter manipulates the other variable, in this case which type font is used.

It is advisable to consult someone who is knowledgeable about relevant statistical tests before doing most experiments, rather than wondering afterwards what to do with the data that is collected.

14.4.1   Variables and conditions

Designs with one independent variable

In order to test a hypothesis, the experimenter has to set up the *experimental conditions* and find ways to control other variables that could influence the test result. So for example, in the experiment in which type font is the independent variable, there are two conditions:

Condition 1 = read screen of text in Helvetica font

Condition 2 = read screen of text in Times New Roman font

It is also helpful to have a *control condition* against which to compare the results of the experiment. For example, in the above test you could set up two control conditions: reading of the same text on printed paper, using Times font and reading of the same text on printed paper, using Helvetica font. The performance measures for both screen conditions could be compared with the paper versions.

Designs with two or more independent variables

Experiments are carried out in user testing usually to compare two or more conditions to see if users perform better in one condition than in the other. For example, we might wish to compare the existing design of a system (e.g., version 5.0) with a redesigned one (e.g., version 6.0). We would need to design a number of tasks that users would be tested on for both versions of the system and then compare their performance across these tasks. If their performance was statistically better in one condition compared with the other, we could say that the two versions were different. Supposing we were then interested in finding out whether the performance of different user groups was affected by the two versions of the system; how could we do this? We could split the users into two groups: those who are beginners and those who are expert users. We would then compare the performance of the two user groups across the two versions of the system. In so doing, we now have two independent variables each with two conditions: the version of the system and the experience of the user.

This gives us a 2×2 design as shown in the table.

Original design	Redesign
Beginners	Beginners
Experts	Experts

Deciding what it means to "perform better" involves determining what to measure; that is, what the dependent variables should be. Two commonly used dependent variables are the time that it takes to complete a task and the number of errors that users make doing the task.

Hypothesis testing can also be extended to include more variables. For example, three variables each with two conditions gives $2 \times 2 \times 2$. In each condition the aim is to test the main effects of each combination and look for any interactions among them.

14.4.2 Allocation of participants to conditions

The discussion so far has assumed that different participants will be used for each condition but sometimes this is not possible because there are not enough participants and at other times it is preferable to have all participants take part in all conditions. Three well-known approaches are used: different participants for all conditions, the same participants for all conditions, and matched pairs of participants.

Different participants

In different participant design a single group of participants is allocated randomly to each of the experimental conditions, so that *different* participants perform in *different* conditions. There are two major drawbacks with this arrangement. The first is making sure that you have enough participants. The second is that if small groups are used for each condition, then the effect of any individual differences among participants, such as differences in experience and expertise, becomes a problem. Randomly allocating the participants and pre-testing to identify any participants that differ strongly from the others helps. An advantage is that there are no *ordering effects*, caused by the influence of participants' experience of one set of tasks on performance on the next, as each participant only ever performs in one condition.

Same participants

In same-participant design, all participants perform in all conditions so only *half* the number of participants is needed; the main reason for this design is to lessen the impact of individual differences and to see how performance varies across conditions for each participant. However, it is important to ensure that the *order* in which participants perform tasks does not bias the results. For example, if there are two tasks, A and B, half the participants should do task A followed by task B and the other half should do task B followed by task A. This is known as *counterbalancing*.

Counterbalancing neutralizes possible unfair effects of learning from the first task, i.e., the *order effect.*

Matched participants

In *matched-participants design*, participants are matched in pairs based on certain user characteristics such as expertise and gender. Each pair is then randomly allocated to each experimental condition. This design is used when participants cannot perform in both conditions. The problem with this arrangement is that other important variables that haven't been taken into account may influence the results. For example, experience in using the web could influence the results of tests to evaluate the navigability of a website. So web expertise would be a good criterion for matching participants.

The advantages and disadvantages of using different experimental designs are summarized in Table 14.3.

Table 14.3   The advantages and disadvantages of different experimental designs

Design	Advantages	Disadvantages
Different participants	No order effects	Many participants needed. Individual differences among participants are a problem. Can be offset to some extent by randomly assigning to groups.
Same participants	Eliminates individual differences between experimental conditions.	Need to counterbalance to avoid ordering effects.
Matched participants	Same as different participants, but the effects of individual differences are reduced.	Can never be sure that subjects are matched across variables.

14.4.3   Other practical issues

Just as in user testing, there are many practical issues to consider and plan, for example where will the experiment be conducted, how will the equipment be setup, how will participants be introduced to the experiment, and what scripts are needed to standardize the procedure? Pilot studies are particularly valuable in identifying potential problems with the equipment or the experimental design.

14.4.4   Data collection and analysis

Data should be collected that measures user performance on the tasks set. These usually include response times, number of errors, and times to complete a task.

Analyzing the data involves knowing what to look for. Do the data sets from the two conditions look different or similar? Are there any extreme atypical values? If so, what do they reflect? Displaying the results on a graph will also help reveal differences.

The response times, errors, etc. should be averaged across conditions to see if there are any marked differences. Simple statistical tests like t-tests can reveal if these are significant. For example, a t-test could reveal whether Helvetica or Times font is slower to read on a screen. If there was no significance then the hypothesis would have to be refuted, i.e., the claim that Helvetica font is easier to read is not true.

Box 14.2 describes an experiment to test whether broad, shallow menu design is preferable to deep menus on the web.

BOX 14.2   An Experiment to Evaluate Structure in Web Page Design

A huge amount of work has been done on exploring the optimal number of items in a menu design, and most studies conclude that breadth is preferable to depth in organizing menu content. By this it is meant having a large number of top level menu items with few levels rather than a small number of top level items with many levels. Around 1997, when the web was still a relatively new phenomenon, there was an assumption that the number of links from a home page to other items should be fewer than 10. Their assumption was based on misapplying Miller's magic number, 7 ± 2. This assumption fails to recognize, however, that users do not need to remember the items, they need only to be able to identify them, which is far easier. A contrary position was that because recognition is easier than recall, it would be better to have a much larger number of links on the home page. This goes against a rule of thumb for information display on paper that advocates the use of white space to prevent confusion and an unpleasing, cluttered design. To solve this controversy Kevin Larson and Mary Czerwinski (1998) from Microsoft Research carried out an experiment and user satisfaction study. The following account outlines the main points of their study.

The goal of the study was to find the optimal depth versus breadth structure of hyperlinks for expertly categorized web content. Three conditions were tested using different link designs for the same web content. Each design had 512 bottom-level nodes.

Condition 1: 8 × 8 × 8 (8 top-level categories, each with 8 sublevels, with 8 content-levels under each)

Condition 2: 16 × 32 (16 top-level categories, each with 32 content-level categories)

Condition 3: 32 × 16 (32 top-level categories, each with 16 content-levels categories)

These conditions were tested by 19 experienced web users, who each performed eight search tasks for each condition, making a total of 24 searches. The eight searches were selected for each participant at random from a bank of 128 possible target items, that were categorized according to content and complexity. Participants were given the same number of items from each category and no one searched for the same item more than once (i.e., there was no duplication of items across conditions).

Reaction times (RT) to complete each search were recorded and the average (Avg.) and standard deviation (SD) for each condition was computed. The results showed that on average participants completed search tasks fastest in the 16 × 32 hierarchy (Avg. RT = 36 seconds, SD = 16), second fastest in the 32 × 16 hierarchy (Avg. RT = 46 seconds, SD = 26), and slowest in the 8 × 8 × 8 hierarchy (Avg. RT = 58 seconds, SD = 23). These results suggest that breadth is preferable to depth for searching web content. However, very large numbers of links on one page may be detrimental to searching performance.

ACTIVITY 14.5

(a) What were the independent and dependent variables in this study?

(b) Write two possible hypothesis statements.

(c) How would you categorize the experimental design?

(d) The participants are all described as "experts." Is this adequate? What else do you want to know about them?

(e) Comment on the description of the tasks. What else do you want to know?

(f) If you know some statistics, suggest what further analysis of the results should be done.

(g) Three other analyses were done on issues that were not mentioned in this description, but that anyone doing this experiment might have looked at. From your knowledge of interaction design, suggest what these analyses might be and say why.

(h) What are the implications of this study for web design?

Comment

(a) The independent variable is menu link structure. The dependent variable is reaction time to complete a search successfully.

(b) Web search performance is better with broad shallow link structures. There is no difference in search performance with different link structures.

(c) All the participants did all the tasks, so this is a same-participant design.

(d) "Expert" could refer to a broad range of expertise. The evaluators could have used a screening questionnaire to make sure that all the participants had reached a basic level of expertise and there were no super-experts in the group. However, given that all the participants did all the conditions, differences in expertise had less impact than in other experimental designs.

(e) Our excerpt contains very little description of the tasks. It would be good to see examples of typical tasks in each task category. How was the similarity and complexity of the tasks tested?

(f) A one-way analysis of variance was used to validate the significance of the main finding. Other tests are also discussed in the full paper.

(g) Participants could be asked to rate their preferences using a subjective rating questionnaire, which is similar to a user satisfaction questionnaire. The researchers also analyzed the paths the participants took to see if any of the conditions caused less optimal searching. They found that the condition with 32 items on the top-level caused a feeling of "lost in hyperspace," though this was not statistically significant. A less obvious analysis examined memory and scanning ability and found that better memory and scanning ability was associated with faster reaction time in the 16×32 hierarchy.

(h) Implications for web design are to avoid deep narrow link hierarchies and very broad shallow ones. However, as the authors emphasize, this is only one study and more research is needed before any generalizations can be made.

14.5 Predictive models

In contrast to the other forms of evaluation we have discussed, predictive models provide various measures of user performance without actually testing users.

This is especially useful in situations where it is difficult to do any user testing. For example, consider companies who want to upgrade their computer support for their employees. How do they decide which of the many possibilities is going to be the most effective and efficient for their needs? One way of helping them make their decision is to provide estimates about how different systems will fare for various kinds of task. Predictive modeling techniques have been designed to enable this.

The most well-known predictive modeling technique in human-computer interaction is GOMS. This is a generic term used to refer to a family of models, that vary in their granularity as to what aspects of a user's performance they model and make predictions about. These include the time it takes to perform tasks and the most effective strategies to use when performing tasks. The models have been used mainly to predict user performance when comparing different applications and devices. Below we describe two of the most well-known members of the GOMS family: the GOMS model and its "daughter," the keystroke level model.

14.5.1 The GOMS model

The GOMS model was developed in the early eighties by Stu Card, Tom Moran and Alan Newell (Card et al., 1983). As mentioned in Chapter 3, it was an attempt to model the knowledge and cognitive processes involved when users interact with systems. The term GOMS is an acronym which stands for *goals, operators, methods and selection rules:*

- *Goals* refer to a particular state the user wants to achieve (e.g., find a website on interaction design).

- *Operators* refer to the cognitive processes and physical actions that need to be performed in order to attain those goals (e.g., decide on which search engine to use, think up and then enter keywords in search engine). The difference between a goal and an operator is that a goal is obtained and an operator is executed.

- *Methods* are learned procedures for accomplishing the goals. They consist of the exact sequence of steps required (e.g., drag mouse over entry field, type in keywords, press the "go" button).

- *Selection rules* are used to determine which method to select when there is more than one available for a given stage of a task. For example, once keywords have been entered into a search engine entry field, many search engines allow users to press the return key on the keyboard or click the "go" button using the mouse to progress the search. A selection rule would determine which of these two methods to use in the particular instance. Below is a detailed example of a GOMS model for deleting a word in a sentence using Microsoft Word.

Goal: delete a word in a sentence

Method for accomplishing goal of deleting a word using menu option:

Step 1. Recall that word to be deleted has to be highlighted

Step 2. Recall that command is "cut"

Step 3. Recall that command "cut" is in edit menu

Step 4. Accomplish goal of selecting and executing the "cut" command

Step 5. Return with goal accomplished

Method for accomplishing goal of deleting a word using delete key:

Step 1. Recall where to position cursor in relation to word to be deleted

Step 2. Recall which key is delete key

Step 3. Press "delete" key to delete each letter

Step 4. Return with goal accomplished

Operators to use in above methods:

Click mouse

Drag cursor over text

Select menu

Move cursor to command

Press keyboard key

Selection Rules to decide which method to use:

1: Delete text using mouse and selecting from menu if large amount of text is to be deleted

2: Delete text using delete key if small number of letters is to be deleted

14.5.2 The Keystroke level model

The keystroke level model differs from the GOMS model in that it provides actual numerical predictions of user performance. Tasks can be compared in terms of the time it takes to perform them when using different strategies. The main benefit of making these kinds of quantitative predictions is that different features of systems and applications can be easily compared to see which might be the most effective for performing specific kinds of tasks.

When developing the keystroke level model, Card et al. (1983) analyzed the findings of many empirical studies of actual user performance in order to derive a standard set of approximate times for the main kinds of operators used during a task. In so doing, they were able to come up with the average time it takes to carry out common physical actions (e.g., press a key, click on a mouse button) together with other aspects of user–computer interaction (e.g., the time it takes to decide what to do, the system response rate). Below are the core times they proposed for these (note how much variability there is in the time it takes to press a key for users with different typing skills).

Operator name	Description	Time (sec)
K	Pressing a single key or button	0.35 (average)
	Skilled typist (55 wpm)	0.22
	Average typist (40 wpm)	0.28
	User unfamiliar with the keyboard	1.20
	Pressing shift or control key	0.08
P	Pointing with a mouse or other device to a target on a display	1.10
P_1	Clicking the mouse or similar device	0.20
H	Homing hands on the keyboard or other device	0.40
D	Draw a line using a mouse	Variable depending on the length of line
M	Mentally prepare to do something (e.g., make a decision)	1.35
R(t)	System response time—counted only if it causes the user to wait when carrying out their task	t

The predicted time it takes to execute a given task is then calculated by describing the sequence of actions involved and then summing together the approximate times that each one will take:

$$T_{execute} = T_K + T_P + T_H + T_D + T_M + T_R$$

For example, consider how long it would take to insert the word *not* into the following sentence, using a word processor like Microsoft Word:

Running through the streets naked is normal.
So that it becomes:
Running through the streets naked is not normal.

First we need to decide what the user will do. We are assuming that he will have read the sentences beforehand and so start our calculation at the point where he is about to carry out the requested task. To begin he will need to think what method to select. So we first note a mental event (M operator). Next he will need to move the cursor into the appropriate point of the sentence. So we note an H operator (i.e., reach for the mouse). The remaining sequence of operators are then: position the mouse before the word normal (P), click the mouse button (P_1), move hand from mouse over the keyboard ready to type (H), think about which letters to type (M), type the letters *n, o* and *t* (3K) and finally press the spacebar (K).

The times for each of these operators can then be worked out:

Mentally prepare (M)	1.35
Reach for the mouse (H)	0.40
Position mouse before the word "normal" (P)	1.10
Click mouse (P_1)	0.20
Move hands to home position on keys (H)	0.40
Mentally prepare (M)	1.35
Type "n" (good typist) (K)	0.22
Type "o" (K)	0.22
Type "t" (K)	0.22
Type "space" (K)	0.22
Total predicted time:	5.68 seconds

When there are many components to add up, it is often easier to put together all the same kinds of operators. For example, the above can be rewritten as:
$2(M) + 2(H) + 1(P) + 1(P_1) + 4 (K) = 2.70 + 0.88 + 1.10 + 0.2 + 0.80 = 5.68$ seconds.

Over 5 seconds seems a long time to insert a word into a sentence, especially for a good typist. Having made our calculation it is useful to look back at the various decisions made. For example, we may want to think why we included a mental operator before typing the letters n, o and t but not one before any of the other physical actions. Was this necessary? Perhaps we don't need to include it. The decision when to include a time for mentally preparing for a physical action is one of the main difficulties with using the keystroke level model. Sometimes it is obvious when to include one (especially if the task requires making a decision) but for other times it can seem quite arbitrary. Another problem is that, just like typing skills vary between individuals, so too do the mental preparation times people spend thinking about what to do. Mental preparation can vary from under 0.5 of a second to well over a minute. Practice at modeling similar kinds of tasks together with comparing them with actual times taken can help overcome these problems. Ensuring that decisions are applied consistently also helps. For example, if comparisons between two prototypes are made, apply the same decisions to each.

ACTIVITY 14.6 As described in the GOMS model above there are two main ways words can be deleted in a sentence when using a word processor like Word. These are:

(a) deleting each letter of the word individually by using the delete key

(b) highlighting the word using the mouse and then deleting the highlighted section in one go

Which of the two methods do you think is quickest for deleting the word "not" from the following sentence:

I do not like using the keystroke level model.

Comment   (a) Our analysis for method 1 is:

Mentally prepare	M	1.35
Reach for mouse	H	0.40
Move cursor one space after the word "not"	P	1.10
Click mouse	P_1	0.20
Home in on delete key	H	0.40
Press delete key 4 times to remove word plus a space (using value for good typist value)	4(K)	0.88

Total predicted time = 4.33 seconds

(b) Our analysis for method 2 is:

Mentally prepare	M	1.35
Reach for mouse	H	0.40
Move cursor to just before the word "not"	P	1.10
Click and hold mouse button down (half a P_1)	P_1	0.10
Drag the mouse across "not" and one space	P	1.10
Release the mouse button (half a P_1)	P_1	0.10
Home in on delete key	H	0.40
Press delete key	K	0.22
(Using value for good typist rate)		

Total predicted time = 4.77 seconds

The result seems counter-intuitive. Why do you think this is? The reason is that the amount of time required to select the letters to be deleted is longer for the second method than pressing the delete key three times in the first method. If the word had been any longer, for example, "keystroke" then the keystroke analysis would have predicted the opposite. There are also other ways of deleting words, such as double clicking on the word (to select it) and then either pressing the delete key or the combination of ctrl+X keys. What do you think the keystroke level model would predict for either of these two methods?

14.5.3 Benefits and limitations of GOMS

One of the main attractions of the GOMS approach is that it allows comparative analyses to be performed for different interfaces or computer systems relatively easily. Since its inception, a number of researchers have used the method, reporting on its success for comparing the efficacy of different computer-based systems. The most well-known is Project Ernestine (Gray et al., 1993). This study was carried out to determine if a proposed new workstation, that was ergonomically designed, would improve telephone call operators' performance. Empirical data collected for a range of operator tasks using the existing system was compared with hypothetical data deduced from doing a GOMS analysis for the same set of tasks for the proposed new system.

Similar to the activity above, the outcome of the study was counter-intuitive. When comparing the GOMS predictions for the proposed system with the empirical data collected for the existing system, the researchers discovered that several tasks would take longer to accomplish. Moreover, their analysis was able to show why this might be the case: certain keystrokes would need to be performed at critical times during a task rather than during slack periods (as was the case with the existing system). Thus, rather than carrying out these keystrokes in parallel when talking with a customer (as they did with the existing system) they would need to do them sequentially—hence the predicted increase in time spent on the overall task. This suggested to the researchers that, overall, the proposed system would actually slow down the operators rather than improve their performance. On the basis of this study, they were able to advise the phone company against purchasing the new workstations, saving them from investing in a potentially inefficient technology.

While this study has shown that GOMS can be useful in helping make decisions about the effectiveness of new products, it is not often used for evaluation purposes. Part of the problem is its highly limited scope: it can only really model computer-based tasks that involve a small set of highly routine data-entry type tasks. Furthermore, it is intended to be used only to predict expert performance, and does not allow for errors to be modeled. This makes it much more difficult (and sometimes impossible) to predict how an average user will carry out their tasks when using a range of systems, especially those that have been designed to be very flexible in the way they can be used. In most situations, it isn't possible to predict how users will perform. Many *unpredictable* factors come into play including individual differences among users, fatigue, mental workload, learning effects, and social and organizational factors. For example, most people do not carry out their tasks sequentially but will be constantly multi-tasking, dealing with interruptions and talking to others.

A dilemma with predictive models, therefore, is that they can only really make predictions about predictable behavior. Given that most people are unpredictable in the way they behave, it makes it difficult to use them as a way of evaluating how systems will be used in real-world contexts. They can, however, provide useful estimates for comparing the efficiency of different methods of completing tasks, particularly if the tasks are short and clearly defined.

14.5.4 Fitts' Law

Fitts' Law (1954) predicts the time it takes to reach a target using a pointing device. It was originally used in human factors research to model the relationship between speed and accuracy when moving towards a target on a display. In interaction design it has been used to describe the time it takes to point at a target, based on the size of the object and the distance to the object. Specifically, it is used to model the time it takes to use a mouse and other input devices to click on objects on a screen. One of its main benefits is that it can help designers decide where to locate buttons, what size they should be and how close together they should be on a screen display. The law states that:

$$T = k \log2(D/S + 0.5), k \sim 100 \text{ msec.}$$

where

    T = time to move the hand to a target

    D = distance between hand and target

    S = size of target

In a nutshell the bigger the target the easier and quicker it is to reach it. This is why interfaces that have big buttons are easier to use than interfaces that present lots of tiny buttons crammed together. Fitts' law also predicts that the most quickly accessed targets on any computer display are the four corners of the screen. This is because of their "pinning" action, i.e., the sides of the display constrain the user from over-stepping the target. However, as pointed out by Tog on his AskTog website, corners seem strangely to be avoided at all costs by designers.

Fitts' Law, therefore, can be useful for evaluating systems where the time to physically locate an object is critical to the task at hand. In particular it can help designers think about where to locate objects on the screen in relation to each other. This is especially useful for mobile devices, where there is limited space for placing icons and buttons on the screen. For example, in a recent study carried out by Nokia, Fitts' Law was used to predict expert text entry rates for several input methods on a 12-key mobile phone keypad. The study helped the designers make decisions about the size of keys, their positioning and the sequences of presses to perform common tasks for the mobile device. Trade-offs between the size of a device, and accuracy of using it were made with the help of calculations from this model.

ACTIVITY 14.7 Microsoft toolbars provide the user with the option of displaying a label below each tool. Give a reason why labeled tools may be accessed faster. (Assume that the user knows the tool and does not need the label to identify it.)

Comment The label becomes part of the target and hence the target gets bigger. As we mentioned earlier bigger targets can be accessed faster.

Furthermore, tool icons that don't have labels are likely to be placed closer together so they are more crowded. Spreading the icons further apart creates buffer zones of space around the icons so that if users accidentally go past the target they will be less likely to select the wrong icon. When the icons are crowded together the user is at greater risk of accidentally overshooting and selecting the wrong icon. The same is true of menus, where the items are closely bunched together.

Assignment

This assignment continues the work you did on the web-based ticketing system at the end of Chapters 7, 8, and 13. The aim of this assignment is again to evaluate the prototypes produced, but this time using user testing. You will then be able to compare the kind of results you got from the heuristic evaluation with those from the user testing. Even though you will be using different prototypes for each evaluation, you should be able to compare the types of problems that each technique reveals.

(a) Based on your knowledge of the requirements for this system, develop a standard task, e.g., booking two seats for a particular performance.

(b) Prepare a short informed consent form, and write an introduction that explains why you are testing this prototype.

(c) Select three typical users, who can be friends or colleagues, and ask them to do the task using your prototype.

(d) `Note the problems that each user encounters. If you can, time their performance. (If you happen to have a video camera you could film each participant.)

(e) Did the kinds of problems that user testing revealed differ from those obtained from a heuristic evaluation? If so, in what ways?

(f) What are the main advantages and disadvantages of each technique?

Summary

This chapter described user testing, which is the core of usability testing. The various aspects of user testing were discussed, including setting up tests, collecting data, controlling conditions and analyzing findings. Experimental design and how experiments differ from user testing was also discussed.

Predicting user performance using the GOMS model, the keystroke level model, and Fitts' Law was presented. These techniques can be useful for determining whether a proposed interface, system or keypad layout will be optimal.

Key points

- User testing is a central component of usability testing which typically also includes observation, user satisfaction questionnaires and interviews.

- Testing is commonly done in controlled laboratory-like conditions, in contrast to field studies that focus on how the product is used in its natural context.

- Experiments aim to answer a question or hypothesis by manipulating certain variables while keeping others constant.

- The experimenter controls independent variable(s) in order to measure dependent variable(s).

- There are three types of experimental design: different participants, same participants, and matched pair participants.

- The GOMS model, keystroke-level model and Fitts' law can be used to predict expert, error-free performance for certain kinds of tasks.

- Predictive models require neither users nor experts, but the evaluators must be skilled in applying the models.

- Predictive models are used to evaluate systems with limited, clearly defined functionality such as data entry applications.

Further reading

DUMAS, J. S., AND REDISH, J. C. (1999) *A Practical Guide to Usability Testing*. Exeter, UK: Intellect. Many books have been written about user testing and usability, but this one is particularly useful because it describes the process in detail and provides many examples.

RUBIN, J. (1994) *Handbook of Usability Testing: How to Plan, Design and Conduct Effective Tests*. New York: John Wiley & Sons. This book also provides good practical advice about preparing and conducting user tests, analyzing and reporting the results.

ROBSON, C. (1994) *Experimental Design and Statistics in Psychology*. Aylesbury, UK: Penguin Psychology. This book provides an introduction to experimental design and basic statistics.

LARSON, K., AND CZERWINSKI, M. (1998) *Web page design: Implications of memory, structure and scent for information retrieval*. Paper presented at CHI 98, Los Angeles. This paper describes the breadth-versus-depth web study outlined in Box 14.2.

CARD, S. K., MORAN, T. P., AND NEWELL, A. (1983) *The Psychology of Human Computer Interaction*. Hillsdale, NJ: Lawrence Erlbaum Associates. This seminal book describes GOMS and the keystroke level model.

MACKENZIE, I. S. (1992) Fitts' law as a research and design tool in human-computer interaction. *Human-Computer Interaction*, 7, 91–139. This early paper by Scott Mackenzie provides a detailed discussion of how Fitts' law can be used in HCI.

INTERVIEW   with Ben Shneiderman

Ben Shneiderman is professor of computer science at University of Maryland, where he was founder and director of the Human-Computer Interaction Laboratory from 1983 to 2000. He is author of the highly acclaimed book *Designing the User Interface: Strategies for Effective Human-Computer Interaction*, now in its third edition. He developed the concept of direct manipulation and created the user interface for the selectable text link that makes the web so easy to use.

JP: Ben you've been a strong advocate of measuring user performance and user satisfaction. Why is just watching users not enough?

BS: Watching users is a great way to begin, but if we are to develop a scientific foundation for HCI that promotes theory and supports prediction, measurement will be important. The purpose of measurement is not statistics but insight.

JP: OK can you give me an example?

BS: Watching users traverse a menu tree may reveal some problems they have, but only when you start to measure the time and number of branches taken can you discover that broader and shallower trees are almost always the winning strategy. This conflict between broader and shallower trees emerged in a conference panel discussion with a leading researcher for a major corporation. She and her colleagues followed up by testing users' speed of performance on searching tasks with two-level and three-level trees.

(Editor's note: You can read about this experiment in Box 14.2).

JP: But is speed of performance always the important measure?

BS: Measuring speed of performance, rate of errors, and user satisfaction separately is important because sometimes users may be satisfied by an elaborate graphical interface even if it slows them down substantially. Finding the right balance among performance, error rates, and user satisfaction depends on whether you are building a repetitive data-entry system, an air-traffic control system, or a game.

JP: Experiments are an important part of your undergraduate classes. Why?

BS: Most computer science and information systems students have had little exposure to experiments. I want to make sure that my students can form lucid and testable hypotheses that can be experimentally tested with groups of real users. They should understand about choosing a small number of independent variables to modify and dependent variables to measure. I believe that students benefit by understanding how to control for biases and perform statistical tests that confirm or refute the hypotheses. My students conduct experimental projects in teams and prepare their reports on the web. For example, one team did a project in which they varied the display size and demonstrated that web surfers found what they needed faster with larger screens. Another group found that bigger mouse pads do not increase speed of performance (www.otal.umd.edu/SHORE2000). Even if students never conduct an experiment professionally, the process of designing experiments helps them to become more effective analysts. I also want my students to be able to read scientific papers that report on experiments.

JP: What "take-away messages" do you want your students to get from taking an HCI class?

BS: I want my students to know about rigorous and replicable scientific results that form the foundation for this emerging discipline of human-computer interaction. Just as physics provides a scientific foundation for mechanical engineering, HCI provides a rigorous foundation for usability engineering.

JP: How do you distinguish between an experiment and usability testing?

BS: The best controlled experiments start with a hypothesis that has practical implications and theoretical results of widespread importance. A controlled experiment has at least two conditions and applies statistical tests such as t-test and analysis of variance (ANOVA) to verify statistically significant differences. The results confirm or refute the hypothesis and the procedure is carefully described so that others can replicate it. I tell my students that experiments have two parents and three children. The parents are "a practical problem" and "a theoretical foundation" and the three children are "help in resolving the practical problem," "refinements to the theory," and "advice to future experimenters who work on the same problem."

By contrast, a usability test studies a small number of users who carry out required tasks. Statistical results are less important. The goal is to refine a product as quickly as possible. The outcome of a usability test is a report to developers that identifies frequent problems and possibly suggests improvements, maybe ranked from high to low priority and from low to high developer effort.

JP: What do you see as the important usability issues for the next five years?

BS: I see three directions for the next five years. The first is the shift from emphasizing the technology to focusing on user needs. I like to say "the old computing is about what computers can do, the new computing is about what users can do."

JP: But hasn't HCI always been about what users can do?

BS: Yes, but HCI and usability engineering have been more evaluative than generative. To clarify, I believe that deeper theories about human needs will contribute to innovations in mobility, ubiquity, and community. Information and communication tools will become pervasive and enable higher levels of social interaction. For example, museum visitors to the Louvre, white-water rafters in Colorado, or family travelers to Hawaii's Haleakala volcano will be able to point at a sculpture, rock, or flower and find out about it. They'll be able to see photos at different seasons taken by previous visitors and send their own pictures back to friends and grandparents. One of our projects allows people to accumulate, organize, and retrieve the many photos that they will take and receive. Users of our PhotoFinder software tool can organize their photos and annotate them by dragging

and dropping name labels. Then they can find photos of people and events to tell stories and reminisce (see figure).

HCI researchers who understand human needs are likely to come up with innovations that help physicians to make better diagnoses, enable shoppers to find what they want at fair prices, and allow educators to create more compelling experiences for students.

JP: What are the other two directions?

BS: The second opportunity is to support universal usability, thereby bringing the benefits of information and communications technology to the widest possible set of users. website designers will need to learn how to attract and retain a broad set of users with divergent needs and differing skills. They will have to understand how to accommodate users efficiently with slow and fast network connections, new and old computers, and various software platforms. System designers who invent strategies to accommodate young and old, novice and expert, and users with varying disabilities will earn the appreciation of users and the respect of their colleagues. Evidence is accumulating that designs that facilitate multiple natural-language versions of a website also make it easy to accommodate end-user customization, convert to wireless applications, support disabled users and speed modifications. The good news is that satisfying these multiple requirements also produces interfaces that are better for all users. Diversity promotes quality.

The third direction is the development of tools to let more people be more creative more of the time. Word processors, painting tools and music-composition software are a good starting point, but creative people need more powerful tools so that they can explore alternative solutions rapidly. Creativity-support tools will speed search of existing solutions, facilitate consultations with peers and mentors, and record the users' history of activity so that they can review or revise their work.

But remember that every positive development also has a potential dark side. One of the formidable challenges for HCI students is to think carefully about how to cope with the unexpected and unintended. Powerful tools can have dangerous consequences.

Chapter 15

Design and evaluation in the real world: communicators and advisory systems

15.1 Introduction
15.2 Key issues
15.3 Designing mobile communicators
    15.3.1 Background
    15.3.2 Nokia's approach to developing a communicator
    15.3.3 Philips' approach to designing a communicator for children
15.4 Redesigning part of a large interactive phone-based response system
    15.4.1 Background
    15.4.2 The redesign

15.1 Introduction

Textbooks about design and usability testing often make the processes sound straightforward and able to be followed in a step-by-step manner. However, in the real world bringing together all the different aspects of a design is far from straightforward. It is only when you become involved in an actual design project that the challenges and multitude of difficult decisions to be made become apparent. Iterative design often involves carrying out different parts of a project in parallel and under tremendous pressure. The need to deal with different sets of demands and trade-offs (e.g., the need for rigorous testing versus the very limited availability of time and resources) is a major influence on the way a design project is carried out.

The aim of this final chapter is to convey what interaction design is like in the real world by describing how others have dealt with the challenges of an actual design project. As you will have noticed, we have written primarily about design in Chapters 6–9 and evaluation in Chapters 10–14. This was to enable us to explain the different techniques and processes involved during a design project. It is important to realize that in the real world these two central aspects are closely integrated. You do not do one without the other. In particular, the main reason for doing an evaluation is to make progress on a design. Conversely, whenever you develop a design you need to evaluate it. Whether you are designing a small handheld device or a large air-traffic control system, a design that takes months to produce or one that spans years of effort, the two processes must be carried out together.

The chapter provides glimpses into the design and evaluation process for quite different types of interactive systems. The first two case studies discuss the design of mobile communicators for different groups of users, showing how the design issues differ for each group. The third case study examines the redesign of a large interactive voice response system. In the original design, the focus was on developing a system where the programmers used themselves as models of the users. Furthermore, the programmers were more concerned with developing elegant programs than with users' needs for easy interaction. As you will see, this caused a mismatch between their design and how users tried to find information. This is a common predicament and interaction designers are often brought in to fix already badly designed systems.

The main aims of this chapter are to:

- Show how design and evaluation are brought together in the development of interactive products.
- Show how different combinations of design and evaluation methods are used in practice.
- Describe the various design trade-offs and decisions made in the real world.

15.2 Key issues

As we have stressed throughout, user-centered approaches to interaction design involve iterative cycles of design-evaluate-redesign as development progresses from initial ideas through various prototypes to the final product. How many cycles need to take place depends on the constraints of the project (e.g., how many people are working on it, how much time is available, how secure the system has to be). To be good at working through these cycles requires a mix of skills involving multitasking, decision-making, team work and firefighting. Many practical issues and unexpected events also need to be dealt with (e.g., users not turning up at testing sessions, prototypes not working, budgets being cut, time to completion being reduced, designers leaving at crucial stages). A design team, therefore, must be creative, well organized, and knowledgeable about the range of techniques that can be brought into play when needed. Part of the challenge and excitement of interaction design is finding ways to cope with the diverse set of problems confronting a project.

A multitude of questions, concerns and decisions come up throughout a design project. No two projects are ever the same; each will face a different set of constraints, demands, and crises. Throughout the book we have raised what we consider to be general issues that are important in any project. These include how to involve users and take their needs into account, how to understand a problem space, how to design a conceptual model, and how to go about designing and evaluating interfaces. In the following case studies, we focus on some of the more practical problems and dilemmas that can arise when working on an actual project.

We present the case studies through a set of questions that draw out a number of key issues for each project. For example, mapping a large number of functions onto a much smaller number of buttons is key for mobile devices; understanding a child's world is key when designing for children; evaluating the current system is key when redesigning any large system.

15.3 Designing mobile communicators

The first two case studies are about the design of mobile communicators. They focus on some of the design decisions and trade-offs that need to be made. We describe example design practices at two companies, Nokia and Philips, highlighting the differences in requirements and design methods for what is seemingly a similar device.

15.3.1 Background

Mobile communicators often combine the functionality of a mobile telephone, a PDA, and a desktop computer. They allow the user to send and receive email and faxes, to make and receive telephone calls, and to keep contact details, diary entries, and other notes. They are an example of new devices that try to push technological boundaries while at the same time being accessible to a wide range of users. A key design challenge, therefore, is how to make such everyday devices usable and affordable to a heterogeneous set of users. Related to this set of usability goals is the decision about which design approach to use. As you are aware, there are many different approaches to choose from, ranging from ethnographic to more analytic methods. Here, we examine the different approaches of the two companies. To put you in a "design" frame of mind, we begin by asking you to consider the requirements for this kind of device.

ACTIVITY 15.1   In Chapter 7, we introduced a number of different kinds of requirements: functional, data, environmental, user, and usability requirements. Which of these is particularly relevant to the design of a communicator?

Comment   All these are relevant in the design of mobile communicators, but one that needs particular attention is environmental requirements. Because the device is aimed at users "on the move" in all kinds of places, the environment in which it should work or its "context of use" is very variable.

Core environmental issues include how to make the device small and light enough to be carried around in a pocket or small handbag. This means the device must be made of light materials and should be physically small, and also the software must be designed to work with a small screen and limited memory. The system must allow for a whole range of situations: noisy or quiet, well lit or poorly lit, hot or cold, wet or dry, vibrating or still, and so on. These constraints have implications for the use of audio, for the levels of display lighting, and for the physical robustness of the device, among other things.

Another consideration in the design of this kind of communication device is what the users are doing when using it. A typical user is likely to be doing something else at the same time as using the communicator. This may be walking around, avoiding obstacles, looking for traffic, etc., or it may be listening for a train announcement or a call from children. So users are trying to combine at least three things: communicating with the device (talking, typing, or whatever), performing the "external" activity (walking, listening, etc.), and operating the device. This creates quite a high cognitive load, so operating the device should occupy as little attention as possible.

Tasks are very likely to be interrupted by external events, so users need to know where in an interaction sequence they are at any time, and be able to restart the sequence after an interruption. For a mobile communicator designed to access the Internet, this raises an interesting design trade-off: how long should a communicator remain connected to the Internet after activity has apparently ceased? A balance is needed between disconnecting so as to minimize connection costs, and remaining connected in a stable state to allow the resumption of an interrupted task. The best option may be to let users set their own time-out period, but this adds to the complexity of operation.

Another implication of the fact that users are likely to be doing other things in parallel with operating the device is that the communicator may need to be operated with one hand, or indeed in a hands-free mode. For example, someone who is walking down the street carrying a bag when the phone rings needs to be able to respond without stopping and putting the bag down, i.e., the operation needs to be one-handed.

For mobile devices in particular, tasks tend to be time-critical, *ad hoc*, triggered by other people or events, relatively brief, low in terms of attention to be applied to the task, and very personal. Because of these characteristics, the flow among tasks must be smooth. It seems that easy transition between contact database, telephone, and calendar is particularly important for mobile devices. The nature of these tasks and the environmental requirements for mobile devices have implications for evaluation, as we discuss in section 15.3.2.

Because this device will be mobile it must be simple to use and not involve much training. It also needs to be robust and reliable, as the user is most likely to be away from any significant technical support.

15.3.2 Nokia's approach to developing a communicator

So how does Nokia deal with these kinds of requirements? And which design and evaluation methods do they use? Here, we look at an example approach of Nokia's, and some of the key decisions in mobile communicator design. A design example of an existing Nokia communicator is illustrated in Figure 15.1. This communicator weighs 244 g, is 158 × 56 × 27 mm, and has a full-color screen. As well

Figure 15.1 The Nokia
9210 communicator.

as email and high-speed WAP connections, it also runs a variety of office applications including word processing, spreadsheets, and presentations.[1]

This case study is based on material from Väänänen-Vainio-Mattila and Ruuska (2000).

What kind of lifecycle does Nokia use?   Nokia follows a user-centered approach to concept development that includes contextual design techniques. They point out that "one clear strength of the methodology is that it makes ethnographic research manageable in a business environment" (Väänänen-Vainio-Mattila and Ruuska, 2000, p. 197). As discussed in Chapter 9, the "rich" descriptions arising from an ethnographic study are often not in a form that can be readily translated into a design specification. Nokia tries to get around this problem by carrying out ethnographic studies in combination with other methods. This enables them to come up with a set of detailed requirements.

Figure 15.2 shows a top-level model of Nokia's approach. It has four main steps:

1. The cycle begins with data gathering. The data is collected through market research studies, data from previous projects, and contextual techniques.

---

[1] Description summarized from information on the Nokia website *www.nokia.com*, as of February 2001.

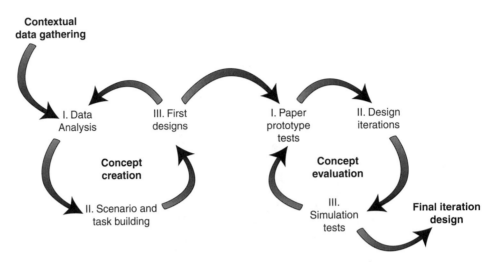

Figure 15.2 The user-centered concept and product development cycle.

2. Scenarios and then task models are built by analyzing the data collected, and initial designs are proposed.

3. Many iterations of design and evaluation are performed before the final design emerges. During this process, it may be found that more data is required, so further data gathering is conducted. The evaluation involves contextual interviews with paper-based prototypes to get feedback on first designs, and usability testing once the design is sufficiently advanced. Evaluation sessions emphasize the most important user tasks, as determined by the data gathering.
   • Once the design is advanced enough, high-fidelity simulations of the design are constructed.
   • Simulation tests are conducted with end users, and expert reviews are performed. Functional prototypes are tested with end users for feedback on long-term acceptability, efficiency, and utility of the concept.

4. During the last iteration phase, the final design is tested with end users and expert usability specialists.

ACTIVITY 15.2   How does this cycle of activities differ from the interaction design model introduced in Figure 6.7?

Comment   This cycle also has a focus on iteration through prototyping and evaluation, which is the basis of the model in Chapter 6. However, this cycle distinguishes between concept creation and concept evaluation. Scenarios and task modeling are used at the concept creation phase but simulation tests are used in the concept evaluation phase.

What challenges does this approach raise?   Nokia is very conscious of the need for iterative design and evaluation in the development of mobile communicators. They also use participatory design to a degree, but they point out that users will not necessarily have the vision of future possibilities that would allow innovative design in the same way as they might if asked to help design a familiar application like a web browser. Nokia is also well aware of the challenges of evaluating an innovative product like a communicator. These include:

- The difficulty of testing in all possible scenarios.
- The difficulty of testing human communication practices, especially when developing innovative products that will encourage novel behavior.
- The difficulty of testing services that cannot all be known beforehand.

What happens when the product is new and there are no users to test? At Nokia, quick and effortless access to critical tasks is a key design driver, and usability tests are used to evaluate the flow of tasks that have been found critical for mobile devices.

In a competitive and innovative market, other evaluation challenges may also arise. For example, consider the original Nokia communicator (the N9000). This was the first of its kind on the market. This had implications for how it could be evaluated because the device could not be shown to people outside the development team for fear of losing the "first-in-the-market" advantage. Thus the first version on the market did not have the benefit of testing with real users. Although extensive paper-based prototyping and simulations were produced, the evaluations were limited to a small group of people.

What methods does Nokia use? Nokia uses a number of methods in its development cycle, in particular "usage scenarios." Usage scenarios are high-level descriptions of uses of the device, based on data collected from representative stakeholders. They differ from the generic scenarios described in Chapter 7 in that they focus specifically on concept creation and high-level design considerations. An example of a usage scenario developed by Nokia is given in Figure 15.3.

What do design teams do next once they have created a set of scenarios? At Nokia, the design teams use the usage scenarios they have developed to identify critical user tasks and their structure. These task descriptions, which are more detailed than the original descriptions provided in the usage scenarios, are then used to consider lower-level design issues. A sample critical user task is shown in Figure 15.4.

ACTIVITY 15.3   To create scenarios, appropriate tasks and stakeholders will need to be identified. Who would the stakeholders be, and what techniques might be used to investigate their needs?

Comment   First, the tasks to be performed and the stakeholders who might be asked about requirements would have to be identified. Stakeholders for a mobile device include users, developers, telephone companies, computer hardware and software vendors, and their shareholders. At least in theory, a user may be almost any member of the population, but in practice, only certain sections of the population are likely to be users. Given the wide functionality of the communicator, the most likely users are professionals.

Example of a Usage Scenario

David works as a legal consultant in an international corporation. He uses a communicator daily for light note taking and communications as well as for his personal organization.

8 A.M. The working day starts with a multiparty conference call to Japan. He uses the communicator as a speakerphone to be able to type notes in it at the same time. At the end of the meeting, he sends everybody a copy of the notes via email directly from the communicator.

1 P.M. At the airport, he downloads all his new email messages to his communicator so that he can start working on them during the flight. On the plane there is always plenty of time to write answers to messages. While downloading, he views the communicator calendar for the day and remembers having promised to send his business card to a potential client. He does this while standing in line for boarding.

At his destination, he switches the communicator phone on, and it automatically starts sending the replies written on the plane. At the same time David can continue reading the rest of the messages.

2:30 P.M. His secretary back in London sends him a calendar reservation for the following week. David checks his calendar in the communicator and accepts the request. His communicator sends the confirmation automatically to the secretary and marks the appointment in David's calendar.

Figure 15.3 An example usage scenario.

If we assume that the user group is professional, then it is necessary to find out more about the tasks they perform. This could be done using questionnaires, interviews and observation, or focus groups, but there would be some other issues to consider. A professional who is constantly on the move will be difficult to track down. However, interviews and questionnaires can be administered in different settings such as at trade fairs where many professionals are all gathered in one place. This would potentially provide a ready audience, reduce travel expenses, and supply immediate responses.

Performing standard observations in an office has its problems, but observing someone on the move, in all the possible locations in which they might use the device, opens up a whole new set of issues. Mobile devices are intended to be used anywhere, so where are observations performed, and how closely can the participants be followed?

What usability and user experience goals are important in designing this kind of device? A mobile communicator would be expected to meet the normal usability goals that we have discussed before. But what about user experience goals? Personalization has been identified as significant in user satisfaction; however, a balance

---

User Tasks: Classification

(1)   Done under pressure: very critical
(2)   Done frequently: critical
(3)   Medium frequency or medium pressure
(4)   Not frequent or not done under pressure

Sample 1: User tasks in person-to-person voice communication

Call-making/in-call

(1)   Making a call to an emergency number

(1)   Answering a call

(1)   Rejecting a call

(2)   Making a call to frequently called numbers (usually 4–10 of them)

(2)   Making a call by manually entering each digit

(2)   Redialing a number/person

(2)   Indication of being busy

(3)   Making a call to semifrequently called numbers (e.g., a vet, hairdresser)

(4)   Making a call to occasionally called numbers (i.e., numbers that are often called only once).

Phone book memory

(1/4) Saving a name and number [1 = very critical during a call]

(2/3) Recalling a name and number and dialing [2 = to a frequently called number]

(4)   Editing a name and number

(4)   Erasing a name and number

(4)   Browsing the contents of a phone book, etc.

Sample 2: User tasks in text messaging

Sending

(4)   Sending a text message to a contact in the phone book

(4)   Setting a message center number, etc.

Receiving

(2)   Reading and replying to a message

(2)   Reading and calling back the sender

(3)   Reading and erasing a message

(4)   Reading and storing a message with a new name, etc.

Figure 15.4 Sample user tasks.

BOX 15.1   Designing an Interface with a Small Number of Keys

What would you do if you had to design a communication device that could accommodate a maximum of only 15 keys? The device has to support numerical and text input. How could you design the mapping of the 15 keys to the various kinds of operations proposed so as to support the range of user tasks identified?

As a minimum, the device will need an on/off switch, a switch for connecting and disconnecting to the network, and a mechanism for entering ten numbers, 26 characters, and the space. You may decide to omit punctuation and capital letters, although this will have implications for the usability of the device. One way in which the functionality can be achieved is described below

One key is a dedicated on/off switch, one key is a dedicated connect/disconnect key, and one function key toggles the keys from numerical to character input. Ten of the keys represent the digits 0–9 and two or three characters each (26 characters plus the space bar means that seven of the keys must represent three characters and three of them must represent two). This uses 13 keys, which means that the keys for on/off, connect/disconnect, and the function button can be made larger to distinguish them from the others.

Alternatively, if you want to include punctuation, then all ten keys could have three characters each (giving room for four punctuation marks), and if you want to include capitals then a 14th key might be used as a "shift" key. Remember, though, that if the device is to be operated with one hand then this must operate more like a "caps lock" key than a shift key.

Some devices let you choose a character using only two keys. One key is repeatedly pressed as it toggles through the character list, and the second is used to accept the choice when the required character is displayed. This design choice requires more time from the user and its suitability depends on the functions the device is to support.

---

must be struck between allowing flexibility and providing sensible default values so that users don't have to customize settings unless they want to.

Mobile communicators are intended to support users wherever they are, so they must be compatible with the users' lifestyles. Designers must therefore understand the design characteristics that make the communicator attractive to different user groups, and those characteristics that will vary from group to group. If we consider the users as business people, then the important user experience goals are likely to include being helpful, motivating, aesthetically pleasing, and rewarding. If we consider children, then entertainment and fun are likely to be more important, while for teenagers its physical appearance might be more significant.

How does Nokia design a communicator's physical aspects?   Deciding how many keys to have and how to map them onto a much larger set of functions is a difficult design challenge in any mobile device (see Box 15.1). For example, in the Nokia 7110 mobile phone, the problem of limited keys and limited space was dealt with by providing softkeys with context-sensitive functions that change depending on where the user is in the interaction sequence. This allows the keys to perform different functions depending on the other contextual issues. The softkeys allow the user to do a variety of things, such as make selections, enter, edit, or delete text. The current label for each softkey is displayed at the bottom of the screen, near the relevant key. There is,

1. Power key. Used for switching the phone on and off. When pressed briefly the user enters the list of profiles (user environments: e.g., *Silent* to turn off all the phone tones).

2. Navi Roller. Used for navigating the Menu and the Phonebook. Navi Roller allows scrolling up and down as well as *selecting, saving*, or *sending* the displayed item by clicking the roller.

3. Two Softkeys. The softkeys are assigned actions that enable the user to manipulate the user interface by making selections and *entering, editing*, and *deleting* text. The name of the action changes according to the state of the phone. Descriptive labels are shown in the lower corner of the display respective to the key underneath.

4. Send key (green receiver). *Send key* is used for call handling, that is, call creation, and also for bringing up the last-called numbers list.

5. End key (red receiver). *End key* is for call termination. It is also an Exit key that can be used as a panic key since it takes the user from any state of the phone to the idle state without saving changes.

6. Numeric keys, with an alphabet according to the ITU-T.161 standard. Used for number and character input. The 1 key also doubles as the Voice Mailbox speed dial key. The # key is used for changing the character case during editing. Nokia 7110 employs a predictive text input method: only one keypress per letter is required, and the entered text string is continually matched with the words in the built-in dictionary.

• The left softkey is basically used as a yes/positive key. It contains options that execute commands and go deeper into the menu structure. In the idle state the left softkey is **Menu** (the hierarchy of phone functions).

• The right softkey is basically used as a no/negative key. It contains options that cancel commands, delete text, and go higher in the menu structure. In the idle state the right softkey is **Names** (the Phonebook).

Figure 15.5 The Nokia 7110 mobile phone.

of course, a balance to be struck between having too many softkeys, each with limited functionality, and having only a few keys that can be overloaded with too many functions. In the end, the Nokia 7110 (Figure 15.5) was designed with just two softkeys that performed multiple functions. (Väänänen-Vainio-Mattila and Ruuska, 2000).

Textual input becomes a major problem when the number of input keys is restricted by the design. Having only a small number means the users must constantly "peck" at a few keys, typically using their thumbs. Trying to place too many keys in a heavily constrained space means that the user is likely to press the wrong key or two keys at once. How was this problem handled by Nokia? They opted for a small number of keys but in combination with a way of speeding up the typing of words, through having the communicator guess what the user is writing. In particular, the Nokia 7110 introduced the T9 predictive text method that allows speedy input of words based on a dictionary. The phone proposes a likely word once the user has typed a few characters. The user then either selects the proposed word and moves on to the next word, or rejects it and continues to enter the current word.

Communicators have also been designed to include a function button to let the user customize the interface to a limited degree, for example by allowing a favorite application to be associated with one of the hard keys.

BOX 15.2   Designing Telephones for the Elderly and Disabled

The British Royal National Institute for the Blind (RNIB), together with the British Department of Trade and Industry and British Telecommunications, have compiled a brochure to explain the different impairments affecting many telephone user groups, together with a set of suggested telephone features that could greatly enhance the accessibility of devices for such user groups. They identify 15 impairments and 44 features that could be added to telephones to make their use more pleasant. The impairments include cognitive impairment, weak grip, limited dexterity, speech impairment, hearing impairment, and hand tremor (Gill and Shipley, 1999). Features that could make a difference to these user groups include:

- Guarded or recessed keys to help prevent pressing the wrong key by mistake.

- Sidetone reduction, which reduces the amount of noise picked up from the environment and mixed with incoming speech at the earpiece.

- Allowing the user to adjust the amount of pressure needed to select a key. Apart from the more obvious consequences of too much or too little pressure, unsuitable key pressure may produce muscle spasms in some users.

- Audio and tactile key feedback to indicate when a key has been pressed.

Is it possible to design consistent interfaces, given the physical constraints of a communicator? A particular problem when developing software for a small display with limited input controls is how to make the interface consistent.

The design dilemma of consistency was addressed in Chapter 1. Consistency is often extolled as a virtue, yet it is sometimes appropriate to be inconsistent. In the design of communicators, the problems of consistency arise again. The device needs to have external consistency, i.e., consistency with users' expectations from their use of other similar tools, and also internal consistency, i.e., consistency with other items of software that the device supports. Sometimes these two design goals are in conflict, and it is appropriate to design a new solution for a particular situation.

The N9000 web browser was developed for the Nokia N9000 communicator. Many design decisions had to be dealt with, especially the problem of consistency (Ketola et al., 2000). Nokia has an internal style guide that all its products must follow in order to maintain internal consistency. External consistency with PC-based products is difficult to achieve because of physical constraints, and because the operating system for the N9000 is not commonly used with a PC. Other constraints on the design were:

1. The N9000 does not have a pointing device. Pointing is therefore done by selection using the scrolling bars. Scrolling down causes selection to jump from one hyperlink to the next; scrolling up causes it to jump to the previous link.

2. In cellular devices, connection rate is limited to 9600 bps, which is slower than the fixed-line rate. Connection can also take up to 30 seconds, considerably slower than the fixed-line equivalent. Web users may be accustomed to slow downloading times, but a long connection time is a new

phenomenon. A progress indicator was included in the design so that users would not become frustrated and start pressing other buttons. This leads to a further external consistency issue: should web pages be made to look the same as on faster desktop machines, or should they be designed for faster downloading?

Specific design decisions and solutions taken under these constraints were as follows:

1. The default page for a desktop web browser is a home page, but because of the connection time and the speed of downloading, the N9000 browser defaults to a list of favorite pages (called the Hotlist) instead. Thus, the default state is offline. This violates external consistency, but proved to be acceptable to users.

2. The functionality of the N9000 browser had to be carefully examined. Because of the Nokia style guide, only three buttons were available for navigating through the function hierarchy, so navigation became a major issue. To cope with the limited availability of command buttons, the N9000 employs the idea of *views*, within which only certain functions are possible. For the web browser, three views were provided: Hotlist view, Document view, and Navigation view. Users can select a document in the Hotlist view and enter the Document view. From here they are able to save, read, disconnect from the network, and close the document. However, they cannot navigate through the document. For this they need to go to the Navigation view. This conceptual shift was difficult for users to come to terms with.

3. The style guide dictated that the fourth command button be used to move upwards in the view hierarchy. It is also a part of the style guide that this button should be called "Back." In other applications this may not be a problem, but in the context of a web browser, a button labeled "Back" is interpreted differently. Internal consistency had to be obeyed here, and so the command that moved back to the previous page in the history list was called "Previous." This caused considerable confusion for users.

4. Optimizing web pages for display on mobile communicators involves the following three issues: content, because it's important to optimize download times; page layout, because of the small size of the screen; and navigation, because it's important to minimize the number of file downloads. User trials showed that, in the mobile context, users are more interested in getting the text information quickly than in downloading the graphics. Downloading unwanted pages also proved to be considered a key aspect of usability. Good link naming and clear, predictable behavior were important because of the long downloading times; locating the wrong page expends much time and cost.

ACTIVITY 15.4 If you are sitting near a desktop computer, study the interface of the piece of software that is running. If you are not near one, then think of the application you run most regularly on a desktop machine. Imagine what this interface would look like if you were to reduce the screen size to a mere 158 mm × 56 mm (the size of the Nokia 9210 communicator). What difficulties can you see? What implications do you think this has for software design, and also for the user who is swapping between desktop systems and mobile systems on a regular basis?

Comment   If the same screen design is carried over to the mobile device then either everything will have to be miniaturized, so that the tool bars, icons and menus will become unreadable, or left at the same size, so that they will take up too much space on the screen. The interface therefore must be designed differently. This has implications for consistency for users who might be using the same application in a desktop environment and on the mobile device.

What kind of user testing does Nokia use?   As mentioned earlier, there were confidentiality problems in testing the first generation of communicators on the intended user population. Hence, user testing could be done only after the product was released on the market. One kind of summative testing Nokia did was to find out what questions people have when first using the communicator. Users were given the device to use for some weeks and were then asked to report on positive and negative features. The results from this study confirmed the developers' concerns about the effects of consistency with other similar applications designed to run on desktop machines. Another study involved sending questionnaires to more critical communicator users whose experience ranged from 0 to 12 months, to find out if their reactions were similar.

As can be seen from this case study, Nokia uses a number of methods to develop their communicators for the general public. Furthermore, many design decisions and problems have to be dealt with, ranging from the lack of real users for testing, to how to let users send text messages with only a few keys and a very confined space.

15.3.3   Philips' approach to designing a communicator for children

We now consider how another company went about designing a mobile communicator aimed at a specific user group, children (mostly girls) aged between 7 and 12. Developing a tool for this user group is quite different from developing a tool for use by the general public, where there is likely to be a huge range of different users. An advantage of designing a device for a smaller set of users is that they are likely to have similar needs and preferences, meaning that the device can be customized much more to their requirements. This case study draws on material reported in Oosterholt et al. (1996).

Which approach did Philips use?   The Philips process of development for this particular communicator made extensive use of prototyping techniques and participatory design. Children were involved from the initial concepts stage right through to final product testing. Each time a prototype was produced, it was shown to children for comment and feedback. A central part of the design process involved developing interface metaphors. Again, when ideas for metaphors were

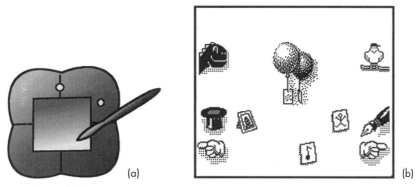

Figure 15.6 (a) The communicator with pen. (b) Product display showing 'the world'.

proposed, the designers turned to the girls in a spirit of participatory design in order to elicit their responses.

What usability and user experience goals were considered important? In the Nokia communicator example we saw the importance of usability goals focusing on effectiveness and efficiency, especially the need to move smoothly among critical tasks. In contrast, Philips focused more on the user experience goals of being enjoyable, entertaining, and fun. Other goals were that it should encourage creativity and provide personal and magical applications. The girls had expressed a specific desire for these.

What functionality did the communicator provide? The communicator was designed to have a touch-sensitive screen, pen input, infrared communications, and audio output (see Figure 15.6(a)). The interface was built on the metaphor of a world in which the users can move around freely, picking things up and starting applications (see Figure 15.6(b)). Available applications include a calendar, alarm clock, photo album, fortune teller, and communicator. The user can also perform tasks such as writing letters, composing tunes, drawing pictures, and sending them to other similar devices (see Figure 15.7).

What methods were used? Development of the product was divided into four phases: initiation, concept creation, specification, and finalization. Whereas Nokia adopted techniques from contextual design, Philips used mainly low-fidelity prototyping techniques for this particular project. Different prototypes were used throughout the development and for different purposes.

During the initiation phase, foam models were used to elicit feedback on the color, shape, size, styles, and robustness of the device, among other things. Using group discussions to encourage the youngsters to express their opinions a lot of feedback was gained from the foam models, even though the models contained no functionality. For example, children liked the idea of protecting the screen when carrying it, so they wanted different bags and cases to be provided for it; privacy was an important aspect, so they did not want it easily accessible by others; the pen should be stored safely within the device rather than underneath it for fear of it

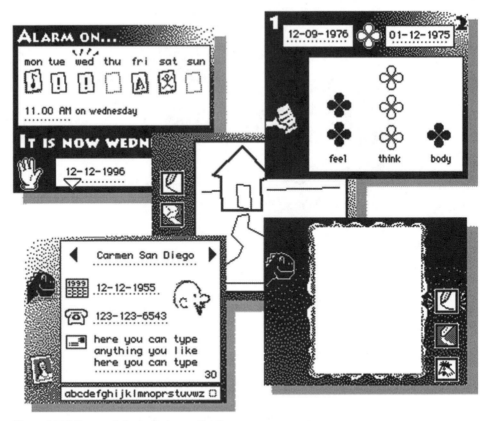

Figure 15.7 Some of the built-in applications.

being lost. One surprising result was that the children did not like the colors. The initial colors were bright (See Figure 15.8 on Color Plate 8), but they wanted dark colors more akin to their parents' hi-fi equipment at home.

The session with the models also provided input for the first user interface design, which was animated using a computer-based tool. This was used to explore navigation, pen-based dialog, types of application, and visual style.

During the concept creation phase, dynamic visualizations, which are like the storyboards described in Chapter 8 but are computer-based, were used to capture the initial ideas about interface and functionality (see Figure 15.9).

During the specification phase, foam models were again used to decide the size of the screen appropriate for writing on while standing up. As well as the size, different display formats were simulated (see Figure 15.10). These prototypes proved to be effective, again eliciting a lot of useful feedback. For example, left-handed users used the upper left part of the product to lean on while writing and the right-handed children used the lower right portion, yielding the design implication that the product should have hand resting places at these two points.

Also during specification, ideas for the interface design were evaluated by youngsters at a fair. There were two main contenders for the interface design.

Figure 15.9 The first dynamic visualizations.

One provided direct access to each of the applications in the device, represented as a static matrix of options. This meant that the visual presentation and size of the applications was limited by the size of the screen. The other interface worked by indirect access, through a navigation model based on the idea of a window moving over a linked list of options.

Prototyping was also used in the finalization phase for market evaluations.

Figure 15.10 Foam models for investigating display size and screen format.

ACTIVITY 15.5   Prototypes are often used to answer specific questions. In this development, what questions were answered by producing and evaluating the foam models?

Comment   Foam models were used at two specific points in the development to answer clear questions. The first set was used to consider the physical design such as size and color. They also elicited comments about storing the pen, covering the display, and having a carrying bag. The second set was used to design the display size and format. This also had the side effect of finding out useful information about where children would rest their hands on the device.

How much did the children participate in the design?   One of the problems with participatory design is knowing how much to involve the users. Trying to involve children too much can be counterproductive, boring them and sometimes making them feel out of their depth. Asking children to participate too little can end up making them feel as if their views and ideas are not being sufficiently taken into account.

The Philips design team involved the children in design and evaluation from the very beginning. The first participatory design session was held during the initiation phase at a local international primary school. The session investigated the social and personal lives of 7 to 12 year-olds. Groups of 8 to 10 children were engaged in discussions and were asked to draw sketches of their ideal product. They were also asked to write stories about the use of the product, so that designers could get some contextual information about how it might be used. From this first session, it was clear that the concept was well received by the children. They particularly liked the communication, the pen-based interface, and its multifunctionality.

There were clear differences between boys, who wanted a broader range of functionality, and girls, who focused on communication. The ability to personalize was important to both groups. For example, one girl wanted the device to cough when a message arrived so that the teacher wouldn't know she was using it during class.

The whole design team was present at participatory design sessions. Spending time to get the children's opinions and to enter their world to understand how they perceive things was important for the success of the product.

One lesson that the designers drew from this exercise echoes a comment by Gillian Crampton Smith in the interview at the end of Chapter 6: users are not designers. In this instance, the children were limited in what they could design by what they knew and what they were used to. Another stakeholder group, parents, expected keyboard input, as they believed this to be more sophisticated than pen input, which was seen as old fashioned.

On the other hand, children are often more imaginative than adults, so involving the children was useful when discussing innovative ideas, or when only partial ideas were available. Working with children like this rather than adults requires a different approach, yet both adults and children need to appreciate each others' strengths and weaknesses. Box 15.3 describes the intergenerational design teams that Druin works with in projects at the University of Maryland.

BOX 15.3 Children and Adults Bring Participant Observation Close to Design

Allison Druin designs innovative technology with intergenerational design teams in which children and adults work together (Druin, 2000). In her teams children and adults observe children interacting with low-fidelity prototyping materials—crayons, pens, paper, glue, scissors, felt, furry cloth, Lego, animal parts, etc. (Figure 15.11), to explore ideas. By keeping more complex technology, such as computers, out of the picture during early observation and brainstorming sessions, adults do not dominate the scene.

Both adult and child members of Druin's teams observe and take notes while other children interact with the prototypes. This enables the team to capture impressions from both child and adult perspectives. Originally observations were recorded on a data-capture form like that

in Figure 15.12 but many children prefer to draw and write simple notes like those in Figure 15.13. Adults, on the other hand, generally prefer to write, so now the team uses both techniques.

A typical observation session includes a pair of observers, an interactor and a child. The interactor's role is to ask questions that initiate discussion about the activities. Without this essential role, the children tend to feel that they are on stage being observed. But when engaged in discussion they are more likely to relax and reveal their real behavior and opinions. It is also important that the interactor does not take notes because this can make the children feel that they are being tested. Therefore, video is used to record observations.

Figure 15.11 Early design ideas for "fluffy-fuzzy" robots using low-tech prototyping materials.

Time	RAW DATA: Quotes	Activities	DATA ANALYSIS: Activity Patterns	Roles	Design Ideas
10.05	E: Can you draw whatever you want? [Gustav: Yes.] E: (To K) A Christmas Tree? K: Yes!	K. takes the mouse rapidly, draws a red tree, takes the yellow crayon	Drawing	Artist	
		draws something in the corner, rubs out, continues	Drawing Erasing	Artist	
		E. tries to take the mouse	Struggling for control of input device	Leader	Multiple input devices
	E: But I want the long one! E: Noo! [Difficult to erase.] E: There. E: But what's this? [Windows Start menu appears.]	E gets the mouse, tries to get the blue crayon, looks irritated when she cannot get the blue one, gets it	Difficulty selecting tools	Frustrated User	Easier way to select tools

Figure 15.12 Excerpt from a data capture form.

Figure 15.13 Sample notes illustrating a child's observation.

ACTIVITY 15.6   Suggest ways of helping adults and children feel comfortable together and gain mutual acceptance.

Comment   Allison Druin asks everyone to dress casually in jeans, sneakers and T-shirts. The group works together at shared tables or on the floor. Snacks are important in creating a relaxed environment, and everyone uses first names. The goal is to create a group in which everyone respects each other's contributions and accepts and welcomes different contributions. Children are used to being controlled by adults and adults are used to being in control, and it takes time to break down these ingrained stereotypes.

What conceptual models did they design?   By the concept creation phase, the importance of four goals for the product and its interface had emerged:

1. to support communication by stimulating social interaction among children
2. to evoke creativity and fantasy
3. to be "alive"—unexpected fun things should happen, surprising and pleasurable to the user, that give the product more character
4. to enhance intimacy—the product is a personal asset containing personal information

Five metaphors were developed by designers based on these values. Each metaphor was represented by a story. Figure 15.14 shows an illustration of one metaphor: the wizard. Specific metaphor workshops were conducted to find out how the girls reacted to the metaphors. They were asked to create a collage to visualize the metaphors, showing what they understood by them. The collages were a combination of drawings, essays, and existing pictures. The metaphor workshop showed that the girls were interested in being able to create, communicate, and organize personal things.

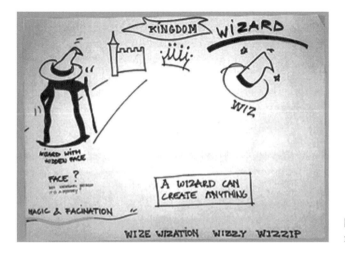

Figure 15.14 One of the metaphors: the wizard.

How did they evaluate the conceptual model?   During the finalization stage, usability evaluations with children were performed to investigate the user interface itself and also to answer specific questions concerned with ideas for games, and writing performance. In most sessions, users were asked to play with the device for a certain period of time before giving feedback.

What lessons were learned from this case study?   Many lessons were learned from developing an innovative product using a combination of participatory design and user testing. Some practical advice offered by Oosterholt and colleagues that can be generalized to the design of other interactive products is:

> *Specify Your User Requirements And Define Milestones* The rationale behind specifying user requirements is not just to develop them, but to make sure that the team agrees on the assumptions and realizes how and when they have been and can be changed.
> *A Product Is Not Designed in a Vacuum* Start thinking about additional and follow-up products at an early stage, so one does not have to change suddenly or add extra functionality in a later phase.
> *Users Are Not Designers* Not all answers can be generated by user or market tests. Users will generally relate any new product concept to existing products.
> *Act Quick And Dirty If Necessary* Often, the purpose of user testing is not to decide whether one interface concept is more usable than an alternative concept, but to discover issues that are important to the children. Small qualitative sessions of user involvement are therefore often appropriate. Furthermore, such sessions provide an opportunity for designers to "enter" the children's world.

15.4 Redesigning part of a large interactive phone-based response system

In this case study, we focus on quite a different kind of system, one being re-designed for a specific application intended to provide the general public with advice about filling out a tax return—and those of you who have to do this know only too well how complex it is. The original product was developed not as a commercial product but as an advisory system to be interacted with via the phone. We report here on the work carried out by usability consultant Bill Killam and his colleagues, who worked with the US Internal Revenue Services (IRS) to evaluate and redesigned the telephone response information system (TRIS).

Although this case study is situated in the US, such phone-based information systems are widespread across the world. Typically, they are very frustrating to use. Have you been annoyed by the long menus of options such systems provide when you are trying to buy a train ticket or when making an appointment for a technician to fix your phone line? What happens is that you work your way through several different menu systems, selecting an option from the first list of, say, seven choices, only to find that now you must choose from another list of five alternatives. Then, having spent several minutes doing this, you discover that you made the wrong choice back in the first menu, so you have to start again. Does this sound familiar? Other problems are that often there are too many options to remember, and that none of the options seems to be the right one for you. In such situations, most users long for human contact, for a real live operator, but of course there usually isn't one.

TRIS provided information via such a myriad of menus, so it was not surprising that users reported many of these problems. Consequently a thorough evaluation and redesign was planned. To do this, the usability specialists drew on many techniques to get different perspectives of the problems and to find potential solutions. Their choice of techniques was influenced by a combination of constraints: schedules, budgets, their level of expertise, and not least that they were working on redesigning part of an already existing system. Unlike new product development, the design space for making decisions was extremely limited by existing design decisions and the expectations of a large existing user population.

15.4.1 Background

Everyone over age 18 living in the US must submit a tax return each year either individually or included in a household. The age varies from country to country but the process is fairly similar in many countries. In the US this amounts to over 100 million tax returns each year. Completing the actual tax return is complex, so the IRS provides information in various forms to help people. One of the most used information services is TRIS, which provides voice-recorded information through an automated system. TRIS also allows simple automated transactions. Over 50 million calls are made to the IRS each year, but of these only 14% are handled by TRIS. This suggested to the designers that something was wrong.

15.4.2 The redesign

How do users interact with the current version of TRIS? The users of TRIS are the public, who get information by calling a toll-free telephone number. This takes them to the main IRS help desk, which is in fact the TRIS. The interface with TRIS is recorded voice information, so output is auditory. Users navigate through this system by selecting choices from the auditory menu that they enter by typing on the telephone keypad. First, the users have to interact with the Auto Attendant portion of the system—a sort of simulated operator that must figure out what the call is about and direct it to the proper part of the system. This sounds simple but there is a problem. Some paths have many subpaths and the way information is classified under the four main paths is often not intuitive to users. Furthermore, some of the functionality available through TRIS is provided by two other independent systems, so users can become confused about which system they are dealing with and may not even know they are dealing with a different system. Users get very few clues that these other systems exist or how they relate to each other, yet suddenly things may be quite different—even the voice they are listening to may change. Navigating through the system, with its lack of visual feedback and few auditory clues, is difficult. Imagine being in a maze with your eyes blindfolded and your hands tied so you can't feel anything, and where the only information you get is auditory. How can you possibly remember all the instructions and construct an accurate mental model in your head to help you?

Once in TRIS, users can take various paths that:

- Provide answers to questions about tax law (provided by one of the two other computer systems accessible through TRIS).

- Allow people to order all the forms and other materials they need to complete their tax return (provided by the two other systems accessible through TRIS).

- Perform simple transactions, such as changing a mailing address, ordering a copy of a tax return, or obtaining answers to specific questions about a person's taxation.

- Reach a live operator if none of the above options are applicable or the user cannot figure out how to use the system.

ACTIVITY 15.7	Why is developing an accurate mental model of TRIS difficult for users?
Comment	Much of TRIS is hidden to the users. Their interaction with it is indirect, through listening to responses from the system and pressing various keys (whose meaning is always context dependent). There is no visual interface and users have only speech output to support their mental model development. Because speech is transient, unlike visual feedback, users must work out the conceptual model without visual cues. The user interface to this system is a series of menus in a tree structure and, since human short-term memory is limited, the structure of the system must also be limited to only a few branches at each point in the tree. Another problem is that TRIS accepts input only from the telephone number keypad, so it's not possible to associate unique or meaningful options with user choices.

What are the main problems identified with the existing version of TRIS? Because one of the main problems users have when using TRIS is developing a mental model of the system it is hard for users to find the information they need. In addition, TRIS was not designed to reveal the mapping of the underlying systems and often did things that made sense from a processing point of view but not from the user's. This is probably because the programmers took a data-oriented view of the system rather than a user-oriented one. For example, TRIS used the same software routine to gather both a social security number and an employee identification number for certain interactions. This may be efficient from a code-development standpoint, since only one code module needs to be designed and tested, but from the user's perspective it presented several problems. The system always had to ask the user which type of number was expected, even though only one of these numbers made sense for many questions being asked. Consequently, many users unfamiliar with employee identification numbers were not sure what to answer, those who knew the difference wondered why the system was even asking, and all users had yet another chance to make an entry error.

What methods did the usability experts use to identify the problems with the current version of TRIS?   To begin with the usability specialists did a general review of the literature and industry standards and identified the latest design guidelines and current industry best practices for interactive voice response (IVR) systems. These guidelines formed the basis for a heuristic evaluation of the existing TRIS user interface and helped identify specific areas that needed improvement. They also used the GOMS keystroke-level modeling technique to predict how well the interface supported users' tasks. Menu selection from a hierarchy of options is quite well suited to a GOMS evaluation, although certain modifications were necessary to estimate values for average performance times.

What did they do with the findings of the evaluation?   Once the analysis of the existing interface and user tasks was complete, the team then followed a set of design guidelines and standards, to develop three alternative interfaces for the Auto Attendant part of TRIS. An expert peer panel then reviewed the three alternatives and jointly selected the one that they considered to have the highest usability. The usability specialists also performed a further GOMS analysis for comparison with the existing system. The analysis predicted that it would only take 216.2 seconds to make a call with the new system, compared with 278.7 seconds with the original system. While this kind of prediction can highlight possible savings, it says little about which aspects of the redesign are more effective and why. The usability specialists, therefore, needed to carry out other kinds of user testing.

ACTIVITY 15.8	Why is it that the results from a GOMS analysis do not necessarily predict the best design?
Comment	The keystroke-level analysis predicts performance time for experts doing a task from beginning to end. Not all of the users of TRIS will be experts, so performance time is not the only predictor of good usability.

The usability specialists did *three iterations* of user testing in which they simulated how the new system would work. When they were confident the new Auto Attendant interface had sufficient usability, they redesigned a subset of the underlying functionality. A new simulation of the entire Auto Attendant portion of TRIS was then developed. It was designed to support two typical tasks that had been identified earlier as problematic, to:

- find out the status of a tax refund
- order a transcript of a tax return for a particular year

These tasks also provide examples of nearly all of the user–system interactions with TRIS (e.g., caller identification, numeric data entry, database lookup, data playback, verbal instructions, etc.). A separate simulation of the existing system was also developed so that the new and existing designs could be compared. The user interaction was automatically logged to make data collection easier and unobtrusive.

What conflicts can arise when suggesting changes for improvement?   When carrying out an evaluation of an existing product, often "jewels in the mud" stick out—glaring usability problems with a system that, if changed, could result in significant improvements. However, conflicts can arise when suggesting such changes, especially if they may decrease the efficient running of the system. The usability specialists quickly became aware that the TRIS system was making too many cognitive demands on users. In particular, the system expected users to select from too many menu choices too quickly. They also realized that immediate usability improvements could be gained by just a few minor changes: breaking menu choices into groups of 3–5 items; making the choices easier to understand; and separating general navigation commands (e.g., repeat the menu or return to the top menu) from other choices with pauses. However, to make these changes would require adding additional menus and building in pauses in the software. This conflicts with the way engineers write their code: they are extremely reluctant to purposely add additional levels to a menu structure and resist purposely slowing down a system with pauses.

ACTIVITY 15.9   The gap between programmers' goals and usability goals is often seen in large systems like TRIS that have existed for some time. How might such problems be avoided when designing new systems?

Comment   It can be hard to get changes made when a system has been in operation for some time, but it is important for interaction designers to be persistent and convince the programmers of the benefits of doing so. Involving users early in design and frequent cycles of 'design-test-redesign' helps to avoid such problems in the design of new systems.

How were the usability tests devised and carried out?   In order to do usability tests, the usability specialists had to identify goals for testing, plan tasks that would satisfy those goals, recruit participants, schedule the tests, collect and analyze data, and report their findings. Their main goals were to:

- evaluate the navigation system of the redesigned TRIS Auto Attendant
- compare the usability of the redesign with the original TRIS for sample tasks

Twenty-eight participants were recruited from a database of individuals who had expressed interest in participating in a usability test. There was an attempt to recruit an equal number of males and females and people from a mixture of education and income levels. The participants were screened by a telephone interview and were paid for their participation. The tests were conducted in a usability lab that provided access to the two simulated TRIS systems (the original design and the redesign). The lab had all the usual features (e.g., video cameras) and a telephone. Timestamps were included in the videotape and the participants' comments were recorded.

The order of the tasks and the order in which the systems were used was counter-balanced. This was done so that participants' experience on one system or task would not distort the results. So, half the participants first experienced the original TRIS design and the other half first experienced the redesigned TRIS system. That way, if a user learned something from one or other system the effects would be balanced. Similarly, the usability specialists wanted to avoid ordering effects from all the participants doing the same task first. Half the participants were therefore randomly allocated to do task A first and the other half to do task B. Taking both these ordering effects into account produced a 4 × 4 experimental design with eight participants for each condition.

ACTIVITY 15.10   Compare the description of this testing procedure with that for HutchWorld in Chapter 10. What differences do you notice and how can they be explained?

Comment   The testing for HutchWorld is more typical. There were fewer participants and only one version of the system was tested at any time. In the TRIS test a larger number of participants were involved and the tests were more like an experiment. TRIS is complex, particularly the mapping between TRIS and the underlying functionality, although the system's purpose is clearly defined. By the time the usability specialists started the tests, they believed that they had fixed the major usability problems because they had responded first to the expert reviewers' feedback and then to the GOMS analysis. They were therefore confident that the new design would be better than the original one, but they had to demonstrate this to the IRS. This style of testing was also possible because there were thousands of potential users and the cost savings over 50 million calls justified the cost of this elaborate testing procedure.

How did they ensure that the participants tested were a representative set of users? In order to get demographic information to make sure the participants were representative, a questionnaire was given to all of them. It revealed a broad range of ethnicity, educational accomplishment, and income among the 18 women and 14 men who took part in the tests. Most had submitted tax returns during the last five years and most were experienced with interactive voice response systems. Eight participants indicated strong negative feelings about IVR systems, saying they were frustrating, time-consuming, and user-unfriendly.

What data was collected during the user testing? A total of 185 subnavigation steps made up the two tasks for the current TRIS. Participants successfully completed 91 steps on their first attempt (49% of the total). This was compared with a similar number of steps for the redesigned system: 187 subnavigation steps made up the same tasks for the redesigned TRIS. Participants were able to complete 117 of the steps on the first attempt (62% of the total), indicating an improvement of over 10%.

The average time to perform tasks was also analyzed. The summary data for the two tasks is shown in Table 15.1. As you can see, performance time on the redesigned system was much better for both tasks.

How was the user's satisfaction with the system assessed? At the end of each task, participants were asked to evaluate how well they thought the system enabled

Table 15.1   Average total task completion time by systems in seconds (s)

Task	Original system (s)	Redesigned system (s)
A	264.3	186.9
B	348.7	218.1

them to accomplish their tasks by completing a user satisfaction questionnaire. The responses again indicated that participants thought the redesign was easier to use and they preferred it. Regardless of the order in which participants used the two systems, the scores on the *redesigned* system were consistently much better than for the *original* system. The questionnaire provided statements that the participants had to rate on a 7-point scale. The difference between the two systems was highly significant, averaging over 3 rating-scale points higher on each statement.

ACTIVITY 15.10   User satisfaction questionnaires like the ones just described enable usability specialists to get answers to questions they regard as important. How can you make sure you collect opinions on all the topics that are most important to users?

Comment   Asking users' opinions informally after pilot testing the questionnaire helps to make sure that you cover everything, but it is not foolproof. Furthermore, you may not want to increase the length of the questionnaire. Two other approaches that could be used separately are to ask users to think aloud and to use open-ended interviews. However, the think aloud method can distort the performance measures, so that is not such a good idea. Open-ended interviews are better, and this was done by the usability specialists in this case.

Participants were also invited to make any additional comments they wanted about the two systems. These were then categorized in terms of how easy the new system was considered to navigate, whether it was less confusing, faster, etc. Specific complaints included that some wording was still unclear and that not being able to return to previous menus easily was annoying. No matter how much usability testing and redesign you do, there is always room for improvement.

Would it have been better to redesign the entire system?   It would have been far too expensive and time-consuming to redesign and test the whole system. A skill that usability specialists need when dealing with this much complexity is how to limit the scope of what they do and still produce useful results.

What other design features could be considered besides improving efficiency? Given that the system is aimed at a diverse set of users, many whose native language is not English, a system that uses different languages would be useful (the Olympic Messaging System used in the Los Angeles games did this very successfully). A range of voices could also be tested to compare the acceptability of different kinds of voices.

This case study has illustrated how to use different techniques in the evaluation and redesign of a system. Expert critiques and GOMS analyses are both useful tools for analyzing current systems and for predicting improvements with a proposed new design. But until the systems are actually tested with users, there is no way of knowing whether the predictions are accurate. What if users can theoretically carry out their tasks faster but in practice the interface is so poor that they cannot use it? In many cases, testing with real users is needed to ensure that the new design really does offer an improvement in usability. In this case study, results from usability testing were able to indicate that not only was the new design faster but users also liked it much better.

Summary

The three case studies illustrate how different combinations of design and evaluation techniques can be used effectively together to arrive at a design for a new product or redesign of an existing system. Quite different demands are placed on the design team when redesigning an existing product compared with designing a new product. Many practical problems and constraints will be encountered in both situations and experience of designing different systems will help you learn how to deal with them.

Key points

- Design involves trade-offs that can limit choices but can also result in exciting design challenges.
- Prototypes can be used for a variety of purposes throughout development, including for marketing presentations and evaluations.
- The design space for making changes when upgrading a product is limited by previous decisions.
- The design space is much greater when building new products.
- Rapid prototyping and evaluation cycles help designers to choose among alternatives in a very short time.
- Simulations are useful for evaluating large systems intended for millions of users when it is not feasible to work on the system directly.
- Piecing together evidence from data from different sources can provide a rich picture of usability problems, why they occur, and possible ways of fixing them.

Further Reading

BREWSTER, S., AND DUNLOP, M. (2000) (eds.) *Personal Technologies*. Special issue on Human Computer Interaction and Mobile Devices, 4, 2&3. This collection of articles discusses many issues in the design of mobile devices and would be a good starting point for anyone interested in pursuing this area.

BERGMAN, ERIC. (2000) (ed.) *Information Appliances and Beyond*. San Francisco, CA: Morgan Kaufmann. This book contains an excellent collection of practical articles describing how different information appliances have been developed, from interactive toys and games to a vehicle navigation system.

KILLAM, H. W. AND AUTRY, M. (2000) IVR interface design standards: A practical analysis. In Proceedings of HFES/IEA 44th Annual Meeting. This paper describes aspects of the TRIS study in more detail.

Reflections from the Authors

To end the book, we each present some of our views about interaction design.

Helen: When I worked as a programmer/analyst in the City of London, during the early 1980s, I was always surprised and impressed by the workarounds that my company's clients devised in order to make the software they used work for them. At the same time, of course, I was also disappointed that the software didn't support them better. The real end users were often not consulted during the development, and had the systems thrust upon them. The situation nowadays is so much better, and I think it's great that the importance of involving users is now so widely recognized.

There have been great technological advances, creating some quite incredible devices, but we also shouldn't forget the more mundane applications of technology, which at times I think we tend to ignore. As Gillian Crampton Smith said in her interview, the software we use has become an environment in which we spend a lot of our time, either at work or in our leisure. These are interactive systems too and deserve our attention to make them more usable.

But for me, one of the most exciting implications of the kinds of advances we are seeing in interaction design is not technological, nor because of the focus on users, but because of the increased need for multidisciplinary teams. Having to work in a multidisciplinary team creates challenges but also great opportunities to learn from other disciplines and to create a much better product. In my research, I have been involved with a variety of different designers, for example software, architectural, knitwear, and electronic. There is so much to learn from each other. I look forward to it!

Jenny: Since the three of us started working together in the early 1990s, the changes in technology have been phenomenal. The web, the Internet, and cell phones have transformed the way we live. Although the usability of these systems has improved, we need to strive to make them even more compact, computationally powerful, universally usable, and attractive.

I'm aware of my good fortune in having access to state-of-the-art technology, but what about people who aren't so privileged? We need low cost products that are faster, do more, and can be used by people of different cultures, ages, abilities, and experiences. Designing fancy web graphics may be fun but if users cannot access them because of slow Internet connections and old machines, what use are they? Designing for universal usability is a challenge and I hope this book will help you to create systems that are more usable by more people, more of the time.

My research is concerned with developing online communities that combine appropriate support for social interaction (i.e., sociability) with well designed software (i.e., usability). These virtual communities enable people to reach out to each other in new ways, but we need a deeper understanding of why some communities fail while others thrive. I hope that more multidisciplinary teams will be inspired to meet this exciting challenge.

Yvonne: Writing this book has made me realize how much and how rapidly the field of interaction design has expanded in the last ten years. When we wrote our first textbook on human-computer interaction in the early '90s, the web hadn't even arrived and mobile and wireless devices were still very much a dream. "WIMP" was very much *the* paradigm which interface designers (*sic*) developed applications for. Now everything has changed. Technology has advanced so rapidly that interaction designers (*sic*) now need to think about a whole host of different issues, besides the way an interface should look and behave. Moreover, there is greater eclecticism, in terms of users, settings, activities, and spaces to design for. For example, interaction designers are now involved in designing interactive products for use both indoors and outdoors (e.g., handheld devices, wearables), for work, home, school, and leisure, for both very large surfaces (e.g., interactive whiteboards) and very small screens (e.g., mobile phone displays)—to name but a few.

What this amounts to is a growing need for new methods and techniques to help in the design and evaluation of this new range of user experiences. As we point out in the book, techniques developed for screen-based systems often do not scale up very well and are inappropriate for other kinds of systems (e.g., very large collaborative virtual environments or "inhabited TV" where there may be thousands of users interacting at the same time). In addition, new theories will also need to be developed to inform the design of user experiences that are enjoyable and meaningful and expand our cognitive and social capabilities. I believe it is a very challenging time for both academic researchers and designers working in the commercial world.

References

ANNETT, J. AND DUNCAN, K. D. (1967) Task analysis and training design, *Occupational Psychology*, 41, 211–21.

APPLE COMPUTER INC. (1993) *Making IT Macintosh: The Macintosh Human Interface Guidelines Companion* (CD-ROM).

APPLE COMPUTER INC., (1987) *Human Interface Guidelines*. Harlow, UK: Addison-Wesley.

ANDREWS, D., PREECE, J., AND TUROFF, M. (2001) A conceptual framework for demographic groups resistant to online community interaction. In *Proceedings of IEEE Hawaiian International Conference on System Science (HICSS)*.

ATKINSON, P., AND HAMMERSLEY, M. (1994) Ethnography and participant observation. In N. K. Denzin and Y. S. Lincoln (eds.) *Handbook of Qualitative Research*. London: Sage.

AUSTIN, J. L. (1962) *How to Do Things with Words*. Cambridge, MA: Harvard University Press.

BAILEY, B. (2000) How to improve design decisions by reducing reliance on superstition. Let's start with Miller's Magic 7±2. *Human Factors International, Inc.* www.humanfactors.com

BAILEY, R. W. (2001) Insights from Human Factors International, Inc. (HFI) Providing consulting and training in software ergonomics. January. www.humanfactors.com/home/

BAINBRIDGE, D. (1999) *Software Copyright Law* (4th ed.). London: Butterworths.

BASILI, V., CALDIERA, G., AND ROMBACH, D. H. (1994) *The Goal Question Metric Paradigm: Encyclopedia of Software Engineering*. New York: John Wiley & Sons.

BATES, J. (1994) The role of emotion in believable characters. *Communications of the ACM*, 37(7), 122–125.

BAUM, F. L., AND DENSLOW, W. (1900) *The Wizard of Oz*. New York: Random House, Inc.

BAYM, N. (1997) Interpreting soap operas and creating community: inside an electronic fan culture. In S. Kiesler (ed.) *Culture of the Internet*. Hillsdale, NJ: Lawrence Erlbaum Associates, 103–119.

BELLOTTI, V AND ROGERS, Y. (1997) From web press to web pressure: multimedia representations and multimedia publishing. In *Proceedings of CSCW'97*, 279–286.

BEN ACHOUR, C. (1999) *Extracting Requirements by Analyzing Text Scenarios*, Thèse de Doctorat de Université Paris-6.

BENFORD, S., BEDERSON, B. B., AKESSON, K. P., BAYON, V., DRUIN, A., HANSSON, P., HOURCADE, J. P., INGRAM, R., NEALE, H., O'MALLEY, C., SIMSARIAN, K. T., STANTON, D., SUNBLAND, Y., AND TAXEN, G. (2000) Designing storytelling technologies to encourage collaboration between young children. In *Proceedings of CHI'2000*, 556–563.

BENNETT, J. (1984) Managing to meet usability requirements. In J. Bennett, D. Case, J. Sandelin, and M. Smith (eds.) *Visual Display Terminals: Usability Issues and Health Concern*. Englewood Cliffs, NJ: Prentice-Hall.

BEWLEY, W. L., ROBERTS, T. L., SCHROIT, D., AND VERPLANK, W. (1990) Human factors testing in the design of Xerox's 8010 'Star' office workstation. In J. Preece and L. Keller (eds.). *Human-Computer Interaction: A Reader*. Hemel Hempstead, UK: Prentice Hall, 368–382.

BERGMAN, E. AND HAITANI, R. (2000) Designing the PalmPilot: a conversation with Rob Haitani. In *Information Appliances*. San Francisco: Morgan Kaufmann.

BEYER, H. AND HOLTZBLATT, K. (1998) *Contextual Design: Defining Customer-Centered Systems*. San Francisco: Morgan Kauffman.

BEYNON-DAVIES, P. (1997) Ethnography and information systems development: ethnography of, for and within IS development. *Information and Software Technology*, 39, 531–540.

BIAS, R. G. (1994) The pluralistic usability walkthrough—coordinated empathies. In J. Nielsen and R. L. Mack (eds.) *Usability Inspection Methods*. New York: John Wiley & Sons.

BLUMBERG, B. (1996) *Old Tricks, New Dogs: Ethology and Interactive Creatures*. PhD Dissertation. MIT Media Lab.

BLY, S. (1997) Field work: is it product work? *ACM Interactions Magazine*, January and February, 25–30.

BØDKER, S. (2000) Scenarios in user-centered design—setting the stage for reflection and action. *Interacting with Computers*, 13(1), 61–76.

BØDKER, S., GREENBAUM, J. AND KYNG, M. (1991) Setting the stage for design as action. In J. Greenbaum and M. Kyng (eds.) *Design at Work: Cooperative Design of Computer Systems*. Hillsdale, NJ: Lawrence Erlbaum Associates, 139–154.

BOEHM B., EGYED A., KWAN, J., PORT, D. SHAH A., AND MADACHY, R. (1998) Using the WinWin spiral model: a case study. *IEEE Computer*, 31(7), 33–44.

BOEHM, B. W. (1988) A spiral model of software development and enhancement, *IEEE Computer*, 21(5), 61–72.

BOGDEWIC, S. P. (1992) Participant observation. In B. F. Crabtree and W. L. Miller (eds.) *Doing Qualitative Research*. Newbury Park, CA: Sage, 45–69.

BORCHERS, J. (2001) *A Pattern Approach to Interaction Design*. Chichester, UK: John Wiley & Sons.

BRAITERMAN, J., VERHAGE, S. AND CHOO, R. (2000) Designing with Users in Internet Time. *ACM Interactions Magazine*, VII.5, 23–27.

BREAZEAL, C. (1999) Kismet: A robot for social interactions with humans. www.ai.mit.edu/projects/kismet/

BRINKLIN, D. (2001) *VisiCalc: Information from its creators*. www.bricklin.com/visicalc.htm

BROWN, B. A., SELLEN, A. J., AND O'HARA, K. P. (2000) *A diary study of information capture in working life*. In *Proceedings of CHI 2000*, The Hague, Holland, 438–445.

BUCHENAU, M. AND SURI, J. F. (2000) Experience prototyping. In *Proceedings of DIS 2000 Design Interactive Systems: Processes, Practices, Methods, Techniques*, 17–19.

BUTTON, G. AND SHARROCK, W. (1994) Occasioned practices in the work of software engineers. In Jirotka, M. and Goguen, J. A. (eds.) *Requirements Engineering: Social and Technical Issues*. San Diego: Academic Press, 217–240.

CARD, S. K., MACKINLEY, J. D., AND SHNEIDERMAN, B. (1999) (eds.) *Readings in Information Visualization: Using Vision to Think*. San Francisco: Morgan Kaufmann.

CARD, S. K., MORAN, T. P. AND NEWELL, A. (1983) *The Psychology of Human-Computer Interaction*. Hillsdale, NJ: Lawrence Erlbaum Associates.

CARROLL, J. M. (2000) Introduction to the special issue on "Scenario-Based Systems Development," *Interacting with Computers*, 13(1), 41–42.

CARROLL, J. M. (1990) *The Nurnberg Funnel*. Cambridge, MA: MIT Press.

CASSELL, J. (2000) Embodied conversational interface agents. *Communications of the ACM,* 43(3), 70–79.

CHENG, L., STONE, L., FARNHAM, S., CLARK, A. M., AND ZANER-GODSEY, M. (2000) HutchWorld: Lessons Learned. A Collaborative Project: Fred Hutchinson Cancer Research Center & Microsoft Research. In *Proceedings of the Virtual Worlds Conference 2000*, Paris, France.

CHI Panel (2000) Scaling for the Masses: Usability Practices for the Web's Most Popular Sites.

CHIN, J. P., DIEHL, V. A., AND NORMAN, K. L. (1988) Development of an instrument measuring user satisfaction of the human-computer interface. *In Proceedings of CHI'88*.

COCKBURN, A. (1995) Structuring use cases with goals. members.aol.com/acockburn/papers/usecases.htm.

COGDILL, K. (1999) *MEDLINEplus Interface Evaluation: Final Report*. College Park, MD: College of Information Studies, University of Maryland.

COMER, E. R. (1997) Alternative lifecycle models. In Merlin Dorfman and Richard H. Thayer (eds.) *Software Engineering*. Piscataway, NJ: IEEE Computer Society Press.

CONKLIN, J. AND BEGEMAN, M. L. (1989) gIBIS: A tool for all reasons. *Journal of the American Society for Information Science*, 40(3), 200-213.

CONSTANTINE, L. L. AND LOCKWOOD, L. A. D. (1999) *Software for use*. Harlow, UK: Addison-Wesley.

COYLE, A. (1995) Discourse analysis. In G. M. Breakwell, S. Hammond, and C. Fife-Schaw (eds.) *Research Methods in Psychology*. London: Sage.

CRAIK, K. J. W. (1943) *The Nature of Explanation*. Cambridge University Press.

CRAMPTON SMITH, G. (1995) The hand that rocks the cradle. *ID Magazine*, May/June, 60–65.

CUSUMANO, M. A. AND SELBY, R. W. (1995) *Microsoft Secrets*. London: Harper-Collins Business.

CUSUMANO, M. A. AND SELBY, R. W. (1997) How Microsoft builds software. *Communications of the ACM*, 40(6), 53–61.

DANIS, C. AND BOIES, S. (2000) Using a technique from graphic designers to develop innovative systems design. In *Proceedings of DIS 2000*, 20–26.

DENZIN, N. K., AND LINCOLN, Y. S. (1994) *Handbook of Qualitative Research*. London: Sage.

DIX, A., FINLAY, J., ABOWD, G., AND BEALE, R. (1993) *Human-Computer Interaction* (2nd ed.). London: Prentice-Hall Europe.

DOURISH, P. AND BELLOTTI, V. (1992) Awareness and coordination in shared workspaces. In *Proceedings of CSCW'92*, 107–114.

DOURISH, P. AND BLY, S. (1992) Portholes: supporting awareness in a distributed work group. In *Proceedings of CHI'92*, 541–547.

DRAY, S. M. AND MRAZEK, D. (1996) A day in the life of a family: an international ethnographic study. In D. Wixon and J. Ramey (eds.) *Field Methods Casebook for Software Design*. New York: John Wiley and Sons, 145–156.

DRUIN, A. (2000) The role of children in the design of new technology. *University of Maryland, Human-Computer Interaction Laboratory Technical Report* 99–23. www.cs.umd.edu/hcil.

DRUIN, A. (1999) *The Design of Children's Software.* San Francisco, CA: Morgan-Kaufmann.

DUMAS, J. S., AND REDISH, J. C. (1999) *A Practical Guide to Usability Testing* (Revised Edition). Exeter, UK: Intellect.

EASON, K. (1987) *Information Technology and Organizational Change.* London: Taylor and Francis.

EBLING, M. R., AND JOHN, B. E. (2000) On the contributions of different empirical data in usability testing. In *Proceedings of ACM DIS 2001*, 289–296.

EDWARDS, A. D. N. (1992) Graphical user interfaces and blind people. In *Proceedings of ICCHP '92*, Vienna: Austrian Computer Society, 114–119.

EHN, P. (1989) *Word-oriented Design of Computer Artifacts* (2nd edn.) Hillsdale, NJ: Lawrence Erlbaum Associates.

EHN, P. AND KYNG, M. (1991) Cardboard computers: mocking-it-up or hands-on the future. In J. Greenbaum and M. Kyng (eds.). *Design at Work*. Hillsdale, NJ: Lawrence Erlbaum Associates.

EICK, S. G. (2001) Visualizing online activity. *Communications of the ACM*, 44(8), 45–50.

ERICKSON, T. D. (1990) Working with interface metaphors. In B. Laurel (ed.). *The Art of Human-Computer Interface Design*. Boston: Addison-Wesley.

ERICKSON, T., SMITH, D. N., KELLOGG, W. A., LAFF, M., RICHARDS, J. T., AND BRADNER, E. (1999) Socially translucent systems: social proxies, persistent conversation and the design of "Babble". In *Proceedings of CHI'99*, 72–79.

ERIKSON, T. D., AND SIMON, H. A. (1985) *Protocol Analysis: Verbal Reports as Data.* Cambridge, MA: The MIT Press.

FETTERMAN, D. M. (1998) *Ethnography: Step by Step* (2nd ed.) Thousand Oaks, CA: Sage.

FISH, R.S. (1989) Cruiser: a multimedia system for social browsing. *SIGGRAPH Video Review* (video cassette) Issue 45, Item 6.

FISKE, J. (1994) Audiencing: cultural practice and cultural studies. In N. K. Denzin and Y. S. Lincoln (eds.) *Handbook of Qualitative Research*. Thousand Oaks, CA: Sage, 189–198.

FITTS, P. M. (1954) The information capacity of the human motor system in controlling amplitude of movement. *Journal of Experimental Psychology*, 47, 381–391.

FITZPATRICK, G., MANSFIELD, T., KAPLAN, S., ARNOLD., D., PHELPS, T., AND SEGALL, B. (1999) Augmenting the workaday world with Elvin. In *Proceedings of the Sixth European Conference on Computer-Supported Cooperative Work*. Dordrecht, The Netherlands: Kluwer, 431–450.

FONTANA, A., AND FREY, J. H. (1994) *Interviewing: The art of science.* In N. Denzin and Y. Lincoln (eds.) *Handbook of Qualitative Research*. London: Sage, 361–376.

FROHLICH, D. AND MURPHY, R. (1999) Getting physical: what is fun computing in tangible form? In *Computers and Fun 2, Workshop*, 20 Dec. York, UK.

GAVER, B., DUNNE, T., AND PACENTI, E. (1999) Cultural probes. *ACM Interactions Magazine*, January and February, 21–29.

GENTNER, D. AND NIELSEN, J. (1996) The anti-Mac interface. *Communictions of the ACM*, 39 (8) 70–82.

GILL, J. AND SHIPLEY, T. (1999) *Telephones–What Features do Disabled People Need?* RNIB.

GOETZ, J. P., AND LECOMPTE, M. D. (1984) *Ethnography and Qualitative Design in Educational Research.* Orlando, FL: Academic Press.

GOUGH, P. A., FODEMSKI, F. T., HIGGINS, S. A., AND RAY, S. J. (1995) Scenarios—an industrial case study and hypermedia enhancements. In *Proceedings of 2nd IEEE Symposium on Requirements Engineering*, IEEE Computer Society, 10–17.

GOULD, J. D., AND LEWIS, C. H. (1985) Designing for usability: key principles and what designers think. *Communications of the ACM*, 28(3), 300–311.

GOULD, J. D., BOIES, S. J., LEVY, S., RICHARDS, J. T., AND SCHOONARD, J. (1987) The 1984 Olympic Message System: a test of behavioral principles of system design. *Communications of the ACM*, 30(9), 758–769. (full paper)

GOULD, J. D., BOIES, S. J., LEVY, S., RICHARDS, J. T., AND SCHOONARD, J. (1990) The 1984 Olympic Message System: a test of behavioral principles of system design. In J. Preece and L. Keller (eds.) *Human-Computer Interaction (Readings)*. Hemel Hempstead, UK: Prentice Hall International Ltd., 260–283.

GRAY, W. D., JOHN, B. E., AND ATWOOD, M. E. (1993) Project Ernestine: validating a GOMS analysis for predicting and explaining real-world performance. *Human-Computer Interaction,* 8(3), 237–309.

GREEN, T. R. G. (1990) The cognitive dimension of viscosity: A sticky problem for HCI. In D. DIAPER, D. GILMORE, G. COCKTON AND B. SHAKEL (eds.) *Human-Computer Interaction—INTERACT'90*. Elsevier Publishers, 79–86.

GREIF, I. (1988) *Computer Supported Cooperative Work: a book of readings*. San Francisco: Morgan Kaufmann.

GRUDIN, J. (1989) The case against user interface consistency. *Communications of the ACM,* 32(10), 1164–1173.

GRUDIN, J. (1990) The computer reaches out: the historical continuity of interface design. In *Proceedings of CHI'90*, 261–268.

GUINDON, R. (1990) Designing the design process: exploiting opportunistic thoughts. *Human-Computer Interaction*, 5(2&3), 305–344.

HALVERSON, C. (1995) Inside the cognitive workplace: new technology and air traffic control. PhD Thesis, Dept. of Cognitive Science, University of California, San Diego.

HAMMERSLEY, M. AND ATKINSON, P. (1983) *Ethnography: principles in practice*. London: Tavistock.

HARPER, R. (2000) The organization of ethnography, In *Proceedings of CSCW 2000*, 239–264.

HARRISON, S., BLY, S. ANDERSON, S. AND MINNEMAN (1997) The media space. In Finn, K. E. Sellen, A. and Wilbur, S. B. (eds.) *Video-Mediated Communication*. Mahwah, NJ: Lawrence Earlbaum Associates, 273–300

HARTFIELD, B. AND WINOGRAD, T. (1996) Profile: IDEO. In T. Winograd (ed.) *Bringing Design to Software*. ACM Press.

HARTSON, H. R. AND HIX, D. (1989) Toward empirically derived methodologies and tools for human-computer interface development. *International Journal of Man-Machine Studies*, 31, 477–494.

HAUMER, P., JARKE, M., POHL, K., AND WEIDENHAUPT, K. (2000) Improving reviews of conceptual models by extended traceability to captured system usage. *Interacting with Computers*, 13(1), 77–95.

HEATH, C. AND LUFF, P. (1992) Collaboration and control: crisis management and multimedia technology in London Underground line control rooms. In *Proceedings of CSCW'92*, 1, 1–2, 69–94.

HEATH, C., JIROTKA, M., LUFF, P., AND HINDMARSH, J. (1993) Unpacking collaboration: the interactional organization of trading in a city dealing room. In *Proceedings of the Third European Conference on Computer-Supported Cooperative Work*. Dordrecht: Kluwer.

HEINBOKEL, T., SONNENTAG, S., FRESE, M., STOLTE, W., and BRODBECK, F. C. (1996) Don't underestimate the problems of user centredness in software development projects—there are many! *Behaviour & Information Technology*, 15(4), 226–236.

HOCHHEISER, H., AND SHNEIDERMAN, B. (2001) Using interactive visualization of WWW log data to characterize access patterns and inform site design. *Journal of the American Society for Information Science*, 52, 4, 331–343.

HOLTZBLATT, K. AND JONES, S. (1993) Contextual Inquiry: a participatory technique for systems design. In D. Schuler, and A. Namioka, (eds.) *Participatory Design: Principles and Practice*, Hillsdale, NJ: Lawrence Erlbaum Associates, 177–210.

HOLTZBLATT, K., AND BEYER, H. (1996) Contextual Design: principles and practice. In D. Wixon and J. Ramey, (eds.) *Field Methods Casebook for Software Design*. New York: John Wiley and Sons, 301–333.

HUGHES, J. A., KING, RANDALL, D. AND SHARROCK (1993) Ethnography for system design: a guide, COMIC working paper, COMIC-LANCS-2-N. More information about COMIC is available from Cooperative Systems Engineering Group, Computing Department, Lancaster University, UK.

HUGHES, J. A., KING, V., RODDEN, T., AND ANDERSEN, H. (1994) Moving out of the control room: ethnography in system design. In *Proceedings of CSCW'94*, Chapel Hill, NC.

HUGHES, J. A., O'BRIEN, J., RODDEN, T. AND ROUNCEFIELD, M. (1997) Designing with Ethnography: a Presentation Framework for Design. In *Proceedings of DIS '97*, 147–159.

HUGHES, J. A., SOMMERVILLE, I., BENTLEY, R. AND RANDALL, D. (1993a) Designing with ethnography: making work visible. *Interacting with Computers*, 5(2), 239–253.

HUTCHINS, E. (1995) *Cognition in the Wild*. Cambridge, MA: MIT Press.

ISENSEE, S., KALINOSKI, K. AND VOCHATZER, K. (2000) Designing Internet appliances at Netpliance. In E. Bergman (ed.) *Information Appliances and Beyond*. San Francisco: Morgan Kaufmann.

ISHII, H. AND ULLMER, B. (1997) Tangible bits: towards seamless interfaces between people, bits and atoms. In *Proceedings of CHI'97*, 234–241.

ISHII, H., KOBAYASHI, M., AND Grudin, J. (1993) Integration of interpersonal space and shared workspace: Clearboard design and experiments. *ACM Transactions on Information Systems*, 11 (4), 349–375.

JACOBSON, I., CHRISTERSON, M., JONSSON, P. AND OVERGAARD, G. (1992) *Object-Oriented Software Engineering—A Use Case Driven Approach*. Harlow, UK: Addison-Wesley.

JOHNSON, M. and LAKOFF, G. (1980) *Metaphors We Live By*. Chicago: The University of Chicago Press.

JOHNSON-LAIRD, P. N. (1983) *Mental Models*. Cambridge: Cambridge University Press.

KAHN, R., AND CANNELL, C. (1957) *The Dynamics of Interviewing*. New York: John Wiley & Sons.

KARAT, C. M. (1993) The cost-benefit and business case analysis of usability engineering. InterChi '93, Amsterdam, Tutorial Notes 23.

KARAT, C.-M. (1994) A comparison of user interface evaluation methods. In J. Nielsen and R. L. Mack (eds.) *Usability Inspection Methods*. New York: John Wiley & Sons.

KARAT, J. (1995) Scenario Use in the Design of a Speech Recognition System. In J. M. Carroll (ed.) *Scenario-based Design*, 109–134. New York: John Wiley & Sons.

KARAT, J. AND BENNET, J. L. (1991) Using scenarios in design meetings. In Karat, J. (ed.) *Taking Design Seriously*. London: Academic Press.

KAY, A. (1969) *The Reactive Engine*. PhD Dissertation, Electrical Engineering and Computer Science, University of Utah.

KEIL, M. AND CARMEL, E. (1995) Customer-developer links in software development. *Communications of the ACM*, 38(5), 33–44.

KEMPTON, W. (1986) Two theories of home heat control. *Cognitive Science*, 10, 75–90.

KETOLA, P., HJELMEROOS, H., AND RAIHA, K.-J. (2000) Coping with consistency under multiple design constraints: The case of the Nokia 9000 WWW browser. *Personal Technologies* 4(2&3), 86–95.

KIM, S. (1990) Interdisciplinary cooperation. In *The Art of Human-Computer Interface Design*. B. Laurel (ed.) Reading, MA: Addison-Wesley.

KOENEMANN-BELLIVEAU, J., CARROLL, J. M., ROSSON, M. B., AND SINGLEY, M. K. (1994) Comparative usability evaluation: critical incidents and critical threads. In *Proceedings of CHI'94*.

KOTONYA, G. AND SOMMERVILLE, I. (1998) *Requirements engineering: processes and techniques*. Chichester, UK: John Wiley & Sons.

KRAUT, R., FISH, R., ROOT, R. AND CHALFONTE, B. (1990) Informal communications in organizations: form, function and technology. In S. Oskamp and S. Spacapan (eds.) *People's Reactions to Technology in Factories, Offices and Aerospace*. The Claremont Symposium on Applied Social Psychology. Thousand Oaks, CA.: Sage Publications, 145–199.

KUHN, S. (1996) Design for people at work. In T. Winograd, (ed.) *Bringing Design to Software*. Boston: Addison-Wesley.

KUJALA, S. AND MÄNTYLÄ, M. (2000) Is user involvement harmful or useful in the early stages of product development? In *CHI 2000 Extended Abstracts*, ACM Press, 285–286.

LAKOFF, G. AND JOHNSON, M. (1980) *Metaphors we Live By*. Chicago: The University of Chicago Press.

LAMBOURNE, R., FEIZ, K., AND RIGOT, B. (1997) Social trends and product opportunities: Philips' Vision of the Future Project. In *Proceedings of CHI'97*, 494–501.

LANSDALE, M. (1988) The psychology of personal information management. *Applied Ergonomics*, 55, 55–66.

LANSDALE, M. AND EDMONDS, E. (1992) Using memory for events in the design of personal filing systems. *International Journal of Human-Computer Studies*, 26, 97–126.

LARSON, K., AND CZERWINSKI, M. (1998) Web page design: implications of memory, structure and scent for information retrieval. In *Proceedings of CHI '98*, 25–32.

LAUREL, B. (1993) *Computers as Theatre*. New York: Addison-Wesley.

LAZAR, J., AND PREECE, J. (1999) Designing and implementing web-based surveys. *Journal of Computer Information Systems, xxxix* (4), 63–67.

LEE, J., KIM, J., AND MOON, JAE YUN (2000) What makes Internet users visit cyber stores again? Key design factors for customer loyalty. In *Proceedings of CHI 2000,* 305–312.

LESTER, J. C., AND STONE, B. A. (1997) Increasing believability in animated pedagogical agent. In *Proceedings of Autonomous Agents'97,* 16–21.

LESTER, J. C., CONVERSE, S. A., STONE, B. A., AND BHOGAL, R. S. (1997) The personal effect: affective impact of animated pedagogical agents. In *Proceedings of CHI'97,* 359–366.

LIDDLE, D. (1996) Design of the conceptual model. In T. Winograd, (ed.) *Bringing Design to Software.* Reading, MA: Addison-Wesley, 17–31.

LUND, A. M. (1994) Ameritech's usability laboratory: from prototype to final design. *Behaviour & Information Technology,* 13(1 & 2), 67–80.

LYNCH, P. J., AND HORTON, S. (1999) *Web Style Guide (Preliminary Version).* New Haven, CT. and London: Yale University Press.

M880 (2000) OSS CD part of *M880 Software Engineering.* Milton Keynes, UK: The Open University.

MACKAY, W. E., RATZER, A. V., AND JANECEK, P. (2000) Video artifacts for design: bridging the gap between abstraction and detail. In *Proceedings of DIS 2000,* 72–82.

MACKENZIE, I. S. (1992) Fitts' law as a research and design tool in human-computer interaction. *Human-Computer Interaction,* 7, 91–139.

MAES, P. (1995) Intelligent software. *Scientific American,* 273(3), 84-86.

MAGLIO, P. P., MATLOCK, T., RAPHAELY, D., CHERNICKY, B., AND KIRSH D. (1999) Interactive skill in Scrabble. In *Proceedings of Twenty-first Annual Conference of the Cognitive Science Society.* Mahwah, NJ: Lawrence Erlbaum Associates.

MAHER, M. L. AND PU, P. (1997) *Issues and Applications of Case-Based Reasoning in Design.* Hillsdale, NJ: Lawrence Erlbaum Associates,

MAIDEN, N. A. M. AND RUGG, G. (1996) ACRE: selecting methods for requirements acquisition. *Software Engineering Journal,* 11(3), 183–192.

MALONE, T. W. (1983) How do people organize their desks? Implications for the design of office information systems. *ACM Transactions on Office Information Systems,* 1(1) 99–112.

MANDLER, R., SALOMON, G. AND WONG, Y. Y. (1992) A 'pile' metaphor for supporting casual organization of information. In *Proceedings of CHI'92,* 627–634.

MANN, S. (1996) Smart clothing: wearable multimedia computing and personal imaging to restore the technological balance between people and their environment. In *Proceedings of ACM Multimedia, 96,* 163–174.

MARCUS, A. (1993) Human communication issues in advanced UIs. *Communications of the ACM,* 101–109.

MARK, G., FUCHS, L. AND SOHLENKAMP, M. (1997) Supporting groupware conventions through contextual awareness. In *Proceedings of the Fifth European Conference on Computer-Supported Cooperative Work.* Dordrecht, The Netherlands: Kluwer, 253–268.

MARMASSE, N. AND SCHMANDT, C. (2000) Location-aware information delivery with ComMotion. In *Proceedings of Handheld and Ubiquitous Computing, Second International Symposium, HUC 2000,* Springer-Verlag, 157–171.

MARSHALL, C., AND ROSSMAN, G. B. (1999) *Designing Qualitative Research* (3rd ed.). Thousand Oaks, CA: Sage Publications.

MARTIN, H. AND GAVER, B. (2000) Beyond the snapshot: from speculation to prototypes in audiophotography. In *Proceedings of DIS 2000,* 55–65.

MATEAS, M., SALVADOR, T., SCHOLTZ, J. AND SORENSEN, D. (1996) Engineering ethnography in the home. *Companion for CHI '96,* ACM, 283–284.

MAYHEW, D. J. (1999) *The Usability Engineering Lifecycle.* San Francisco: Morgan Kaufmann.

MCLAUGHLIN, M., GOLDBERG, S. B., ELLISON, N. AND LUCAS, J. (1999) Measuring Internet audiences: patrons of an online art museum. In S. Jones (ed.) *Doing Internet Research: Critical Issues and Methods for Examining the Net.* Thousand Oaks, CA: Sage, 163–178.

Microsoft Corporation (1992) *The Windows Interface, An Application Design Guide.* Microsoft Press.

MILLER, G. (1956) The Magical Number Seven, Plus or Minus Two: Some Limits on our Capacity for Processing Information. *Psychological Review,* 63, 81–97.

MILLER, L.H. AND JOHNSON, J. (1996) The Xerox Star: an influential user interface design. In M. Rudisill, C. Lewis, P. G. Polson, and T. D. McKay, (eds.) *Human-Computer Interface Design.* San Francisco: Morgan-Kaufmann.

MILLINGTON, D. AND STAPLETON, J. (1995) Special report: developing a RAD standard. *IEEE Software*, 12(5), 54–6.

MONTEMAYOR, J., DRUIN, A. AND HELANDER, J. (2000) PETS: A personal electronic teller of stories.' In C.A. Druin & J. Helander (eds.) *Robots for Kids*. San Francisco: Morgan Kaufmann.

MORAN, T. P., AND R. J. ANDERSON (1990) The workaday world as a paradigm for CSCW design. In *Proceedings of the CSCW '90*, 381–393.

MORIKAWA, O. AND MAESAKO, T. (1998) HyperMirror: towards pleasant-to-use video mediated communication system. In *Proceedings of CSCW'98*, 149–158.

MOSIER, J. N. AND TAMMARO, S. G., (1997) When are group scheduling tools useful? In *Proceedings of CSCW '97*, 6, 53–70.

MULLER, M. J. (1991) PICTIVE—An exploration in participatory design. In *Proceedings of CHI '91*, 225–231.

MULLER, M. J., TUDOR, L. G., WILDMAN, D. M., WHITE, E. A., ROOT, R. W., DAYTON, T., CARR, R., DIEKMAN, B., AND DYKSTRA-ERICKSON, E. (1995) Bifocal tools for scenarios and representations in participatory activities with users. In J. M. Carroll (ed.) *Scenario-Based Design*. New York: John Wiley & Sons, 135–163.

MULLET, K. AND SANO, D. (1995) *Designing Visual Interfaces*. Mountain View, CA: Prentice-Hall.

MYERS, B. A. (1995) State of the Art in User Interface Software Tools. In R. Baecker, J. Grundin, W. Buxton, and S. Greenberg (eds.) *Readings in Human-Computer Interaction: Toward the Year 2000* (2nd ed.) San Francisco: Morgan Kaufmann, 344–356.

MYERS, B., HUDSON, S. E., AND PAUSCH, R. (2000) Past, present and future of user interface software tools. *ACM Transactions on Computer-Human Interaction*, 7(1), 3–28.

NARDI, B. A., AND O'DAY, V. L. (1999) *Information Ecologies: Using Technology with a Heart*. Cambridge, MA: The MIT Press.

NELSON, T. (1980) Interactive Systems and the design of Virtuality. *Creative Computing,* Nov.–Dec., 1980.

NELSON, T. (1990) The right way to think about software design. In B. Laurel, (ed.) *The Art of Human-Computer Design*. Reading, MA: Addison-Wesley.

NEWMAN, W. AND LAMMING, N. (1995) *Interactive System Design*. Harlow, UK: Addison-Wesley.

NIE, N. H., AND EBRING, L. (2000) *Internet and Society. Preliminary Report*. Stanford, CA: The Stanford Institute for the Quantitative Study of Society.

NIELSEN, J. (1992) Finding usability problems through heuristic evaluation. In *Proceedings of CHI'92*, 373–800.

NIELSEN, J. (1993) *Usability Engineering*. San Francisco: Morgan Kaufmann.

NIELSEN, J. (1994a) Heuristic evaluation. In J. Nielsen and R. L. Mack (eds.) *Usability Inspection Methods*. New York: John Wiley & Sons.

NIELSEN, J. (1994b) Enhancing the explanatory power of usability heuristics. In *Proceedings of ACM CHI'94*, 152–158.

NIELSEN, J. (1999) www.useit.com

NIELSEN, J. (2000) *Designing Web Usability*. Indianapolis: New Riders Publishing.

NIELSEN, J. (2001) Ten Usability Heuristics. www.useit.com/papers/heuristic

NIELSEN, J., AND MACK, R. L. (1994) *Usability Inspection Methods*. New York: John Wiley & Sons.

NODDER, C., WILLIAMS, G., AND DUBROW, D. (1999) Evaluating the usability of an evolving collaborative product—changes in user type, tasks and evaluation methods over time. In *Proceedings of GROUP'99*, 150–159.

NOLDUS (2000) *The Observer Video-Pro*. www.noldus.com/products/observer/obs_spvta30.html.

NONNECKE, B., AND PREECE, J. (2000) Lurker demographics: counting the silent. In *Proceedings of CHI 2000*, 73–80.

NORMAN, D. (1983) Some observations on mental models. In Gentner, D. and A. L. Stevens (eds.) *Mental Models*. Hillsdale, NJ: Lawrence Earlbaum Associates.

NORMAN, D. (1988) *The Design of Everyday Things*. New York: Basic Books.

NORMAN, D. (1990) Four (more) issues for Cognitive Science. *Cognitive Science Technical Report No. 9001*, Dept. of Cognitive Science, UCSD, USA.

NORMAN, D. (1993) *Things That Make Us Smart*. Reading, MA: Addison-Wesley.

NORMAN, D. (1999) Affordances, conventions and design. *ACM Interactions Magazine,* May/June 1999, 38–42.

NYGAARD, K. (1990) The origins of the Scandinavian school, why and how? *Participatory Design Conference 1990 Transcript*, Computer Professionals for Social Responsibility.

OLSON, J. S., AND MORAN, T. P. (1996) Mapping the method muddle: guidance in using methods for user interface design. In M. Rudisill, C. Lewis, P. B. Polson and T. D. McKay (eds.) *Human-Computer Interface Design: Success Stories, Emerging Methods, Real-World Context,* San Francisco: Morgan Kaufmann, pp. 269–300.

OOSTERHOLT, R., KUSANO, M., AND DEVRIES, G. (1996) Interaction design and human factors support in the development of a personal communicator for children. In *Proceedings of CHI '96,* 450–465.

OPPENHEIM, A. N. (1992) *Questionnaire Design, Interviewing and Attitude Measurement.* London: Pinter Publishers.

OREN, T., SALOMON, G., KREITMAN, K. AND DON, A. (1990) Guides: characterizing the interface. In B. Laurel (ed.) *The Art of Human-Computer Interface Design.* Reading, MA: Addison-Wesley, 367–381.

PAGE, S. R. (1996) User-centered Design in a commercial software company. In D. Wixon and J. Ramey, (eds.) *Field Methods Casebook for Software Design.* New York: John Wiley & Sons, 197–213.

PAYNE, S. (1991) A descriptive study of mental models. *Behaviour and Information Technology,* 10, 3–21.

Penpoint *hci.stanford.edu/cs147/notes/penpoint.html*

PICARD, R. W. (1998) *Affective Computing.* Cambridge, MA: MIT Press.

PLOWMAN, L., ROGERS, Y. AND RAMAGE, M. (1995) What are workplace studies for? In *Proceedings of the Fourth European Conference on Computer-Supported Cooperative Work,* Dordrecht: The Netherlands, Kluwer, 309–324.

POTTER, J. AND WETHERELL, M. (1987) *Discourse and Social Psychology.* London: Sage.

PREECE, J. (2000) *Online Communities: Designing Usability, Supporting Sociability.* Chichester, UK: John Wiley & Sons.

PREECE, J., ROGERS, Y., SHARP, H., BENYON, D., HOLLAND, S., AND CAREY, T. (1994) *Human-Computer Interaction.* Wokingham, UK: Addison-Wesley.

PRESSMAN, R. (1992) *Software Engineering: A Practitioner's Approach.* New York: McGraw-Hill.

QUINTANAR, L. R., CROWELL, C. R., AND PRYOR, J. B. (1982) Human-computer interaction: a preliminary social psychological analysis. *Behavior Research: Methods and Instrumentation,* 13, (2), 210–220.

REEVES, B., AND NASS, C. (1996) *The Media Equation: How People Treat Computers, Television, and New Media like Real People and Places.* Cambridge: Cambridge University Press.

RETTIG, M. (1994) Prototyping for tiny fingers. *Communications of the ACM,* 37(4), 21–27.

RHODES, B., MINAR, N. AND WEAVER, J. (1999) Wearable computing meets ubiquitous computing: reaping the best of both worlds. In *Proceedings of the Third International Symposium on Wearable Computers (ISWC '99),* San Francisco, 141–149.

RIBA (1988) *Architect's Job Book: Volume 1, Job Administration* (5th edition), London: RIBA Publications.

ROBERTSON, S. AND ROBERTSON, J. (1999) *Mastering the Requirement Process.* Boston: Addison-Wesley.

ROBINSON, J. P., AND GODBEY, G. (1997) *Time for Life: The Surprising Ways that Americans Use Their Time.* University Park, PA: The Pennsylvania State University Press.

ROBSON, C. (1993) *Real World Research.* Oxford, UK: Blackwell.

ROBSON, C. (1994) *Experimental Design and Statistics in Psychology.* Aylesbury, England: Penguin Psychology

ROGERS, Y. (1993) Coordinating computer-mediated work. *Computer Supported Cooperative Work,* 1, 2995–3315.

ROGERS, Y. AND SCAIFE, M. (1998) How can interactive multimedia facilitate learning? In J. Lee (ed.) *Intelligence and Multimodality in Multimedia Interfaces: Research and Applications.* Menlo Park, CA: AAAI Press.

ROSE, A., SHNEIDERMAN, B., AND PLAISANT, C. (1995) An applied ethnographic method for redesigning user interfaces. In *Proceedings of DIS 95,* 115–122.

ROTH, I. (1986) An introduction to object perception. In I. Roth and J.B. Frisby (eds.) *Perception and Representation: A Cognitive Approach.* Milton Keynes: Open University.

RUBIN, J., (1994) *Handbook of Usability testing: How to Plan, Design and Conduct Effective tests.* New York: John Wiley & Sons.

RUBINSTEIN, R. AND HERSH, H. (1984) *The Human Factor: Designing Computer Systems for People.* Woburn, MA: Digital Press.

RUDD, J., STERN, K. R. AND ISENSEE, S. (1996) Low vs. High-fidelity Prototyping Debate. *ACM Interactions Magazine,* January, 76–85.

RUDMAN, C. AND ENGELBECK, G. (1996) Lessons in choosing methods for designing complex graphical user interfaces. In M. Rudisill, C. Lewis, P. B. Polson and T. D. McKay (eds.). *Human-Computer Interface Design: Success Stories, Emerging Methods, Real-World Context.* San Francisco: Morgan Kaufmann, 198–228.

SACKS, H., SCHEGLOFF, E., AND JEFFERSON, G. (1978) A simplest systematics for the organization of turn-taking for conversation. *Language,* 50, 696–735.

SCAIFE, M. AND ROGERS, Y. (1996) External cognition: how do graphical representations work? *International Journal of Human-Computer Studies,* 45, 185–213.

SCAIFE, M., AND ROGERS, Y. (2001) Informing the design of virtual environments. *International Journal of Human-Computer Systems,* 55(2), 115–143.

SCAIFE, M., ROGERS, Y., ALDRICH, F., AND DAVIES, M. (1997) Designing for or designing with? Informant design for interactive learning environments. In *Proceedings of CHI '97,* 343–350.

SCHANK, R. C. (1982) *Dynamic Memory: a Theory of Learning in Computers and People.* Cambridge, UK: Cambridge University Press.

SCHÖN, D. (1983) *The Reflective Practitioner: How Professionals Think in Action.* New York: Basic Books.

SCHRAGE, M. (1996) Cultures of prototyping. In T. Winograd (ed.) *Bringing Design to Software.* Boston: Addison-Wesley.

SEARLE, J. (1969) *Speech Acts.* Cambridge: Cambridge University Press.

SEGALL, B., AND ARNOLD, D. (1997) Elvin has left the building: A publish/subscribe notification service with quenching. In *Proceedings of AUUG Summer Technical Conference,* Brisbane, Australia.

SHACKEL, B. (1990) Human factors and usability. In J. Preece and L. Keller (eds.) *Human-Computer Interaction: Selected Readings.* Hemel Hempstead, UK: Prentice-Hall, 27–41.

SHAPIRO, D. (1995) Noddy's guide to . . . ethnography and HCI. *HCI Newsletter* 27, 8–10.

SHARF, B. F. (1999) Beyond netiquette: the ethics of doing naturalistic discourse research on the Internet. In S. Jones (ed.) *Doing Internet Research: Critical issues and methods for examining the net.* Thousand Oaks, CA: Sage Publications, 243–256.

SHARP, H. C., ROBINSON, H. M., AND WOODMAN, M. (1999) The role of culture in successful software process improvement. In *EUROMICRO '99, Proceedings of 25th EUROMICRO Conference.* Piscataway, NJ: IEEE Press, II, 170–176.

SCHEGLOFF, E. A., AND SACKS, H. (1973) Opening up closings. *Semiotica,* 7, 289–327.

SHNEIDERMAN, B. (1983) Direct manipulation: a step beyond programming languages. *IEEE Computer,* 16(8), 57–69.

SHNEIDERMAN, B. (1998) *Designing the User Interface: Strategies for Effective Human-Computer Interaction* (3rd ed.). Reading, MA: Addison-Wesley.

SHNEIDERMAN, B. (1998a) Relate-Create-Donate: A teaching philosophy for the cyber-generation. *Computers in Education,* 31(1), 25–39.

SILFVERBERG, M., MACKENZIE, I. S., AND KORHONEN, P. (2000) *Predicting text entry speed on mobile phones.* In *Proceedings of* CHI'2000, 9–16.

SMITH, D., IRBY, C., KIMBALL, R., VERPLANK, B. AND HARSLEM, E. (1982) Designing the Star user interface. *Byte,* 7(4), 242–82.

SMITH, S. L. AND MOSIER, J. N. (1986) *Guidelines for Designing User Interface Software.* Report ESD-TR-86–278, Electronic Systems Division, Bedford, MA: The Mitre Corporation.

SOMMERVILLE, I. (2001) *Software Engineering* (6th ed.) Boston and Harlow, UK: Addison-Wesley.

SPENCER, R. (2000) The streamlined cognitive walkthrough method: working around social constraints encountered in a software development company. In *Proceedings of CHI 2000,* 253–359.

SPIEGEL, D., BLOOM, J. R., KRAEMER, H. C., AND GOTTHEIL, E. (1989) Effect of psychosocial treatment on survival of patients with metastatic breast cancer. *The Lancet,* October 4, 888–891.

SPREENBERG, P., SALOMON, G., AND JOE, P. (1995) Interaction design at IDEO product development. In *Proceedings of ACM CHI'95 Conference Companion,* 164–165.

SPROULL, L., SUBRAMANI, M. M., KIESLER, S., WALKER, J. H., AND WATERS, K. (1996) When the interface is a face. *Human-Computer Interaction,* 11, 97–124.

STROMMEN, E. (1998) When the interface is a talking dinosaur: learning across media with ActiMates Barney. In *Proceedings of CHI'98,* 288–295.

SUCHMAN, L. A. (1983) Office procedures as practical action: models of work and system design. *ACM Transactions on Office Information Systems,* 1(4), 320–328.

SUCHMAN, L. A. (1987) *Plans and Situated Actions.* Cambridge: Cambridge University Press.

SULLIVAN, K. (1996) Windows 95 user interface: A case study in usability engineering. In *Proceedings of CHI '96*, 473–480.

TAYLOR, A. (2000) IT projects: sink or swim. *The Computer Bulletin*, January, 24–26.

TEASLEY, B., LEVENTHAL, L., BLUMENTHAL, B., IN-STONE, K., AND STONE, D. (1994) Cultural diversity in user interface design. *SIGCHI Bulletin*, 26(1), 36–40.

THIMBLEBY, H. (1990) *User Interface Design.* Harlow, UK: Addison Wesley.

TRACTINSKY, N. (1997) Aesthetics and apparent usability: empirically assessing cultural and methodological issues. In *Proceedings of CHI'97*, 115–122.

TUDOR, L. G. (1993) A participatory design technique for high-level task analysis, critique and redesign: The CARD method. In *Proceedings of the Human Factors and Ergonomics Society 1993 Meeting*, Seattle, October 1993, 295–299.

VÄÄNÄNEN-VAINIO-MATTILA, K. AND RUUSKA, S. (2000) Designing mobile phones and communicators for consumers' needs at Nokia. In E. Bergman (ed.) *Information Appliances and Beyond.* San Francisco: Morgan Kaufmann, 169–204.

VEEN, J. (2001) *The Art and Science of Web Design.* Indianapolis: New Riders Publishing.

VERPLANK, B. (1989) Tutorial Notes. In *Proceedings of CHI'89 Conference.*

VERPLANK, B. (1994) Interview with Bill Verplank. In PREECE, J., ROGERS, Y., SHARP, H., BENYON, D., HOLLAND, S., AND CAREY, T., *Human-Computer Interaction.* Wokingham, UK: Addison-Wesley, 467–468.

VILLER, S. AND SOMMERVILLE, I. (1999) Coherence: an approach to representing ethnographic analyses in systems design. *Human-Computer Interaction*, 14.

WALKER, J., SPROULL, L., AND SUBRAMANI, R. (1994) Using a human face in an interface. In *Proceedings of CHI'94*, 85–91.

WEBB, B. R. (1996) The role of users in interactive systems design: when computers are theatre, do we want the audience to write the script? *Behaviour and Information Technology*, 15(2), 76–83.

WEISER, M. (1991) The computer for the 21st Century. *Scientific American*, 265 (3), 94–104.

WELLNER, P. (1993) Interacting with paper on the digital desk. *Communications of the ACM*, 36(7), 86–96.

WHARTON, C., RIEMAN, J., LEWIS, C., AND POLSON, P. (1994) The cognitive walkthrough method: a practitioner's guide. In J. Nielsen and R. L. Mack (eds.), *Usability Inspection Methods.* New York: John Wiley & Sons.

WHITESIDE, J., BENNETT, J. AND HOLTZBLATT, K. (1988) Usability engineering: our experience and evolution. In *Handbook of Human-Computer Interaction.* Helander, M. (ed.) Amsterdam: Elsevier Science Publishers, 791–817.

WHITTAKER, S., AND SCHWARTZ, H. (1995) Back to the future: pen and paper technology supports complex group coordination. In *Proceedings of CHI'95*, 495–502.

WILLIAMS, F., RICE, R. E., AND ROGERS, E. M. (1988) *Research Methods and the New Media.* New York: The Free Press, Macmillan Inc.

WITMER, D. F., COLMAN, R. W., AND KATZMAN, S. L. (1999) From paper-and-pencil to screen-and-keyboard. In S. Jones (ed.) *Doing Internet Research: Critical Issues and Methods for Examining the Net.* Thousands Oaks, CA: Sage, 145–161.

WINOGRAD, T. (1988) A language/action perspective on the design of cooperative work. *Human-Computer Interaction*, 3, 3–30.

WINOGRAD, T. (1994) Categories, disciplines, and social coordination. *Computer Supported Cooperative Work*, 2, 191–197.

WINOGRAD, T. (1996) (ed.) *Bringing Design to Software.* Reading, MA: Addison-Wesley.

WINOGRAD, T. (1997) From computing machinery to interaction design. In P. Denning and R. Metcalfe (eds.) *Beyond Calculation: the Next Fifty Years of Computing.* Amsterdam: Springer-Verlag, 149–162.

WINOGRAD, T. AND FLORES, W. (1986) *Understanding Computers and Cognition.* Norwood, NJ: Addison-Wesley.

WIXON, D., AND WILSON, C. (1997) The usability engineering framework for product design and evaluation (Chapter 27). In M. G. Helander, T. K. Landauer, and P. V. Prabju (eds.) *Handbook of Human-Computer Interaction.* Amsterdam, Holland: Elsevier, 653–688.

WOOD, J. AND SILVER, D. (1995) *Joint Applications Development* (2nd ed.) New York: John Wiley & Sons.

Credits

Chapter 1

Figure 1.1: after Gillian Crampton Smith, The hand that rocks the cradle, *ID Magazine*, May/June 1995; Figure 1.2 (on Color Plate 1) (i): gif from www.electrolux.com/screenfridge/start.html, reproduced by permission of AB Electrolux; Figure 1.2(ii): gif from http://houns54.clearlake.ibm. com / solutions/media/medpub.nsf/ebrcs/Ask_a_Question? OpenDocument reproduced by permission of IBM; Figure 1.2(iii): gif from http://www.research. philips.com/pressmedia/pictures/passw3.html, copyright © Philips Research, reproduced by permission of Philips Research; Figure 1.4: figure under section heading 32.1 Interdisciplinary Cooperation, Chapter by S.Kim in *The Art of Human Interface Design*, edited by B. Laurel (1990), Addison Wesley; Figure 1.5: gif from www.ideo. com/studies/scout.htm, reproduced by permission of IDEO; Figure 1.6(a) and (b): screenshots from www.qualcomm.com/eudora reproduced by permission of QUALCOMM Eudora Products; Figure 1.8: screenshot of Photoshop™ menu reproduced by permission of Adobe Systems Incorporated; Table 1: reproduced by permission of www.useit.com/papers/heuristic/heuristic-list.html, copyright © Jakob Nielsen. All Rights Reserved. Fig 1. Interview: reproduced by permission of IDEO.

Chapter 2

Figure 2.1 (on Color Plate 2): gif from www.ai.mit. edu/projects/medical-vision/surgery/ surgical_navigation.html reproduced by permission of Michael E. Leventon; Figure 2.6(a): gif from http://vibes.cs.uiuc.edu/Project/VR/Virtue/VirtueOve rview.htm, reproduced by permission of Dr Daniel A. Reed (University of Illinois at Urbana-Champaign) from work on the Collaborative Virtual Environments for Direct Software Manipulation research project, supported in part by the Defense Advanced Research Projects Agency under contract numbers DABT63-94-C0049, F30602-96-C-0161, DABT63-96-C0027, N66001-97-C-8532, in part by the National Science Foundation under grants CDA 94-

01124 and ASC 97-20202, and in part by the Department of Energy under contracts B-341494, W-7405-ENG-48, and 1-B-333164; Figure 2.5: The Finder Desktop from *Apple Human Interface Guidelines*, Apple Computer Inc. (1987), Addison Wesley; Figure 2.6(b) (on Color Plate 3): gif from http://www.evl.uic.edu/pape/projects/crayoland/big/, copyright © 1997 Dave Pape, image courtesy of the Electronic Visualization Laboratory, University of Illinois at Chicago; Figure 2.7: gif of annotated screen dump for Visicalc™ used with permission of Lotus Development Corporation—Visicalc is a trademark of Lotus Development Corporation; Figure 2.8: Johnson, J. *et al.*, The Xerox "Star": a retrospective, in *IEEE Computer*, copyright © 1989 IEEE, reproduced by permission of IEEE; Figure 2.9: Figure 1.10 (page 16) from *The Psychology of Everyday Things*, by Donald A. Norman, copyright © 1988 by Donald A. Norman, reprinted by permission of Basic Books, a member of Perseus Books, L.L.C.; Figure 2.10: Figure 32 (page 33) from *Designing Visual Interfaces* by K. Mullett and D. Sano © 1995 reprinted by permission of Pearson Education, Inc., Upper Saddle River, NJ 07458; Figure 2.11(i): gif from http://tangible.media.mit.edu/papers/ Tangible_Bits_CHI97.html, Ishii, H. and Ullmer, B. (1997) Tangible Bits: towards seamless interfaces, in *CHI'97 Proceedings*, reprinted by permission of Association for Computing Machinery, Inc.; Figure 2.11(ii) gif from www.almaden.ibm.com/cs/blueeyes/ reproduced by permission of IBM; Figure 2.11(iii): gif from www.parc.xerox.com/red/members/richgold/ livingdoc/slide6. html, reproduced by permission of Rich Gold of PARC Communications; Figure 2.12: gif from www.mbay.net/~brendah/articles/PDA. Mar.95/ reproduced by permission of General Magic, Inc.; Figure 2.13(b): gif from http://thesims.ea.com/us/ reproduced by permission of Electronic Arts Inc. © 2001 Electronic Arts Inc., all rights reserved; Figure 2.14 (on Color Plate 2): gif from http://graphics. stanford.EDU/projects/iwork/ reproduced by permission of Professor Terry Winograd; *Cartoon:* Copyright © CartoonStock, www.CartoonStock.com.

Chapter 3

Figures 3.2(a) and (b): two screenshots of lodging information reproduced by permission of T. S. Tullis from his Ph.D. Dissertation *Predicting the Usability of Alphanumeric Displays*, Rice University, Houston, Texas, USA; Figures 3.3 and 3.10: screenshots of Google search engine reproduced by permission of Google Inc.; Figure 3.4: summarized text from page 192 from *Designing Visual Interfaces* by K. Mullett and D. Sano © 1995 reprinted by permission of Pearson Education, Inc., Upper Saddle River, NJ 07458; Figure 3.6: Lonsdale and Edmunds (1992) *International Journal of Human Computer Studies*, 26, 97–126, Figure 3, reproduced by permission of Academic Press Ltd; Figure 3.8: Mander, R., Salomon, G. and Wong, Y. (1992) Figure 6 (page 631) in *CHI'92 Proceedings*, reprinted by permission of Association for Computing Machinery, Inc.; Figure 3.9 (on Color Plate 4): gif of a transparent phone reproduced by permission of Lazerbuilt Limited; Figure 3.11: redrawn and adapted from Barber, P. (1988) *Applied Cognitive Psychology*, Figure 3.1 (page 63) published by Routledge and reproduced by permission of ITPS Ltd; Figure 3.12: Card, S., Moran, T. and Newell, A. (1983) *The Psychology of HCI*, Figure 2.1, page 26, reproduced by permission of Lawrence Erlbaum Associates, Inc.; Figure 3.13: reproduced courtesy of Lucent Technologies Inc. © [1997] Lucent Technologies Inc., all rights reserved; *Cartoon:* Reproduced by permission of Randy Glasbergen.

Chapter 4

Figure 4.1 (on Color Plate 5): Three gifs of BowieWorld from www.worlds.com/bowie reproduced by permission of worlds.com; Figure 4.2: reprinted from *Decision Support Systems*, 5(2), Nunamaker, J. *et al.*, Experiences at IBM with group support systems, 183–196, Figure 2 © 1989, with permission from Elsevier Science; Figure 4.3: gif of Willow Tree ACTIVboard reproduced by permission of Promethean Ltd.; Figure 4.4(a): photograph of an early model of a videophone (prototype) by courtesy of BT Archives; Figure 4.4(b): photograph of the VP-210 VisualPhone reproduced by permission of Kyocera Corporation, © 1999 Kyocera Corporation; Figure 4.5: illustration of the Video Window System in use from Kraut, R. E., Root, R. W. and Chalfonte, B. L. (1990) Informal communication in

organisations (pages 145–199) in Oskamp, S. and Spacapan, S. (eds.) *People's Reactions to Technologies in Factories, Offices and Aerospace–The Claremont Symposium on Applied Psychology* copyright © 1990 Sage Publications, reprinted by permission of Sage Publications Inc.; Figure 4.7: Morikawa, O., Yamashita, J. and Fukui, Y. (2000) The sense of physically crossing paths, Figure 1 (page 183) in *CHI 2000 Proceedings*, reprinted by permission of Association for Computing Machinery, Inc.; Figure 4.8: *Computer Supported Cooperative Work Journal*, 1, 303, Rogers, Y. Figure 3, reproduced with kind permission from Kluwer Academic Publishers; Figure 4.10: reproduced by permission of the Xerox Research Centre Europe; Figure 4.11: *ECSCW* (1999) 438, Augmenting the workaday world, Fitzpatrick, G. *et al.*, Figures 4 and 5, reproduced with kind permission from Kluwer Academic publishers and the authors; Figure 4.12: Erickson, T. *et al.* (1999) Socially translucent systems, Figure 2 (page 74) in *CHI'99 Proceedings*, reprinted by permission of Association for Computing Machinery, Inc.; Figure 4.13: Winograd, T. and Flores, W. (1986) *Understanding Computers and Cognition*, Figure 5.1 (page 65), Addison Wesley: Figure 4.14: Winograd, T. (1988) Where the action is, Table A (page 257) in *BYTE*, reproduced by permission of CMP Media LLC and Byte.com; Figure 4.15: after Halverson, C., Inside the cognitive workplace: new technology and air traffic control. PhD Thesis, U. of California, San Diego (1995); Figure 4.16: Preece, J. and Keller, L. (1994) *Human-Computer Interaction*, Figure 3.5 (page 70) © Selection and editorial material, the Open University, reprinted by permission of Pearson Education Ltd.; *Cartoon:* Copyright © CartoonStock, www.CartoonStock.com.

Chapter 5

Figure 5.1: gif from www.ai.mit.edu/projects/humanoid-robotics-group/kismet/kismet-html reproduced by permission of Peter Menzel Photography; Figure 5.2: Figure 40 (page 40) from *Designing Visual Interfaces* by K. Mullett and D. Sano © 1995 reprinted by permission of Pearson Education, Inc., Upper Saddle River, NJ 07458; Figure 5.3 (on Color Plate 6) (i): photograph of an iMac from www.apple.com/hardware reproduced by permission of Mark Laita; Figure 5.3(ii): screenshot of a Nokia mobile phone from www.nokia.com/phones reproduced by permission of Nokia Corporation; Figure 5.3(iii): gif from www.ideo.com/studies/bbc.htm reproduced by permission of IDEO; Figures 5.4(a) and (b): Marcus, A. (1993) Human communication in advanced uIs, Figures 2 and 4 (pages 106 and 107) in *Communications of the ACM*, 36(4), 101–109, reprinted by permission of Association for Computing Machinery, Inc.; Figure 5.5: Figure 7.2 (page 147) in *Bringing Design to Software*, edited by Winograd, T. (1996), Addison Wesley; Figure 5.8: from Oren, T., Salomon, G. *et al.*, Guides: Characterizing the Interface, Figure 6 (page 370) in *The Art of Human Interface Design* edited by Laurel, B. (1990), Addison Wesley; Figure 5.9 (on Color Plate 6) (i): gif of Aibo from www.newscast.co.uk reproduced by permission of Sony Corporation; (ii): screenshot of www.ananova.com showing *Ananova*, the virtual news presenter, © Ananova Ltd. 2001, reproduced by permission of Ananova Ltd., all rights reserved; (iii): screenshot from www.e-cyas.com of E-cyas avatar reproduced by permission of I-D Media Ltd.; Figure 5.10: gifs from alive.www.media.edu/projects/alive reproduced by permission of Professor Bruce Blumberg; Figure 5.11: gif from www.csc.ncsu.edu/eos/users/1/lester/www/imedia/DAP.html reproduced by permission of Professor James Lester; Figure 5.12 (on Color Plate 7): gif from www.cs.cmu.edu/afs/cs.cmu.edu/project/oz/web/woggles_clr.html reproduced by permission of Joseph Bates, Zoesis Studios; Figure 5.13 (on Color Plate 8): gif from http://gn.www.media.mit.edu/groups/gn/projects/humanoid/ reproduced by permission of Professor Justine Cassell; Figure 5.14: Figure 2 (page 365) in *The Art of Human Interface Design* edited by Laurel, B. (1991), Addison Wesley.

Chapter 6

Figures 6.2–6.4: reproduced by permission of IDEO, photographs by Jorge Davies; Figures 6.5 and 6.6: Cusumano, M. and Selby, R. (1997) How Microsoft builds software, Figures 2 and 3 (pages 56 and 57) in *Communications of the ACM*, 40(6) reprinted by permission of Association for Computing Machinery, Inc.; Figure 6.9: Boehm, B. W. A spiral model of software development and enhancement, *IEEE Computer*, 21 (5), Figure 2 (page 64) reproduced by permission of IEEE C1988 IEEE; Figures 6.11 and 6.12: Isensee, S. *et al.* Designing internet appliances

at Netpliance from *Information Appliances and Beyond* (2000) edited by Bergman, E., Figures 3.2 (page 58) and 3.6 (page 71) reproduced by permission of Academic Press Inc.; Figure 6.13: Hartson, H. R. and Hix, D. (1989) How Microsoft builds software, *International Journal of Man-Machine Studies*, 31, 477–494, the Star lifecycle model, reproduced by permission of Academic Press Ltd.; Figure 6.14: The usability engineering lifecycle figure in *The Usability Engineering Lifecycle* by Mayhew, D. J. (1999) reproduced by permission of Academic Press Inc.; *Cartoon:* Copyright © CartoonStock, www.CartoonStock.com.

Chapter 7

Figures 7.1 and 7.5: Robertson, S. and Robertson, J. (1999) *Mastering the Requirements Process*, Figures 10.3 (page 184) and 1.3 (page 9) © Pearson Education Ltd 1999, reprinted by permission of Pearson Education Ltd.; Figure 7.2: Bergman, E. and Haitani, R. (2000) Designing the PalmPilot: a conversation with Rob Haitani, from *Information Appliances and Beyond*, (edited by Bergman, E.) Figure 4.3 (page 86) reproduced by permission of Academic Press Inc.; Figure 7.3(a) photograph of the KordGrip reproduced by permission of WetPC Pty. Ltd., Australia 2605; Figure 7.3(b) (on Color Plate 8): photograph of the KordGrip being used under water by permission of the Australian Institute of Marine Science; Figure 7.4: Gaver, B., Dunne, T. and Pacenti, E. (1999) Cultural probes, Figure 1 (page 22) in *Interactions* (January/February) reprinted by permission of Association for Computing Machinery, Inc.; Figure 7.7: screenshot reproduced by permission from Symbian–http://www.symbian.com; figure in Suzanne Robertson's interview reproduced by permission of The Atlantic Systems Guild Ltd.; *Cartoon:* © The 5th Wave, www.the5thwave.com.

Chapter 8

Figure 8.1: reprinted with kind permission of Sigil Khwaja; Figure 8.2: Figure 8.2 (page 169) in *Bringing Design to Software*, edited by Winograd, T. (1996), Addison Wesley; Figure 8.5: Buchenau, M. and Suri, J. F. (2000) Experience prototyping, Figure 1, in Boyarski, D. and Kellogg, W. (eds.) *DIS 2000–Design Interactive Systems, Processes, Practices, Methods, Techniques, Conference Proceedings*, reprinted by permission of Association for Computing Machinery,

Inc.; Figure 8.6(a) and (b): photographs reproduced by permission of ICE Ergonomics Ltd., Loughborough, UK; Figure 8.7: text quoted from Mayhew, D. (1999) *The Usability Engineering Lifecycle*, pages 212–214, reproduced by permission of Academic Press Inc.; Figure 8.8: reprinted from *Interacting with Computers*, 13 (1) Bodker, S. Scenarios in user-centred design–setting the stage for reflection and action, Figure 2 (page 70), © 2000 with permission from Elsevier Science; Figure 8.12: an excerpt from BS-EN-ISO 9241 concerning how to group items in a menu reproduced by permission of the British Standards Institute; Figure 8.14: screenshot of "arrange a meeting" icon from http://www.palm.net/Registration/RegistrationAdd.jsp reproduced by permission of Palm, Inc.; Figure 8.15: reproduced by permission of New Riders Publishing, copyright © 2001 Jeffrey Veen, from the book *The Art and Science of Web Design* by Jeffrey Veen; Figure 8.16: screenshot of the front web page of the Aftonbladet Newspaper from http://www.aftonbladet.se reproduced by permission of Aftonbladet Nya Medier; *Cartoon:* Copyright © CartoonStock, www.CartoonStock.com.

Chapter 9

Figures 9.1–9.3: Tables 1–3 (pages 7,8), Tables 4–7 (pages 9,10), Table 9 (page 15) from Viller, S. and Somerville, I. (1999) Coherence: an approach to representing ethnographic analyses in systems design, *Human-Computer Interaction*, 14 (special issue on representations in interactive systems and development) reproduced by permission of Lawrence Erlbaum Associates, Inc.; Figures 9.4–9.8: Figure 11.5 (page 206), Figure 17.4 (page 315), Figure 17.5 (page 316), Figure 17.2 (page 312), Figure 17.3 (page 313) from Wixon, D. and Ramey, J. (eds.) *Field Methods Casebook for Software Design*, © 1996 John Wiley & Sons, Inc., reprinted by permission of John Wiley & Sons, Inc.; Figure 9.9: Beyer, H. and Holtzblatt, K. (1998) *Contextual Design*, Figure 9.1 (page 155) reproduced by permission of Academic Press, Inc.; Figure 9.10: Ehn, P. and Kyng, M. (1991) Cardboard computers: mocking-it-up or hands-on the future, sort machine mock-up (page 175) in *Design at Work: Cooperative Design of Computer Systems* (Greenbaum, J. and Kyng, M., eds.) reproduced by permission of Lawrence Erlbaum Associates, Inc.; Figure 9.11: Muller, M. J. (1991) PICTIVE–an

exploration in participatory design, Figures 1 and 2 (page 26) in *CHI'91 Proceedings*, reprinted by permission of Association for Computing Machinery, Inc.; Figure 9.12: Muller, M. J. *et al.* (1995) Bifocal tools for scenarios and representations in participatory activities with users, Figure 6.3 (page 149) in *Scenario-based Design* (Carroll, J., ed.) © John Carroll, reproduced by permission of John Carroll, Virginia Tech.; *Cartoon:* Reproduced by permission of Randy Glasbergen.

Chapter 10

Figures 10.1 and 10.2: Gould, J. D. *et al.* (1990) The 1984 Olympic Message System–a test of behavioral principles of system design, in Preece, J. and Keller, L. (eds.) *Human-Computer Interaction (Readings)* Figures 12.4 (page 265) and 12.1 (page 263) © Selection and editorial material, the Open University, reprinted by permission of Pearson Education Ltd.; Figures 10.3–10.8: Figure 1 (page 6), Appendix A of Usability study, Figure 3 (page 10), Appendix B (pages 14, 15) of Usability study, Table 3 (page 6) of Usability study, Summary (page 8) of Usability study from Cheng, L. *et al.* (2000) Hutchworld: lessons learned. A collaborative project: Fred Hutchsinson Cancer Research Center and Microsoft Research, *Virtual Worlds Conference 2000, Paris, France* © Springer-Verlag GmbH & Co., reproduced by permission of Springer-Verlag GmbH & Co. and the author.

Chapter 11

Cartoon: Reproduced by permission of Randy Glasbergen.

Chapter 12

Figures 12.1 and 12.2: screenshots from http://www.northernlight.com reproduced by permission of Northern Light Technology, Inc.; Figure 12.3: Figure 5 (pages 7 and 8) from Hochheiser, H. and Shneiderman, B. (2001) Using interactive visualizations of WWW log data to characterize access patterns and inform site design, *Journal of the American Society for Information Science* (in press) reproduced by permission from University of Maryland, Human-Computer Interaction Lab; *Cartoon:* HERMAN ® is reprinted with permission from LaughingStock Licensing Inc., Ottawa, Canada, all rights reserved.

Chapter 13

Figure 13.1: screenshot from http://ananova.com ©
Ananova Ltd. 2001, reproduced by permission of
Ananova Ltd., all rights reserved; Figure 13.3: B.
Shneiderman (1998) *Designing the User Interface:
Strategies for Effecive Human-Computer Interaction*,
Third Edition, Table 4.1, Part 3 (page 136), Addison
Wesley; Figure 13.4: from Andrews *et al.*, A
Conceptual Framework framework for demographic
groups resistant to online community interaction. In
*Proceedings of IEEE Hawaiian International
Conference on System Science (HICSS)*, 2001; Figure
13.5: Nielsen, J., Finding Usability Problems through
Heuristic Evaluation. In *Proceedings of CHI'92*,
373–800; Figure 13.6: Adapted from Appendix G.
page 204 (2001) Ph.D. Thesis by Dorine C. Andrews,
'Computer-Supported Social Networks: Audience-
Centric Online Community Implementation.'
Communications Design. University of Baltimore,
Maryland; Figure 13.7: Figure 2.2 (page 33) from
Nielsen, J. and Mack, R. L. (1994) *Usability
Inspection Methods*, © 1994, John Wiley & Sons, Inc.,
reprinted by permission of John Wiley & Sons, Inc.;
Figures 13.7–13.9: Figures 1–3 (pages 11, 12 and 14)
from Cogdill, K. (1999) *MEDLINEplus Interface
Evaluation: Final Report*, reproduced by permission
of Professor Keith Cogdill, College of Information
Studies, University of Maryland; Figure 13.10:
screenshot from http://REI.com reproduced by
permission of Recreational Equipment, Inc.; *Cartoon:*
© The 5th Wave, www.the5thwave.com.

Chapter 14

Figure 14.1: Figure 1 (page 11) from Cogdill, K.
(1999) *MEDLINEplus Interface Evaluation: Final
Report*, reproduced by permission of Professor Keith
Cogdill, College of Information Studies, University
of Maryland; Figure 14.2: Figure 2, pages 67–80,
from Lund, A.M. Ameritech's usability laboratory:
from prototype to final design, *Behaviour and
Information Technology*, 13, 1–2 (1994)
(http://www.tandf.co.uk/journals) reproduced by
permission from Taylor & Francis Ltd.; Figure 14.3:
Nodder, C., Williams, G. and Dubrow, D. (1999)
Evaluating the usability of an evolving collaborative
product–changes in user type, tasks and evaluation
methods over time, Figure 6 (page 156) in
GROUP'99, Phoenix, Arizona, USA, reprinted by
permission of Association for Computing Machinery,
Inc.; Figure 14.4: Larson, K. and Czerwinski, M.
(1998) Web page design: implications of memory,
structure and scent for information retrieval, Figure 1
(page 28), in *CHI'98 Proceedings*, reprinted by
permission of Association for Computing Machinery,
Inc.; *Cartoon:* From *The Wall Street Journal*—
Permission, Cartoon Features Syndicate.

Chapter 15

Figure 15.1: screenshot of the Nokia 9210
Communicator from http://www.nokia.com/press/
photo/phones/jpeg/9210_09.jpg reproduced by
permission of Nokia Corporation; Figures 15.2–15.5;
Figure 7.11 (page 195), an example usage scenario
(page 181), Figures 7.6 and 7.7 (pages 183 and 186)
from Vaananen-Vainio-Mattila, K. and Ruuska, S.
(2000) Designing mobile phones and communicators
for consumers' needs at Nokia, *Information
Appliances and Beyond* (Bergman, E., ed.)
reproduced by permission of Academic Press, Inc.;
Figures 15.6–15.10, including Figure 15.8 (on Color
Plate 8) and 15.14: Oosterholt, R., Kusano, M. and
de Vries, G. (1996) Interaction design and human
factors support in the development of a personal
communicator for children, Figures 1, 2, 3, 5, 9, 10
and 7 in *CHI'96 Proceedings*, reprinted by permission
of Association for Computing Machinery, Inc.,
communicator concept development and execution
by Philips Design, Eindhoven, The Netherlands;
Figures 15.11–15.13: Figure 19 (page 28), Table 2
(pages 24 and 25) and Figure 16 (page 25) from
Montemayor, J. *et al.* (2000) PETS: A personal
electronic teller of stories, *Robots for Kids* (Druin,
C.A. and Helander, J., eds.) reproduced by
permission of Academic Press, Inc. and the authors,
Institute for Advanced Computer Studies, University
of Maryland.

*The publisher has made every attempt to obtain
permission to reproduce material in this book from the
appropriate source. If there are any errors or
omissions please contact the publisher, who will make
suitable acknowledgement when the book is reprinted.*

Index

Page references followed by italic t indicate material in tables. Page references followed by italic n indicate material in footnotes.

abstraction
  dynalinking for learning, 87
  loss of information, 293
  realism contrasted, 66–67
access, to websites, 415–416
ACM Code of Ethics, 351–352
ACRE (ACquisition REquirements), 219
ActiMates, 154
ACTIVBoard, 114t
activities, of people interacting with products, 4–5
activity-based conceptual models, 41–51, 250, 252
activity-based planning, 184, 282
activity theory, 136, 382
actors, 226–230
aesthetics, 27, 409
  user experience goal, 18, 19
affective aspects, 141–142
  and anthropomorphism, 153–157
  expressive interfaces, 143–147
  user frustration, 147–153
affective computing, 142
affinity diagrams (Contextual Design method), 304, 305
affordance, 25–26, 29
agents
  for conversation-based conceptual models, 46–47, 50
  design, 160–162
  friendly interface agents, 144, 146
  types of, 157–160
Aibo, 157
alternative designs
  choosing among, 179–182
  conceptual models, 254
  generation, 12, 166, 169, 174–179
  and lifecycle model, 186
  and prototyping, 241

Amazon.com
  cognitive walkthrough of book purchase, 421–422
  one-click purchasing, 14, 179
animated agents, 46–47, 158
animation, 143
  avoiding gratuitous use on websites, 416
annotating, 98–100
  shared external representations, 121
ANOVA (analysis of variance), 457
Ananova (virtual newscaster), 392–394
anthropomorphism, 153–157
apologies, by computers, 153
appearance
  of interfaces, user frustration with, 152
  of virtual characters, 160–161
Apple Macintosh, *See* Macintosh
architectural design, 168
artifact model (Contextual Design method), 301, 305
artifacts, collection in field studies, 342
artist-design
  approach to users, 212–213
  relation to interaction design, 8
Ask Jeeves, 155
Ask Jeeves for Kids, 44–45
asynchronous communication, 327
  computer-mediated, 112–113t
atomic requirements, 236–237
attention, 75–76
  design implications, 77
attentive environments, 62, 63, 257
audio recording. *See also* interviews
  data analysis, 381–385
  interaction logging with, 378
  in observation, 365, 369, 374, 376t
  in requirements identification, 218
augmented reality, 36, 63
autistic communication-support device, 241–242
Auto Attendant interface, TRIS, 485, 486
automated phone-based systems, 45

awareness mechanisms, in collaboration, 124–126

Babble, 128
back channeling, 106, 108
Barney, example of anthropomorphism, 154
biases
  in evaluation data, 355–356
  in interview questions, 391
  in questionnaires, 406
BlueEyes, 61, 63
BlueTooth, 57
Bly, Sara, interview with, 387–388
Bob (friendly interface agent), 144, 146
body-area network, 60
body language, 106, 108
bookmarking, 80
  problem space definition, 37–38
book metaphor, problems of using, 59
branding, web pages, 273
browsers, *See* web browsers
browsing-based conceptual models, 41, 49
bulletin boards
  conversational analysis, 354
  discourse analysis, 384
  usage tracking, 378

CARD (Collaborative Analysis of Requirements and Design), 307, 309–311
case-based reasoning, 175
CASE (Computer-Aided Software Engineering), 259
CD-ROM tutorials, 16
cell phones, 38–39, 463. *See also* mobile communicators
  culture change required for, 173
  evaluation, 322
  physical design, 265–266
  transparency of functioning, 95
chatrooms, 110, 112t
  conversational analysis, 354
  discourse analysis, 384

check boxes, in questionnaires,
    400–401
children
    computerized toy evaluation,
        419–420
    participant observation,
        479–480
chunks, of memory, 82
ClearBoard, 115, 118
Clippy, 49, 144, 146
closely-knitted teams, 125–126
cluster analysis, 407
COG, 142
cognition, 74–75, 286. *See also*
    memory
    and attention, 75–76, 77
    distributed, 98, 133–136
    external, 98–101
    information processing, 96–98
    and learning, 86, 87
    and memory, 76, 78–85
    mental models, 92–95, 101
    and perception, 76–78
    and problem solving, reasoning,
        and decision making, 88–89
    and reading, speaking, and
        listening, 86–88, 89
cognitive engineering, relation to
    interaction design, 9
cognitive ergonomics, relation to
    interaction design, 9
cognitive science, relation to
    interaction design, 9
cognitive tracing, 98–100
cognitive walkthroughs, 420–423
coherence method, 293–295, 310t
cohort, 401
collaboration and communication,
    105
    awareness mechanisms, 124–126
    conversational mechanisms,
        106–110
    coordination mechanisms,
        118–122
    difficulties with in design,
        198–199
    distributed cognition approach,
        130–133
    ethnographic studies, 129
    language/action framework
        approach, 130–133
    and physical design, 267
    for user involvement, 281
collaborative technologies, 105
    designing to support awareness,
        126–128

designing to support coordination,
    122–124
designing to support social
    conversation, 110–118
collaborative virtual environments,
    110–111, 112t
color, avoiding gratuitous use on
    websites, 416
command-based interfaces, 42, 50
    memory aspects, 79–80
command-based programming
    languages, 7
commercial style guides, 267
communication, *See* collaboration
    and communication
component systems, 276
computational offloading, 99, 100
computer conferencing, 110
computerized toys, 419–420
computer-mediated communication,
    111, 115–118
    types, 112–114t
computer science, relation to
    interaction design, 9
computer-supported cooperative
    work, relation to interaction
    design, 9
conceptual design, 239, 249–250
    iterative nature of, 250, 265
    and physical design, 265
    prototypes in, 262–265
    scenarios in, 259–262
conceptual models, 39–41,
    249–250
    activity-based, 41–51, 250, 252
    for collaboration and
        communication, 130–136
    expanding, 257–259
    hybrid, 54–55
    and interaction modes, 40–55,
        250–253
    and interaction paradigms, 40,
        60–64, 257
    and interface metaphors, 40,
        55–60, 253–257
    from model to physical design,
        64–68
    object-based, 51–53, 250, 253
    Philips mobile communicator,
        481–482
    process- *vs.* product-oriented, 253,
        254–255
    user understanding of, 54
consistency, 408
    design principle, 24–25, 29, 266,
        412

Nokia mobile communicators,
    472–473
    usability principle, 27
consolidation (Contextual Design
    method), 296
constraints, 21–23
    support tools designed to
        maintain, 276
construction, 248–249
content analysis, 342
    described, 383
context-free grammars, 276
context of use, 207. *See also*
    environmental requirements
    mobile communicators, 463
    and user-centered development,
        286
context-sensitive information, 94,
    100
Contextual Design method, 250,
    310t
    described, 295–300, 313–315
    Nokia mobile communicators,
        465–466
    for office products design, 297–298
contextual inquiry process
    (Contextual Design method),
    296, 298–300, 313
contextualized observations, 372
controlled environment studies, *See*
    laboratory studies
convenience sampling, 406
conventions
    for collaborative meetings, 121
    reasons for not following, 122
conversational analysis, 342, 384
conversational mechanisms, in
    collaboration, 107–110
conversation-based conceptual
    models, 41, 44–47
conversations for action (CfA),
    130–131
coordination mechanisms, in
    collaboration, 118–122
Coordinator System, 131–133
coping strategies, in physical world,
    90–91
copyright, 179
corporate style guides, 267
counterbalancing, 445–446
Crampton Smith, Gillian, interview
    with, 198–199
creativity
    enhancing in design process, 175
    user experience goal, 18, 19, 141
    and user involvement, 247–248

creativity-support tools, 459
Creatures, 157
critical incident analysis, 382
critical mass, 327
critical user tasks, 467, 469
crit reports, 347*t*
Cruiser, 117
cues, in conversation, 107, 108
cultural constraints, 22–23
cultural diversity, 173, 350
cultural model (Contextual Design method), 301–302, 305
cultural probes, 212

Dangling String, 61
data-flow diagrams, 220
data gathering
  in evaluation, 344*t*
  in experiments, 446–448
  MEDLINEplus user testing, 435–436
  in observation, 363, 365, 371–377, 376*t*
  props with, 210
  in requirements activity, 202–203, 210–218, 213*t*
  in TRIS redesign, 487
data interpretation and analysis
  in evaluation, 355–356
  in experiments, 446–448
  in interviews, 392, 398
  MEDLINEplus user testing, 436–438
  in observation, 365, 372, 376*t*, 379–385, 387
  in questionnaires, 407
  in requirements activity, 202–203, 219–221
data requirements, 206–207
DECIDE evaluation framework, 348–356
  observation application, 379
  user testing application, 438–443
decision making, 88–89
defibrillator, chest-implanted automatic, 251
dependent variables, 444
design, 166. *See also* interaction design
The Design of Everyday Things (Norman), 21, 25
design principles
  described, 20–27
  level of guidance and terms used with, 28
  for physical design, 268

design room (Contextual Design method), 306
desktop paradigm, 60, 257
dialog boxes, 267, 413
  design for closure, 266
  expressive interfaces, 144, 145
diaries, 377
different participant design, of experiments, 445, 446*t*
digital butler, 50
digital desk, 63
direct manipulation interfaces, 47–49, 50
  and learning through doing, 86
discount evaluation, 410
discourse analysis, 342
  described, 383–384
distributed awareness systems, 127–128
distributed cognition, 98
  and collaboration, 133–136
Distributed Systems Technology Center, 117
documentation, 180
  as usability principle, 27
  use in requirements activity, 213*t*, 214–215
drop-down menus, 268
dynalinking, 77, 87
dynamic icons, 143
Dynamic Systems Development Method (DSDM), 190
dynamic visualization, 476, 477
dyslexics, 88

ecological validity, of evaluation, 356
e-commerce
  culture change required for, 173
  efficiency, 14
educational software, 7
effectiveness, usability goal, 14
efficiency
  usability criteria, 18
  usability goal, 14
  usability principle, 27
e-jacket, 60
electronic calculator, 167–168, 175
electronic commerce, *See* e-commerce
electronic ink, 5
electronic meeting rooms, 113*t*
electronic whiteboards, 124
Elvin, 127–128
email, 110
  conversational analysis, 354

email questionnaires, 405
embodied conversational interface agents, 159–160
emoticons, 146–147, 147*t*
  for online patient support community, 322
emotional agents, 158–159
emotional fulfillment, user experience goal, 18, 19, 141
emulation, of physical world knowledge, 90–91
engineering, 6
  relation to interaction design, 9
enjoyment, user experience goal, 18, 19, 141
entertainment, user experience goal, 18, 19
entity-relationship diagrams, 221
environmental requirements, 207.
  See also context of use
  mobile communicators, 463–464
ergonomics, relation to interaction design, 9
error handling, 266
error messages, 147, 148–150
  design, 149, 266
error prevention, 27, 266, 408, 413
error recovery, 27, 408
essential use cases, 229–231
  and functional requirements, 258
e-tailing, *See* e-commerce
ethical issues
  in evaluation, 352–355
  in observation, 378
  in unstructured interviews, 392
  in user testing, 443
Ethnograph, 381, 398
ethnography. *See also* field studies
  adapting to fit development process, 373
  coherence method, 293–295, 310*t*
  of communication, 129
  contextual Design method, 250, 295–300, 310*t*, 313–315
  example, 289–290
  goals, 360
  of home technology use, 291
  Nokia mobile communicators, 465
  in observation, 361, 363, 364, 380–381
  and participant observation, 364, 370–373
  in user-centered development, 279, 288–306, 310*t*
ethnomethodology, 136

Eudora, safe and unsafe menus, 15
evaluation, 12, 169–170, 317–318. *See also* DECIDE evaluation framework; field studies; predictive evaluation; usability testing; user testing
  ethical issues, 352–355
  formative and summative, 323
  goals, 360–361
  HutchWorld case study, 318, 324–336, 440
  insider *vs.* outsider, 342, 361–364
  integration with design, 461–462
  and lifecycle model, 186
  mobile communicators case study, *See* mobile communicators
  Nokia mobile communicators, 466–467
  Philips mobile communicator, 482
  phone-based response system redesign case study, 482–489
  pilot studies, 356
  practical issues, 350–351
  reasons for, 319–323
  terminology, 340, 345
  what to evaluate, 318–319
  when to evaluate, 323–324
  when to stop, 334
evaluation paradigms, 340, 341–345, 344t
  choosing in DECIDE framework, 349
  techniques used with, 347t
evaluation techniques, 345–347
  choosing in DECIDE framework, 349
event languages, 276
evolutionary prototyping, 248, 249
expectation management, and user involvement, 280–281
experiential cognition, 74
experimental conditions, 444
experiments, 430, 431, 443–444
  allocation of participants to conditions, 445–446
  data collection and analysis, 446–448
  usability testing contrasted, 457–458
  variables and conditions, 430, 443–445
  website design structure, 447
expert crit, 410

expert opinions, 346, 347t
  HutchWorld case study, 325
  in quick and dirty evaluation, 341
  in TRIS redesign, 485, 488
exploration-based conceptual models, 41, 49
expressive interfaces, 143–147
external cognition, 98–101
externalization, of memories, 98–99

facial expressions, 106
feedback
  design and usability principles for, 20–21
  in evaluation paradigms, 344t
  interview-like, 397
  and iterative design, 170
  in observation, 376t
field studies, 341. *See also* ethnography
  challenges, 388
  described, 342
  goals, 360
  observation, 359, 363–364, 368–370
  techniques applied, 347t
  user screening, 350
file locking, for coordinating collaborative technologies, 122
file management systems, 81, 83
  and pile phenomenon, 91
film industry, relation to interaction design, 9
Fitts' Law, 454–455
flaming, 113t, 153
flexibility, 409
  of observation data-collection techniques, 376t
  usability principle, 27
flight strips, 296
flow chart diagrams, for constraining, 22
focus groups
  use in evaluation, 396–397
  use in requirements activity, 213t, 214, 217
formal communication, 110
formal language-based tools, 276
formative evaluations, 323
Fred Hutchison Cancer Research Center, 324–325, 334
friendly interface agents, 144, 146
fun, user experience goal, 18, 19
functional requirements, 205, 206
  analysis, 220–221
  and conceptual model, 258–259

gesturing, 106, 108
gIBIS, 114t
gimmicks, user frustration with, 148
GOMS model (goals, operators, methods, and selection rules), 102, 231, 346
  benefits and limitations, 453–454
  described, 449–450
  in TRIS redesign, 485, 488
Google, 22, 77
  background information on operation, 95
graphical user interfaces, 7, 42, 60
  and affordance, 25–26
  and learning through doing, 86
  memory aspects, 79–80
  memory load reduction, 101
  shading for menu item deactivation, 21–22
graphic design, 416
  relation to interaction design, 8
graphics, avoiding gratuitous use on websites, 416
group interviews, 390
  described, 396–397
GroupSystem, 113t
groupware, 105. *See also* collaborative technologies
GUIs, *See* graphical user interfaces
GVU survey, 406

Hawthorne effect, 356
HCI Bibliography Project, xxii, xxiii
hearing, 77
help, 409
  as usability principle, 27
helpfulness, user experience goal, 18, 19
Herman the Bug, 158
heuristic evaluation, 26, 341, 343
  adapting to Web, 248–249
  described, 408–410
  MEDLINEplus, 412–416, 432
  of online communities, 417–419
  problems with, 411
  process of, 410–412
  walkthroughs, 210, 420–423
  of websites, 412–417
heuristics, 26–27, 28, 408–409, 419–420. *See also* usability principles
  for predictive evaluation, 343
  for website evaluation, 412–413
Hierarchical Task Analysis, 231–233

high-fidelity prototyping, 245–246, 246*t*, 263
high-level programming languages, 7
Holtzblatt, Karen, interview with, 313–315
HOME RUN heuristic, 409
horizontal prototyping, 248
human-computer interaction, 458–459
  design patterns, 272
  and ethnography, 342
  lifecycle models in, 192–196
  relation to interaction design, 9
human factors, relation to interaction design, 9
Hutchinson Cancer Research Center, 324–325, 334
HutchWorld case study, 318, 324–336, 440
hyperlinks, 273–274
HyperMirror, 118
hypertext, 274, 276
hypotheses, 443–444, 445

IBM usability laboratory, 441
icons, 268
  design, 270–271
IDEO Scout, 12
IDEO TechBox, 176–178
incidents, analyzing in observational data, 381–382
independent variables, 444
index cards, prototyping with, 244
indirect observation, 377–379
industrial design, relation to interaction design, 9
informal communication, 110
informatics, relation to interaction design, 9
information appliances, 9
information architects, 11
information design, of websites, 416
information display design, 274–275
information processing, 96–98
information retrieval, 81, 83
information visualization, 7, 101
informed consent, 352–353, 354, 365
  unstructured interviews, 392
infrared sensing, 7
innovation
  and prototyping culture, 247–248
  and user involvement, 247–248
insider evaluation, 342, 361–364

inspections, 407–408. *See also* heuristic evaluation; walkthroughs
  walkthroughs, 210, 420–423
instruction-based conceptual models, 41, 42–44
interaction design. *See also* affective aspects; cognition; conceptual design; conceptual models; interaction design process; lifecycle models; physical design; requirements; usability goals; user experience goals; specific types of interfaces
  aim of, 1–2
  and anthropomorphism, 153–157
  in business, 10–12
  defined, 6–12, 166–168
  emulation of physical world knowledge, 90–91
  good and poor contrasted, 2–6
  history, 7–8
  and human-computer interaction, 8
  integration with evaluation, 461–462
  iterative nature of, *See* iterative design
  mobile communicators case study, *See* mobile communicators
  multidisciplinary teams for, 9–10, 282
  notation for, 222
  and other approaches, 9
  phone-based response system redesign case study, 482–489
  realism or abstraction?, 66–67
  relation of other approaches, 8
  terminology, 11
  from theory to practice, 100–101
  trade-offs, 166
  what to design: activities supported, 4–6
interaction design process, 12–13, 165–170. *See also* alternative designs; lifecycle models; prototyping
  activities associated with, 168–170
  building interactive design versions, 12, 169
  practical issues, 170–182
interaction logs, 354, 365
  described, 377–379
interaction modes, 40–55, 250–253
interaction paradigms, 40
  and conceptual design, 257
  types of, 60–64

interaction styles, 41, 250
interactive development environment, 422
interactive graphical tools, 276
interactive/interaction designers, 11
interactive learning environments, 7
interactive pets, 157
interactive phone-based response system redesign, 482–489
interactive products, 1–2. *See also* conceptual models; evaluation
  defined, 2*n*
  interaction paradigms, 40, 60–64
  interface metaphors, 40, 55–60
  problem space, 36–39
interactive toys, 5
interactive voice response systems, 485
interface designers, 11
interface metaphors, 40, 55–60, 253–257
  Philips mobile communicators, 474–475
intergenerational design teams, 479
internal consistency, 413, 414
internal locus of control, 266, 413
Internal Revenue Service, TRIS redesign (telephone response information system), 443, 482–489
inter-research reliability rating, 383
interrupt-driven tasks, 319
interviews. *See also* semi-structured interviews; structured interviews; unstructured interviews
  believability of responses, 397
  data analysis, 398
  in evaluation pilot studies, 356
  field studies technique, 342
  HutchWorld case study, 330
  planning for, 391
  question development, 390–391
  in requirements activity, 210, 211, 213*t*, 214, 215, 217
  retrospective, 372
  types of, 392–397
  usability testing technique, 340, 341
  for user opinion solicitation, 346
i-opener, 191
ISO 9241, 268, 269
ISO 13407, 268
ISO 14915, 268
iterative design, 64–65, 68
  in conceptual design, 250, 264
  and feedback, 170

iterative design, (*Continued*)
  in physical design, 265
  in prototyping, 239, 247, 248
  real world pressure, 461
  in requirements activity, 203
  and user-centered development,
    285, 462
  in user need identification, 203
IT project failure, 203

jargon, avoiding in interviews, 391
Java, 57
Java Beans, 276
Joint Application Development
  (JAD) workshops, 190, 214

keystroke level method, 102, 346
  described, 450–453
  scope, 356
KidPad, 114*t*
Kismet, 142
knowledge
  circulation in social circles, 106
  emulation of physical world's,
    90–91
Knowledge Navigator, 161
KordGrip (WetPC), 208

laboratory studies, 345
  ecological validity, 356
  observation, 359, 363, 365–368
  user screening, 350
language/action framework, 130–133
laptop computers, in observation,
    369, 374
large interactive screens, 9
learnability
  usability criteria, 18
  usability goal, 14, 16–17
learning, 86
  design implications, 87
  resistance to time spent, 94
library catalog, 252, 256
  task description and analysis,
    222–234
lifecycle models, 182–186
  in human-computer interaction,
    192–196
  Nokia mobile communicators,
    465–467
  in software development, 187–192
Likert scales, 401–403
listening, 86–88
design implications, 89
listserver discussion groups, lurking
    behavior, 378

liveboards (ubiquitous computing
    device), 61, 62
logical constraints, 22–23
London Underground, 125–126,
    361
low-fidelity prototyping, 243–245,
    246*t*, 249, 263
  for rapid feedback, 250
lurking behavior, 378

Macintosh
  direct manipulation as conceptual
    model, 47–49
  expressive interface: smiling and
    sad Macs, 143
  garbage can, user confusion with,
    49, 58
  pile approach used by, 91
Macromedia Director, for
    prototyping, 245
Magic Cap, 66
manipulation-based conceptual
    models, 41, 47–49
mapping, 23
marble phone answering machine,
    example of good design, 3–4
matched-participants design, of
    experiments, 446, 446*t*
measurement, 285. *See also* user
    testing
  importance of, 457
  in usability testing, 341–342
media spaces, 110, 111, 112*t*
MEDLINEplus
  heuristic evaluation, 412–416,
    432
  user testing, 432–438
MeetingMaker, 120
meetings, 290
MEMOIRS, 83
memorability
  usability criteria, 18
  usability goal, 14, 17, 19
memory, 78–85
  design implications, 85, 266,
    268, 413
  externalizing to reduce load,
    98–99
  and information processing,
    97
  and perception, 76
  seven chunks theory, 82
mental models, 92–95, 101
menus, 268
  design, 268–270
messaging, 110, 112*t*

Microsoft Corporation. *See also*
    Windows environment
  HutchWorld involvement, 324,
    326, 328
  synch and stabilize software
    design process, 183, 184–185
  usability laboratory, 441, 442
  user involvement, 282
Microsoft Office 4.0, usability
    testing, 282
Microsoft Windows, *See* Windows
    environment
Microsoft Word 2001, sorting
    operation, 24–25
minimalist design, 27, 409
minus scenarios, 260–261
mnemonics, 81
mobile communicators, 463–464
  Nokia's approach to design,
    464–474
  Philips' approach to design,
    474–482
mobile computing, 7
mobile telephones, *See* cell phones
mobile usability laboratories, 365, 442
mockup and text with customers
    (Contextual Design method),
    296
mockups, 240–241, 307
monitors (visual display units), 7
MOOs, 111
motivation, user experience goal, 18,
    19, 141
MUDs, 111, 112*t*
multidisciplinary teams, 9–10
  user involvement with, 282
multimedia applications, 5, 7
  dynalinking, 87
MUMMS (Measuring the Usability
    of Multi-Media Systems), 407
musical playing devices, 23

naturalistic observation, 279. *See
    also* field studies
  use in requirements activity, 213*t*,
    214, 217
natural-language-based systems, 44,
    88
navigation, 415
navigation-based conceptual
    models, 41, 47–49
need identification, *See* user need
    identification
NetMeeting, 442
Netpliance, 173
  spiral development cycle, 191–192

networked classrooms, 114t
networked clothing, 5
networking, 7
Nielson, Jakob, interview with, 426–427
Nokia, mobile communicator design approach, 464–474
Nokia 9000 communicator, 467
Nokia 9210 communicator, 465
Nokia 7110 mobile phone, 470–471
Nokia 9000 web browser, 472–473
nonfunctional requirements, 205, 206
non-verbal communication, 106, 119
Northernlight, 365–367
note taking
    in observation, 365, 369, 370, 374, 376t
    in requirements identification, 218
noticeboards, 121
NUDIST, 381, 382, 383, 398

object-based conceptual models, 51–53, 250, 253
objective evaluations, 345
object-oriented programming, 276
object-oriented software engineering, 195, 259
Object Oriented Software Engineering, 226
observation. See also naturalistic observation
    approaches to, 363–364
    in controlled environments, 365–368
    data gathering, 363, 365, 371, 372, 373–377, 376t
    data interpretation and analysis, 365, 372, 376t, 379–385, 387
    described, 345–346, 347t
    ethical issues, 378
    in field studies, 342, 368–370
    framework for, 368–369
    goals, 360–361
    HutchWorld case study, 327
    indirect, 377–379
    trend toward real world observation, 319
    usability testing technique, 340, 341
    what and when to observe, 361–363
    when to stop, 372
Observer Video-Pro, 382–383
Olympic Messaging System (1984), 285, 319, 323, 336
    described, 320–321
online communities, heuristic evaluation, 417–419

online interviews, 397
online patient support communities evaluation, 322
    HutchWorld case study, 318, 324–336
online questionnaires, 405–407
online tutorials, 16
open-ended interviews, See unstructured interviews
open-ended problem spaces, 39
order effects, 446
ordering effects, 445
organizational environment, 207
orphan pages, 415
outsider evaluation, 342, 361–364
overhearing, 125–126
overseeing, 125–126
ownership, and user involvement, 280, 281

pads (ubiquitous computing device), 61, 62
PalmPilot, 60, 63
    requirements activity, 205–206
    wooden prototype, 241
paradigms, 183n. See also evaluation paradigms; interaction paradigms; lifecycle models
PARC Media Space project, 387
participant observation, 342, 361, 363. See also observation
    with children and adults, 479–480
    described, 364, 370–373
    Philips mobile communicator, 478
participatory design, 306–311, 310t
participatory prototyping, 210
patenting, 179
patterns
    analyzing in observational data, 381–382
    analyzing in questionnaires, 407
    design, 272
PDAs, 463
perception, 76–78
design implications, 78
Perl, 276
personalization
    Nokia mobile communicator, 468
    Philips mobile communicator, 478
personal workstations, 7
pervasive computing, 60, 257
Phil, Knowledge Navigator agent, 160–161
Philips, mobile communicator design approach, 474–482

Philips Vision of the Future Project, 10
phone answering system (marble answering machine), as example of good design, 3–4
phone banking, 83–85
phone-based response system redesign, 482–489
photocopiers, 179–180
    problems with, 1
PhotoFinder, 458–459
physical constraints, 22
    and evaluation, 340
    Nokia mobile communicators, 470–473
physical design, 239, 265–266
    from conceptual model to, 64–68
    guidelines and standards, 266–267, 268
    icons, 270–271
    information displays, 274–275
    menus, 267–270
    screens, 271–272, 274
physical limitations, 286
physical model (Contextual Design method), 302, 303, 305
physical/virtual integration, 63
PICTIVE (Plastic Interface for Collaborative Technology Initiatives through Video Exploration), 307–309
pilot studies
    in evaluation, 356
    for refining structured interview questions, 394
    in requirements identification, 217
pleasure factors, See user experience goals
plug-and-play interfaces, 96
plug-ins, user frustration with, 151–152
pluralistic walkthroughs, 420, 423
plus scenarios, 260–261
Pokemon, 157
POLITeam workspace system, 135
pop-up menus, 268
portal website, conceptual model, 56
Portholes, 126–127, 127
predictive evaluation, 449. See also GOMS model; keystroke level method
    benefits and limitations, 453–454
    defined, 343, 344t
    Fitts' Law, 454–455
    techniques applied, 347t
predictive models<\#208>455

Presence Project, 212
primary users, 171
privacy protection
    in evaluation, 351–352, 353, 354
    in observation, 378
probes, in semi-structured
    interviews, 394
problem solving, 88–89
    design implications, 89
problem space, of interactive
    products, 36–39
process, of interaction design, *See*
    interaction design process
process models, 183*n*. *See also*
    lifecycle models
process-oriented conceptual models,
    253, 254–255
product design, relation to
    interaction design, 8
product-oriented conceptual
    models, 253, 254–255
Project Ernestine, 453–454
project failure, reasons for, 203
project management systems, 123
prompting, in semi-structured
    interviews, 394
props, with data-gathering
    techniques, 210
prototyping, 64–65, 169
    compromises in, 246–248
    in conceptual design, 262–265
    and construction, 248–249
    defined, 180, 240–241
    evolutionary, 248, 249
    high-fidelity, 245–246, 246*t*, 263
    horizontal and vertical, 248
    HutchWorld case study, 325–326
    iterative nature of, 239, 247, 248
    low-fidelity, 243–245, 246*t*, 249, 263
    notation formality of software, 222
    observation for evaluation, 345
    participatory, 210
    Philips mobile communicators,
        474–478
    rapid, 195
    reasons for doing, 241–242
    role-playing walkthroughs, 210
    scenarios as scripts for user
        evaluation, 261
    and spiral lifecycle model, 188
    throw-away, 248–249
    and Usability Engineering
        Lifecycle model, 195
    user involvement, 284
    value of, 181
prototyping cultures, 247–248

proxy-users, 280
psychology, 6
    relation to interaction design, 8
putting it into practice (Contextual
    Design method), 296
Python, 276

qualitative evaluations, 345
    importance of, 387
quality, for choosing between
    alternative designs, 180–181
quantitative evaluations, 345
Questionnaire for User Interaction
    Satisfaction (QUIS), 402, 404,
    435
questionnaires
    administering, 404
    data analysis, 407
    design, 399–400
    in evaluation pilot studies, 356
    HutchWorld case study, 330
    MEDLINEplus user testing, 435,
        438
    online, 405–407
    question and response format,
        400–403
    in requirements activity, 211, 213*t*,
        215, 217
    usability testing technique, 340,
        341, 342
    for user opinion solicitation, 346
    user screening, 350
quick and dirty evaluation
    defined, 341, 344*t*
    goals, 360
    HutchWorld case study, 336
    observation, 363, 364
    techniques applied, 347*t*
    user testing, 431
Quicken, 53
QUIS (Questionnaire for User
    Interaction Satisfaction), 402,
    404, 435

radio-frequency tags, 9
ranges, in questionnaires, 400–401
Rapid Application Development
    (RAD), 187, 188–190
rapid prototyping, 195
Razor Freestyle Scooter, 67
Rea, 159
reading, 86–88
    design implications, 89
realism, abstraction contrasted, 66–67
reasoning, 88–89
    design implications, 89

recognition, preferred to recall, 27,
    408
recycle bins, 57–58
redesign, phone-based response
    system case study, 482–489
reflective cognition, 74
REI.com, 416–417, 422
reliability
    of evaluation data, 355
    of observation data, 376*t*, 383
requirements activity, 64, 201–202
    balancing conflicting, 166
    data gathering, 202–203, 210–218,
        213*t*
    data interpretation and analysis,
        202–203, 219–221
    defined, 204–208, 236
    essential use cases, 229–231
    iterative nature of, 203
    and lifecycle models, 186–188, 195
    mobile communicators, 463–464
    for new Internet appliances, 191
    and prototyping, 241
    scenarios, 211, 223–226
    task analysis, 231–234
    task description, 222–231
    types of requirements, 205–208
    use cases, 226–229
    what, how, and why of, 202–204
requirements analysis, 204
requirements engineering, 204
requirements specification template,
    238
retrospective interviews, 372
reviews, 408
rewarding activities, user experience
    goal, 18, 19
rich descriptions, 380
risk analysis, and spiral lifecycle
    model, 188
Robertson, Suzanne, interview with,
    236–238
role-playing prototyping
    walkthroughs, 210
Royal National Institute for the
    Blind, telephone design
    guidelines, 472
rules
    for collaborative meetings, 121
    level of guidance and terms used
        with, 28
    for physical design, 268

safety, usability goal, 14–16
Salomon, Gitta, interview with,
    31–33

same-participant design, of experiments, 445–446, 446t
satisfaction, user experience goal, 18, 19
scenarios. *See also* prototyping
  in conceptual design, 259–262
  and functional requirements, 258
  interviews for eliciting, 211
  in pluralistic walkthroughs, 423
  plus and minus, 260–261
  in requirements activity, 223–226
  usage, 467–468
schedules, for meetings, 119–120
scope
  of evaluation, 356
  of redesign, limiting, 489
Scout Modo, 12
screen design, 271–272, 274
scripting languages, 276
scrollbar, conceptual model, 56
search engines, 89
  background information on operation, 95
  as interface metaphor, 55
secondary users, 171
Sellen, Abigail, interview with, 138–140
semantic differential scales, 401–403
semi-structured interviews, 211
  described, 394–396
sequence model (Contextual Design method), 301
seven chunks theory, 82
shared calendars, 120, 121, 252, 256
  card-based prototype, 263–265
  physical design, 269–271, 275
  task description and analysis, 222–234, 258, 259
shared external representations, 121–122, 123
shared feedback, 127
Sherlock, 84
Shneiderman, Ben, interview with, 457–459
shortcuts, 266, 413
Shredit, 114t
Silas The Dog, 157–158, 161
simplicity, design principle, 27
Sims World, 67
single-dialog menus, 268
situated action and common ground theory, 136
sketching, for prototyping, 244

Smalltalk
  programming manual efficiency observation, 381–382
  for prototyping, 245
smart (intelligent) fridges, 5, 62
Smith, Gillian Crampton, interview with, 198–199
soap opera online community, 371–372
social environment, 207
social mechanisms
  in collaboration, 106–128
  in patient support communities, 325, 334–335
social sciences, 6
  relation to interaction design, 8
software bots, 155
software development
  ethnographic studies, 288
  heuristic evaluation, 343
  lifecycle models in, 187–192
  Microsoft's synch-and-stabilize process, 183, 184–185
  prototyping in, 241, 245–246, 248
  prototyping vs. specification cultures, 247–248
  relation to interaction design, 6, 8
  requirements, 205
software inspections, 346
software reviews, 346
software upgrades
  evaluation, 323
  evolutionary vs. revolutionary, 102
  user frustration with, 150, 152
sounds, 143
spaghetti code, 248
speaking, 87–89
  design implications, 90
specification culture, 247
speech act theory, 130
speech recognition, 88
  scenario applications, 262
spiral lifecycle model, 187, 188
spoken messages, 143
spreadsheets, 51–53
stakeholders
  conflict resolution, 236–237
  defined, 171–172
  discussing ideas with, 241, 250
  needs identification, 203
  prototypes for discussing ideas with, 241
  and quality of design, 181
  and requirements activity, 214, 215, 216–217

scenario construction, 223, 259–260
  and WinWin spiral lifecycle model, 188
standards, 408
  for evaluation, 323
  for physical design, 268
  usability principle, 27
Star interface, 53, 55, 430, 431
Star lifecycle model, 192–193
state charts, 221
statistical analysis
  experiments, 431, 457–458
  observation, 381
  questionnaires, 407
Steelcase showroom, 32
stock exchange dealers, 290
storyboards, 64, 243–244
  for incident analysis, 382–383
  as prototypes, 241, 243–245
structured interviews, 211
  data analysis, 398
  described, 394
structured tasks, HutchWorld case study, 328, 331–333
style guides, 267, 268
subjective evaluations, 345
SUMI (Software Usability Measurement Inventory), 407
summative evaluations, 323
Swim Interaction Design Studio, 11, 31
synch-and-stabilize process (Microsoft), 183, 184–185
synchronous communication, computer-mediated, 112t
synthetic characters, 157–158
system status visibility, 27

tabs (ubiquitous computing device), 61, 62
talking, 107–110
tangible bits, 61, 62, 63, 257
task allocation, 258
task analysis, 231–234, 259
  early focus on, 285, 286
  mobile communicators, 464
  and screen design, 271
task description, 222–231
technical environment, 207
telephone design guidelines, 472
telephone interviews, 211, 397
templates
  for diaries, 377
  for requirements identification, 204–205, 219

ten-minute rule, 16
tertiary users, 171
thick descriptions, 380
think-aloud technique, 365–368
    data analysis, 381
Third Age suit, 251–252
3D games
    conceptual model, 49
    realism in, 67
3D rendering, 66–67
throw-away prototyping, 248–249
Tickertape, 127–128
ticket machines, 44
Tognazzini, Bruce, 219, 321
tool support, 275
toolbars, 268
    conceptual model, 56
touch, 77
training
    for ethnographic studies, 291,
        293
    for expectation management,
        280–281
    of experts to be evaluators, 411
training simulators, 7
transcription, of observational
        notes, 374
transparency, 94–95
transparent computing, 62
travel metaphor, problems of using,
        59
triangulation, 335
TRIS redesign (IRS telephone
        response information system),
        443, 482–489
T-test, 457
typeface, 267

ubiquitous computing, 60, 62, 257
underwater PCs, 208
undo key, 266
universal usability, 459
Unix pipe symbol, 57
unstructured interviews, 211
    data analysis, 392, 398
    described, 392–394
    ethical issues, 392
upgrades, See software upgrades
URLs, avoiding complex, 415–416
usability
    aim of interaction design, 2
    business case for good, 318
    design principles, 20–27
    and evaluation, 317–318
    future issues, 458

terms used with, 28
    trade-offs, 29, 65
usability criteria, 18
usability engineering, 181–182, 193,
        195
    and evaluation, 323, 342
Usability Engineering Lifecycle,
        193–196
usability engineers, 11
usability goals
    clarifying, 37
    described, 14–18
    and evaluation, 319–322, 339
    identification in design process,
        170
    level of guidance and terms used
        with, 28
    Nokia mobile communicators,
        469, 470
    overlooking, 36
    Philips mobile communicators,
        475
    and requirements activity, 208
usability laboratories, 441–442
    mobile, 365, 442
usability principles, 26–27
    level of guidance and terms used
        with, 28
usability requirements, 207–208
usability testing, 323
    defined, 341–342, 344t
    experiments contrasted, 457–458
    HutchWorld case study, 328–334
    observation, 359, 363
    techniques applied, 340, 347t
    in TRIS redesign, 486–487
    user screening, 350
usage scenarios, Nokia mobile
        communicators, 467–468
use cases, 226–229
    essential, 229–231, 258
    and functional requirements, 258
use-oriented scenarios, 262
user abilities, 172–173, 207. See also
        cognition
    and user-centered development,
        286
user-centered development, 165, 279
    CARD approach, 307, 309–311
    defined, 285–287
    ethnography applications,
        288–306
    iterative nature of, 285, 462
    methods compared, 210t
    participatory design, 306–311

PICTIVE approach, 211, 307–309
    and requirements activity,
        203–204
user characteristics, See user
        abilities
user control, 27, 408
user environment design
        (Contextual Design method),
        296
user-experience designers, 11
user experience goals
    clarifying, 37
    described, 18–20
    and evaluation, 322, 339
    identification in design process,
        170
    level of guidance and terms used
        with, 28
    Nokia mobile communicators,
        469, 470
    Philips mobile communicators,
        475
    and requirements activity, 208
user experiences, 6, 319
    understanding, 251–252
user freedom, 27, 408
user frustration, 147–153
user interface builders, 276
user interface management tools
        (UIMs), 276
user interfaces. See also graphical
        user interfaces; interaction
        design
    early history of, 7
    with small number of keys, 470
user interface tools, 275–276
user involvement
    evaluation practical issues, 350
    importance of, 280–285
    negative effects of, 284
    participatory design, 306–311
    in user-centered development,
        279, 285–287
user need identification, 12, 169,
        202
    iterative nature of, 203
    and lifecycle model, 186
user needs, 172–173
    and evaluation, 340
    identifying, 12, 169, 202
user observation, See naturalistic
        observation; observation
user opinions, 346, 347t
    HutchWorld case study, 325, 336
    in quick and dirty evaluation, 341

user profile, 207
user requirements, 207. *See also*
    requirements activity
user roles, 230
users
    artist-design approach to,
        212–213
    as codesigners, 279
    on design team, 199, 281
    early focus on, 285
    identifying, 171–172
    as project team leaders, 282
user skills, 172–173, 207
user studies, 340
    described, 138–140
user task performance modeling,
        102. *See also* task analysis
    described, 346, 347*t*
    scope, 356
    in usability testing, 342
user tasks, *See* task analysis
user testing. *See also* experiments
    described, 346, 347*t*, 429–431
    ethical issues, 443
    with heuristic evaluation, 426
    HutchWorld case study,
        327–334
    MEDLINEplus, 432–438
    Nokia mobile communicators,
        474
    number of users, 433, 441
    origins of, 431
    process of, 438–443
    reasons for investing in, 321
    in TRIS redesign, 443, 485,
        487–488
    usability testing technique, 340,
        342
utility, usability goal, 14, 16
UTOPIA Project, 306–307

validity, of evaluation data, 355
variables, 430, 443–445
V-Chat, 326, 327
VCRs
    problems with, 1, 17
    using with Observer Video-Pro,
        382–383
vending machines, 42–43, 44
verbal communication, 106, 119
vertical prototyping, 248
videoconferencing, 110, 112*t*
videophones, 110, 112*t*, 115

video recording
    data analysis, 381–385
    interaction logging with, 378
    in observation, 365, 369, 374–377,
        376*t*
    in requirements identification, 218
VideoWindow System, 116–117
virtual assistants, 155, 157
virtual bartenders, 157
virtual calculator, 58
virtual newscasters, 157, 392–394
virtual pop stars, 157
virtual reality, 7
    direct manipulation in, 48
    physical/virtual integration, 63
virtual talk-show hosts, 157
virtual worlds, 47
    discourse analysis, 384
visibility, of system status, 21, 408
VisiCalc, 51–53
vision, 76–77
Vision of the Future Project, 9–10
Visual Basic, 276
    for prototyping, 245
voice intonation, 106
voice mail systems, as example of
        poor design, 2–3
voice-recognition menu-driven
        systems, 44
Volere requirements shell, 204–205,
        219
Volere Requirements Specification
        Template, 238

walkthroughs, 420
    cognitive, 420–423
    pluralistic, 420, 423
    role-playing prototyping, 210
waterfall lifecycle model, 187–188
wearable computing, 60, 62–63, 257
web-based questionnaires, 404–407
web browsers
    bookmarking, 37–38, 80
    conceptual model, 49
    interface metaphors, 60
    Nokia 9000 browser, 472–473
web designers, 11
WebLog, 378, 379
websites
    counters, 378
    design, 273–274
    design structure evaluation
        experiment, 447

future developments in, 427
heuristic evaluation, 412–417
optimizing for mobile
    communicators, 473
for selling clothes, 322
Webtrends, 378
web usage logging, 354, 378–379
WetPC, 208
whiteboards, 124
widgets, 268
WIMP interfaces (windows, icons,
    mouse, and pull-down menus),
    60, 257
window managers, 276
Windows 95, 184
    design, 175
Windows environment
    conceptual model, 49
    friendly interface agents, 143–144,
        146
    style guide, 267
    toolbars, 143–144, 146
    Windows 95 design, 175
Winograd, Terry, interview with,
    70–71
WinWin spiral lifecycle model, 188
wireless phones, *See* cell phones
Wizard of Oz (prototyping method),
    245
Woggles, 159
WordPerfect, Contextual Design
    application, 297–298
word-processing applications
    consistency of button design, 24
    Contextual Design application,
        297–298
    evaluation, 322
    evolution of, 174
Workaday World, 62, 64, 257
work-flow charts, 221
work flow model (Contextual
    Design method), 300
work modeling (Contextual Design
    method), 296, 300–306
work redesign (Contextual Design
    method), 296
workshops, use in requirements
    activity, 213*t*, 214, 217
World Wide Web, *See* websites

Xerox Star interface, 53, 55, 430,
    431